BACTERIOLOGY RESEARCH DEVELOPMENTS

LACTOBACILLUS

CLASSIFICATION, USES AND HEALTH IMPLICATIONS

BACTERIOLOGY RESEARCH DEVELOPMENTS

Additional books in this series can be found on Nova's website
under the Series tab.

Additional E-books in this series can be found on Nova's website
under the E-book tab.

MICROBIOLOGY RESEARCH ADVANCES

Additional books in this series can be found on Nova's website
under the Series tab.

Additional E-books in this series can be found on Nova's website
under the E-book tab.

BACTERIOLOGY RESEARCH DEVELOPMENTS

LACTOBACILLUS

CLASSIFICATION, USES AND HEALTH IMPLICATIONS

ALBA I. PEREZ CAMPOS
AND
ARTURO LEON MENA
EDITORS

NOVA BIOMEDICAL

New York

Library of Congress Cataloging-in-Publication Data

Lactobacillus : classification, uses and health implications / editors, Alba I. Perez Campos and Arturo Leon Mena.
p. cm.
Includes index.
ISBN 978-1-62081-151-1 (hardcover)
1. Lactobacillus--Classification. 2. Lactobacillus--Health aspects. I. Campos, Alba I. Perez. II. Mena, Arturo Leon.
QR82.L3L32 2012
579.3'7--dc23

2012005424

Published by Nova Science Publishers, Inc. † New York

Contents

Preface **vii**

Chapter I *Lactobacillus Plantarum:* An Overview with Emphasis in Biochemical and Healthy Properties **1**
Guiomar Melgar-Lalanne, Yadira Rivera-Espinoza and Humberto Hernández-Sánchez

Chapter II Characterization and Evaluation of *Lactobacillus plantarum* Probiotic Potential **35**
Rafael Chacon Ruiz Martinez, Antônio Diogo Silva Vieira, Karina Maria Olbrich dos Santos, Bernadette Dora Gombossy de Melo Franco and Svetoslav Dimitrov Todorov

Chapter III Bacteriocin-Producing Lactic Acid Bacteria for Biopreservation: Example of Application in Raw and Processed Salmon **65**
Svetoslav Dimitrov Todorov, Jean Guy LeBlanc, Bernadette Dora Gombossy de Melo Franco and Manuela Vaz-Velho

Chapter IV Resistance of Spoilage *Lactobacillus* Spp. to Food Processing Technologies **93**
Fernando Sampedro

Chapter V Probiotic and Health Effects of *Lactobacillus* Strains in Humans **119**
Ana Belén Rey Morán, María del Pilar González Abad, María Sol Pérez Rodríguez, Eva María Martínez Vázquez, Ana Isabel Tizón Varela and Nelson Pérez Guerra

Chapter VI Lactic Acid Bacteria in Meat and Fish: New Approaches for Traditional Applications **147**
Silvina Fadda, Lucila Saavedra, Carolina Belfiore and Graciela Vignolo

Chapter VII Strategies for Low-Cost Production and Modeling of Highly Concentrated Cultures of *Lactobacillus Casei* CECT 4043 **177**
Nelson P. Guerra, Paula Fajardo Bernárdez and Lorenzo Pastrana Castro

Chapter VIII Unraveling Genomics of Lactic Acid Bacteria and Flavor Formation in Dairy Products **203**
Ana Paula do Carmo, Arnaldo Chaer Borges, Célia Alencar de Moraes and Antônio Fernandes de Carvalho

Chapter IX Beneficial Lactobacilli for Improving Respiratory Defenses: The Case of *Lactobacillus Rhamnosus* CRL1505 **223**
Julio Villena, Susana Salva, Martha Núñez, Josefina Corzo, René Tolaba, Julio Faedda, Graciela Font and Susana Alvarez

Chapter X Selection of LAB Strains Based on Species- and Strain-Specific Typing for Probiotic Applications **241**
Palmiro Poltronieri, Oscar Fernando D'Urso, Alla V. Belyakova and Alexei B. Shevelev

Chapter XI Lactobacillus in Lacto-Fermented Vegetables **257**
Zsolt Zalán and Anna Halász

Chapter XII Lactobacilli – Functional Starter Cultures for Meat Applications **273**
Nevijo Zdolec

Chapter XIII Environmental Applications of *Lactobacillus* for Protein Recovery and Biodegradation of Recalcitrant Chemical Compounds **291**
Phisit Seesuriyachan, Charin Techapun and Ken Sasaki

Chapter XIV Use of Probiotics and Prebiotics on Functional Dairy Products: The Health Benefits **309**
Svetoslav Dimitrov Todorov and Ricardo Pinheiro de Souza Oliveira

Chapter XV Lactobacillus Reuteri ATCC 55730 and L22 Display Probiotic Potential In Vitro and Protect against Salmonella-Induced Pullorum Disease in a Chick Model of Infection **325**
Zhang Dexian, LI Rui and LI Jichang

Chapter XVI Probiotics in Pediatric Diarrheal Diseases **343**
Tapas K. Sabui and Sudipta Misra

Index **351**

Preface

Lactobacillus plantarum is a widespread lactic acid bacterium commonly found in fermented foods as well as in the human gastrointestinal tract. Application of Lactobacillus plantarum and its probiotic properties has been widely used over the past several years. The authors of this book present topical research in the study of the classification, uses and health implications of lactobacillus. Topics include the biochemical and genetic characteristics of L. plantarum; the probiotic potential of different strains of Lactobacillus plantarum; beneficial lactobacilli for improving respiratory defenses; the implementation of lactobacilli in meat applications; environmental applications of lactobacillus; and probiotics in pediatric diarrheal diseases.

Chapter I - *Lactobacillus plantarum* is a widespread lactic acid bacterium commonly found in fermented foods as well as in the human gastrointestinal tract (GIT). Their use as probiotic has increased during the last years. *L. plantarum* is a facultative anaerobic bacterium which in the absence of oxygen is able to carry on fermentations and turn sugars into lactic acid. Its genome has been fully sequenced for some strains. This analysis confirmed that *L. plantarum* has the encoded capacity for the uptake and utilization of many different sugars, uptake of peptides and formation of most amino acids. This bacterium is commonly used for food fermentations like dairy products (fermented milk and cheeses), vegetable (pickles, table olives, sauerkraut, sourdough, etc.) and meat and fish sausages. In some strains of *L. plantarum* the ability to survive along the human gastrointestinal tract has been proved aside of their capacity to adhere to the epithelium cells of the small intestine where benefic actions can take place. Some strains are used as a treatment for irritable bowel syndrome and some clinical evidence suggests effects in reducing pain, abdominal distention and flatulence. The intake of *L. plantarum* is shown to reduce certain gastrointestinal symptoms during treatment with antibiotics and the colonization of *Clostridium difficile* in ill patients treated with antibiotics Additionally, it has been shown that *L. plantarum* can protect epithelial cells from *E. coli*-induced damage by preventing changes in host cell morphology, monolayer resistance and macromolecular permeability. These results show that different strains of *L. plantarum* have a great potential to be used as a probiotic bacterium.

Chapter II - *Lactobacillus plantarum* belongs to the group of lactic acid bacteria (LAB) and has been fundamentally studied in the last few decades. The microorganism is isolated from different sources including dairy products, meat, fish, fruits, vegetables and cereal products and has a well accepted GRAS status. Application of *Lactobacillus plantarum* and its probiotic properties has been subject of several studies. In this review, the authors describe

Lactobacillus plantarum main characteristics, the biochemical and molecular methods used for its identification and investigate the probiotic potential of different strains of *Lactobacillus plantarum*, including management of gastrointestinal disorders, enhancement of gut barrier function, immunomodulatory effects, maintenance of oral health, potential treatment of burns, potential role in prevention/treatment of cardiovascular disease, cholesterol-lowering effect, and potential antiobesity effect.

Chapter III - The concept of bio-preservatives has gained significant interest during the last decades based on consumer's requests for more natural and healthier food. Lactic acid bacteria produce different antimicrobial compounds, including bacteriocins, lactic acid, hydrogen peroxide, benzoic acid, fatty acids, diacetyl and other low molecular weight compounds. Bacteriocins produced by lactic acid bacteria have been reported to be active against a wide range of pathogenic microorganisms commonly found in foods. This chapter will discuss the potential of various bacteriocin producing LAB and particularly those from the genera *Lactobacillus* and their potential in the bio-preservation of fish products. These bacteriocins, produced by beneficial microorganisms, can prevent or at least inhibit the growth of pathogens such as *Listeria monocytogenes*, thus increasing the shelf-life without affecting the physical and sensorial qualities of the end-product. Different methods of application of these LAB strains or of the purified bacteriocins will also be described. Traditional methods for preservation of food products need to be re-evaluated in view on the potential application of bacteriocinogenic LAB.

Chapter IV - Food processing preservation requisites are based on achieving an adequate food safety and prolonged shelf-life while maintaining the fresh character and quality properties of food. To preserve food and prolong the shelf-life, thermal pasteurization is the most common processing technology employed, but thermal processing can damage the original characteristics of food. Recently, some emerging non-thermal technologies (high pressure processing, pulsed electric fields, high pressure carbon dioxide, high pressure homogenization) along with new applications of non-conventional thermal technologies such as microwave irradiation have appeared in the market due to the consumers demand for food with fresh-like characteristics. To establish the optimum treatment conditions of a new technology requires not only studies on the influence of the various process parameters on the death of the microorganism and quality factors, but also the choice of a reference microorganism on which to carry out the studies. *Lactobacillus* spp. such as *Lactobacillus plantarum*, *Lactobacillus brevis*, *Lactobacillus sporogenes*, *Lactobacillus hilgardii* and *Lactobacillus viridescens* are considered common spoilage microflora encountered in fruit juices, vegetable-based products, wine and meat. *Lactobacillus* bacteria are capable of growing over a wide pH range and spoiling minimally processed foods owing to its aciduric nature, producing quality detrimental effects. The high resistance to different processing technologies and acidic conditions make *Lactobacillus* genus a target microbial factor to be considered when designing a new processing treatment.

Chapter V - The main purpose of this work is to give an overview on the positive and negative effects of some lactobacilli in human health. Firstly, the main characteristics of the most commonly used probiotic strains as well as their use for treatment of human diseases are exposed. Secondly, the manuscript describes the positive results obtained in a number of studies dealing with the evaluation of the potential of different *Lactobacillus* strains in preventing infections by pathogenic bacteria in humans. These positive effects are commonly associated to the capacity of lactobacilli for surviving the stressful environment of the stomach (low pH and bile), for adhering to host epithelial tissue and colonizing the gut, production of ant-adhesion factors, their ability for producing antimicrobial products (lactic acid, hydrogen peroxide, bacteriocins) antagonistic to pathogen growth, and perhaps immune modulation or signalling effects. Evidence of some infections (such as infectious endocarditis, bacteraemia and localized infections) caused by lactobacilli, especially in severely ill or immunocompromised patient, and the ineffectiveness of lactobacilli treatment for the prevention or treatment of human diseases are also showed.

In addition, examples of different commercial dietary supplements and food products containing probiotic lactobacilli in combination or not with other probiotic bacteria are given, and the major problems limiting the use of these commercial products as probiotic supplements for human and animals are mentioned and discussed.

Chapter VI - Lactic acid bacteria (LAB) are the main characters during fermentation of muscle-based foods, guaranteeing safety and sensorial quality of the products. The diversity of LAB resulting from the prevailing environmental conditions in processed meat and fish as well as the results in which LAB are involved as functional cultures are discussed. Particularly, the role of bacteriocinogenic *Enterococcus* during meat fermentation and the effect of curing agents on the antimicrobial activity is here described. In addition, LAB adaption to stressful conditions is analyzed by post-genomic technologies.

Chapter VII - Two re-alkalized fed-batch cultures with the potential probiotic strain *Lactobacillus casei* subsp. *casei* CECT 4043 were carried out in whey by using different mixtures of feeding substrates (concentrated whey plus a concentrated solution of lactose and mussel processing wastes plus a concentrated solution of glucose). Mass balance equations were carried out in the two cultures to calculate the volumes of feeding substrates, the concentrations of nutrients (lactose, glucose, proteins, nitrogen and phosphorous) added to the fermentation medium, as well as the concentrations of biomass and products in the fermentor justly after each feeding.

The data obtained was used to develop mathematical models to describe the growth and product formation by the CECT 4043 strain. The growth models were based on Monod kinetics and related to the phases of nitrogen consumption observed in each culture. The antibacterial activity produced in these fermentations was described by using a modified form of the Luedeking and Piret model that includes a term for the specific effect of pH drop on the antibacterial activity synthesis. Production of lactic acid and acetic acid was modeled by using the Luedeking and Piret model without modifications. A reasonable description of the growth and product formation over the time by the *Lactobacillus casei* strain was provided by the developed models, which appear to be useful as a control tool in the large scale production of a potentially probiotic *L. casei* CECT 4043 culture.

Chapter VIII - Lactic acid bacteria (LAB) are industrially important organisms and have been used as starter cultures in various food-fermentation processes. The improvement of gene annotation has led to a better prediction of flavor producing pathways in specific lineages. LAB of the genus *Lactobacillus* are widely used in the dairy industry. The most recent phylogenetic classification of this genus includes 20 completely sequenced genomes. Thus, it was possible to distinguish three groups: NCFM, GG and WCFS; this nomenclature is derived from the reference strains of each group, *Lactobacillus acidophilus* NCFM, *Lactobacillus rhamnosus* GG and *Lactobacillus plantarum* WCFS1, respectively. *Lactobacillus delbrueckii* species can be represented by three main subspecies, *delbrueckii*, *bulgaricus* and *lactis*. These subspecies are evolutionary distinct, colonizing different habitats and also presenting distinct fermentation patterns. While the subspecies *bulgaricus* and *lactis* are almost exclusively associated with milk and dairy products, the subspecies *delbrueckii* colonizes plant sources. Flavor compounds can be produced from the degradation of three principal constituents: lactose (glycolysis and Leloir pathway), lipids (lipolysis) and protein (proteolysis), the latter being the most complex and important process for the formation of flavor compounds in fermented foods. The proteolytic systems of LAB can be divided into specific components that degrade proteins into amino acids by sequence reactions involving cell-wall proteinases, peptide transport systems and different enzymes. These include intracellular proteinases (endopeptidases and aminopeptidases), whose degradation reactions usually involve release of flavor compounds. The introduction of probiotic LAB in starter cultures for manufacturing fermented products, in addition to meeting current demands of consumers who seek innovative and differentiated products while at the same time providing health benefits to the organism, has provided significant economic return, driving a multimillion-dollar market worldwide. *Lactobacillus delbrueckii* UFV H2b20 was isolated from faecal material of a breast fed newborn child in the first week of life, and since then has been the focus of numerous studies to demonstrate promising characteristics for its use as a probiotic. The operon that encodes ribosomal rRNAs in *L. delbrueckii* UFV H2b20 was partially characterized. Data suggests that *L. delbrueckii* UFV H2b20 is more closely related to *L. delbrueckii* subsp. *delbrueckii* than *L. delbrueckii* subsp. *lactis*. On the other hand, this bacterium can be compared to *L. delbrueckii* subsp. *lactis*, *L. delbrueckii* subsp. *bulgaricus* and *L. acidophilus* by a phenotypical aspect, since both of them are able to degrade lactose and therefore can grow on milk. In previous studies, viable or inactivated cells of *L. delbrueckii* UFV H2b20 were concentrated in stabilized Minas Frescal cheese whey, from which it was possible to manufacture dairy products containing this bacterium. Subsequently, an efficient method for dehydration and maintenance of cell viability in the dehydrated state was evaluated. Thus, a probiotic starter culture with promising characteristics for use in dairy products was produced at the laboratory scale. Recent studies have demonstrated that *L. delbrueckii* UFV H2b20 possesses several proteins needed for an efficient proteolytic system that supports growth on milk. Based on the genes identified so far, it was possible to propose a schematic model to explain how these proteolytic enzymes are involved in flavor formation. Moreover, this strain has some unexpected metabolic characteristics which suggest that *L. delbrueckii* UFV H2b20 might be considered as a new and unusual subspecies among *L. delbrueckii* subspecies, the first one identified as a probiotic. Since *L. delbrueckii* UFV H2b20 has been extensively characterized and because it is a probiotic, this review includes some more detailed information about this bacterium. Additionally, current scientific researches about genomics of LAB and flavor formation in dairy products are thoroughly discussed.

Chapter IX - Lactic acid bacteria (LAB) are technologically and commercially important and have various beneficial effects on human health. Several studies have demonstrated that certain LAB strains can exert their beneficial effect on the host through their immunomodulatory activity. Although most research concerning LAB-mediated enhanced immune protection is focused on gastrointestinal tract pathogens, recent studies have centred on whether these immunobiotics might sufficiently stimulate the common mucosal immune system to provide protection to other mucosal sites as well. In this sense, LAB have been used for the development of probiotic foods with the ability to stimulate respiratory immunity, which would increase resistance to infections, even in immunocompromised hosts.

This article examines the current scientific literature dealing with the use of probiotic strains to prevent respiratory infections. More specifically, the present work reviews the work of the authors' laboratory in the use of *Lactobacillus rhamnosus* CRL1505 for the improvement of respiratory immunity in adult immunocompetent host and the early recovery of the immune mechanisms against respiratory pathogens in malnourished hosts. Moreover, considering that recent reviews suggest common respiratory infectious diseases continue to be a major cause of death among preschool children in developing countries and that the identification of feasible strategies to combat this disease burden is an important public health need, the authors have evaluated the efficacy of *L. rhamnosus* CRL1505 in preventing respiratory infections and severe illnesses, in children aged 2–6 years. The research from the authors' laboratory demonstrates that *L. rhamnosus* CRL1505 represent a promising resource for the development of prevention strategies against respiratory infections that could be effective tools for medical application.

Chapter X - Methods for screening putative probiotic strains are based on selection of a high number of strains isolated from foods, human or animal sources. This preliminary selection can be followed by molecular analyses focusing on gene products, metabolites and proteome analysis. Candidate strains to be screened for potential probiotic properties are required to pass several tests to assess some beneficial effect, either in the gut environment, such as the binding to cell receptors on epithelial cells, competing for attachment of pathogenic strains, either for the production of specific compounds with known beneficial property. *Lactobacillus* strains have to be isolated from the environments and niches to which these strains are adapted to live, and then are needed to be molecularly assigned to species and clusters according to genomic sequences and phylogenetic analysis. Recently, methods were set up to select LAB strains resistant to upper gastro-intestinal tract, not isolated with traditional culture methods due to other major microbial populations in natural sources (environmental samples, dairy and fermented products). After isolation of strains surviving low pH, trypsin, pancreatin, and in 0.1% up to 0.3% bile salts, strains can be screened for other beneficial effects. In this chapter a review is presented on the selection of LAB strains such as the screening of *Lactobacillus* strains based on binding to CACO-2 cells or adhering to specific substrates, protein ligands as fibronectin, collagen, or immune cell surface receptors. Additional assays may include the displacement of pathogenic species from binding to cultured cells. Finally, a multiplex PCR exploring the structure of specific domains in extracellular proteins linked to strains adapted to the GI tract may provide the basis to identify the candidate strains as belonging to gastrointestinal tract niche-adapted strains. *Lactobacillus* strains thus selected may be exploited for probiotic applications in foods, or may be transformed for the expression of heterologous proteins, immunotherapy or tolerance-regulating cytokines.

Chapter XI - The lactic acid fermentation of vegetables is an ancient practice. Wide range of fermented vegetable products is presented all over the world and takes a specific part in the human diet. The role of the lactic acid bacteria in this process is not only the extension of the shelf life but also make the raw material more digestible and enhance the organoleptical properties of the end product. The members of the genera of *Lactobacillus* play an important role in the vegetable fermentation and often they are the predominant microorganisms in these products. They produce beside the organic acids several other antimicrobial metabolites (e.g. hydrogen peroxide, alcohol, carbon dioxide, bacteriocins), which ensure the microbial safety of the product and in this way follow out the biopreservation. Lactobacilli produce also several aromatic compounds and by their activity can form bioactive components to enhance the nutritional value of the product. The lacto-fermented vegetables combine the vitamins, minerals and fibre of vegetables with the potential probiotic properties of *Lactobacillus*. Several products are fermented by spontaneously, which can form although unique food, but it is unrepeatable. For the constant and appropriate quality the application of selected starter *Lactobacillus* culture with good fermentation properties is necessary.

In this chapter is collected the knowledge about the lacto-fermented vegetables, report about the authors' studies and the role and the usability of the *Lactobacillus* in this process.

Chapter XII - Lactic acid bacteria, primarily lactobacilli are known beneficial microbes with wide agri-food applications. Their use in animal production as feed/water additives could contribute to better growth performance, feed conversion or gut health by reducing colonization of pathogens. Today, when antibiotics are banned as growth promoters in food animals, lactobacilli and other natural antimicrobial agents are promising tool for improving microbial safety in agri-food chain. Furthermore, beneficial properties of lactobacilli are also employed in processing of food for human consumption, mainly fermented meat and dairy products. Naturally fermented foods with autochthonous „wild" microbial population are rich source of lactobacilli with favorable technological and hygienic properties for food applications. For selection of potential functional starter cultures the strains must be phenotypically and genotypically characterized. Selection criteria should include technological, safety and probiotic features of the strain. Results of in vitro studies at given laboratory conditions could be promising, but performance of the culture in real food fermentation may not be relevant. This chapter summarizes results of current studies regarding selection and implementation of lactobacilli in (non)fermented meats with particular emphasis on technological and health benefits.

Chapter XIII - In the past few decades concern about environmental issues has increased considerably and the implementation of water management has shifted towards the incorporation of new technology for the protection of ecosystems and sustainable development. Treatment of contaminated wastewater using biological processes has been widely applied. At this point, a combination of conventional knowledge and advances in technology has generated many new biological processes for wastewater treatment.

In this chapter, a biological treatment for dairy wastewater using lactic acid bacteria (LAB), *Lactobacillus* in particular, causing solid-liquid separation via biocoagulation and flocculation was introduced. A method of applying fermented lactic acid to settle protein in dairy effluent prior to the secondary treatment was demonstrated and a significant lowering in the organic load of these effluents was achieved. It was found that *Lactobacilli* are considerably more tolerant to low cytoplasmic pH than other groups of lactic acid bacteria. Sustained lactic fermentation was achieved by means of an ingenious fill-react-drain-idle sequence. This is a highly efficient new semi-continuous system, with a micro-aerobic sequencing batch reactor (micro-aerobic SBR), lactic fermentation was achieved when *Lactobacillus* was used as a biocatalyst for dairy wastewater protein recovery and biodegradation of chemically recalcitrant compounds. This is alternative to other more costly chemical and biological treatments currently in use. Additionally, protein recovered from this process can be used as a probiotic feed replacing feed antibiotics and promoting health worldwide.

This chapter gives the understanding of lactic acid bacteria in environmental applications by providing data on its potential for biodegrading recalcitrant compounds and toxic substances, such as azo dyes, by *Lactobacillus*.

Chapter XIV - In recent years, the number of functional dairy products enriched with live probiotic microorganisms, like bifidobacteria and lactobacilli, has increased. According to FAO/WHO (2002), probiotics are live microorganisms which, when administered in adequate amounts, confer health benefits on the host. In fact, these therapeutic effects are the reason for which most probiotics are used in yoghurts, fermented milks, ice creams and pharmaceutical products. Currently, to enhance health benefits, dairy foods usually contain members of probiotic genus in association with prebiotics, i.e. non-digestible oligosaccharides that resist hydrolysis and absorption in the upper gastrointestinal tract and are metabolized selectively by at least one type of probiotic in the colon.

Chapter XV - *Lactobacillus reuteri* ATCC 55730 (*L. reuteri* ATCC 55730) and *Lactobacillus reuteri* L22 (*L. reuteri* L22) were studied for their probiotic potential. These two strains were able to produce an antimicrobial substance, termed reuterin, the maximum production of reuterin by these two strains was detected in the late logarithmic growth phase (16h in MRS and 20h in LB broths). These two strains could significantly reduce the growth of *S. pullorum* ATCC 9120 in MRS broth, *L. reuteri* ATCC 55730 with a reduction of 48.2±4.15% (in 5 log) and 89.7±2.59% (in 4 log) respectively, at the same time, *L. reuteri* L22 was 69.4±3.48% (in 5 log) and 80.4±3.22% respectively. *L. reuteri* ATCC 55730 was active against the majority of the pathogenic species, including *S. pullorum* ATCC 9120 and *E. coli* O_{78}, while *L. reuteri* L22 was not as effective as *L. reuteri* ATCC 55730. The two potential strains were found to survive variably at pH 2.5 and were unaffected by bile salts, while neither of the strains was haemolytic. Moreover, *L. reuteri* ATCC 55730 exhibited variable susceptibility towards commonly used antibiotics; but *L. reuteri* L22 showed resistant to most antibiotics in this study. *L. reuteri* ATCC 55730 consequently was found to significantly increase survival rate in a *Salmonella*-induced pullorum disease model in chick. To conclude, strain *L. reuteri* ATCC 55730 possesses desirable probiotic properties, such as antimicrobial activity and immunomodulation *in vitro*, which were confirmed *in vivo* by the use of animal models.

Chapter XVI - Probiotics have recently gained popularity as therapeutic agents in different gut and non-gut related disease conditions to prevent or to reduce the severity or to gain effective early control of the disease. However, the cost benefit analyses of probiotics treatment have not been done. The authors have reviewed the role of probiotic in acute and persistent diarrhea in children in this article.

Meta-analysis and evidence-based reviews suggest modest beneficial effect of probiotics, especially LGG, in infectious diarrhea in children. In a recent meta analysis that included 63 RCTs with majority (56) pediatric studies, the authors concluded that probiotics appeared to be safe and had beneficial effects in shortening the duration and reducing stool frequency in acute infectious diarrhea. The studies done in developing countries failed to show any significant benefit. The response again is dose and strain specific. LGG was found to be most effective. However, the clinical benefit of reducing duration of mild to moderate diarrhea by a day was not clear to justify the cost of probiotics. The use of probiotic for prevention of diarrhea is not recommended with the currently available data.

Meta-analysis of 4 trials of probiotics in persistent diarrhea with a total number of 464 participants showed that probiotics reduced the duration of diarrhea (mean difference 4.02

days, 95% CI 4.61 to 3.43 days, n=324, 2 trials). Stool frequency was reduced with probiotics in two trials. One trial with a small patient population reported a significantly shorter hospital stay. No adverse events were reported. Authors concluded that there is limited evidence suggesting probiotics may be effective in treating persistent diarrhea in children.

The key factor in the pathogenesis of antibiotic associated diarrhea (AAD) is a disturbance in normal intestinal microflora. Therefore, the probiotics may be effective in preventing AAD. A review of six RCTs revealed that there was a moderate beneficial effect of LGG, B. Lactis, S. thermophilus and S.boulardii in preventing AAD.As AAD is uncommon in children, mostly mild in severity and responds well to discontinuation of antibiotics; its routine administration with antibiotic is not recommended.

In: Lactobacillus: Classification, Uses and Health Implications ISBN: 978-1-62081-151-1
Editors: Alba I. Perez Campos and Arturo Leon Mena

Chapter I

Lactobacillus Plantarum: An Overview with Emphasis in Biochemical and Healthy Properties

Guiomar Melgar-Lalanne, Yadira Rivera-Espinoza and Humberto Hernández-Sánchez
Depto. Graduados e Inv. Alimentos, Escuela Nacional de Ciencias Biológicas,
Instituto Politécnico Nacional, Carpio y Plan de Ayala, México, DF, Mexico

Abstract

Lactobacillus plantarum is a widespread lactic acid bacterium commonly found in fermented foods as well as in the human gastrointestinal tract (GIT). Their use as probiotic has increased during the last years. *L. plantarum* is a facultative anaerobic bacterium which in the absence of oxygen is able to carry on fermentations and turn sugars into lactic acid. Its genome has been fully sequenced for some strains. This analysis confirmed that *L. plantarum* has the encoded capacity for the uptake and utilization of many different sugars, uptake of peptides and formation of most amino acids. This bacterium is commonly used for food fermentations like dairy products (fermented milk and cheeses), vegetable (pickles, table olives, sauerkraut, sourdough, etc.) and meat and fish sausages. In some strains of *L. plantarum* the ability to survive along the human gastrointestinal tract has been proved aside of their capacity to adhere to the epithelium cells of the small intestine where benefic actions can take place. Some strains are used as a treatment for irritable bowel syndrome and some clinical evidence suggests effects in reducing pain, abdominal distention and flatulence. The intake of *L. plantarum* is shown to reduce certain gastrointestinal symptoms during treatment with antibiotics and the colonization of *Clostridium difficile* in ill patients treated with antibiotics Additionally, it has been shown that *L. plantarum* can protect epithelial cells from *E. coli*-induced damage by preventing changes in host cell morphology, monolayer resistance and macromolecular permeability. These results show that different strains of *L. plantarum* have a great potential to be used as a probiotic bacterium.

1. Introduction

1.1. Lactobacilli Overview

Lactic Acid Bacteria (LAB) are typically low G+C, gram-positive, non-sporulating, catalase-negative, acid tolerant and facultative anaerobic organisms wide spread in a broad variety of ecological niches (Mayo et al., 2010). Most of LAB are non-pathogenic and generally recognized as safe microorganisms (GRAS). Under a biochemical perspective LAB include both homo-fermentative, producing mainly lactic acid, and hetero-fermentative which apart from lactic acid yield a large variety of fermentation products such as acetic acid, ethanol, carbon dioxide and formic acid (Kleerebezem and Hugenholtz 2003). LAB are able to grow in a wide range of temperatures, salt concentrations and pH (Doyle and Mena, 2006) (Table 1).

Typical LAB species belong to the genera *Lactobacillus, Leuconostoc, Pediococcus and Streptococcus.* Other genera recently proposed as LAB comprise *Aerococcus, Alloiococcus, Carnobacterium, Dolosigranulum, Enterococcus, Globicatella, Lactococcus, Oenococcus, Tetragenococcus, Vagococcus, and Weissella* (Khalid, 2011). Many LAB are actually recognized as probiotic microorganisms which when administered in adequate amounts confer health benefits on the host (FAO, 2002).

Lactobacilli genus includes 175 recognized species (November 17, 2011; http://www.bacterio.cict.fr/l/lactobacillus.html) that are traditionally divided into three groups based on their fermentation characteristics: A) obligatory homo-fermentative, B) facultative hetero-fermentative and C) obligatory hetero-fermentative (Claesson et al., 2008). Many *Lactobacilli* are associated with food and feed products, mainly because they contribute to conservation due to acidification, but also because of their capacity to contribute to their unique flavor and texture characteristics.

Table 1. Characteristics of LAB at the genus level based on morphology and physiology (Doyle and Mena., 2006)

	Rods	Cocci				
	Lactobacillus	*Lactococcus*	*Streptococcus*	*Pediococcus*	*Entero-coccus*	*Leuco-nostoc*
Characteristics						
CO_2 from glucose [(a)]	Homo /hetero	hetero	hetero	hetero	hetero	Homo
Growth:						
at 10°C	±	+	-	±	+	+
at 45°C	±	+	±	±	+	-
in 6.5% NaCl	±	+	-	±	+	±
at pH 4.4	±	+	-	+	+	±
at pH 9.6	±	+	-	-	+	-
Lactic acid [(b)]	D, L; DL	L	L	L, DL	L, DL	D

+ Positive, - negative, ± responses varies between species.

(a) Test for homo or hetero-fermentation of glucose :homo (homo-fermentation), hetero (hetero-fermentation).

(b) Configuration of Lactic acid produced by glucose.

Table 2. *Lactobacillus* species commonly detected in human feces, saliva and food (Walter 2008)

Specie	Feces	Oral cavity	Food
L. acidophilus	+	+-	
L. crispatus	+ *(P)*		
L. gasseri	+ *(P)*	+	
L. johnsonii	+		+
L. salivarius	+ *(P)*	+	
L. ruminis	+ *(P)*		
L. casei	+	+	
L. paracasei	+	+	+
L. rhamnosus	+	+	+
L. plantarum	+	+	+
L. reuteri	+ *(P)*		+
L. fermentum	+	+	+
L. brevis	+	+	+
L. delbrueckii	+		+
L. sakei	+		+
L. vaginalis	+	+	
L. curvatus	+		+

P indicates species that were reported to endure in some human subjects.

The natural habitat of *Lactobacilli* ranges from dairy, meat, and plant material to the oral cavity, genital and gastrointestinal tracts of humans and animals (Mayo and van Sinderem., 2010). Taxonomically, the genus *Lactobacilli* belong to the phylum *Firmicutes*, class *Bacilli*, order *Lactobacillales* and family *Lactobacillaceae*.

They are nutritionally fastidious and require rich media to grow (carbohydrates, aminoacids, peptides, fatty acid esters, salts and nucleic acid derivatives) (Leeber et al., 2008). *Lactobacilli* seem to be part of only a minor proportion of the human adult fecal microbiota (around 0.01 % to 0.6% of total bacterial counts) but, unlike adult microbiota, infant microbiota contains important amounts (10^5 CFU/g feces in neonates and 10^6-10^8 CFU/g feces in one month infants). The presence of *Lactobacilli* is more prominent in the female genitourinary tract where they often dominate the microbiota (Leeber et al., 2008).

1.2. *Lactobacillus plantarum*: An Overview

Lactobacillus plantarum is a rod and facultative hetero-fermentative *Lactobacilli* metabolically flexible and versatile, encountered in many environmental niches with broad applications.

It has been found in a wide variety of fermented foods (Table 3) and it is part of the normal microbiota of human and other animals (Table 4).

Table 3. Food products where *Lactobacillus plantarum* was found

Product	Reference
Amasi	Todorov et al., 2008
Argentine Cheeses	Zago et al., 2011
Ayurvedic medicine (Kutajarista)	Kumar et al., 2011
Barley beer	Todorov et al., 2004
Belgian soft cheese	Burns et al., 2011
Bulgarian Salami	Todorov and Vaz-Velho, 2008
Camel cheese milk	Nanda et al., 2011
Cheddar cheese	Lynch et al., 1999
Cider	Yin et al., 2008
Cocoa beans	Camu et al., 2007
Corn silage	Tallon et al., 2003
Dry fermented sausage	Müller et al., 2009
Ewe cheese	Zhou et al., 2005
Fermentation coffee bean	Avallone et al., 2001
Fermented milk from Zimbabwe	Todorov et al., 2007
Fermented sourdough	Pepe et al., 2004
Fermented vegetables	Mäkimattila et al., 2011
Fish	Jónsson et al., 1983
Flemish artisan gouda-type cheese	Van Hoorde et al., 2008
Grape must	Rojo-Bezares et al., 2007
Greek graviera cheese	Samelis et al., 2011
Home-made cheese	Strahniic et al., 2010
Iranian cheeses	Edalatian et al., 2011
Italian ewe cheeses	De Angelis et al., 2001
Kummis	Koleva et al., 2009
Meat dry sausage	Enan et al., 1995
Molasses	Todorov et al., 2005
Mozarella cheese	De Angelis et al., 2008
Olive fermentation	Leal-Sánchez et al., 2002
Olive products	Landete et al., 2008
Olive wastes	Ayed and Hamdi, 2002
Olives	Maldonado et al., 2008
Papaya	Todorov et al., 2011
Pickled cabbage	Osawa et al., 2000
Pickled cucumber	Daeschel et al., 1990
Raw milk cheese	Feld et al., 2009
Raw turkey meat	Cho et al., 2010
Raw vegetables and fruits	Di Cagno et al., 2010
Ready to eat-salad	Franz et al., 1998
Red Gingseng	Kim et al., 2010
Salted meat	Essid et al., 2009
Sauerkraut	Makimattitla et al., 2011
Sourdough	Todorov et al., 1999; Lavermicocca et al., 2000; Siezen et al., 2010
Spanish Majorero cheese	Nespolo et al., 2010
Spanish Manchego cheese	Nieto-Arribas et al., 2009
Spoiled olive brine	Todorov and Dicks, 2006

Product	Reference
Swiss cheese	Sumner et al., 1985
Tenerife cheese	Hernández et al., 2005
Tibetan qula cheese	Duan et al., 2008
Traditional African Fermented cereal (ragi)	Desai et al., 2006
Traditional African fermented food Pearl-millet based (Ben-saalga)	Omar et al., 2006
Traditional Chinese pickled	Pan et al., 2011
Traditional Coreal fermented vegetable (Kimchi)	Hong and Pyun, 1999
Traditional dairy product from China (Koumiss)	Xie et al., 2010
Traditional fermented cream from China (Jiaoke)	Gong et al., 2010
Traditional fermented food from Philippines	Kaneko et al., 2000
Traditional fermented foods from Nigeria	Sanni et al., 2002
Traditional fermented Japanese meat product (Aji-narezuski)	Kuda et al., 2010
Traditional fermented Japanese vegetable product (kaburazushi)	Kuda et al., 2010
Traditional Indonesian fermented milk (Dadih)	Surono et al., 2008; Collado et al., 2007
Traditional Italian cheese products	Siragusa et al., 2011
Traditional Japanese fermented food	Kawashima et al., 2011
Traditional Kenya fermented milk (Kule naoto)	Matahara et al., 2008
Traditional Mexican alcoholic beverage (Mezcal)	Escalante-Minakate et al., 2008
Traditional Mexican cheese	Morales et al., 2011
Traditional Mexican fermented beverage (Pulque)	Escalante-Minakata et al., 2008
Traditional Philippines fermented foods from rice and fish	Kim et al., 2008
Traditional pork products from Portugal (Beloura and Chouriço)	Todorov et al., 2010
Traditional Spanish fermented meat (Botillo)	García-Fontán et al., 2007
Traditional Taiwan food (Fu-Tsan)	Liu et al., 2011
Traditional Thai fermented fruits and vegetables	Tanganurat et al., 2009
Traditional Thai meat sausage (Mhom)	Samappito et al., 2011
Traditional Thai style fermented pork sausage	Jaichumjai et al., 2010
Traditional Tiber kefir	Wang et al., 2010
Traditional Tunisian salted meat	Essid et al., 2009
Traditional Turkey dry fermented sausage (sucuk)	Kaban and Koya, 2009
Wine	Spano et al., 2005; López et al., 2008

Table 4. Human and other animals mucous where *Lactobacillus plantarum* were found

Mucous	Reference
Bovine fecal samples	Kagkli et al., 2007
Breast milk	Albesharat et al., 2011; Jara et al., 2011
Breast-fed infant feces	Wang et al., 2010; Albesharat et al., 2011
Breast-fed infants	Khalil et al., 2007;
Feces from breast-fed mothers	Albesharat et al., 2011
Feces from young dairy calves	Abu-Tarboush et al., 1996
Feces of herbivores, omnivores and carnivores	Endo et al., 2010
Human feces	Löonnemark et al., 2012; Vizoso-Pinto et al., 2006
Human healthy rectal mucosa	Ahrné et al., 1998

Table 4. (Continued)

Mucous	Reference
Human Intestinal Tract	Wang et al., 2009
Human saliva	Kleerebezem et al., 2003
Infant feces	Nguyen et al., 2006
Oral cavity	Lönnermark et al., 2012; Badet et al., 2008; Piwat et al., 2011
Oral mucosa	Ahmé et al., 1998
Pig feces	De Angelis et al., 2006
Vagina	Kmet and Lucchini, 1998; Vásquez et al., 2002, Pavlova et al., 2002; Martínez et al., 2008; Martín et al., 2007

2. Biochemical and Genetics Characteristics of *L. Plantarum*

2.1. Biochemical Properties of *L. plantarum*

Lactobacillus plantarum was first named as *Streptobacterium plantarum* by Orla-Jennsen in 1919 and renamed as *Lactobacillus plantarum* by Pederson (1936) who described this species due to some biochemical and morphological characteristics. *Lactobacillus plantarum* usually ferments hexoses via the EMP (Embden-Meyerhoff-Parnas) metabolic pathway resulting in the formation of D and L- lactic acids. Besides, pentoses are fermented to form lactic and acetic acid in the presence of inducible phosphoketolase (Todorov et al., 2010).

The species *Lactobacillus plantarum*, *L. pentosus* and *L. paraplantarum* are genotypically closely related and show highly similar phenotypes and are easily mistaken. The genetic heterogeneity of the *L. plantarum* group has been demonstrated by Dellaglio et al., (1975) on the basis of DNA-DNA hybridization data. Serological studies have shown that most *L. plantarum* strains belong to serological group B, as they contain ribitol and teichoic acid in their cell walls (Sharpe, 1955; Knox and Wicken,. 1972). *L. plantarum* has specific lactic acid dehydrogenases (LDHs) that forms L (+) and D (-) lactate (Garvie, 1980). Normal anaerobic glycolysis would give L (+) lactic acid whereas D (-) lactic acid could be synthesized from the glyoxalase system. The function of the D (-) lactic acid dehydrogenase may be to convert D (-) - lactic acid to pyruvic acid, which could be utilized to synthesize some other compounds such as alanine (Dennis and Kaplan, 1960).

L. plantarum ferments carbohydrates as amygdalin, cellobiose, esculin, gluconate, mannitol, melezitose, melibiose, raffinose, ribose, sorbitol, sucrose and xylose in more than 90%, and arabinose and xylose between 11 to 89% (Bergey and Boone, 2009). Sugar probes have shown that sugar fermentation was principally homo-lactic with some hetero-lactic activity (acetate production) (Plumed-Ferrer et al., 2008). Organic acids like malic, acetic and ethanol can be partially metabolized resulting in the production of carbon dioxide, lactic and acetic acid (Plumer-Ferrer et al., 2008). *L. plantarum* has shown diverse responses for some stress factor as heat shock (55°C, 10 min), bile (0.5% oxgall®), oxidative stress (0.1% H_2O_2), low pH (2.5), ethanol (10%) salt (7.5% NaCl) and detergent (0.05% sodium dodecyl sulfate) (Parente et al., 2010).

Bacteriocins are biologically active proteins that exhibit bactericidal activity against species closely related to the producer strain and may help to reduce the utilization of chemical preservatives and physical treatments in food industry. (Todorov, 2009). Several *L. plantarum* bacteriocins (most of them called plantaricins) have been partially characterized and only in a few of them their aminoacid sequence had been completely sequenced (Todorov, 2009).

2.2. Genetical Characteristics

Nowadays, over 2,000 genes have been identified from different strains of *L. plantarum*, most of them encoding proteins with homologues in other LAB. Of the remaining 300 genes with no homologues in sequenced LAB, 121 genes were found to be present in 42 *L. plantarum* strains analyzed by Siezen et al., (2010) suggesting that these 121 are marker genes from *L. plantarum*. The large majority of these specific genes seem to encode proteins. 54 genes were found to be unique for *L. plantarum* WCFS1 strain and were absent in all other 41 strains analyzed by Comparative genomic hybridization (CGH). More specifically, three gene clusters resulted exclusive to WCFS1 strain and were presumably acquired recently in evolution.

2.2.1. Genetic Sequences of Some L. Plantarum Strains

Lactobacillus plantarum sequence currently shows a genome size around 3.2 Mbp, with between 1 to 84 contigs, GC% 44.0 to 45.3, CDS ~3,000 and with 1 to 3 plasmids (Siezen et al., 2011). *Lactobacillus plantarum* WCFS1 found in human saliva was the first strain to be completely sequenced. It consists of a 3.3 Mbp chromosome, still the largest of any sequenced LAB analyzed to date, has three plasmids of 1.9 kbp, 2.3 kbp and 36.1 kbp, one contig a GC% 44.5% and a CDS (identification of coding sequences) of 3007 (Kleebenzem et al., 2003). It contains 3,052 predicted protein-encoding genes. Putative biological functions could be assigned to 2,120 of the predicted proteins. The genome sequence resulted consistent with the classification of *L. plantarum* as a facultative hetero-fermentative LAB. The genome encodes all enzymes required for glycolisis and phosphoketolase pathways and encodes a large pyruvate-dissipating potential, leading to various fermentation products. The chromosome encodes 200 extracellular proteins, many of which are predicted to be bound at the cell envelope. A large proportion of the genes encoding sugar transport and utilization, as well as genes encoding extracellular functions, appear to be clustered in a 600-kbp region near the origin of replication. Many of these genes display deviation of nucleotide composition, consistent with a foreign origin. These findings suggest that genes mentioned above provide an important part of the interaction of *L. plantarum* with its environment (Kleerebezen et al., 2003).

Other two *L. plantarum* strains were completely sequenced: *L. plantarum* JDMI, from a Chinese commercial LAB with probiotic functions (Zhang et al., 2009) and *L. plantarum* ST-III, isolated from Kimchi, a traditional Korean fermented vegetable product (Wang et al., 2010). *L. plantarum* ATCC 14917, isolated from pickled cabbage, are partially studied (Iwamoto et al., 2008; Kook et al., 2010) and *L. plantarum* KCA-1, (data not published) (http://www.ncbi.nlm.nih.gov/genome/?term=lactobacillus+plantarumandsubmit=Go).

In silico genome analysis of *L. plantarum* WCFS1 reveals that this strains have at least three systems for the uptake and biosynthesis of osmoprotectants glycine-betains /carnitin / choline (Kleerebezem et al., 2003). Besides, *L. plantarum* WCFS1 genome encodes a large number of proteins which are involved in the survival of oxidative stress, including catalases, peroxidases, thioredoxins, NADH oxidases and glutathione reductases (O`Flaherly et al., 2009). This strain has 31 genes bile -induced (Bron et al., 2004) and four putative bile salt hydrolase (BSH) genes but only *bsh1* encodes a functional BSH, whereas the remaining BSH genes encode proteins exhibiting penicillin acylase activity (Lambert et al., 2008, O`Flaherly and Klaenhammer, 2010). Cell surface factors which interact between the host GIT and the probiotics that may promote pathogen exclusion, mucosal integrity, and host immunomodulation, have been recently studied in *L. plantarum* WCFS1 and a mucin-binding protein has been found and recognized as a lectin like mannose-specific adhesin, *msa* (Pretzer et al., 2005). Major cell surface structures such as surface layer proteins (SLPs), teichoic acids (TA), and lipoteichoic acids (LTA) have also been implicated as mediators of adherence and immunomodulation and clusters for encoded surface layer proteins have been found in *L. plantarum* WCFS1 (Siezen et al., 2006).

3. Safety of *L. Plantarum* Strains

Although the use of probiotics is generally recognized as safe in healthy people, clinical evaluation of safety aspects in probiotics should be addressed especially in at risk population. Currently, there is not a legislation about probiotic safety although some recommendations have been done by FAO/WHO expert consultation in food health and safety, FDA (Food and Drug Administration, United States of America) and ESPGHAN (European Society for Paediatric Gastroenterology, Hepatology and Nutrition) (Lara – Villoslada 2010).

L. plantarum has a long story of natural occurrence and safe utilization in a variety of food products and it is a normal host in human GIT (tables 3 and 4). On the other hand, it is known that *Lactobacilli* can generate some infections in humans usually immune-depressed whereas, only a few of these *Lactobacilli* infections could be completely characterized. In a review of 200 cases of *Lactobacilli* infections (Cannon et al., 2005) only 14 were related to *Lactobacillus plantarum* (2 bacteremias, 11 endocarditis and one localized infection). The most important concern about *L. plantarum* infection is with endocarditis because some strains are able to coagulate blood *in vitro* by aggregation of human platelets However, this *in vitro* trait may not be reflecting the *in vivo* danger, since a large number of lactic acid bacteria appear to share this property (de Vries et al., 2006).

3.1. *In Vitro* Test to Ensure Safety

Antibiotic resistance can occur in two ways in a bacterial population: mutation of an endogenous gene or acquisition of a resistance gene from an exogenous source (vertical and horizontal gene transfer) (Liu et al., 2009). Many LAB are naturally resistant to some antibiotics. These resistance attributes are often intrinsic and usually not transferable to opportunistic pathogens in the human gastrointestinal tract (Zhou et al., 2005; Jacobsen et al.,

2007; Liu et al., 2009). On the other hand, intrinsically antibiotic resistant probiotic strains may benefit patients whose normal intestinal microbiota has become unbalanced or greatly reduced in numbers due to the administration of antibiotic (Salminen et al., 1998; Zhou et al., 2005). Most of *L. plantarum* strains showed to be susceptible to cephalotin, clindamicin, chloramphenicol, cloxacilin, co-amoxyclav and novobiocin and resulted to be resistant to eritromin, fusidic acid, gentamicin, kanamycin, nalidixic acid, neomycin, ofloxacin, and vancomycin (Table 5). In general, *L. plantarum* is susceptible to the β- Lactam antibiotic family but some *L. plantarum* strains are resistant to vancomycin. Vancomycin resistance is a great concern because is one of the last antibiotics broadly efficacious against clinical infections caused by multidrug-resistant pathogens; such resistance is usually intrinsic, being chromosomally encoded and non transmissible (Zhou et al., 2005).

Table 5. *L. plantarum* strains antibiotic resistance

antibiotic	μg/ disk	17 Strains	HN045	LPL1	LPL2	13 Strains	12 Strains
Reference		Essid et al., 2009	Zhou et al., 2005	Jamaly et al., 2011	Jamaly et al., 2011	Cebecci et al., 2003	Karasu et al., 2010
Amoxicilin	25	nd	nd	nd	nd	2 R, 4 MS	nd
Ampicillin	10	9 R	nd	R	S	1R, 3 MS	12 S
Amykacin	30	nd	nd	nd	nd	13 R	nd
Aztreomicin	30	nd	nd	nd	nd	13 R	nd
Bacitracin	0.04	nd	nd	nd	nd	13 R	11 R
Cefoperazone	75	nd	nd	nd	nd	2 R, 8 MS	nd
Cefriaxone	30	nd	nd	nd	nd	4 R, 3 MS	nd
Cephalotin	30	nd	S	nd	nd	4 MS	nd
Cephoxitin	30	nd	nd	nd	nd	13 R	nd
Cephuroxime	30	nd	nd	nd	nd	6 R, 2 MS	nd
Clindamicin	2	nd	nd	nd	nd	2 MS	nd
Cloramphenicol	30	nd	S	nd	nd	nd	12 S
Cloxacilin	1	nd	S	S	S	nd	nd
Co- amoxyclav	30	nd	nd	nd	nd	1MS	nd
Co- trymozazole	25	nd	nd	nd	nd	4 R, 1 MS	nd
Erytromycin	15	15 R	nd	nd	nd	2 MS	0 R
Fusidic acid	10	nd	R	nd	nd	13 R	nd
Gentamycin	10	nd	R	S	S	3 R	12 R
Kanamycin	30	nd	R	S	S	nd	12 R
Naldixic acid	30	nd	R	nd	nd	13 R	nd
Neomycin	30	nd	R	nd	nd	nd	11 R
Netilmycin	30	nd	nd	nd	nd	12 R, 1 MS	nd
Nitrofurantozin	300	nd	nd	nd	nd	13 R	nd
Novobiocin	5	nd	S	nd	nd	nd	nd
Ofloxacin	5	nd	nd	nd	nd	13 R	nd

Table 5. (Continued)

antibiotic	μg/ disk	17 Strains	HN045	LPL1	LPL2	13 Strains	12 Strains
Reference		Essid et al., 2009	Zhou et al., 2005	Jamaly et al., 2011	Jamaly et al., 2011	Cebecci et al., 2003	Karasu et al., 2010
Penicillin	10	9 R	nd	nd	nd	3 R, 10 MS	8 R
Pipieracilin	100	nd	nd	nd	nd	1 R	nd
Polymyxin B	300	nd	R	nd	nd	5 R, 1 MS	12 R
Rifampicin	5	14 R	nd	nd	nd	7 R, 3 MS	12 S
Streptomycin		nd	R	nd	nd	nd	12 R
Tetracyclin	30	17 R	nd	S	S	10 R, 1MS	2 R
Tobramycin	10	nd	nd	nd	nd	13 R	nd
Vancomycin	30		,R	nd	nd	11 R	11 R

S sensible, R resistant, MS moderate Sensible, nd: not determined.

Table 6. Some antibiotic resistance genes located in different strains of *L. plantarum*

strain	Gene	Antibiotic resistance	Location	Reference
DG 522	Tet (M)	Tetracycline	plasmid plasmid	Gevers et al., 2003
DG 507	Tet (M)	Tetracycline	plasmid	Gevers et al., 2003
DG 507	Erm(B)	Erytromycine	Plasmid	Gevers et al., 2003
VTT E-042708	Tet(M)	Tetracycline	Plasmid	Egervärn et al., 2009
VTT E-042709	Tet(W)	Tetracycline	Plasmid	Egervärn et al., 2009
JDM1	Van(X)	Vancomycin	D-ala.D-ala dipeptidase	Zhang et al., 2011
JDM1	Van(H)	Vancomycin	Essential dehydrogenase	Zhang et al., 2011
JDM1	Van(D)	Vancomycin	D-alanine-D-lactate ligase	Zhang et al., 2011
JDM1	Van(R)	Vancomycin	Response regulator	Zhang et al., 2011
JDM1	Van(S)	Vancomycin	Histidine kinase	Zhang et al., 2011
E14	Tet(M)	Tetracycline	Plasmid	Rojo-Bezares et al., 2006
C709	aac(6′)-aph(2″)	Gentamycin and Kanamycin	nd	Rojo-Bezares et al., 2006
J62	ant(6)	Streptocmycin	nd	Rojo-Bezares et al., 2006
J65	aac(6′)-aph(2″)	Kanamycin	nd	Rojo-Bezares et al., 2006
J75	aac(6′)-aph(2″)	Gentamycin and Kanamycin	nd	Rojo-Bezares et al., 2006
J77	aac(6′)-aph(2″), aph (3′)-IIIa	Gentamycin and Kanamycin	nd	Rojo-Bezares et al., 2006
34	tet(M)	Tetracycline	Plasmid	Toomey et al., 2010
LP1	van(X)	Vancomycin	D-ala-D-ala dipeptidase	Liu et al., 2009
LP2	van(X)	Vancomycin	D-ala-D-ala dipeptidase	Liu et al., 2009
M345	Erm(B)	Erytromycine	Plasmid	Feld et al., 2009

nd: not data.

Tetracyclin, erithromycin, gentamycin, kanamycin and vancomycin resistance genes have been isolated and located in *some L. plantarum* strains (Table 6). Some of these genes are present in plasmid fractions so, hypothetically, a horizontal gene transfer of antibiotic resistance would be possible. Jacobsen et al., (2006) demonstrated that *in vivo* antibiotic resistance gene transfer can take place from a wild type of *L. plantarum* isolated from food to *Enterococcus faecalis,* a natural inhabitant of the human gut and a potentially pathogenic species to a gnotobiotic rats.

Not all the antibiotic resistance is associated to genes. Ouoba et al., (2008) analyzed the presence of 28 genes associated to antibiotic resistance in a *L. plantarum* strain from human origin and found no relation.

To prevent undesirable transfer of antibiotic resistance to pathogenic endogenous bacteria, LAB such as *L. plantarum* should not carry resistance genes to other species into the gastrointestinal tract (Marthur and Singht, 2005).

No references were found about positive haemolytic potential (α or β haemolysis) in *L. plantarum* strains. All strains analyzed resulted γ-haemolytic (or not haemolytic) and rarely cause illness (Ruiz-Moyano et al., 2009; Anas et al., 2008; Jamaly et al., 2011, Zoumopoulou et al., 2008; Mourad and Nour-Eddine., 2006).

Although nisin is an antimicrobial peptide produced by *Lactotoccus lactis* widely used as food preservative, nisin tolerance of *L. plantarum* strains been poorly analyzed. In an analysis of 64 LAB isolated from wine (of which 22 were *L. plantarum* strains) nisin tolerance changed greatly depending on the content (between 0.4 and 400 μg / mL) leading to the conclusion that appropriate combinations of nisin and other antimicrobial methods could control the growth of spoilage bacteria in wine (Rojo – Bezares et al., 2007).

4. *In Vitro* Survival of L. Plantarum in Gastrointestinal Tract Simulation

After oral ingestion, bacteria encounter a number of human defense systems associated with different secretions through the GIT. The first defense system is the presence of lysozime and α- amylase in the oral cavity which display antimicrobial activity. After that, bacteria suffer the presence of low pH (between 2.0 to 3.0) and proteolytic enzymes like pepsin. In the small intestine pH increases to 8.0 and bile salt and pancreatic juices are secreted (de Vries et al., 2006).

Only a few authors have investigated lysozyme tolerance of *L. plantarum* strains. Zago et al., (2011) analyzed 27 *L. plantarum* strains and found high resistance to lysozyme from 15 strains (≥ 68% of survival rate) and Golowyc et al., (2010) found a survival rate close to 100% in *L. plantarum* CIDCA83114. Many strains of *L. plantarum* showed high tolerance when exposed to hydrochloric acid (pH 3.0) in the presence or in absence of pepsin (0.3%) and most of them exhibited a great decrease in survival rate with pH 2.0 with and without pepsin. Zago et al., (2011) found no significant decrease in population in 27 strains when the pH decreased from 5.0 to 2.5 in 60 min, but only two strains survived with no significant decrease in population when the pH dropped from 5.0 to 2.2 in 90 min in the presence of 0.6% pepsin. In general, most studies reported a high death rate of *L. plantarum* in the presence of pepsin at the same pH probably due to the hydrolysis of the peptidoglycan present in the cell wall (Zhu et al., 2006). The intestinal survival rate of different strains of *L. plantarum* is close to 100% at concentrations of up to 0.5% bile salt (Ox-gall®) for 4 h. Some strains survived at bile stress conditions (1.0% bile salt, pH 8.0) during 24 h (Jamaly et al., 2011; Wang et al., 2010; Zago et al., 2011). The presence of pancreatin (1% with bile salts) seemingly did not have a significant effect on the survival rate of *L. plantarum* strains (Botes et al., 2008; Michida et al., 2006; Jiménez- Pranteda et al., 2011).

Microencapsulation is a technology designed for the improvement of probiotic survival (both during the shelf life of a product and gastrointestinal transit), because it is aimed to the protection of probiotic bacteria offering a great potential in the delivery of viable cells. In the case of *L. plantarum*, only a few studies have been reported and results are strongly dependent on the polysaccharide and protein matrixes used as encapsulation material. Alginate–chitosan and pectin–alginate microcapsules did not show differences in survival rate in comparison with unencapsulated *L. plantarum* BL011 (Brinques et al., 2011). The same phenomenon was found using a jamilan-gellan gum matrix with *L. plantarum* CRL1815 (Jiménez – Pranteda et al., 2011). However a xanthine – gellan gum showed a slight survival increase at pH 2.0 (Jiménez – Pranteda et al., 2011). An alginate matrix coated with whey proteins showed a better survival rate in an *in vitro* gastrointestinal simulation with *L. plantarum* 299v (Gbassi et al., 2009). The nature of the food matrix in where probiotics are delivered seems to improve the viability of other *Lactobacillus* strains as well (Burgain et al., 2011). There is only one study employing a mixture of malt extracts with *L. plantarum* NCJMB8826 (Michida et al., 2006), where the presence of these extracts significantly improved the gastric tolerance of *L. plantarum* and the sole presence of malt extract improved the cell population in the small intestine.

5. *In Vitro* Adhesion to Human Gastrointestinal Tract

The intestinal mucosa, which covers the intestinal tract, forms a biological barrier between the external and internal environment of human body. Mucosa consists of polarized epithelial cells that form a single layer of columnar cells named enterocytes. Scattered between these cells are specialized globet cells that synthesize and excrete mucus. Mucus is a gel layer covering the epithelial lining which has the main function to protect the epithelium. Adhesion to the intestinal mucosa is a prerequisite for colonization and it is an important characteristic related to the ability of strains to interact with the host. Due to all mentioned above is important to consider a selection criteria for new probiotics strains (Collado et al., 2010).

Major components of mucus are mucins, a group of polymeric glycoproteins that have been widely used for *in vitro* adhesion tests. For example, in a study with 31 *L. plantarum* strains, mucin adherence varied from 15 ± 3 to 20,088 ± 3442 CFU/wall (Tallon et al., 2006). Other studies have been done with human intestinal mucus (Gueimonde et al., 2006; Kinoshita et al., 2008) and *L. plantarum* strains showed good adherence properties.

Although the action mechanisms are not fully understood, it is generally accepted that the ability of a strain to auto-aggregate and to co-aggregate with other probiotic or pathogens strains is a desired property because they are harder to remove from the intestinal environment (Darilmaz et al., 2011). Co-aggregation with potential pathogens allows probiotics strains to produce antimicrobial substances in a very close proximity *per se* inhibiting the growth of pathogen strains (Botes et al., 2008). Auto-aggregation in *L. plantarum* depends on the strains (Table 7) and in general *L. plantarum* strains showed medium and high levels of auto-aggregation (from 31% to 95.4%). Few studies are available on co-aggregation with other probiotic or pathogen strains. Vaz – Velho et al., (2011) analyzed co-aggregation of *L. plantarum* ST16Pa with different *Listeria* spp. showing a co-aggregation of around 30% and Collado et al., (2008a) studied the Lp-115 strain co-aggregated with *S. aureus* (21.4%), *Enterobacter sakazaki* (2.2%), *Bacteroides vulgatus* (13.7%) and *Clostridium histolyticum* (21.5%). Best abilities to co-aggregate with pathogens (*B. vulgatus*, *C. histolyticum* and *difficile*, *St. aureus*, *E. sakazakii*, and *E. coli)* were shown by *L. plantarum* IS – 20506 (Collado et al., 2008b). *L. plantarum* 423 showed effective co-aggregation rate with *Salmonella* sp. (12.9%), *L. monocytogenes* (51.3%) and *E. coli* (43.6%) (Botes et al., 2008). *L. plantarum* ST194BZ, ST414BZ and ST664BZ showed high co-aggregation rates to *Listeria innocua, L. sakei, Enterococcus faecium* and *Lactococcus lactis subs. lactis* (Todorov et al., 2008). The ability of *L. plantarum* stains to generate co-aggregates is better with gram positive strains probably due to the cell wall composition.

Adhesion is a complex process involving non –specific (hydrophobic) and specific ligand – receptors mechanisms. Cell surface hydrophobicity of bacteria may contribute with the capability to adhere to epithelial cells and extracellular protein matrix. Different *L. plantarum* strains show different hydrophobicity levels (Table 7). Correlation between aggregation and hydrophobicity to different hydrocarbons is not well defined.

Table 7. Auto-aggregation (%) and hydrophobicity (%) of different *L. plantarum* strains

Strain	Auto-aggregation (%)	Hydrophobicity (%)	Reference
LP-115	44.5 ± 16.2	44.2 ± 18.7 (a)	Collado et al., 2008
2012	13.7 ± 0.9	1.6 ± 0.2 (a)	Katmzamanidis et al., 2010
ST194BZ	~ 78	~ 30 (b)	Todorov et al., 2008
ST414BZ	~ 85	~ 60 (b)	Todorov et al., 2008
ST664BZ	~ 72	~ 70 (b)	Todorov et al., 2008
2022	15.7 ± 1.5	13.5 ± 0.3 (a)	Kotzamandis et al., 2010
2035	44.3 ± 1.9	61.3 ± 2.3 (a)	Kotzamandis et al., 2010
Lp9	31 ± 1.0	37.7 ± 1.3 (a)	Kaushik et al., 2011
Lp 590	nd	17.64 ± 2.32 (a)	Liu et al., 2011
Lp ASI 2986	nd	16.97 ±1.618 (a)	Liu et al., 2011
ATCC 8014	nd	16 (a)	Ouwehand et al., 1999
8 different strains	74.3 to 95.4	Nd	Todorov and Dickins 2008
Lp ST16Pa	37.05	68.7 (b)	Van Velho et al., 2011
27 different strains	nd	2.19 to 30.83% (a)	Zago et al., 2011
IS-10506	9.0 ± 0.0	76.3 ± 16.7% (a)	Collado et al., 2008b
Is-20506	0.5	10.3 ± 6.3 (a)	Collado et al., 2008b
423	52	50 (b)	Botes et al., 2008
76 strains	nd	5 to 80 (a)	Mathara et al., 2008
BMS2	nd	1.50 (b)	Samot et al., 2011

(a) Hydrophobicity to xylene.
(b) Hydrophobicity to n-hexadecane.

Collado et al. (2008) found that auto-aggregation properties are strongly correlated with xylene hydrophobicity in some strains but other authors did not find a correlation between them (Ouwerhand et al., 1999). Besides xylene and n-hexadecane, other organic compounds are used to measure hydrophobicity like chloroform (a monopolar and acid organic solvent) and ethyl acetate (a monopolar and basic organic solvent). *L. plantarum* L4 had low xylene hydrophobicity (fully hydrophilic) and a strong affinity for chloroform (an acid organic solvent) so the presence of polysaccharides in its cell surface may be assumed (Kos et al., 2003). In the case of *L. plantarum* BMS2 the same pattern was reported (Samot et al., 2011). To confirm the presence of a bacterial sugar – binding proteins, *in vitro* agglutination tests with yeasts are conduced. Yeast cell surface express a variety of carbohydrates, which can form a cross-linked binding body with bacterial lectins and finally lead to aggregation. For this reason, the yeast cell is usually used as the host cell in an *in vitro* test. Bacterial adhesion to the intestinal mucosa includes a mucus binding ability and an intestinal epithelial cell adhesion (Zago et al., 2011; Liu et al., 2011; Liu et al., 2011).

To date, the most common *in vitro* adhesion test is adhesion to cell culture lines. The most used cells are HT-29 (which is a human colonic cancer line) and Caco-2 (another human colonic cancer cell line). In the case of *L. plantarum* strains, most researchers have found a low or medium adhesion rate to both cell lines. Besides, Ouwerhand et al., (1999) did not find any correlation between adhesive ability and cell surface hydrophobicity to n-hexadecane with *L. plantarum* strains. Furthermore, Van Tasssel (2011) claims that there is not a clear correlation between *in vitro* and *in vivo* adhesion probiotic tests. Other cell lines employed to investigate adhesion were IPEC-J2, a porcine intestinal cell line, which showed adhesion with *L. plantarum* Q47 (Larssen et al., 2009), and rat ileum epithelial cells showed adhesion with *L. plantarum* LPL2 (Jamaly et al., 2011). Adhesion of *L. plantarum* strains depends on the model as well as the strain used. Chemical and enzymatic pretreatments applied to bacterial cell suggested that lectins such as adhesins and other proteinaceus cell- surface structures are involved in adhesion mechanisms (Tallon et al., 2006).

Adhesins are cell – surface components of bacterium which facilitate bacterial adhesion to other cells or to inanimate surfaces. Ramiah et al., (2007) found two adhesion related genes in *L. plantarum* 4232: *mub* (mucus binding protein gene) and *mapA* (mucus adhesion-promoting protein) and an elongation factor *EF-Fu* (adhesion like protein) and Sánchez et al., (2011) found a *cbp* gene (chitin- binding protein) in *L. plantarum* BMCM12. *L. plantarum* WCFS1 genes: lp-1229, lp_1643 and lp_3144 genes were predicted *in sillico* to encode proteins containing mucus-binding domains (Boekhorst et al., 2006). Also, lp_1229 was identified as a gene encoded mannose-specific adhesin of some *L. plantarum* strains with the ability to agglutinate the yeast *S. cerevisiae* (Pretzer et al., 2005).

Electrical resistance enhancement in Caco-2 and HT-29 cells by *L. plantarum* was proposed as a dependent mechanism of modulation of cytoskeleton and tight junction protein phosphorylation (Remus et al., 2011*)*. *L. plantarum* CGMCC No.1258 prevents enteroinvasive *E. coli* to induce tight junction protein changes in intestinal epithelial cells (Caco-2 cell line), causing damage of the epithelia monolayer barrier function by preventing changes in host cell morphology (Qin et al., 2009). On the other hand *L. plantarum* DSMZ 12028 showed, in an *in vitro* model with Caco-2 cell line, a proinflammatory reducing response against *E. coli* K4 just at the hold level, which could prevent an inflammatory outcome while inducing a higher state of alertness in the defense system of the host intestinal epithelial cells (Cammarota et al., 2009). An *in vivo* study with *L. plantarum* WCSF1 administered to healthy volunteers showed a regulation of the epithelial tight junction proteins by the LAB *in vivo* and also a protective effect on the epithelial barrier (Karczewski et al., 2010).

6. *In Vitro* Probiotic Potential of *L. Plantarum* Strains

Antibacterial activity of *L. plantarum* strains toward different pathogens and food spoilage bacteria had been widely analyzed (Table 8) and depends on *L. plantarum* strains and pathogen strains as well as culture conditions. There are several molecules and mechanisms involved in the interrelationship between probiotics and pathogens such organic acids, hydrogen peroxide, ethanol and bacteriocin production (Margolles et al., 2009).

Only a few studies have dealt with potential antiviral activity: one related with herpes simplex (HSF) strain F (Todorov et al., 2008) and other against influenza infection in mice (Takeda et al., 2011). Antifungal activity against molds and yeasts are widely reported, most of them in dough and bakery products as well as in some cheeses (Dal Bello et al., 2007; Voulgari et al., 2010; Djossou et al., 2011; Yang and Chang, 2010).

β Galactosidase (β-D-galactohydrolase, EC 3.2.1.23) or lactase is an enzyme that hydrolyzes D-galactosyl residues from polymers, oligosaccharides or secondary metabolites. Polysaccharide specific β-galactosidases include β-galactanases, which attack pectic polymers and β-galactosidases that attack xyloglucans. These enzymes have two main applications: the removal of lactose from milk products for lactose intolerant consumers and production of *galactosylated* products. It is widely used in food industry to improve sweetness, solubility, flavor and digestibility of dairy products.

Most of *L. plantarum* strains are capable to produce β galactosidase *in vitro* in medium or high amounts, however, some β galactosidase-negative strains have also been found. Zago et al. (2011) found β galactosidase activity (Millet Units) in the range from 12.69 to 1062.14 for 27 strains of *L. plantarum*. α- galactosides are carbohydrate reserves in many plant tissues and seeds which are not digested in the duodenum by human and other monogastric animals. They include one or several galactose units, linked together or to the glucose moiety of sucrose through α – 1, 6 linkages. Several LAB, like some *L. plantarum* strains, are able to produce α – galactosidase. The digestion of α- galactosides in duodenum could reduce the flatulence problem and increase the metabolizable energy from vegetable products rich in non-digestible carbohydrates (Silvestroni et al., 2002). Only a few studies had been done to confirm the presence of this enzyme in *L. plantarum* strains (Songré – Outtera et al., 2008; Essid et al., 2009). The genes that encode this enzymatic activity are galM, and LacLM in *L. plantarum* ATCC8014 (Silvestroni et al., 2002).

Bile salts play an essential role in lipid digestion. They act as a detergent that emulsifies and solubilizes dietary lipids and lipid soluble vitamins. Usually, species of the intestinal microbiota, including *Lactobacilli* produce bile salt hydrolases (BSH) which are able to deconjugate the aminoacid moiety from the bile salt in the intestine. BSH is delivered to play an important role in host physiology, as the gatekeeping reaction in further oxidation and dehydroxylation steps of primary bile salts into secondary bile salts (Lammbert et al., 2008).

The microbial BSH role is not clear enough and it is related to bacterial gastrointestinal persistence, host cholesterol lowering and with a possible cancer activation in the host.

Table 8. Antimicrobial activity of *L. plantarum* strains to pathogens and food spoilage bacteria

Strain	Antibacterial activity	Reference
2.9	*Eliminate B. cereus, E. coli O57:H7* and *S. enterica CHC4300* in MRS broth and in a malted millet flour slurry.	Valenzuela et al., 2008
BFE5092	Protective effect against *L. monocytogenes* EGDe in turkey meat under low temperatures	Cho et al., 2010
ACA-DC-287	No inhibits against 507 indicator strains, pathogenic and food spoilage bacteria (443 gram positive and 64 gram negative) in plat agar.	Zompopoulou et al., 2008
6 strains	Inhibits growth of *E. faecium* HKLHS, *L. sakei* DSM 20017, *Lc. Lactis sub.lactis* HV219, *L. innocua* LMG 13568, L. inoccua UWC N27 *and L. ivanovii subsp ivanovii* ATCC 19119 in well diffusion method.	Todorov and Dicks, 2008
4.1	Inhibits against *E. coli K12*, *S. thyphimurium* ATCC 6994, *Cl. perfrigens* type A strain 22G, *L. innocua* DSM 20649 and *S. aureus* ATCC 25923. No inhibits against *Brachispora hyadisenteriae* in MRS broth	De Angelis et al., 2006
LpL1 and LpL2	Inhibits against *L. innocua* LMHAE-LI 107, *E. coli,* LMHAE-SA and *E. coli* ATCC25922, *S. aureus* LMHAE-SA 105, *E. faecium* ATCC 25212, *Streptococcus* D and *Klebsiella pneumonia* CIP 53153 in plat agar	Jamaly et al., 2011
27 strains	No one inhibit *E. coli* V517, *S. enteroditis* OM-Ca, *L. monoctytogenes* ATCC 15313 and *S. aureus* 76	Zago et al., 2011
4 strains	Inhibit againt *E. coli* ATCC 25922, *S. aureus* ATCC 25928, *S. aureus* ATCC 11778, *P. aeurogeinosa* ATCC 27853, S. *thyphi, S. sonnei* and *C. albicans*	Sirilum et al., 2010
NBIM CC241S	Inhibits growth of *E. coli* ATCC 25922, *E. coli* ATCC 8739, *Proteus vulgaricus* G, *Salmonella* sp., *S. abony* NTCC 6017, *S. aureus* ATCC 25093, *S. aureus* ATCC6538, *L. monocytogenes* and *L. monocytogenes* I.	Nedelcheva et al., 2010
30 strains	7 inhibit *L. innova* DSM 20649 0 inhibit *L. monocytogenes* DSM 20600 2 inhibit *B. cereus* CCH2010 12 inhibit *S. aureus* S1 5 inhibit *S. mutants* DSM 6138 3 inhibit *Klebsiella pneumonia* subsp. *Pneumonia* BFE 147 1 inhibit *E. cloacae* BFE 282 7 inhibit *E. agglomerans* BFE 154 7 inhibit *Pseuomonas aeroginosa* BFE 162	Tamang et al., 2009
Lp9	Antibacterial activity against *E. coli, L. monocytogenes, S. thyphi, S. aureus* and *B. cereus* in agar well diffusion assay	Kauskih et al., 2009
ACA-DC287	Inhibits growth of *Salmonella enteric server. Thyphimurium* SL1344 *in vitro* and *in vivo*	Fayol- Messaoudi et al., 2006
17 strains	Inhibit Salmonella arizonae ATCC 25922, S. aureus ATCC 25923, Pseudomonas aeuroginosa MBA and E. coli DH5a, IPT	Essid et al., 2009
ITM21B	Inhibit *Bacillus subtilis* ATCC 8473 in bread making products	Valerio et al., 2008
FST 1.7	Inhibits growth of *Bacillus subtilus, Ciobacter fieondis, E. faecalis, E. coli, L. innocua, Micrococcus luteus, Proteus vulgaris and S. aureus* in wheat bread	Dal Bello et al., 2007

However *Lactobacilli* are not capable of dehydroxylating deconjugated bile salt and so the majority of the breakdown products of BSH activity may be precipitated and excreted in feces (Bergley et al., 2006) lowering serum cholesterol (Nguyen et al., 2007). BSH genes were analyzed by Lambert et al. (2008), who found four genes: *bsh1, bsh2, bsh3* and *bsh4* that encode the enzymatic activity. *In vitro* BSH activity had been studied in *L. plantarum* strains (Table 9).

Table 9. *In vitro* BSH activity in *L. plantarum* strains

strain	BSH activity	Reference
98 strains	27 strain showed high activity	Zago et al., 2011
6 strains	6 strain showed high activity	Vinderola et al., 2006
PH04	High activity	Nguyen et al., 2007
423	No activity	Botes et al., 2008
NR74 and 66	High activity	Lee et al., 2010
Lp9	High activity	Kauskih et al., 2011

A higher BSH activity was detected during the stationary phase than during the exponential phase of growth and was induced by conjugated bile salts but not by deconjugated bile salts and was greater in the presence of glycine – conjugated than in taurine conjugated bile salts (Nguyen et al., 2007).

In an *in vivo* animal model, the oral administration of *L. plantarum* PH04 lowered total serum cholesterol levels in mice without any side effects and without bacteria translocation. Serum triglycerides were also lowered as a result of the treatment, without affecting the structure and relative weight of the liver. This suggests that the hypolipoid effect of the bacteria may not be due to a redistribution of lipids from the plasma to the liver, but rather to a decreased intestinal absorption of lipids or to an increment in the lipid catabolism (Nguyen et al., 2007).

Besides, *L. plantarum* KCTC3928 coated with polysaccharides and proteins showed a hypocholesterolemic effect in mice fed with a high – fat diet due to induction of fecal bile acid secretion followed by an increased degradation of hepatic cholesterol (Jeun et al., 2010). In other tests with rats fed a high cholesterol diet, *L. plantarum* MA2 was able to reduce serum cholesterol, low density lipoprotein cholesterol and triglyceride levels (Wang et al., 2009).

Cytokines are key regulators of inflammation in intestinal bowel disease (IBD), and several pro-inflammatory and immune regulatory cytokines are dysregulated in the mucosa of IBD patients. It is important then to know the cytokine anti inflammatory profile of these patients. It is possible to predict the *in vivo* protective capacity of potential probiotics based on the cytokine profile established by *in vitro* tests that may serve as a primary indicator to narrow down the number of candidate strains to be tested in animal models for their anti inflammatory potential (Foligne et al., 2007).

Several studies have revealed that some specific strains of *Lactobacillus* can induce pro-inflammatory cytokines such as interleukin (IL) -1, IL – 6 and IL -12, as well as the anti inflammatory cytokine IL – 10 (Paolillo et al., 2009). Many components of gram positive bacterial cell wall (capsular polysaccharides, peptidoglycans and lipoteichoic acids) have been involved in cytokine induction by macrophage cells (Haza et al., 2004). *L. plantarum*

CBL /J isolated from ewe`s milk cheese induced *in vitro* production of several cytokines by human macrophages. (Haza et al., 2004). *L. plantarum* L8 induced high levels of the Th – 1 cytokines TNF – α snd IL -12 but induced a low level of IL -10 secretion from PBMC, suggesting a pro-inflammatory effect. This may suggest a possible anti-inflammatory regulation of immune system components by this particular potential probiotic strain.

The simultaneous induction of some pro-inflammatory and regulatory cytokines could be beneficial for the maintenance of a chronic and immunologically balanced intestinal inflammation response (Wang et al., 2009).

Some studies with heat-killed *L. plantarum* cells have been done in relation to immunity factors. Heat-killed *L. plantarum* L-137 enhanced protection against influenza virus infection by stimulating type I interferon production in mice (Maeda et al., 2009).

Oral administration of boiled *L. plantarum* O6CC2 was significantly effective in preventing weight loss of mice infected with influenza virus, reducing virus yields in the lungs and prolonging survival times without toxicity (Takeda et al., 2011).

7. *In Vivo* Probiotic Findings Around *L. Plantarum* Strains

7.1. *Lactobacillus plantarum* 299v

L. plantarum 299v after oral administration in humans has been found in high numbers on the rectal mucosa and in feces (Nobeck *et al.,* 2000) increasing the viable count of total *Lactobacilli* in feces (Goossens et al., 2005). Besides, it is capable to adhere to the tonsillar mucosa directly after oral intake (Stjernquist-Desatnik *et al.,* 2000). *L. plantarum* 299v was used for togwa production (a cereal based lactic acid fermented beverage).

This product was given to children under five years old once a day for 13 days. The presence of fecal enteropathogenic bacteria were evaluated resulting in a decrease in fecal counts in the group that consumed towga fermented with *L. plantarum* 299v.

L. plantarum 299v lowers total cholesterol concentration and LDL cholesterol as well as fibrinogen level of serum after consumption (Naruszewicz et al., 2002).

Other study with smokers showed a decrement in the level of fibrinogen but also F2-isoprostanes and IL-6 which are other inflammatory markers and affected positively the systolic blood pressure (Naruszewicz et al., 2002). *L. plantarum* 299v have been administrated to patients with irritable bowel syndrome in two, double blinded, placebo controlled studies, and in both of them a reduction of symptoms like pain and flatulence was detected (Nobaek et al., 2000; Niedzielin et al., 2001).

Administration of *L. plantarum* 299v reduces the negative effects of antibiotics on colonic fermentation and could provide an additional benefit in patients with recurrent *Clostridium difficile* associated diarrhea (Wullt et al., 2007) as well as in critically ill patients (Klarin et al., 2008). Intake of *L. plantarum* 299v could have a preventive effect on milder gastrointestinal symptoms during treatment with antibiotics (Loönnermark et al., 2010).

7.2. Other *L. plantarum* Strains

There are only a few clinical trials with *L. plantarum* strains different from 299v. The administration of strain PL02 did not reduce significantly the diarrhea rate in children or the frequency of stools per day (Szymanski et al., 2008). A mixture of *Lactobacillus plantarum* CECT 7315 and CECT 7316 enhanced the systemic immunity in elderly subjects in a dose-response, double-blind, placebo-controlled, randomized pilot trial. A study with *L. plantarum* LP01 and *Bifidobacterium breve* BRO showed a significant reduction in pain and inflammation in patients with irritable bowel syndrome (Saggioro, 2010). A treatment with *Bifidobacterium bifidum* and *L. plantarum* 8PA3 for 5 days was compared with a standard therapy in 66 patients with alcoholic psychosis and liver damage.

Conclusions, Perspectives and Future

L. plantarum is a widely spread LAB both in food products and in human intestinal tract with a great probiotic potential. There is enough evidence of its use in irritable bowel syndrome and in *Clostridium difficile* diarrhea and shows remarkable prospects of being successfully employed in controlling hypercholesterolemia but more researches are necessary to clarify the whole mechanisms. Currently, only three *L. plantarum* strains are completely sequenced and other two are in study: *L. plantarum* NC8 and *L. plantarum* P8. It is necessary to complete these sequences as well as to study the genome sequence of *L. plantarum* strains actually sold as probiotics. It is necessary to compare more *in silico* findings with *in vitro* and *in vivo* research to assure not only genes presence but their genic expression. Even though there are international guidelines for the evaluation of probiotics in foods (FAO, 2002), the lack of standardized *in vitro* tests makes comparison, review and meta analysis studies among different strains of probiotics a difficult task. For that, we suggest an international effort to standardize probiotic *in vitro* tests in order to simplify the evaluation of probiotic potential. Although *L. plantarum* is traditionally related to a great variety of food and feed fermentations, it is necessary to further analyze some technological aspects around *L. plantarum* strains such as the use of encapsulating agents, technologies and their addition in different food matrixes because this would enhance their survival in food products and in the GIT. Human trials around *L. plantarum* strains are limited and the quantity of subjects analyzed is usually reduced. The most well-known strain is *L. plantarum* 299v, which is present actually in foods, beverages and food supplements (Proviva ®). We did not find any meta-analysis about *L. plantarum* potential health benefits and only a few about probiotics in general. We strongly recommend more human trials about *L. plantarum* and other probiotic strains in order to know better their clinical uses.

References

Abu-Tarboush, H.M., Al-Saiady, M.Y., Keir El-Din, A.H. 1996. Evaluation of diet containing *Lactobacilli* on performance, fecal coliform, and *Lactobacilli* of young dairy calves. *Animal Feed Science and Technology* 57, 39-49.

Ahrné, Nobaek, Jeppsson, Adlerberth, Wold, Molin. 1998. The normal *Lactobacillus* flora of healthy human rectal and oral mucosa. *Journal of Applied Microbiology* 85, 88-94.

Albesharat, R., Ehrmann, M.A., Korakli, M., Yazaji, S., Vogel, R.F. 2011. Phenotypic and genotypic analyses of lactic acid bacteria in local fermented food, breast milk and faeces of mothers and their babies. *Systematic and Applied Microbiology* 34, 148-155.

Anas M, Jamal H, Mebrouk E, Mebrouk K. 2008. Antimicrobial activity of *Lactobacillus* species isolated from Algerian Raw goat`s milk against *Staphilococcus aureus. World Journal of Dairy and Food Sciences*. 3 (2), 39-49.

Avallone, S., Guyot, B., Brillouet, J.-M., Olguin E. and Guiraud, J.-P. 2001. Microbiological and Biochemical Study of Coffee Fermentation. *Current Microbiology* 42, 252-256.

Ayed, L., Hamdi, M. 2002. Culture conditions of tannase production by *Lactobacillus plantarum. Biotechnology Letters* 24, 1763-1765.

Badet, C., Thebaud, N.B. 2008. Ecology of *Lactobacilli* in the oral cavity: a review of literature. *The open microbiology journal* 2, 38-48.

Begley, M., Hill, C., Gahan, C.G.M., 2006. Bile Salt Hydrolase Activity in Probiotics. *Applied and Environmental Microbiology* 72, 1729-1738.

Boekhorst, J., Wels, M., Kleerebezem, M., Siezen, R.J., 2006. The predicted secretome of *Lactobacillus plantarum* WCFS1 sheds light on interactions with its environment. *Microbiology* 152, 3175-3183.

Botes, M., van Reenen, C.A., Dicks, L.M.T. 2008. Evaluation of *Enterococcus mundtii* ST4SA and *Lactobacillus plantarum* 423 as probiotics by using a gastro-intestinal model with infant milk formulations as substrate. *International Journal of Food Microbiology* 128, 362-370.

Brinques, G.B., Ayub, M.A.Z. 2011. Effect of microencapsulation on survival of *Lactobacillus plantarum* in simulated gastrointestinal conditions, refrigeration, and yogurt. *Journal of Food Engineering* 103, 123-128.

Bron, P.A., Grangette, C., Mercenier, A., de Vos, W.M., Kleerebezem, M. 2004. Identification of *Lactobacillus plantarum* Genes That Are Induced in the Gastrointestinal Tract of Mice. *Journal of Bacteriology* 186, 5721-5729.

Burns, P., Cuffia, F., Milesi, M., Vinderola, G., Meinardi, C., Sabbag, N. and Hynes, E. 2011. Technological and probiotic role of adjunct cultures of non-starter lactobacilli in soft cheeses. *Food Microbiology. In press.*

Cammarota, M., De Rosa, M., Stellavato, A., Lamberti, M., Marzaioli, I., Giuliano, M. 2009. In vitro evaluation of *Lactobacillus plantarum* DSMZ 12028 as a probiotic: Emphasis on innate immunity. *International Journal of Food Microbiology* 135, 90-98.

Camu, N., De Winter, T., Verbrugghe, K., Cleenwerck, I., Vandamme, P., Takrama, J.S., Vancanneyt, M. and De Vuyst, L. 2007. Dynamics and biodiversity of populations of lactic acid bacteria and acetic acid bacteria involved in spontaneous heap fermentation of cocoa beans in ghana. *Applied and Environmental Microbiology* 73 (6), 1809-1824.

Cannon, J.P., Lee, T.A., Bolanos, J.T., Danziger, L.H. 2005. Pathogenic relevance of *Lactobacillus*: a retrospective review of over 200 cases. *European Journal of Clinical Microbiology and Infectious Diseases* 24, 31-40.

Cesaro, C., Tiso, A., Del Prete, A., Cariello, R., Tuccillo, C., Cotticelli, G., del Vecchio Blanco, C., Loguercio, C., 2011. Gut microbiota and probiotics in chronic liver diseases. *Digestive and Liver Disease* 43, 431-438.

Cho, G.-S., Hanak, A., Huch, M., Holzapfel, W., Franz, C., 2010. Investigation into the Potential of Bacteriocinogenic *Lactobacillus plantarum* BFE 5092 for Biopreservation of Raw Turkey Meat. *Probiotics and Antimicrobial Proteins* 2, 241-249.

Claesson M. J., van Sinderen D and O'Toole P. W. 2008. *Lactobacillus* phylogenomics – towards a reclassification of the genus. International *Journal of Systematic and Evolutionary Microbiology*. 58 (12): 2945-2954.

Collado MC, Gueimonde M, Salminen S. 2010. Probiotics in adhesion of pathogens: mechanisms of actions. Bioactive foods in promoting health: Probiotics and probiotics. Edited by Watson RR, Preedy VR. Academic Press/Elsevier. 13, 353-370.

Collado, M., Meriluoto, J., Salminen, S. 2008. Adhesion and aggregation properties of probiotic and pathogen strains. *European Food Research and Technology* 226, 1065-1073.

Collado, MC., Surono, I., Meriluoto, J., Salminen, S. 2007. Indigenous Dadih Lactic Acid Bacteria: Cell-Surface Properties and Interactions with Pathogens. *Journal of Food Science* 72, M89-M93.

Daeschel, M.A., McKenney, M.C., McDonald, L.C. 1990. Bacteriocidal activity of *Lactobacillus plantarum* C-11. *Food Microbiology* 7, 91-98.

Dal Bello, F., Clarke, C.I., Ryan, L.A.M., Ulmer, H., Schober, T.J., Ström, K., Sjögren, J., van Sinderen, D., Schnürer, J., Arendt, E.K. 2007. Improvement of the quality and shelf life of wheat bread by fermentation with the antifungal strain *Lactobacillus plantarum* FST 1.7. *Journal of Cereal Science*. 45, 309-318.

Darilmaz, D.O., Beyatli, Y., Yuksekdag, Z.N. 2011. Aggregation and Hydrophobicity Properties of 6 Dairy Propionibacteria Strains Isolated from Homemade Turkish Cheeses. *Journal of Food Science*, no-no.

De Angelis, M., Corsetti, A., Tosti, N., Rossi, J., Corbo, M. R., Gobbetti, M. 2001. Characterization of Non-Starter lactic acid bacteria from italian ewe cheeses based on phenotypic, genotypic, and cell wall protein analyses. *Applied and Environmental Microbiology*. 67 (5), 2011-2020.

De Angelis, M., de Candia, S., Calasso, M.P., Faccia, M., Guinee, T.P., Simonetti, M.C., Gobbetti, M. 2008. Selection and use of autochthonous multiple strain cultures for the manufacture of high-moisture traditional Mozzarella cheese. *International Journal of Food Microbiology*. 125, 123-132.

De Angelis, M., Siragusa, S., Berloco, M., Caputo, L., Settanni, L., Alfonsi, G., Amerio, M., Grandi, A., Ragni, A., Gobbetti, M. 2006. Selection of potential probiotic lactobacilli from pig feces to be used as additives in pelleted feeding. *Research in Microbiology*. 157, 792-801.

de Vries, M.C., Vaughan, E.E., Kleerebezem, M., de Vos, W.M. 2006. Lactobacillus plantarum—survival, functional and potential probiotic properties in the human intestinal tract. *International Dairy Journal*. 16, 1018-1028.

Dellaglio, F., Bottazzi V., Vescovo, M. 1975. Deoxyribonucleic Acid Homology Among *Lactobacillus* Species of the Subgenus *Streptobacterium* Orla-Jensen. *International Journal of Systematic Bacteriology*. 25, 160-172.

Dennis, D., Kaplan, N.O. 1960. D- and L-lactic acid dehydrogenases in *Lactobacillus plantarum*. *The Journal of biological chemistry*. 235, 810-818.

Desai, K.M., Akolkar, S.K., Badhe, Y.P., Tambe, S.S., Lele, S.S. 2006. Optimization of fermentation media for exopolysaccharide production from *Lactobacillus plantarum* using artificial intelligence-based techniques. *Process Biochemistry*. 41, 1842-1848.

Di Cagno, R., Cardinali, G., Minervini, G., Antonielli, L., Rizzello, C.G., Ricciuti, P., Gobbetti, M. 2010. Taxonomic structure of the yeasts and lactic acid bacteria microbiota of pineapple (Ananas comosus L. Merr.) and use of autochthonous starters for minimally processing. *Food Microbiology*. 27, 381-389.

Djossou, O., Perraud-Gaime, I., Lakhal Mirleau, F., Rodriguez-Serrano, G., Karou, G., Niamke, S., Ouzari, I., Boudabous, A., Roussos, S. 2011. Robusta coffee beans post-harvest microflora: *Lactobacillus plantarum* sp. as potential antagonist of *Aspergillus carbonarius*. Anaerobe. *In press.*

Doyle P and Meng J. 2006. Bacteria in Food and Beverage Production. Chapter 3.5. Prokaryotes. 1: 797:811.

Duan Y, Tan Z, Wang Y, Li Z, Li Z, Qin, Huo Y and Cai Y. 2008. Identification and characterization of lactic acid bacteria isolated from Tibetan Qula cheeses. *The Journal of General and Applied Microbiology*. 54, 51-60.

Edalatian, M., Najafi, M., Mortazavi, S., Alegría, Á., Nassiri, M., Bassami, M., Mayo, B. 2012. Microbial diversity of the traditional Iranian cheeses Lighvan and Koozeh, as revealed by polyphasic culturing and culture-independent approaches. *Dairy Science and Technology* 92, 75-90.

Egervärn, M., Roos, S., Lindmark, H. 2009. Identification and characterization of antibiotic resistance genes in *Lactobacillus reuteri* and *Lactobacillus plantarum*. *Journal of Applied Microbiology* 107, 1658-1668.

Enan, G., El-Essawy, A.A., Uyttendaele, M., Debevere, J. 1996. Antibacterial activity of Lactobacillus plantarum UG1 isolated from dry sausage: characterization, production and bactericidal action of plantaricin UG1. *International Journal of Food Microbiology* 30, 189-215.

Endo, A., Futagawa-Endo, Y., Dicks, L.M.T. 2010. Diversity of *Lactobacillus* and *Bifidobacterium* in feces of herbivores, omnivores and carnivores. *Anaerobe*. 16, 590-596.

Escalante-Minakata, P., Blaschek, H.P., Barba de la Rosa, A.P., Santos, L., De León-Rodríguez, A., 2008. Identification of yeast and bacteria involved in the mezcal fermentation of Agave salmiana. *Letters in Applied Microbiology*. 46, 626-630.

Essid, I., Medini, M., Hassouna, M., 2009. Technological and safety properties of Lactobacillus plantarum strains isolated from a Tunisian traditional salted meat. *Meat Science*. 81, 203-208.

FAO.2002. Guidelines for the Evaluation of Probiotics in Food. Report of a Joint FAO/WHO Working Group on Drafting Guidelines for the Evaluation of Probiotics in Foods. London Ontario, Canada. April 30 and May 1, 2002.

Fayol-Messaoudi, D., Coconnier-Polter, M.H., Moal, V.L.-L., Atassi, F., Berger, C.N., Servin, A.L. 2007. The *Lactobacillus plantarum* strain ACA-DC287 isolated from a Greek cheese demonstrates antagonistic activity in vitro and in vivo against *Salmonella enterica* serovar *Typhimurium*. *Journal of Applied Microbiology*. 103, 657-665.

Feld, L., Bielak, E., Hammer, K., Wilcks, A. 2009. Characterization of a small erythromycin resistance plasmid pLFE1 from the food-isolate *Lactobacillus plantarum* M345. *Plasmid*. 61, 159-170.

Foligne B, Grangette C, Pot B. 2005. Probiotics in IBD: Mucosal and systemic routes of administration may promote similar effects. *Gut.* 54:727–728.

Franz, Du, T., Olasupo, Schillinger, Holzapfel. 1998. Plantaricin D, a bacteriocin produced by *Lactobacillus plantarum* BFE 905 from ready-to-eat salad. *Letters in Applied Microbiology*. 26, 231-235.

García Fontán, M.C., Lorenzo, J.M., Martínez, S., Franco, I., Carballo, J. 2007. Microbiological characteristics of Botillo, a Spanish traditional pork sausage. LWT - *Food Science and Technology*. 40, 1610-1622.

Gbassi, G.K., Vandamme, T., Ennahar, S., Marchioni, E. 2009. Microencapsulation of Lactobacillus plantarum spp in an alginate matrix coated with whey proteins. *International Journal of Food Microbiology* 129, 103-105. Burgain, J., Gaiani, C., Linder, M., Scher, J., 2011. Encapsulation of probiotic living cells: From laboratory scale to industrial applications. *Journal of Food Engineering*. 104, 467-483.

Gevers, D., Danielsen, M., Huys, G., Swings, J., 2003. Molecular Characterization of tet(M) Genes in *Lactobacillus* Isolates from Different Types of Fermented Dry Sausage. *Applied and Environmental Microbiology*. 69, 1270-1275.

Golowczyc, M.A., Silva, J., Teixeira, P., De Antoni, G.L., Abraham, A.G., 2011. Cellular injuries of spray-dried Lactobacillus spp. isolated from kefir and their impact on probiotic properties. *International Journal of Food Microbiology*. 144, 556-560.

Gong, H.S., Meng, X.C., Wang, H., 2010. Plantaricin MG active against Gram-negative bacteria produced by *Lactobacillus plantarum* KLDS1.0391 isolated from "Jiaoke", a traditional fermented cream from China. *Food Control*. 21, 89-96.

Goossens, D., Jonkers, D., Russel, M., Thijs, A., Bogaard, A.v.d., Stobberingh, E., Stockbrügger, R. 2005. Survival of the probiotic, *L. plantarum* 299v and its effects on the faecal bacterial flora, with and without gastric acid inhibition. *Digestive and Liver Disease*. 37, 44-50.

Gueimonde, M., Jalonen, L., He, F., Hiramatsu, M., Salminen, S. 2006. Adhesion and competitive inhibition and displacement of human enteropathogens by selected *Lactobacilli*. *Food Research International*. 39, 467-471.

Haza, A.I., Zabala, A., Morales, P., 2004. Protective effect and cytokine production of a *Lactobacillus plantarum* strain isolated from ewes' milk cheese. *International Dairy Journal*. 14, 29-38.

Hernández, D., Cardell, E., Zárate, V., 2005. Antimicrobial activity of lactic acid bacteria isolated from Tenerife cheese: initial characterization of plantaricin TF711, a bacteriocin-like substance produced by *Lactobacillus plantarum* TF711. *Journal of Applied Microbiology*. 99, 77-84.

Hong, S.I., Pyun, Y.R., 1999. Inactivation Kinetics of *Lactobacillus plantarum* by High Pressure Carbon Dioxide. *Journal of Food Science*. 64, 728-733.

Iwamoto, K., Tsuruta, H., Nishitaini, Y., Osawa, R. 2008. Identification and cloning of a gene encoding tannase (tannin acylhydrolase) from *Lactobacillus plantarum* ATCC 14917T. *Systematic and Applied Microbiology*. 31, 269-277.

Jacobsen, L., Wilcks, A., Hammer, K., Huys, G., Gevers, D., Andersen, S.R., 2007. Horizontal transfer of tet(M) and erm(B) resistance plasmids from food strains of *Lactobacillus plantarum* to *Enterococcus faecalis* JH2-2 in the gastrointestinal tract of gnotobiotic rats. *FEMS Microbiology Ecology*. 59, 158-166.

Jaichumjai, P., Valyasevi, R., Assavanig, A., Kurdi, P., 2010. Isolation and characterization of acid-sensitive *Lactobacillus plantarum* with application as starter culture for Nham production. *Food Microbiology.* 27, 741-748.

Jamaly N, Benjouad A, Bouksaim M. 2011. Probiotic potentital of *Lactobacillus* strains isolated from known popular traditional Moroccan dairy products. *British microbiology research Journal.* 1(4) 79-94.

Jeun, J., Kim, S., Cho, S.-Y., Jun, H.-j., Park, H.-J., Seo, J.-G., Chung, M.-J., Lee, S.-J., 2010. Hypocholesterolemic effects of *Lactobacillus plantarum* KCTC3928 by increased bile acid excretion in C57BL/6 mice. *Nutrition.* 26, 321-330.

Jiménez-Pranteda, M.L., Poncelet, D., Náder-Macías, M.E., Arcos, A., Aguilera, M., Monteoliva-Sánchez, M., Ramos-Cormenzana, A. 2011. Stability of *Lactobacilli* encapsulated in various microbial polymers. *Journal of Bioscience and Bioengineering. In press.*

Jónsson, S., Clausen, E., Raa, J. 1983. Amino Acid Degradation by a *Lactobacillus plantarum* Strain from Fish. *Systematic and Applied Microbiology.* 4, 148-154.

Kaban, G., Kaya, M., 2009. Effects of *Lactobacillus plantarum* and *Staphylococcus xylosus* on the Quality Characteristics of Dry Fermented Sausage "Sucuk". *Journal of Food Science.* 74, S58-S63.

Kagkli, D.M., Vancanneyt, M., Hill, C., Vandamme, P., Cogan, T.M., 2007. *Enterococcus* and *Lactobacillus* contamination of raw milk in a farm dairy environment. *International Journal of Food Microbiology.* 114, 243-251.

Kaneko, Y., Kobayashi, H., Kiatpapan, P., Nishimoto, T., Napitupulu, R., Ono, H., Murooka, Y., 2000. Development of a host-vector system for Lactobacillus plantarum L137 isolated from a traditional fermented food produced in the Philippines. *Journal of Bioscience and Bioengineering.* 89, 62-67.

Karasu, N., Şimşek, Ö., Çon, A., 2010. Technological and probiotic characteristics of *Lactobacillus plantarum* strains isolated from traditionally produced fermented vegetables. *Annals of Microbiology.* 60, 227-234.

Karczewski, J., Troost, F.J., Konings, I., Dekker, J., Kleerebezem, M., Brummer, R.-J.M., Wells, J.M., 2010. Regulation of human epithelial tight junction proteins by *Lactobacillus plantarum in vivo* and protective effects on the epithelial barrier. *American Journal of Physiology - Gastrointestinal and Liver Physiology.* 298, G851-G859.

Kaushik, J., Kumar, A., Duary, R., Mohanty, A., Grover, S., Batish, V., 2009. Functional and Probiotic Attributes of an Indigenous Isolate of *Lactobacillus plantarum. PLoS ONE* 4, e8099.

Kawashima, T., Hayashi, K., Kosaka, A., Kawashima, M., Igarashi, T., Tsutsui, H., Tsuji, N.M., Nishimura, I., Hayashi, T., Obata, A., 2011. *Lactobacillus plantarum* strain YU from fermented foods activates Th1 and protective immune responses. *International Immunopharmacology.* 11, 2017-2024.

Khalid, K. 2011. An overview of lactic acid bacteria. *International Journal of Biosciences* (IJB). 1 (3), 1-13.

Khalil R, El-Hlafawy K, Mahrous H, Kamaly K, Frank J, El Solda M. 2007. Evaluation of the probiotic potential of lactic acid bacteria isolated from faeces of breast-fed infants in Egypt. *African Journal of Microbiology.* 6 (7), 939-949.

Kim, B.-G., Choi, S.-Y., Kim, M.-R., Suh, H.J., Park, H.J., 2010. Changes of ginsenosides in Korean red ginseng (Panax ginseng) fermented by *Lactobacillus plantarum* M1. *Process Biochemistry*. 45, 1319-1324.

Kim, J.-H., Sunako, M., Ono, H., Murooka, Y., Fukusaki, E., Yamashita, M. 2008. Characterization of Gene Encoding Amylopullulanase from Plant-Originated Lactic Acid Bacterium, *Lactobacillus plantarum* L137. *Journal of Bioscience and Bioengineering* 106, 449-459.

Kinoshita, H., Wakahara, N., Watanabe, M., Kawasaki, T., Matsuo, H., Kawai, Y., Kitazawa, H., Ohnuma, S., Miura, K., Horii, A., Saito, T., Cell surface glyceraldehyde-3-phosphate dehydrogenase (GAPDH) of Lactobacillus plantarum LA 318 recognizes human A and B blood group antigens. *Research in Microbiology*. 159, 685-691.

Klaenhammer T, Altermann E, Arigoni F, Bolotin A, Breidt F, Broadbent J, Cano R, Chaillou S, Deutscher J, Gasson M, van de Guchte M, Guzzo J, Hartke A, Hawkins T, Hols P, Hutkins R, Kleerebezem M, Kok J, Steele J, O'Sullivan D, de Vos W, Weimer B, Zagorec M and Siezen R. 2002. Discovering lactic acid bacteria by genomics. *Antonie van Leeuwenhoek*. 82 (1), 29-58.

Klarin, B., Wullt, M., Palmquist, I., Molin, G., Larsson, A., and Jeppsson, B. 2008. Lactobacillus plantarum 299v reduces colonisation of Clostridium difficile in critically ill patients treated with antibiotics. *Acta Anaesthesiologica Scandinavica, 52*(8), 1096-1102.

Kleerebezem, M., Hugenholtz, J. 2003. Metabolic pathway engineering in lactic acid bacteria. *Curr. Opin. Biotechnol.* 14 (2), 232-237.

Kmet, V., Lucchini, F., 1997. Aggregation-promoting factor in human vaginal Lactobacillus strains. *FEMS Immunology and Medical Microbiology*. 19, 111-114.

Knox, K.W., Wicken, A.J., 1972. Serological Studies on the Teichoic Acids of Lactobacillus plantarum. Infection and Immunity. 6, 43-49.

Koleva P, Georgieva R, Nikolova D, Danova S. 2009. Lactic acid microflora of bulgarian milk products from mountain regions. *Biotechnoliogy and biotechnological equipment*. 23 (2), 856-860.

Kook M-C, Seo M-J, Cheigh C-I, Lee S-J, Pyun Y-R, Park H. 2010. Enhancement of c-amminobutyric acid production by *Lactobacillus sakei* B2–16 expressing glutamate decarboxylase from *Lactobacillus plantarum* ATCC 14917. *J. Korean Soc. Appl. Biol. Chem.* 53(6), 816–820.

Kos, B., Šušković, J., Vuković, S., Šimpraga, M., Frece, J., Matošić, S., 2003. Adhesion and aggregation ability of probiotic strain *Lactobacillus acidophilus* M92. *Journal of Applied Microbiology*. 94, 981-987.

Kotzamanidis, C., Kourelis, A., Litopoulou-Tzanetaki, E., Tzanetakis, N., Yiangou, M., 2010. Evaluation of adhesion capacity, cell surface traits and immunomodulatory activity of presumptive probiotic *Lactobacillus* strains. *International Journal of Food Microbiology*. 140, 154-163.

Kuda, T., Kaneko, N., Yano, T., Mori, M., 2010. Induction of superoxide anion radical scavenging capacity in Japanese white radish juice and milk by *Lactobacillus plantarum* isolated from aji-narezushi and kaburazushi. *Food Chemistry*. 120, 517-522.

Kumar, H., Rangrez, A., Dayananda, K., Atre, A., Patole, M., Shouche, Y. 2011. *Lactobacillus plantarum* (VR1) isolated from an Ayurvedic medicine (Kutajarista) ameliorates in vitro cellular damage caused by Aeromonas veronii. *BMC Microbiology*. 11, 1-11.

Lambert, J.M., Bongers, R.S., de Vos, W.M., Kleerebezem, M., 2008. Functional Analysis of Four Bile Salt Hydrolase and Penicillin Acylase Family Members in *Lactobacillus plantarum* WCFS1. *Applied and Environmental Microbiology*. 74, 4719-4726.

Landete, J.M., Curiel, J.A., Rodríguez, H., de las Rivas, B., Muñoz, R., 2008. Study of the inhibitory activity of phenolic compounds found in olive products and their degradation by *Lactobacillus plantarum* strains. *Food Chemistry*. 107, 320-326.

Lara-Villoslada F, Olivares M, Xaus J. 2010. Safety of probiotic bacteria. Bioactive foods in promoting health: probiotics and prebiotics. Edited by Watson RR and Preedy VR. Academic Press. 4, 47-58.

Larsen, N., Michaelsen, K.F., Pærregaard, A., Vogensen, F.K., Jakobsen, M., 2009. A comparative study on adhesion and recovery of potential probiotic strains of *Lactobacillus* spp. by in vitro assay and analysis of human colon biopsies. *Microbial Ecology in Health and Disease*. 21, 95-99.

Lavermicocca, P., Valerio, F., Evidente, A., Lazzaroni, S., Corsetti, A., Gobbetti, M., Sep. 2000. Purification and characterization of novel antifungal compounds from the sourdough *Lactobacillus plantarum* strain 21B. *Applied and Environmental Microbiology*. 66 (9), 4084-4090.

Leal-Sánchez, M. V., Jiménez-Díaz, R., Maldonado-Barragán, A., Garrido-Fernández, A., Ruiz-Barba, JL. 2002. Optimization of bacteriocin production by batch fermentation of lactobacillus plantarum LPCO10. *Applied and Environmental Microbiology*. 68 (9), 4465-4471.

Lebeer, S., Vanderleyden, J., De Keersmaecker, S. C J. 2008. Genes and molecules of *Lactobacilli* supporting probiotic action. *Microbiology and Molecular Biology Reviews* 72 (4), 728-764.

Liu, C., Zhang, Z.-Y., Dong, K., Yuan, J.-P., Guo, X.-K., 2009. Antibiotic Resistance of Probiotic Strains of Lactic Acid Bacteria Isolated from Marketed Foods and Drugs. *Biomedical and Environmental Sciences*. 22, 401-412.

Liu, H., Xu, W., Luo, Y., Tian, H., Wang, H., Guo, X., Yuan, Y., Huang, K. 2011. Assessment of tolerance to multistresses and *in vitro* cell adhesion in genetically modified *Lactobacillus plantarum* 590. *Antonie van Leeuwenhoek*. 99, 579-589.

Liu, Y.-W., Su, Y.-W., Ong, W.-K., Cheng, T.-H., Tsai, Y.-C., 2011. Oral administration of *Lactobacillus plantarum* K68 ameliorates DSS-induced ulcerative colitis in BALB/c mice via the anti-inflammatory and immunomodulatory activities. *International Immunopharmacology*. 11, 2159-2166.

Lönnermark, E., Friman, V., Lappas, G., Sandberg, T., Berggren, A., Adlerberth, I. 2010. Intake of *Lactobacillus plantarum* Reduces Certain Gastrointestinal Symptoms During Treatment With Antibiotics. *Journal of Clinical Gastroenterology*. 44, 106-112.

Lönnermark, E., Nowrouzinan, F., Adlerberth, I., Ahrné, S., Wold, A., Friman, V., 2012. Oral and faecal lactobacilli and their expression of mannose-specific adhesins in individuals with and without IgA deficiency. *International Journal of Medical Microbiology* 302, 53-60.

López I, López R, Martínez MP, Torres C, Ruiz-Larrea F. 2008. Performance of malolactic fermentation by inoculation of selected *"Lactobacillus plantarum"* and "*Oenococcus oeni"* strains isolated from Rioja red wines. Vitis: *Journal of Grapevine Research*. 47 (2), 123 -129.

Lucchini, F., Kmet, V., Cesena, C., Coppi, L., Bottazzi, V., Morelli, L. 1998. Specific detection of a probiotic *Lactobacillus* strain in faecal samples by using multiplex PCR. *FEMS Microbiology Letters*. 158, 273-278.

Lynch, C.M., Muir, D.D., Banks, J.M., McSweeney, P.L.H., Fox, P.F. 1999. Influence of Adjunct Cultures of *Lactobacillus paracasei ssp. paracasei* or *Lactobacillus plantarum* on Cheddar Cheese Ripening. *Journal of Dairy Science*. 82, 1618-1628.

Maeda, N., Nakamura, R., Hirose, Y., Murosaki, S., Yamamoto, Y., Kase, T., Yoshikai, Y. 2009. Oral administration of heat-killed *Lactobacillus plantarum* L-137 enhances protection against influenza virus infection by stimulation of type I interferon production in mice. *International Immunopharmacology*. 9, 1122-1125.

Mäkimattila, E., Kahala, M., Joutsjoki, V. 2011. Characterization and electrotransformation of *Lactobacillus plantarum* and *Lactobacillus paraplantarum* isolated from fermented vegetables. *World Journal of Microbiology and Biotechnology*. 27, 371-379.

Margolles A, Mayo B, Ruas-Madievo P. 2009. Screening, identification and characterization of *Lactobacillus* and *Bifidobacterium* strains. Handbook of probiotics and probiotics. Edited by Lee YK, Salminen S. John Wiley and Sons. Inc. 1 (1.2), 4-24.

Martín, R., Heilig, G.H.J., Zoetendal, E.G., Smidt, H., Rodríguez, J.M., 2007. Diversity of the *Lactobacillus* group in breast milk and vagina of healthy women and potential role in the colonization of the infant gut. *Journal of Applied Microbiology*. 103, 2638-2644.

Martinez, R., Franceschini, S., Patta, M., Quintana, S., Nunes, Á., Moreira, J., Anukam, K., Reid, G., De Martinis, E. 2008. Analysis of Vaginal *Lactobacilli* from Healthy and Infected Brazilian Women. *Applied and Environmental Microbiology*. 74, 4539-4542.

Mathara, J.M., Schillinger, U., Guigas, C., Franz, C., Kutima, P.M., Mbugua, S.K., Shin, H.K., Holzapfel, W.H., 2008. Functional characteristics of *Lactobacillus* spp. from traditional Maasai fermented milk products in Kenya. *International Journal of Food Microbiology*. 126, 57-64.

Mathur, S., Singh, R., 2005. Antibiotic resistance in food lactic acid bacteria—a review. *International Journal of Food Microbiology*. 105, 281-295.

Mayo B, Aleksandrzak-Piekarczyk T, Fernández F, Kowalczyk M, Álvarez-Martín P, and Bardowski J. 2010. Updates in the Metabolism of Lactic Acid Bacteria, Biotechnology of Lactic Acid Bacteria. Novel applications. Edited by Fernanda Mozzi, Raúl R. Raya and Graciela M. Vignolo. Blackwell Publishing. Chapter 1: 1-32.

Michida, H., Tamalampudi, S., Pandiella, S.S., Webb, C., Fukuda, H., Kondo, A. 2006. Effect of cereal extracts and cereal fiber on viability of *Lactobacillus plantarum* under gastrointestinal tract conditions. *Biochemical Engineering Journal*. 28, 73-78.

Morales, F., Morales, J., Hernández, C., Hernández-Sánchez, H., 2011. Isolation and Partial Characterization of Halotolerant Lactic Acid Bacteria from Two Mexican Cheeses. *Applied Biochemistry and Biotechnology*. 164, 889-905.

Mourad, K. and K. Nour-Eddine, 2006. *In vitro* Preselection Criteria For Probiotic *Lactobacillus Plantarum* Strains Of Fermented Olives Origin. *Int. J. Prob. and Preb*. 1: 27-32.

Müller, D.M., Carrasco, M.S., Tonarelli, G.G., Simonetta, A.C. 2009. Characterization and purification of a new bacteriocin with a broad inhibitory spectrum produced by *Lactobacillus plantarum* lp 31 strain isolated from dry-fermented sausage. *Journal of Applied Microbiology*. 106, 2031-2040.

Nanda, D. K., Tomar, S. K., Singh, R., Mal, G., Sing, P., Arora, D. K., Joshi, B. K., Chaudarty, R. and Kumar, D. 2011. Phenotypic and genotypic characterisation of *Lactobacilli* isolated from camel cheese produced in India. *International Journal of Dairy Technology*. 64:437–443.

Naruszewicz, M., Johansson, M.-L., Zapolska-Downar, D., Bukowska, H. 2002. Effect of *Lactobacillus plantarum* 299v on cardiovascular disease risk factors in smokers. *The American Journal of Clinical Nutrition*. 76, 1249-1255.

Nawaz M, Wang J, Zhou A, Ma Ch, Wu X, Xu J. 2011. Screening and characterization of new potentially probiotic lactobacilli from breast .fed healthy babies in Pakistan. *African Journal of Microbiology Research*. 5 (12), 1428-1436.

Nedelcheva, P., Denkova, Z., Denev, P., Slavchev, A., Krastanov, A., 2010. Probiotic Strain *Lactobacillus Plantarum* NBIMCC 2415 with Antioxidant Activity as a Starter Culture in the Production of Dried Fermented Meat Products. *Biotechnology and Biotechnological Equipment*. 24, 1624-1630.

Nespolo, C.R., Brandelli, A., 2010. Production of bacteriocin-like substances by lactic acid bacteria isolated from regional ovine cheese. *Brazilian Journal of Microbiology*. 41, 1009-1018.

Nguyen, T.D.T., Kang, J.H., Lee, M.S. 2007. Characterization of *Lactobacillus plantarum* PH04, a potential probiotic bacterium with cholesterol-lowering effects. *International Journal of Food Microbiology*. 113, 358-361.

Niedzielin, K., Kordecki, H., Birkenfeld, B.e. 2001. A controlled, double-blind, randomized study on the efficacy of *Lactobacillus plantarum* 299V in patients with irritable bowel syndrome. *European Journal of Gastroenterology and Hepatology*. 13, 1143-1147.

Nieto-Arribas, P., Poveda, J.M., Seseña, S., Palop, L., Cabezas, L. 2009. Technological characterization of *Lactobacillus* isolates from traditional Manchego cheese for potential use as adjunct starter cultures. *Food Control*. 20, 1092-1098.

Nishitani, Y., Sasaki, E., Fujisawa, T., Osawa, R.o., 2004. Genotypic Analyses of Lactobacilli with a Range of Tannase Activities Isolated from Human Feces and Fermented Foods. *Systematic and Applied Microbiology*. 27, 109-117.

Nobaek, S., Johansson, M.-L., Molin, G., Ahrne, S., Jeppsson, B. 2000. Alteration of intestinal microflora is associated with reduction in abdominal bloating and pain in patients with irritable bowel syndrome. *Am. J. Gastroenterol*. 95, 1231-1238.

O`Flaherty S; Yong Jun G, Klaenhammer TR. 2009. Genomics of probiotic bacteria. Prebiotics and Probiotics Science and Technology, by Charalampopoulos, Dimitris; Rastall, R A.Springer-Verlag. 17, 681-723.

O'Flaherty, S., Klaenhammer, T.R. 2010. The role and potential of probiotic bacteria in the gut, and the communication between gut microflora and gut/host. *International Dairy Journal*. 20, 262-268.

Omar, N.B., Abriouel, H., Lucas, R., Martínez-Cañamero, M., Guyot, J.-P., Gálvez, A. 2006. Isolation of bacteriocinogenic Lactobacillus plantarum strains from ben saalga, a traditional fermented gruel from Burkina Faso. *International Journal of Food Microbiology* 112, 44-50.

Osawa R, Kuroiso K, Goto S, Shimizu S. 2000. Isolation of tannin-Degrading Lactobacilli from Human and Fermented Foods. *Appl. Envioon. Microbiol*. 66 (7), 3093-3097.

Ouoba, L.I.I., Lei, V., Jensen, L.B., 2008. Resistance of potential probiotic lactic acid bacteria and bifidobacteria of African and European origin to antimicrobials: Determination and transferability of the resistance genes to other bacteria. *International Journal of Food Microbiology*. 121, 217-224.

Ouwehand, A.C., Kirjavainen, P.V., Grönlund, M.M., Isolauri, E., Salminen, S.J., 1999. Adhesion of probiotic micro-organisms to intestinal mucus. *International Dairy Journal*. 9, 623-630.

Pan, Q., Zhang, L., Li, J., Chen, T., Chen, W., Wang, G., Yin, J., 2011. Characterization of pLP18, a novel cryptic plasmid of Lactobacillus plantarum PC518 isolated from Chinese pickle. *Plasmid*. 65, 204-209.

Paolillo, R., Romano Carratelli, C., Sorrentino, S., Mazzola, N., Rizzo, A., 2009. Immunomodulatory effects of *Lactobacillus plantarum* on human colon cancer cells. *International Immunopharmacology*. 9, 1265-1271.

Parente, E., Ciocia, F., Ricciardi, A., Zotta, T., Felis, G.E., Torriani, S., 2010. Diversity of stress tolerance in *Lactobacillus plantarum, Lactobacillus pentosus* and *Lactobacillus paraplantarum:* A multivariate screening study. *International Journal of Food Microbiology*. 144, 270-279.

Pavlova, S.I., Kilic, A.O., Kilic, S.S., So, J.S., Nader-Macias, M.E., Simoes, J.A., Tao, L., 2002. Genetic diversity of vaginal *lactobacilli* from women in different countries based on 16S rRNA gene sequences. *Journal of Applied Microbiology*. 92, 451-459.

Pepe, O., Blajotta, G., Anastasio, M., Moschetti, G., Ercolini, D., Villani, F., 2004. Technological and Molecular Diversity of *Lactobacillus plantarum* Strains Isolated from Naturally Fermented Sourdoughs. *Systematic and Applied Microbiology*. 27, 443-453.

Piwat, S., Teanpaisan, R., Dahlén, G., Thitasomakul, S., Douglas, C.W.I., 2011. Acid production and growth by oral *Lactobacillus* species *in vitro*. *Journal of Investigative and Clinical Dentistry*. 2, 1-6.

Plumed-Ferrer, C., Koistinen, K.M., Tolonen, T.L., Lehesranta, S.J., Kärenlampi, S.O., Mäkimattila, E., Joutsjoki, V., Virtanen, V., von Wright, A. 2008. Comparative Study of Sugar Fermentation and Protein Expression Patterns of Two *Lactobacillus plantarum* Strains Grown in Three Different Media. *Applied and Environmental Microbiology*. 74, 5349-5358.

Pretzer, G., Snel, J., Molenaar, D., Wiersma, A., Bron, P.A., Lambert, J., de Vos, W.M., van der Meer, R., Smits, M.A., Kleerebezem, M. 2005. Biodiversity-Based Identification and Functional Characterization of the Mannose-Specific Adhesin of Lactobacillus plantarum. *Journal of Bacteriology*. 187, 6128-6136.

Qin, H., Zhang, Z., Hang, X., Jiang, Y. 2009. *L. plantarum* prevents Enteroinvasive *Escherichia coli* -induced tight junction proteins changes in intestinal epithelial cells. *BMC Microbiology*. 9, 1-9.

Ramiah, K., van Reenen, C.A., Dicks, L.M.T. 2007. Expression of the mucus adhesion genes Mub and MapA, adhesion-like factor EF-Tu and bacteriocin gene plaA of *Lactobacillus plantarum* 423, monitored with real-time PCR. *International Journal of Food Microbiology*. 116, 405-409.

Remus, D.M., Kleerebezem, M., Bron, P.A. 2011. An intimate tête-à-tête — How probiotic *lactobacilli* communicate with the host. *European Journal* of Pharmacology. 668, Supplement 1, S33-S42.

Ridlon, J.M., McGarr, S.E., Hylemon, P.B., 2005. Development of methods for the detection and quantification of 7α-dehydroxylating clostridia, *Desulfovibrio vulgaris,*

Methanobrevibacter smithii, and *Lactobacillus plantarum* in human feces. *Clinica Chimica Acta*. 357, 55-64.

Rojo-Bezares, B., Sáenz, Y., Navarro, L., Zarazaga, M., Ruiz-Larrea, F., Torres, C., 2007. Coculture-inducible bacteriocin activity of *Lactobacillus plantarum* strain J23 isolated from grape must. *Food Microbiology*. 24, 482-491.

Rojo-Bezares, B., Sáenz, Y., Poeta, P., Zarazaga, M., Ruiz-Larrea, F., Torres, C., 2006. Assessment of antibiotic susceptibility within lactic acid bacteria strains isolated from wine. *International Journal of Food Microbiology*. 111, 234-240.

Ruiz-Moyano, S., Martín, A., Benito, M.J., Casquete, R., Serradilla, M.J., Córdoba, M.d.G., 2009. Safety and functional aspects of pre-selected *lactobacilli* for probiotic use in Iberian dry-fermented sausages. *Meat Science*. 83, 460-467.

Saggioro, A., 2004. Probiotics in The Treatment of Irritable Bowel Syndrome. *Journal of Clinical Gastroenterology*. 38, S104-S106.

Salminen, S., von Wright, A., Morelli, L., Marteau, P., Brassart, D., de Vos, W.M., Fondén, R., Saxelin, M., Collins, K., Mogensen, G., Birkeland, S.-E., Mattila-Sandholm, T. 1998. Demonstration of safety of probiotics — a review. *International Journal of Food Microbiology*. 44, 93-106.

Samappito W, Leenanon B, Levin RE. 2011. Microbiological Characteristics of ''Mhom'', a Thai Traditional Meat Sausage. *The open Food Science Journal*. 5, 31-36.

Samelis, J., Bleicher, A., Delbès-Paus, C., Kakouri, A., Neuhaus, K., Montel, M.-C. 2011. FTIR-based polyphasic identification of lactic acid bacteria isolated from traditional Greek Graviera cheese. *Food Microbiology*. 28, 76-83.

Samot, J., Lebreton, J., Badet, C., 2011. Adherence capacities of oral *lactobacilli* for potential probiotic purposes. *Anaerobe*. 17, 69-72.

Sánchez, B., González-Tejedo, C., Ruas-Madiedo, P., Urdaci, M.C., Margolles, A., 2011. Lactobacillus plantarum Extracellular Chitin-Binding Protein and Its Role in the Interaction between Chitin, Caco-2 Cells, and Mucin. *Applied and Environmental Microbiology*. 77, 1123-1126.

Sanni, A.I., Morlon-Guyot, J., Guyot, J.P. 2002. New efficient amylase-producing strains of Lactobacillus plantarum and L. fermentum isolated from different Nigerian traditional fermented foods. *International Journal of Food Microbiology*. 72, 53-62.

Sharpe, M.E., 1955. A Serological Classification of Lactobacilli. *Journal of General Microbiology*. 12, 107-122.

Siezen, R., Boekhorst, J., Muscariello, L., Molenaar, D., Renckens, B., Kleerebezem, M., 2006. *Lactobacillus plantarum* gene clusters encoding putative cell-surface protein complexes for carbohydrate utilization are conserved in specific gram-positive bacteria. *BMC Genomics*. 7, 1-13.

Siezen, R., van Hylckama Vlieg, J. 2011. Genomic diversity and versatility of *Lactobacillus plantarum* a natural metabolic engineer. *Microbial Cell Factories*. 10, 1-13.

Siezen, R.J., Tzeneva, V.A., Castioni, A., Wels, M., Phan, H.T.K., Rademaker, J.L.W., Starrenburg, M.J.C., Kleerebezem, M., Molenaar, D., Van Hylckama Vlieg, J.E.T. 2010. Phenotypic and genomic diversity of *Lactobacillus plantarum* strains isolated from various environmental niches. *Environmental Microbiology*. 12, 758-773.

Silvestroni, A., Connes, C., Sesma, F., de Giori, G.S., Piard, J.-C. 2002. Characterization of the melA Locus for α-Galactosidase in *Lactobacillus plantarum*. *Applied and Environmental Microbiology*. 68, 5464-5471.

Siragusa, S., Fontana, C., Cappa, F., Caputo, L., Cocconcelli, P.S., Gobbetti, M., De Angelis, M., 2011. Disruption of the gene encoding glutamate dehydrogenase affects growth, amino acids catabolism and survival of *Lactobacillus plantarum* UC1001. *International Dairy Journal*. 21, 59-68.

Sirilun S., Chaiyasut C., Kantachote D., Luxananil P. (2010). Characterization of non human origin probiotic *Lactobacillus plantarum* with cholesterol-lowering property. *African Journal of Microbiology Research*. 4, 994-1000.

Songré-Ouattara, L.T., Mouquet-Rivier, C., Icard-Vernière, C., Humblot, C., Diawara, B., Guyot, J.P. 2008. Enzyme activities of lactic acid bacteria from a pearl millet fermented gruel (ben-saalga) of functional interest in nutrition. *International Journal of Food Microbiology*. 128, 395-400.

Spano, G., Rinaldi, A., Ugliano, M., Moio, L., Beneduce, L., Massa, S. 2005. A β-glucosidase gene isolated from wine *Lactobacillus plantarum* is regulated by abiotic stresses. *Journal of Applied Microbiology*. 98, 855-861.

Stjernquist-Desatnik, A., 2000. Persistence of *Lactobacillus plantarum* DSM 9843 on Human Tonsillar Surface after Oral Administration in Fermented Oatmeal Gruel: A Pilot Study. *Acta Oto-laryngologica*. 120, 215-219.

Strahinic, I., Kojic, M., Tolinacki, M., Fira, D., Topisirovic, L., 2010. The presence of prtP proteinase gene in natural isolate *Lactobacillus plantarum* BGSJ3–18. *Letters in Applied Microbiology*. 50, 43-49.

Sumner, S.S., Speckhard, M.W., Somers, E.B., Taylor, S.L. 1985. Isolation of histamine-producing *Lactobacillus buchneri* from Swiss cheese implicated in a food poisoning outbreak. *Applied and Environmental Microbiology*. 50, 1094-1096.

Szymański, H., Armańska, M., Kowalska-Duplaga, K., Szajewska, H., 2008. *Bifidobacterium longum* PL03, *Lactobacillus rhamnosus* KL53A, and *Lactobacillus plantarum* PL02 in the Prevention of Antibiotic-Associated Diarrhea in Children: A Randomized Controlled Pilot Trial. *Digestion*. 78, 13-17.

Takeda, S., Takeshita, M., Kikuchi, Y., Dashnyam, B., Kawahara, S., Yoshida, H., Watanabe, W., Muguruma, M., Kurokawa, M. 2011. Efficacy of oral administration of heat-killed probiotics from Mongolian dairy products against influenza infection in mice: Alleviation of influenza infection by its immunomodulatory activity through intestinal immunity. *International Immunopharmacology*. 11, 1976-1983.

Tallon, R., Arias, S., Bressollier, and Urdaci, M.C. 2007. Strain- and matrix-dependent adhesion of Lactobacillus plantarum is mediated by proteinaceous bacterial compounds. *Journal of Applied Microbiology*. 102, 442-451.

Tamang, J.P., Tamang, B., Schillinger, U., Guigas, C., Holzapfel, W.H., 2009. Functional properties of lactic acid bacteria isolated from ethnic fermented vegetables of the Himalayas. *International Journal of Food Microbiology*. 135, 28-33.

Tanganurat, W., Quinquis, B., Leelawatcharamas, V., Bolotin, A., 2009. Genotypic and phenotypic characterization of *Lactobacillus plantarum* strains isolated from Thai fermented fruits and vegetables. *Journal of Basic Microbiology*. 49, 377-385.

Todorov SD, van Reenen CA and Theodore Dicks LM. 2004. Optimization of bacteriocin production by *Lactobacillus plantarum* ST13BR, a strain isolated from barley beer. *The Journal of General and Applied Microbiology*. 50, 149-157.

Todorov SD. 2008. Bacteriocin production by Lactobacillus plantarum AMA-K isolated from Amasi, a Zimbabwean fermented milk product and study of the adsorption of bacteriocin AMA-K to Listeria sp. *Braz. J. Microbiol.* 39 (1), 178-187.

Todorov, S., Onno, B., Sorokine, O., Chobert, J.M., Ivanova, I., Dousset, X. 1999. Detection and characterization of a novel antibacterial substance produced by *Lactobacillus plantarum* ST 31 isolated from sourdough. *International Journal of Food Microbiology*. 48, 167-177.

Todorov, S.D., 2009. Bacteriocins from *Lactobacillus plantarum* production, genetic organization and mode of action: produção, organização genética e modo de ação. *Brazilian Journal of Microbiology*. 40, 209-221.

Todorov, S.D., Dicks, L.M.T., 2005. *Lactobacillus plantarum* isolated from molasses produces bacteriocins active against Gram-negative bacteria. *Enzyme and Microbial Technology*. 36, 318-326.

Todorov, S.D., Dicks, L.M.T., 2006. Effect of medium components on bacteriocin production by *Lactobacillus plantarum* strains ST23LD and ST341LD, isolated from spoiled olive brine. *Microbiological Research*. 161, 102-108.

Todorov, S.D., Ho, P., Vaz-Velho, M., Dicks, L.M.T. 2010. Characterization of bacteriocins produced by two strains of *Lactobacillus plantarum* isolated from Beloura and Chouriço, traditional pork products from Portugal. *Meat Science*. 84, 334-343.

Todorov, S.D., Nyati, H., Meincken, M., Dicks, L.M.T. 2007. Partial characterization of bacteriocin AMA-K, produced by *Lactobacillus plantarum* AMA-K isolated from naturally fermented milk from Zimbabwe. *Food Control* 18, 656-664.

Todorov, S.D., Prévost, H., Lebois, M., Dousset, X., LeBlanc, J.G., Franco, B.D.G.M., 2011. Bacteriocinogenic *Lactobacillus plantarum* ST16Pa isolated from papaya (Carica papaya) — From isolation to application: Characterization of a bacteriocin. *Food Research International*. 44, 1351-1363.

Todorov, S.D., Vaz-Velho, M. 2008. Isolation and characterization of plantaricin ST8SH, a bacteriocin produced by *Lactobacillus plantarum* ST8SH, strain isolated from Bulgarian salami. *Journal of Biotechnology*. 136, Supplement, S735.

Todorov, SD., Dicks, L. 2008. Evaluation of lactic acid bacteria from kefir, molasses and olive brine as possible probiotics based on physiological properties. *Annals of Microbiology*. 58, 661-670.

Toomey, N., Bolton, D., Fanning, S. 2010. Characterisation and transferability of antibiotic resistance genes from lactic acid bacteria isolated from Irish pork and beef abattoirs. *Research in Microbiology*. 161, 127-135.

Valenzuela, A.S., Ruiz, G.D., Omar, N.B., Abriouel, H., López, R.L., Cañamero, M.M., Ortega, E., Gálvez, A. 2008. Inhibition of food poisoning and pathogenic bacteria by Lactobacillus plantarum strain 2.9 isolated from ben saalga, both in a culture medium and in food. *Food Control*. 19, 842-848.

Valerio, F., De Bellis, P., Lonigro, S.L., Visconti, A., Lavermicocca, P., 2008. Use of *Lactobacillus plantarum* fermentation products in bread-making to prevent *Bacillus subtilis* ropy spoilage. *International Journal of Food Microbiology*. 122, 328-332.

Van Hoorde, K., Verstraete, T., Vandamme, P., Huys, G., 2008. Diversity of lactic acid bacteria in two Flemish artisan raw milk Gouda-type cheeses. *Food Microbiology*. 25, 929-935.

Van Tassell, Maxwell L.; Miller, Michael J. 2011. *Lactobacillus* Adhesion to Mucus. *Nutrients*. 3 (5), 613-636.

Vásquez, A., Jakobsson, T., Ahrné, S., Forsum, U., Molin, G. 2002. Vaginal *Lactobacillus* Flora of Healthy Swedish Women. *Journal of Clinical Microbiology*. 40, 2746-2749.

Vaz- Velho M, Todorov SD. 2011. Potential probiotic evaluation of bacteriocin producing *Lactobacillus plantarum ST16PA* isolated from Papaya *(Carica papaya)*. *The 12th* ASEAN food conference 2011. 16-18 June, 2011.BITEC Bangna Bangkok, Thailande. 88-95.

Vizoso Pinto, M.G., Franz, C.M.A.P., Schillinger, U., Holzapfel, W.H., 2006. Lactobacillus spp. with in vitro probiotic properties from human faeces and traditional fermented products. *International Journal of Food Microbiology*. 109, 205-214.

Voulgari, K., Hatzikamari, M., Delepoglou, A., Georgakopoulos, P., Litopoulou-Tzanetaki, E., Tzanetakis, N. 2010. Antifungal activity of non-starter lactic acid bacteria isolates from dairy products. *Food Control*. 21, 136-142.

Wang, Y., Li, C., Liu, P., Ahmed, Z., Xiao, P., Bai, X., 2010. Physical characterization of exopolysaccharide produced by *Lactobacillus plantarum* KF5 isolated from Tibet Kefir. *Carbohydrate Polymers* 82, 895-903.

Wang, Y., Xu, N., Xi, A., Ahmed, Z., Zhang, B., Bai, X., 2009. Effects of *Lactobacillus plantarum* MA2 isolated from Tibet kefir on lipid metabolism and intestinal microflora of rats fed on high-cholesterol diet. *Applied Microbiology and Biotechnology*. 84, 341-347.

Xie, Y., An, H., Hao, Y., Qin, Q., Huang, Y., Luo, Y., Zhang, L. 2011. Characterization of an anti-Listeria bacteriocin produced by *Lactobacillus plantarum* LB-B1 isolated from koumiss, a traditionally fermented dairy product from China. *Food Control*. 22, 1027-1031.

Yang, E.J., Chang, H.C., 2010. Purification of a new antifungal compound produced by *Lactobacillus plantarum* AF1 isolated from kimchi. *International Journal of Food Microbiology*. 139, 56-63.

Yin, S., Hao, Y., Zhai, Z., Li, R., Huang, Y., Tian, H., Luo, Y. 2008. Characterization of a cryptic plasmid pM4 from *Lactobacillus plantarum* M4. *FEMS Microbiology Letters*. 285, 183-187.

Zago M, Fornasari M. E., Carminati D, Burns P, Suàrez V, Vinderola G, Reinheimer J and Giraffa G. 2011.Characterization and probiotic potential of *Lactobacillus plantarum* strains isolated from cheeses. *Food Microbiology*. 28 (5), 1033-1040.

Zhang, Z.-Y., Liu, C., Zhu, Y.-Z., Zhong, Y., Zhu, Y.-Q., Zheng, H.-J., Zhao, G.-P., Wang, S.-Y., Guo, X.-K. 2009. Complete Genome Sequence of *Lactobacillus plantarum* JDM1. *Journal of Bacteriology*. 191, 5020-5021.

Zhou, J.S., Pillidge, C.J., Gopal, P.K., Gill, H.S., 2005. Antibiotic susceptibility profiles of new probiotic *Lactobacillus* and *Bifidobacterium* strains. *International Journal of Food Microbiology*. 98, 211-217.

Zhu, H., Hart, C.A., Sales, D., Roberts, N.B., 2006. Bacterial killing in gastric juice – effect of pH and pepsin on Escherichia coli and Helicobacter pylori. *Journal of Medical Microbiology*. 55, 1265-1270.

Zoumpopoulou, G., Foligne, B., Christodoulou, K., Grangette, C., Pot, B., Tsakalidou, E., 2008. *Lactobacillus fermentum* ACA-DC 179 displays probiotic potential in vitro and protects against trinitrobenzene sulfonic acid (TNBS)-induced colitis and Salmonella infection in murine models. *International Journal of Food Microbiology*. 121, 18-26.

In: Lactobacillus: Classification, Uses and Health Implications ISBN: 978-1-62081-151-1
Editors: Alba I. Perez Campos and Arturo Leon Mena

Chapter II

Characterization and Evaluation of *Lactobacillus plantarum* Probiotic Potential

***Rafael Chacon Ruiz Martinez*[1], *Antônio Diogo Silva Vieira*[2,3], *Karina Maria Olbrich dos Santos*[3], *Bernadette Dora Gombossy de Melo Franco*[1] *and Svetoslav Dimitrov Todorov*[1,*]**

[1]Department of Food Science and Experimental Nutrition, Faculty of Pharmaceutical Sciences, Sao Paulo University, Sao Paulo, SP, Brazil
[2]Biochemical and Pharmaceutical Technology Department, Faculty of Pharmaceutical Sciences, Sao Paulo University, Sao Paulo, SP, Brazil
[3]EMBRAPA Goats and Sheep, Fazenda Três Lagoas, Estrada Sobral Groairas, Sobral, CE, Brazil

Abstract

Lactobacillus plantarum belongs to the group of lactic acid bacteria (LAB) and has been fundamentally studied in the last few decades. The microorganism is isolated from different sources including dairy products, meat, fish, fruits, vegetables and cereal products and has a well accepted GRAS status. Application of *Lactobacillus plantarum* and its probiotic properties has been subject of several studies. In this review, we describe *Lactobacillus plantarum* main characteristics, the biochemical and molecular methods used for its identification and investigate the probiotic potential of different strains of *Lactobacillus plantarum*, including management of gastrointestinal disorders, enhancement of gut barrier function, immunomodulatory effects, maintenance of oral

* Corresponding author. Tel.: +55-11-3091 2191; fax: +55-11-3815 4410. E-mail address: slavi310570@abv.bg (SD Todorov).

health, potential treatment of burns, potential role in prevention/treatment of cardiovascular disease, cholesterol-lowering effect, and potential antiobesity effect.

1. Lactic Acid Bacteria and *Lactobacillus plantarum* – Taxonomy and Biochemical and Physiological Characteristics

Lactic acid bacteria (LAB) are microorganisms of great interest to food industries, being present in several dairy products and fermented foods. These microorganisms constitute a heterogeneous group, although their members share several physiological characteristics. LAB can be found not only in food, but also in the gastrointestinal tract (GIT) and urogenital tract of human beings (Kleerebezem and de Vos, 2011; Todorov et al., 2011).

LAB receives this designation due to their ability to ferment carbohydrates and produce mainly lactic acid as end-products by homofermentative or heterofermentative metabolism (Settanni and Moschetti, 2010). This group of microorganisms comprises Gram-positive, non-sporulating bacteria, with cocci or rod morphology; LAB are generally non-motile [with the exception of *Lactobacillus agilis, Lactobacillus ghanensis* (Nielsen *et al.*, 2007) and *Lactobacillus capillatus* (Chao *et al.*, 2008)], display catalase and oxidase negative activities, are not able to reduce nitrate to nitrite, although can use lactate, and are tolerant to acid environments. LAB exhibit growth under completely absence of oxygen or in the presence of small amounts of that molecule (Wolf and Hammes, 1988; Mares *et al.*, 1994; Holzapfel *et al.*, 2001; Singh *et al.*, 2007).

Nowadays, LAB are used in the production of many types of drinks, yogurts and several products with functional claiming, which are intended to have potential benefits on consumers' health (Kleerebezem and de Vos, 2011).

Taxonomically, seven genera of microorganisms are found in LAB group, including *Lactobacillus* sp. and *Carnobacterium* sp. (rods), *Lactococcus* sp., *Enterococcus* sp., *Streptococcus* sp., *Leuconostoc* sp. and *Pediococcus* sp. (cocci) (Wessels *et al.*, 2004). Among them, the most important group for food industry is *Lactobacillus* sp. (Singh *et al.*, 2009; Todorov and Franco, 2010).

Lactobacilli normally present rod morphology; however, depending on the conditions of the environments where they are present, other morphologies can be found as spirals or cocci-rod. These microorganisms are genetically different among each other, and the genus *Lactobacillus* comprises more that 140 species (Singh *et al.*, 2009). Among them, *Lactobacillus plantarum* is one of the most studied microorganisms, due to its ability to survive in acid environments and in the presence of high salt contents, such as fermented vegetables, including olives, cucumber, and sauerkraut (Kuratsu *et al.*, 2010).

Lactobacillus plantarum is a versatile bacterium that can be found in a variety of ecosystems (Boekhost *et al.*, 2004), and this characteristic can be explained by its relatively large genome, which presents several proteins involved in regulation and transportation roles, besides the microorganism exhibits high metabolic potential (Kleerbezem *et al.*, 2003).

Originally, *Lactobacillus plantarum* had been taxonomically classified as *Streptobacterium plantarum* by Orla-Jensen (1919) and Holland (1920). Additionally, the bacterium was described by Pederson (1957) as a LAB capable to use vegetables as its substrate. Based only in biochemical features, until a few decades ago, *Lactobacillus pentosus, Lactobacillus arabinosus* (Fred *et al.*, 1921)*, Lactobacillus rudensis* (Davies, 1997) and *Lactobacillus plantarum* subsp. *mobilis* (Harrison, 1950) were considered as identical species, due to their high phenotypic similarity. Nowadays, the taxonomic group of *Lactobacillus plantarum* includes five species, to know: *Lactobacillus plantarum*, subsp. *plantarum, Lactobacillus plantarum* subsp. *argentoratensis, Lactobacillus paraplantarum, Lactobacillus pentosus* and *Lactobacillus fabifermentans* (De Bruyne *et al.*, 2009).

Lactobacillus plantarum shows rod morphology with dimensions of approximately 0.9-1.2 x 1.0-8.0 μm. The microorganism can be found alone or in small chains, is heterofermentative, facultative anaerobic, ferments hexoses by the Embden Meyerhof Parnas (EMP) metabolic pathway, leading to the production of D- and L-lactic acid. By use of other metabolic pathways, *Lactobacillus plantarum* is able to degrade pentoses and produce lactic and acetic acid in the presence of fosfoacetolase (Kandler and Waiss, 1986).

Lactobacillus plantarum exhibits mesophilic characteristic, can grow in temperatures ranging from 15ºC to 45°C, and the optimal growth condition is reached when the microorganism is cultivated in the presence of 4 % to 6 % of sodium chloride (NaCl) and at pH zone between 4 and 9 (Kandler and Waiss, 1986; Tanasupawat *et al.*, 1992). *Lactobacillus plantarum* is able to ferment several carbohydrates, including hexoses and hilolosides (Table 1). A large variety of organic acids (including malic, tartaric, and acetic acids) can be metabolized by *Lactobacillus plantarum* and transformed into carbon dioxide, lactic and acetic acid. It is important to highlight here that the ability of some *Lactobacillus plantarum* strains to use malic acid and produce carbon dioxide is useful for wine industry (Wibowo *et al.*, 1985). Some studies have demonstrated that many *Lactobacillus plantarum* strains belong to B serological group, since they contain ribitol and teichoic acid in the wall cell, whereas other strains were assessed as belonging to other groups (Sharpe, 1981).

Table 1. Some of *Lactobacillus plantarum* main physiological, biochemical, and genetic characteristics

Production of lactic acid from:	Amygdalin, arabinose, cellobiose, fructose, galactose, glucose, gluconate, lactose, maltose, mannitol, mannose, melibiose, raffinose, ribose, salicin, sorbitol, sucrose, and trehalose
Lactic acid are not formed from:	Xylose, sorbose, inulin, inositol, starch, and glycerol
Esculin hydrolysis	+
Lactic acid isomers	D and L
Type teichoic acids	Ribitol or glycerol
Growth with 4% taurocholate	+
Growth in presence of 10% NaCl	+
Hydrolysis of arginine	-
Production of dextran from saccharose	-
Mol% G+C*	44-46
Growth facto requirements	Calcium pentothenate, niacin, and nicotinic acid
Antibiotic resistance to	Kanamycin, gentamicin, neomycin, streptomycin, polimixin B, and colistin

* G: Guanine; and C: Cytosine (Nucleotides of DNA).
Adapted from Todorov and Franco (2010).

2. *Lactobacillus plantarum*: Genetic Identification

Traditionally, morphological and physiological analyzes (such as cultivation in specific culture media and performance of biochemical tests) have been used for the identification of species belonging to the group of *Lactobacillus plantarum* (Todorov and Franco, 2010; Huang *et al.*, 2011). These characteristics can be combined with the use of more appropriate methods, such as the characterization of peptideoglycan structure, DNA homology, species-specific polymerase chain reaction (PCR) based on the evaluation of fractions of 16S rRNA and 23S rRNA, RAPD (randomly amplified polymorphic DNA)-PCR, PFGE (Pulse field gel electrophoresis), and enzymatic restriction analysis (Kandler and Waiss, 1986; Siezen *et al.*, 2004). As gold standard, sequencing of 16S rDNA is necessary for an adequate identification of LAB. For this purpose, amplification of 16S rDNA and confirmation of the identification may be obtained by amplification of genomic DNA using primers F8 and R1512, according to Felske *et al.* (1997). Genotypic methods are frequently recommended for the identification of LAB, since several microbial species may display similar phenotypical features, which cannot be distinguished by traditional methods (Vandamme *et al.*, 1996).

For the molecular characterization of *L. plantarum* strains, methods such as species-specific PCR and partial sequencing of 16S rDNA are commonly used. Different primers for species-specific PCR have been designed and successfully used for that purpose (Todorov and Franco, 2010). Rantsiou *et al.* (2006) used primers Lp (5'-ATG AGG TAT TCA ACT TAT G-3') and 16S (5'-GCT GGA TCA CCT CCT TTC-3'), according to what was previously described by Berthier and Ehrlich (1998), to identify *Lactobacillus plantarum* strains isolated from traditional fermented Greek sausage. Other researchers have used species-specific primers for identification of the microorganism. Omar *et al.* (2006) used primers planF (5'-CCG TTT ATG CGG AAC ACC TA-3') and pREV (5'-TCG GGA TTA CCA AAC ATC AC-3') for the identification of bacteriocin-producing *Lactobacillus plantarum*, isolated from a traditional fermented gruel from Burkina Faso. Furthermore, primers F (5'-AAT TGA GGC AGC TGG CCA-3') and R (5'-GAT TAC GGG AGT CCA AGC-3') were successfully used for the identification of *Lactobacillus plantarum* strains isolated from wines (Quere *et al.*, 1987; Rojo-Bezares *et al.*, 2006). On the other hand, Chagnaud *et al.* (2001) used primers Plant1 (5'-ATC ATG ATT TAC ATT TGA GTG-3') and Lowlac (5'-CGA CGA CCA TGA ACC ACC TGT-3') for the identification of *Lactobacillus plantarum* strains obtained from different origins.

Kleerbezem *et al.* (2003) sequenced the entire genome of *L. plantarum* WCFS1 and observed that the microorganism has the ability to use different sugar sources, to identify peptides and to produce different amino acids. Several works have been carried out on the basis of the results obtained with the genome sequencing of *Lactobacillus plantarum* WCFS1. Bove *et al.* (2011) studied the inactivation of *ftsH* gene from *Lactobacillus plantarum* WCFS1 with quantitative real time PCR (RT-PCR) method, and evaluated its effects on growth, stress tolerance, cell surface proteins, and biofilm formation. Those authors concluded that both transcript profiling and the phenotypic characterization confirmed that in *Lactobacillus plantarum* FtsH is needed for adaptation to several stresses and the gene also plays a major role in survival to heat shock, even when this is limited to a short period.

Duary *et al.* (2010) evaluated the expression of *atpD* gene in probiotic strains of *Lactobacillus plantarum* Lp 9 and Lp 91 under different *in vitro* pH conditions which closely mimic the physiological environment prevalent in the human gut by use of RT-PCR technique. Although both strains were able to survive the stressful conditions tested, Lp91 exhibited relatively greater acid tolerance, as revealed by 4.7-fold up-regulation of the *atp*D gene, as well as higher populations assessed at pH 2.5 after 90 min.

Zago *et al.* (2011) characterized the probiotic potential of *Lactobacillus plantarum* strains isolated from different cheeses, and also evaluated different genes that could be involved in bile salts hydrolysis, adhesion to intestinal cells, and resistance to antibiotic agents, including erythromycin, and tetracycline, among others. Huang *et al.* (2011) identified several strains belonging to *Lactobacillus plantarum* group by use of a novel technique, known as SNaPshot minisequencing, and concluded that the method was efficient and not time-consuming.

3. *Lactobacillus Plantarum*: Resistance to Antibiotics

The susceptibility of LAB to antibiotics, especially for *Lactobacillus* spp., has not received special attention by scientific community for a long time. However, the detection of LAB isolates that are resistant to several antimicrobial agents used in human and animal medicine practice has rapidly increased the interest in this research field, especially due to the possibility of transference of resistant genes to other microorganisms, including pathogens (Charteris *et al.*, 1998; Danielsen, 2002; Delgado *et al.*, 2007).

This lack of attention given to the potential of *Lactobacillus* spp. strains and other LAB members to be resistant to antibiotics could be related to their status of non-pathogenic microorganisms recognized for decades, in combination with their reputation as health-promoting microbes, especially in the GIT. However, these issues have been questioned in the last years, since some *Lactobacillus* and other LAB strains have been linked with different human infections, including cavities, endocarditis, septicaemia, urogenital infections, and pneumonia, among other diseases (Charteris *et al.*, 1998; Franz *et al.*, 2005).

Bacterial strains intended to be used as probiotics in food systems need to be carefully checked for their susceptibility to antibiotics, in order to avoid the dissemination of resistant microorganisms. The possibility of interchange of plasmids containing genes for antibiotic resistance among microorganisms is considered an important issue especially in food and drug industries (Danielsen, 2002; Delgado *et al.*, 2007).

It is believed that LAB present in fermented products could act as a reservoir of resistant genes for antibiotics, and those could be transferred to pathogenic microorganisms found in food matrix or in the GIT (Flórez *et al.*, 2007). Given this fact, several studies have been performed to evaluate the susceptibility of *Lactobacillus* spp. and other LAB strains to antimicrobial agents, and also to assess the presence and possible transference of resistance plasmids to pathogenic strains.

Charteris *et al.* (1998) determined the susceptibility of 46 strains of *Lactobacillus* spp. to 44 antibiotics, including inhibitors of cell wall synthesis, protein synthesis, and nucleic synthesis, among others. Interestingly, those authors verified that all strains studied were resistant to at least 14 antibiotics, highlighting the importance of the determination of the susceptibility profile to antibiotics of strains normally considered as nonhazardous.

In a work done by Danielsen and Wind (2003), 62 strains of *Lactobacillus* spp. were evaluated for the determination of the susceptibility to 25 antibiotics by use of E-test and the authors observed that the results were species-dependent.

Danielsen (2002) characterized the tetracycline resistance plasmid pMD5057 from *Lactobacillus plantarum* 5057, revealing its structural composition. Essid *et al.* (2009) evaluated the safety for use and technological properties of 17 strains of *Lactobacillus plantarum* isolated from traditional Tunisian sausage and verified that all the microorganisms were resistant to tetracycline. Feld *et al.* (2009) studied the erythromycin resistance plasmid pLFE1 from *Lactobacillus plantarum* M345 and noticed that it was transferred to *Lactobacillus rhamnosus*, *Lactococcus lactis* and *Listeria innocua* and also to pathogenic strains such as *Enterococcus faecalis* and *Listeria monocytogenes.* Jacobsen *et al.* (2007) used in *in vivo* model to study the horizontal transfer of tetracycline [*tet*(M)] and erythromycin [*erm*(B)] resistance plasmids from *Lactobacillus plantarum* isolated from foods to *Enterococcus faecalis* JH2-2 and observed that both plasmids were transferred, although the highest transference rates were assessed for *erm*(B) plasmid.

The resistance to antibiotics in *Lactobacillus plantarum* strains is not restricted to tetracycline (Danielsen, 2002; Cebeci and Gürakan, 2003; Flórez *et al.*, 2007; Jacobsen *et al.*, 2007; Essid *et al.*, 2009) and erythromycin (Cebeci and Gürakan, 2003; Jacobsen *et al.*, 2007; Essid *et al.*, 2009; Feld *et al.*, 2009). Actually, others researchers verified the resistance of the microorganisms to different antibiotics, including fusidic acid, penicillin G, vancomycin, amoxilin, piperacilin, rifampicin, ampicilin, cephuroxine, cephoxitin, ceftadizime, cefoperazone, aztreonam, gentamicin, moxifloxacin, metronidazole, streptocin, among others (Charteris *et al.*, 1998; Cebeci and Gürakan, 2003; Choi *et al.*, 2003; Danielsen and Wind, 2003; Zhou *et al.*, 2005; Delgado *et al.*, 2007; Flórez *et al.*, 2007; Klare *et al.*, 2007; Essid *et al.*, 2009).

4. Probiotic Potential of *Lactobacillus Plantarum*

Probiotics are microorganisms that, when used in large amounts in the preparation of foods, are able to survive the passage through the upper digestive tract and adhere to intestinal cells, helping in the intestinal balance. In particular, probiotics used in functional dairy products belong to the genera *Lactobacillus*, *Bifidobacterium*, *Streptococcus* and *Saccharomyces*. To produce the desired benefits, these bacteria should be present in the product in viable counts during their whole shelf-life (7-9 Log CFU/mL); however, their viability in commercial preparations is affected by several factors, among which the presence of other microorganisms (Kailasapathy and Rybka, 1997).

The main health benefits associated with the ingestion of probiotic microorganisms have been described by many researchers and can be summarized as: alleviation of lactose intolerance, prevention and reduction of diarrhoea symptoms, treatment and prevention of allergy, reduction of the risk associated with mutagenicity and carcinogenicity, hypocholesterolemic effect, inhibition of *Helicobacter pylori* and intestinal pathogens, prevention of inflammatory bowel disease and modulation of the immune system (Mattila-Sandholm *et al.*, 1992; Goldin, 1998; Holzapfel *et al.*,1998; Ouwehand *et al.*,2003; Akin *et al.*, 2007).

4.1. Management of Gastrointestinal Disorders

Dicks and ten Doeschate (2010) investigated the ability of *Lactobacillus plantarum* 423 and *Enterococcus mundtii* ST4SA to alleviate symptoms of *Salmonella* infection in Wistar rats. Those authors evaluated four groups of animals that were supplemented during 10 days with 200 µl ($1x10^8$ CFU) of *Enterococcus mundtii* ST4SA, 200 µl ($1x10^8$ CFU) of *Lactobacillus plantarum* 423, 200 µl of a 1:1 combination of *Enterococcus mundtii* ST4SA and *Lactobacillus plantarum* 423 ($1x10^8$ CFU) or 200 µl of sterile physiological water (placebo). On the 8th and 9th days, the animals were infected with 200 µl ($2x10^8$ CFU) of *Salmonella enterica* serovar Typhimurium. Dicks and ten Doeschate (2010) observed slightly lower levels of endotoxin in blood samples obtained from rats supplemented with *Lactobacillus plantarum* 423 and *Enterococcus mundtii* ST4SA. Interestingly, those researches also verified that although both strains were able to ameliorate the symptoms of *Salmonella* infection, *Lactobacillus plantarum* 423 administered alone was more efficient than *Enterococcus mundtii* ST4SA.

Fayol-Messaoudi *et al.* (2007) analyzed the ability of *Lactobacillus plantarum* ACA-DC287 (a strain isolated from a Greek cheese) to colonize the GIT of germ-free mice and also evaluated the antagonistic effect of the LAB in conventional mice infected with *Salmonella typhimurium* SL1344. For the determination of lactobacilli populations in the different segments of the GIT of germ-free mice fed with 10^8 CFU of *Lactobacillus plantarum* ACA-DC287, those authors used cultivation method and verified that the microorganism colonized the GIT and became abundant in all samples examined, ranging from 4.5 to 7.0 log CFU/g of tissue evaluated. Fayol-Messaoudi *et al.* (2007) also found that conventional C3H/He/Oujco mice infected with *Salmonella typhimurium* SL1344 and supplemented with *Lactobacillus plantarum* ACA-DC287 displayed lower levels of the pathogen detected in both intestinal tissues and intestinal contents, in comparison to control infected mice (without supplementation with the LAB). According to the results obtained, those authors concluded that *Lactobacillus plantarum* ACA-DC287 exerted anti-*Salmonella* activity similar to those observed for well-established probiotic strains.

Vandeplas *et al.* (2009) assessed the ability of the probiotic strain *Lactobacillus plantarum* CWBI-B659 combined with a xylanase to reduce the effects of *Salmonella* Thyphimurium infection in broiler chickens. Chicks were fed with four types of diet, including a wheat-based diet (C+) supplemented with 0.1g/kg of xylanase (E) or 10^6 CFU/g of *Lactobacillus plantarum* CWBI-B659 (P) or both (PE). The animals were infected with 10^8 or 10^5 CFU of *Salmonella* Typhimurium on their third day of life. As negative control, uninfected animals were fed with diet C. Growth performance and feed conversion rate (FCR) were assessed weekly. Those authors observed that after challenge with the pathogen, daily weight gain was significantly reduced, while FCR was increased by 1.0 to 19.7%; furthermore, chickens fed PE displayed improved performance compared to the other groups evaluated. Those researchers also verified that PE diet tended to restore a microflora similar to that of non-infected chicks; besides, at slaughter age, *Salmonella* contamination was diminished by 2.0 and 1.85 log CFU in E and PE groups, respectively.

Several studies have addressed the ability of *Lactobacillus plantarum* 299v in the management of gastrointestinal disorders. Sen *et al.* (2002) conducted a double-blind, placebo-controlled, cross-over, four-week trial in which *Lactobacillus plantarum* 299V was administered to 12 previously untreated patients with irritable bowel disease (IBD), which is the most common functional digestive disorder, affecting 11-20% of the adult population in industrialized countries (Bixquert Jiménez, 2009). Those authors verified that the LAB did not improve abdominal pain in the participants, compared to placebo group; however, there was a significant reduction in the production of colonic gas in subjects given probiotics.

Klarin *et al.* (2008) performed a double-blind, placebo-controlled clinical trial with patients diagnosed with *Clostridium difficile*-associated disease (CDAD) to determine the impact of administration of *Lactobacillus plantarum* 299v on several parameters, such as its presence in feces, gut permeability, inflammation and infection. In that study, 22 patients consumed a fermented oatmeal gruel containing 8×10^8 CFU/ml of *Lactobacillus plantarum* 299v (ProbiAB), whereas 22 participants received the same gruel without addition of the LAB (control). The products were administered as bolus doses: commencing with six doses of 100ml each at 12h intervals, followed by 50ml twice a day during the period the patient remained in the intensive care unit. Those authors verified that among the parameters evaluated, the enteral administration of *Lactobacillus plantarum* 299v to CDAD patients reduced the detection of *Clostridium difficile* in feces, as determined at 19% (4/21) for subjects allocated in placebo group, compared to none in the probiotic group ($P<0.05$).

Niedzielin *et al.* (2010) performed a double-blind, placebo-controlled, randomized study to assess the efficacy of *Lactobacillus plantarum* 299v in patients diagnosed with IBD. Patients enrolled in the study were randomly given 200 ml (twice a day) for daily use of either an active preparation containing *Lactobacillus plantarum* 299v (5×10^7 CFU/ml) or an inactive preparation without the microorganism (control), during four weeks. Patients were evaluated at baseline and at the end of the study. Subjects that participated in the study reported their symptoms on a weekly basis, including level of abdominal pain, stool frequency and consistency, and frequency of flatulence. Among 20 patients treated with the LAB, all them reported resolution in abdominal pain in comparison to only 11 (from 20) patients allocated in the placebo group ($P<0.05$). Concerning all symptoms evaluated, an improvement was observed in 95% of patients that received *Lactobacillus plantarum* 299v in comparison to only 15% of subjects in the placebo group ($P<0.05$), demonstrating a beneficial effect of the LAB in patients with the IBD.

Oudhuis *et al.* (2011) evaluated the efficacy of probiotic *versus* antibiotic in the decontamination of the digestive tract (SDD), which is a procedure used to decrease the infection rate and mortality of patients in intensive care units (ICU). Those authors studied a group of 254 ICU patients that were given four times daily an oral paste, containing polymyxin E, gentamicin, and amphotericin B, an enteral solution with the same antibiotics; intravenous cefotaxime (during the first four days) or two times daily a solution (through a nasogastric tube) with *Lactobacillus plantarum* 299v in a dose of 5×10^9 CFU. Oudhuis *et al.* (2011) observed no statistical differences in infection and mortality rates between both groups ($P>0.05$); furthermore, no increase in antibiotic resistance was assessed during and after SDD or LAB treatments. Those researches concluded that, according to their data, there was no increased ICU mortality risk in the LAB group.

A study conducted by Lönnermark *et al.* (2010) aimed to assess the ability of *Lactobacillus plantarum* to prevent gastrointestinal side effects related to use of antibiotic agents in patients. Those authors performed a double-blind, placebo-controlled trial, in which subjects treated for infections received either a fruit drink containing blueberries and 5% oats gruel with *Lactobacillus plantarum* (10^{10} CFU a day) or a similar drink without the microorganism (placebo), until a week after completion of antibiotic regimen. Those researchers observed that the overall risk of developing loose or watery stools was significantly lower in individuals that consumed *Lactobacillus plantarum*, as was the occurrence of nausea ($P<0.05$). However, no significant differences were assessed in terms of diarrhea or carriage of toxin producing *Clostridium difficile* in fecal samples when probiotic and placebo groups were compared ($P>0.05$). Lönnermark *et al.* (2010) concluded that ingestion of *Lactobacillus plantarum* could exert a preventive effect on milder gastrointestinal symptoms during treatment with antimicrobial agents.

Del Piano *et al.* (2010) performed a double-blind, randomized, placebo-controlled trial to assess the effectiveness of two probiotic preparations containing either *Lactobacillus plantarum* LP01 (LMG P-21021) and *Bifidobacterium breve* BR03 (DSM 16604) (I) or *Bifidobacterium animalis* subsp. *lactis* BS01 (LMG P-21384) (II) in the management of evacuation and intestinal discomfort. Those authors evaluated during five years a group of 300 healthy volunteers with evacuation disorders and hard stools that were allocated in one of the following groups: 80 received treatment I, 110 received treatment II, and 110 were given placebo (III) during 30 days. Del Piano *et al.* (2010) observed that individuals administered preparation I reported a significant improvement in the number of weekly bowel movements and in feces consistency and easy of expulsion. Furthermore, both groups I and II displayed relevant improvement in several discomfort items, such as abdominal bloating and anal itching, burning or pain, in comparison to group III, demonstrating that treatment with mixed *Lactobacillus plantarum* LP01 and *Bifidobacterium breve* BR03 was able to relieve important symptoms associated with malfunctioning of the GIT.

Higashikawa *et al.* (2010) evaluated the effects of different yogurts on the gastrointestinal system. Those authors conducted a double-blind, randomized trial in which 68 patients displaying complaints with regard to intestinal health, such as constipation and diarrhea, were given one of three types of yogurt, during six weeks: (A) a yogurt made by *Lactobacillus plantarum* SN35N; (B) a yogurt containing *Lactobacillus plantarum* SN13T, and (C) a yogurt made by *Lactococcus lactis* A6 and *Streptococcus thermophilus* 510. Higashikawa *et al.* (2010) noticed that subjects supplemented with yogurts types A and B, but not type C, have shown improvement in constipation, suggesting that the LAB have potential as a functional food to promote human health.

4.2. Enhancement of Gut Barrier Function

Once the intestinal barrier function is compromised, it can become "leaky" and, therefore, allow that pathogens and toxins enter the body. It is known that the intestinal barrier function is impaired in human conditions including Inflammatory Bowel Disease (Bruewer *et al.*, 2006), IBD (Barbara, 2006), and due to food-borne infections (Guttman *et al.*, 2006). Furthermore, stress (Hart and Kamm, 2002) and aging process (Mullin et al., 2002) can also compromise the gut barrier function. An impaired intestinal barrier function is also linked with illnesses in other parts of the body (Liu *et al.*, 2005; Maes, 2008; Vaarala *et al.*, 2008).

A critical component of the intestinal barrier is the intercellular junction complexes between adjacent intestinal epithelial cells that create a semi-permeable diffusion barrier (Mitic *et al.*, 2000). Some probiotic strains have been shown to improve intestinal permeability induced by pathogens *in vitro*, while other probiotics are believed to enhance tight junction integrity between intestinal epithelial cells that are not compromised (Klingberg *et al.*, 2005; Sherman *et al.*, 2005; Ewaschuk *et al.*, 2008; Putaala *et al.*, 2008). More recently, Anderson *et al.* (2010a) performed a study using gene expression analysis and fluorescent microscopy and demonstrated the ability of *Lactobacillus plantarum* MB452 to enhance the function of the intestinal barrier by increasing the expression levels of genes involved in tight junction formation.

Zhou *et al.* (2010) demonstrated in *in vitro* experiments with Caco-2 cells and by use of different techniques including confocal laser scanning microscopy, immunohistochemistry, Western blotting and real-time quantitive PCR, that *Lactobacillus plantarum* was able to inhibit the intestinal epithelial barrier dysfunction induced by unconjugated bilirubin.

Anderson *et al.* (2010b) screened several *Lactobacillus plantarum* strains and human oral isolates for their ability to enhance tight junction integrity in Caco-2 cells by measurement obtained with the use of a transepithelial electrical resistance (TEER) method. Those authors used the strain *Lactobacillus rhamnosus* HN001 as standard and observed that the isolate *Lactobacillus plantarum* DSM 2648 showed the most promising results, significantly increased TEER, in comparison with control ($P<0.05$). Furthermore, those researchers verified that *Lactobacillus plantarum* DSM 2648 was able to reduce the harmful effects of enteropathogenic *Escherichia coli* O127:H6 on TEER and adherence as up to 98.75 % and 80.18 %, respectively, during simultaneous or prior coculture compared with the pathogen incubated alone.

A study performed by Ko *et al.* (2007) demonstrated that *Lactobacillus plantarum* ATCC 8014 was able to counteract the induction of epithelial barrier disfunction and IL-8 secretion promoted by TNF-α, as determined in an *in vitro* model using Caco-2 cells. Those authors measured the electrical resistance across the monolayers at different time-intervals, quantified cytokines production and analyzed main proteins secreted and concluded that *Lactobacillus plantarum* ATCC 8014 could be useful to preserve the epithelial barrier function and might inhibit the inflammatory response by altering the signal transduction pathway.

Mangell *et al.* (2002) demonstrated that pre-treatment during one week by free access to *Lactobacillus plantarum* 299v protected Sprague-Dawley rats against *Escherichia coli* - induced increase in intestinal permeability. Those authors also verified that the administration of the LAB alone had no negative effect on that same parameter, demonstrating its potential beneficial effect in the GIT. Similar positive outcomes with this same strain were observed by McNaught *et al.* (2002). Those authors evaluated a group of 129 patients undergoing elective major abdominal surgery that randomly received an oral preparation containing *Lactobacillus plantarum* 299v or placebo for at least one week before surgical procedure and also in the postoperative period. The researchers aimed to investigate the effect of the LAB on bacterial translocation, gastric colonization with enteric organisms, and septic complications among the subjects studied, and observed no significant difference between the two groups of patients ($P>0.05$) for all the parameters studied.

According to a study performed by Wang *et al.* (2011), pretreatment of rats with *Lactobacillus plantarum* L2 during 14 days was able to prevent ischemia/reperfusion-induced bacterial translocation and intestinal barrier dysfunction, resulting in a recover in the microflora and the mucosal integrity, in comparison to a group of animals not supplemented with the LAB. Those authors concluded that *Lactobacillus plantarum* L2 could exhibit promising effects in the GIT.

4.3. Immunomodulatory Effects

Bujalance *et al.* (2007) evaluated the immunomodulation capacity of a potential probiotic strain of *Lactobacillus plantarum* on intact and cyclophosphamide-treated BALB/c mice and observed that the microorganism was not able to modify the cyclophosphamide-induced leucopenia; however, it partially restored the proliferation of spleen cells from cyclosphophamide-treated animals in response to lipopolysaccharide. The authors concluded that the microorganism may be useful to improve the immune function damaged by immunosuppressive agents.

Peral *et al.* (2010) studied a group of 34 patients diagnosed with chronic venous ulcer who received once-daily, during 10 days, a topical application of *Lactobacillus plantarum* ATCC 10241 (10^5 UFC/ml). The lesions were clinically monitored and evaluated weekly by the plastic surgeon. Among the patients, 14 suffered from moderately controlled type 2 diabetes mellitus and 20 remaining were non-diabetic. After 30 days of treatment, a reduction of more than 90 % of the lesion area was observed by the authors in 43 % and 50 % of the diabetics and non-diabetic patients, respectively ($P<0.001$). Furthermore, no significant differences were observed between the diabetic and non-diabetic groups, highlighting the importance of the microorganism in wound healing. Those authors also investigated the effect on interleukin (IL)-8 production by cells from ulcer bed and neutrophils isolated from peripheral blood previously challenged *in vitro* with *Pseudomonas aeruginosa* and *Lactobacillus plantarum*. The authors observed that cells from ulcer beds collected after treatment with *Lactobacillus plantarum* for 10 days showed a reduction in the percentage of polymorphonuclear, apoptotic and necrotic cells and an enhancement of IL-8 production. Furthermore, the basal values for IL-8 assessed in diabetic and ulcer patients were higher than normal ($P<0.001$) and were increased by *Pseudomonas aeruginosa* infection in normal, diabetics ($P<0.001$) and non-diabetics with ulcers ($P<0.01$), whereas preincubation with *Lactobacillus plantarum* decreased IL-8 production in non-diabetic and diabetic patients with ulcers ($P<0.001$).

Cammarota *et al.* (2009) analyzed the *in vitro* probiotic potential of *Lactobacillus plantarum* DSMZ 12028 using the pathogen *Escherichia coli* and a probiotic strain (*Lactobacillus plantarum* F19) as controls. Those authors evaluated the expression of innate immunity receptors, the toll-like receptors (TLR), by use of semi-quantitative PCR technique on an epithelial/bacteria co-culture, whereas real-time PCR was used to monitor the expression of TLRs and cytokines in amonocytic cell line (THP-1) following bacterial exposure. Cammarota *et al.* (2009) observed that *Lactobacillus plantarum* DSMZ 12028 down-regulated TLR mRNAs except for TLR2, while *Lactobacillus paracasei* F19 increased the level of TLR2 and *Escherichia coli* up-regulated TLR4 ($P<0.05$). By use of THP-1 the authors demonstrated an augmentation of the proinflammatory cytokine TNFα in the presence of the potential probiotic strain, whereas *Escherichia coli* led to a strong up-regulation of IL8, IL1β, and IL23. Those researches concluded that *Lactobacillus plantarum* DSMZ 12028 induced a proinflammatory response that was insufficient to result in an inflammatory outcome, but induced a higher state of host defense system alertness in the intestinal epithelial cells.

Wang et al. (2009a) evaluated five *Lactobacillus* strains (*Lactobacillus plantarum* L2, *Lactobacillus plantarum* L8, *Lactobacillus plantarum* L9, *Lactobacillus rhamnosus* GG and *Lactobacillus acidophilus* L050103) for their ability to induce the secretion of pro- and anti-inflammatory cytokines by human peripheral blood mononuclear cells (PBMCs). Those authors observed that, among the tested strains, *Lactobacillus plantarum* L2 was able to induce a considerable reduced level of IL-10 and displayed low levels of pro-inflammatory cytokines tested (TNF-α, INF-γ, and IL-12). That microorganism was selected and given daily to rats during 28 days ($2x10^9$ CFU) and after this period, the animals were not fed any *Lactobacillus*. Interestingly, the researchers observed a significant increase ($P<0.05$) in CD19-positive cells in the ileum of the rats after intake of *Lactobacillus plantarum* L2, which indicated a significant alteration of the immune response in the gastrointestinal mucosa.

According to a study performed by Paolillo *et al.* (2009), *Lactobacillus plantarum* may play an important role in the control of inflammation in the gut and may also contribute for the activation of systemic immunity through the induction of TLR-2.

Liu *et al.* (2011) studied the effects of exopolysaccharides produced by *Lactobacillus plantarum* NTU 102 (102EP) and *Lactobacillus paracasei* subsp. *paracasei* NTU 101 (101EP) on antioxidant activity and immunomodulation *in vitro*. The authors used a raw 264.7 cell line and demonstrated that the production of cytokines IL-6, TNF-α, and IL1-β was induced by the presence of both microorganisms in a dose-dependent manner. Furthermore, 102EP and 101 EP exhibited potential antioxidant properties. Those researchers concluded that the microorganisms might be mild immune modulators for macrophages.

Mañé *et al.* (2011) conducted a double-blind, placebo-controlled, randomized pilot trial to determine the effects of a mixture of two new probiotic strains *Lactobacillus plantarum* CECT 7315 and 7316 on systemic immunity in elderly. The subjects were daily treated for 12 weeks with one of the following protocols: $5x10^8$ CFU of *Lactobacillus plantarum* CECT 7315/7316 (low probiotic dose), $5x10^9$ CFU of the probiotic mixture (high probiotic dose), or placebo. Among others, the leukocyte populations, IL-1, IL-10, and TGF-β1 levels were assessed in venous blood at baseline, end of treatment (week 12), and end of follow-up period (week 24). Those researches observed that depending on the dose, *Lactobacillus plantarum* CECT 7315/7316 displayed different immune-enhancing effects in elderly subjects, since in the end of the treatment, patients given high probiotic dose demonstrated significant increases in the percentages of activated potentially T-suppressor and NK cells, whereas patients fed with low probiotic dose increased populations of activated T-helper lymphocytes, B lymphocytes, and antigen presenting cells. Interestingly, plasma TGF-β1 concentration significantly decreased by use of both probiotic regimens, and even 12 weeks after probiotic discontinuation most of the parameters evaluated remained unaltered.

In a study performed by Puertollano *et al.* (2008), the authors treated balb/c mice daily with $5x10^7$ CFU of *Lactobacillus plantarum* or PBS (phosphate-buffered saline) (control) during four weeks, and after that time, the animals were intravenously infected with *Listeria monocytogenes*. The survival of the mice was monitored every eight hours for eight days after the pathogen challenge. Those authors observed that *Lactobacillus plantarum* did not significantly increase the survival ($P>0.05$) of infected mice. Moreover, the levels of IL-1β and IL-6 from sera of orally administered *Lactobacillus plantarum* drastically reduced at 0.4 ($P<0.01$), and 6 days after *Listeria monocytogenes* infection, while TNF-α production was unaltered. Those authors also hypothesized that *Lactobacillus plantarum* could exert anti-inflammatory effects, which may represent an important model to attenuate inflammatory disorders.

A study performed by Hisbergues *et al.* (2007) evaluated *in vitro* and *in vivo* the potential of *Lactobacillus plantarum* NCIMB 8826 to reduce the allergic response induced by Der p 1, the major house dust mite allergen of *Dermatophagoides pteronyssinus*. Those authors observed that the co-administration of *Lactobacillus plantarum* and Der p 1 induced a T-helper type 1 (Th1)-biased allergen-specific IgG response, absence of IgE response and favored the production of IFN-γ upon allergen re-stimulation. *Lactobacillus plantarum* also stimulated a high production of IL-2 and moderate production of IL-10 in bone marrow-derived dendritic cells. Those researchers concluded that the potential probiotic strain *Lactobacillus plantarum* NCIMB 8826 could modulate different immune parameters required for the development of an allergic reaction and that the intranasal co-administration of the micro-organism and *n*Der p 1 might represent an efficient vaccine strategy to minimize its harmful effects.

Snel *et al.* (2011) performed a double-blind, placebo-controlled study to investigate the immunomodulatory properties of five *Lactobacillus* strains in birch-pollen-allergic subjects. Patients received daily, during four weeks, a yogurt drink containing 10^{10} CFU of one of the following micro-organisms: *Lactobacillus casei* Shirota, *Lactobacillus plantarum* WCFS1, *Lactobacillus plantarum* NIZO3400, *Lactobacillus plantarum* NIZO2577 and *Lactobacillus plantarum* CBS 125632 or placebo. Those authors collected blood samples at the start and in the end of the study and determined several parameters in serum and in peripheral blond mononuclear cell cultures (PBMC) derived from the subjects enrolled. Snel *et al.* (2011) verified a reduction in birch-pollen specific IgE for four probiotic strains. Besides, *Lactobacillus plantarum* CBS 125632 decreased several immune markers related to allergy such as IgE, IL-5, and IL-13, and for this reason those researchers concluded that the micro-organism might have the potential to alleviate the severity of seasonal allergy symptoms.

Herías *et al.* (1999) evaluated the immunomodulatory properties of *Lactobacillus plantarum* 299v in gnotobiotic rats fed with the LAB and challenged with *Escherichia coli* 06:K13:H1. Those authors observed that animals given both microorganisms displayed significantly higher levels of total serum IgA and marginally higher contents of IgM and IgA antibody against the pathogen in comparison to those administered only *Escherichia coli* (control). Herías *et al.* (1999) concluded that *Lactobacillus plantarum* 299v competed with the pathogen for intestinal colonization and could have influenced immunity at both intestinal and systemic levels.

Pathmakanthan *et al.* (2004) also evaluated the role of *Lactobacillus plantarum* on its immunomodulatory properties. Those authors studied the pattern of cytokine secretion and their cellular sources in mononuclear cells isolated from normal and ulcerative colitis (UC) colonic mucosa, challenged with *Lactobacillus plantarum* 299v and pathogenic microorganisms, including *Salmonella dublin* ATCC 39184, *Escherichia coli* ATCC 11303, *Bacteroides vulgatus* ATCC 4456, and enteropathogenic *Escherichia coli*. Those researchers observed that the pathogens tested increased the levels of IL-1β, TNF-α, and IFN-γ secreted by the cells at different extents, whereas no considerable changes were observed for those same cytokines when normal or inflamed mucosa was challenged only with the LAB. Furthermore, incubation of *Lactobacillus plantarum* 299v with mononuclear cells isolated from UC mucosa displayed a significant augmentation in the levels of IL-10 detected, in comparison with unstimulated control cells ($P<0.05$). Intracellular cytokine staining also confirmed T-cell and macrophage IL-10 synthesis after stimulation with *Lactobacillus plantarum* 299v. Pathmakanthan *et al.* (2004) concluded that the tested LAB could be useful to reduce excessive inflammation at mucosal level.

Maeda *et al.* (2009) investigated the effect of oral administration of *Lactobacillus plantarum* L-137 (HK-LP) on protection against influenza virus infection in mice, a condition related to high-morbidity and significant mortality (Gillim-Ross and Subbarao, 2006). Those authors tested female 7-week-old C57BL/6 mice orally given lyophilized HK-LP at doses of 5-100 mg/kg/day before and after infection with influenza virus A/FM/1/47 (H1N1, a mouse-adapted strain). Maeda *et al.* (2009) noticed that the survival time was significantly higher in mice treated with HK-LP in comparison to mice treated with PBS as controls. Furthermore, the viral titers in the lung were significantly lower in mice treated with HK-LP than controls at the early stage after the virus infection. Furthermore, no interferon (INF)-β production was detected in controls after influenza infection, whereas it was assessed in considerable levels in the serum of mice treated with the LAB. Since HK-LP is as a potent INF-β inducer, those authors concluded that the microorganism could be beneficial in the prevention against influenza virus infection.

Berggren *et al.* (2011) evaluated the effect of consumption of a product containing *Lactobacillus plantarum* HEAL 9 (DSM 13434) and *Lactobacillus paracasei* 8700:2 ($1x10^9$ CFU/day) on reduction of the risk of common cold episodes, number of days with common cold symptoms, frequency and severity of symptoms, and cellular immune response in healthy subjects. Those researchers conducted a randomized, double-blind, placebo-controlled study in which 135 subjects were supplemented daily with the probiotic product and 137 participants were given a control product, both administered during 12 weeks. Berggren *et al.* (2011) observed that the incidence of acquiring one or more common cold episode significantly decreased from 67 % in the control group to 55 % in the probiotic group. Furthermore, the number of days with cold common symptoms was 8.6 days in the control group in comparison to 6.2 in the probiotic group ($P<0.05$) during the period of the study. The total symptom score was reduced during the study from a mean of 44.4 for the control group to 33.6 for the probiotic group. Additionally, higher counts ($P<0.05$) of peripheral B lymphocytes were determined in control group in comparison to probiotic group. Those authors concluded that *Lactobacillus plantarum* HEAL 9 and *Lactobacillus paracasei* 8700:2 contributed to the body's defense against common cold infections.

4.4. Maintenance of Oral Health

Lactobacilli are important for the oral ecosystem and can be associated with health of disease oral status (Badet and Thebaud, 2008). It is known that naturally occurring *Lactobacillus* species, including *Lactobacillus plantarum*, may inhibit strains of mutans streptococci and subject's autologous mutans streptococci *in vitro* (Simark-Mattsson *et al.*, 2007). According to de Carvalho *et al.* (2006), *Streptococcus mutans* is significantly associated with early childhood caries. On the other hand, *Candida albicans* is a persistent member of the oral microbiota in children with caries, displaying a substantial growth response to sucrose exposure (de Carvalho *et al.*, 2006; Sissons *et al.*, 2007).

Hasslöf et al. (2010) tested eight probiotic lactobacilli (*Lactobacillus acidophilus* La5, *Lactobacillus plantarum* 299v, *Lactobacillus plantarum* 931, *Lactobacillus rhamnosus* GG ATCC 53103, *Lactobacillus rhamnosus* LB21, *Lactobacillus paracasei* F19, *Lactobacillus reuteri* PTA 5289 and *Lactobacillus reuteri* ATCC 55730) for growth inhibition on five strains of mutans spretococci, including both laboratory reference strains (*Streptococcus mutans* NCTC 10449, *Streptococcus mutans* Ingbritt, and *Streptococcus sobrinus* OMZ176) and clinical isolates (*Streptococcus mutans* P1:27 and *Streptococcus mutans* P2:29) and five strains of *Candida* spp., including both reference strains *Candida albicans* ATCC 28366 and *Candida albicans* ATCC 10231, and three clinical isolates of the yeast (*Candida albicans* 1957, *Candida albicans* 3339 and *Candida albicans* GDM8). Hasslöf et al. (2010) observed that at higher concentrations (10^5 to 10^9 CFU/ml) all lactobacilli strains (with the exception of *Lactobacillus acidophilus* La5) inhibited completely the growth of mutans streptococci. At the lowest cell concentration (10^3 CFU/ml) only *Lactobacillus plantarum* 299v and *Lactobacillus plantarum* 931 showed a total growth inhibition. Also, all the tested lactobacilli strains reduced *Candida* spp. growth but *Lactobacillus plantarum* strains and *Lactobacillus reuteri* ATCC 55730 displayed the strongest inhibition on *Candida albicans*. Interestingly, no significant differences were observed between the reference strains and the clinical isolates. Those authors concluded that although the selected probiotic strains displayed a significant ability to inhibit growth of oral pathogens in vitro, further clinical studies are needed to determine the probiotic potential in vivo in such a complex environment, represented by the oral biofilm.

Yang *et al*. (2008) studied the probiotic characteristics of *Lactobacillus plantarum* HO-69 in oral cavity by measurement of adhesion ratios of the microorganism to teeth and dentine surfaces, determination of hydrophobicity and surface charges by microbial adhesion to hydrocarbons (MATH) method and evaluation of the inhibitory activity spectrum using the agar well diffusion method.

Those authors observed that the adhesion ratios to hard tissue were low, whereas the hydrophobicity and surface charges were high. Furthermore, *Lactobacillus plantarum* HO-69 demonstrated inhibitory activity against several pathogens. Yang *et al.* (2008) concluded that it is very unlikely that *Lactobacillus plantarum* HO-69 is able to initiate or accelerate dental caries and since the microorganism displays broad-spectrum of antimicrobial activity it can be considered as a potential probiotic for oral cavity.

Kõll et al. (2008) studied 67 lactobacilli isolates belonging to 10 species (*Lactobacillus acidophilus, Lactobacillus crispatus, Lactobacillus delbrueckii, Lactobacillus fermentum, Lactobacillus gasseri, Lactobacillus oris, Lactobacillus paracasei, Lactobacillus plantarum, Lactobacillus rhamnosus* and *Lactobacillus salivarius*) obtained from human healthy humans for antimicrobial activity against oral pathogens, tolerance of low pH and bile content. Furthermore, the lysozyme tolerance and antibiotic susceptibility of 22 potential probiotic strains were determined. Those authors observed that the majority of the lactobacilli strains tested suppressed the growth of the oral pathogens, with the exception of *Candida albicans* 048 (wild-type).

The lowest pH tolerated by lactobacilli after 4h of incubation was pH 2.5, although no growth of the microorganism was verified under this condition. All lactobacilli strains tested tolerated a high concentration (5% [v/v]) of human bile. Furthermore, four *Lactobacillus plantarum* and two *Lactobacillus oris* strains expressed resistance to tetracycline and/or doxycycline. Kõll et al. (2008) concluded that *Lactobacillus plantarum*, *Lactobacillus paracasei, Lactobacillus salivarius* and *Lactobacillus rhamnosus* have potential to be used for oral health; however, the absence of transferable antibiotic-resistance genes in *Lactobacillus plantarum* strains remains to be confirmed.

4.5. Burns Treatment

Burns produce disruption of the mechanical integrity of the skin and generalized immune suppression that allows micro-organisms to multiply freely (Peral *et al.*, 2009). The most prevalent pathogens isolated from burns are *Staphylococcus aureus*, detected in 75 % of wounds, followed by *Pseudomonas aeruginosa* (25 %), *Streptococcus pyogenes* (20 %) and several coliforms (5 %). Other microorganisms, including streptococci, anaerobic organisms and fungi can also cause infections (Edwards-Jones and Greenwood, 2003). These bacteria normally produce biofilms, a complex community embedded in a polysaccharide matrix, which makes them extremely resistant to antibiotics, antiseptics and to the host immune response (Percival *et al.*, 2011). Furthermore, the use of antibiotics in these patients is associated with the risk of adverse effects (Edwards-Jones *et al.*, 2003).

According to *in vitro* experiments, Valdez *et al.* (2005) observed that *Lactobacillus plantarum* was able to inhibit the activity of N-acyl-homoserine and also interfered with the formation of biofilm and production of elastase by *Pseudomonas aeruginosa.* Those researchers also performed *in vivo* tests to evaluate the role of *Lactobacillus plantarum* in murine burns infected with *Pseudomonas aeruginosas* and observed that the LAB inhibited the colonization of the pathogen, modified the inflammatory response and led to tissue repair. Those researchers concluded that *Lactobacillus plantarum* and/or its metabolites could be considered as potential therapeutic agents for the local treatment of *Pseudomonas aeruginosa* burn infections.

More recently, Peral *et al.* (2009, 2010) performed a clinical trial with *Lactobacillus plantarum* to assess its effectiveness for burn treatment, in comparison with silver sulphadiazine (SD-Ag), which is considered the gold standard in topical burn treatment (Atiyeh *et al.*, 2007). Peral *et al.* (2009) evaluated 80 burned patients which were grouped into infected (delayed) second- and third-degree and non infected (early) third-degree burns treated with *Lactobacillus plantarum* ATCC 10 241 (steeped in a gauze pad applied directly to the burn and changed every day by using freshly-cultivated LAB) or SD-Ag for ten days. Those researchers did not observe any differences between the two groups in routine blood tests and found no allergic or systemic symptoms for both treatments. Those authors also observed that the proportion of patients with delayed second-degree burns was 0.71 for *Lactoabcillus plantarum* and 0.73 for SD-Ag with respect to the decrease in bacterial load, promotion of granulating tissue wound bed and healing, whereas in early third-degree burns, the figures were 0.75 for *Lactobacillus plantarum* and 0.84 for SD-Ag in preventing wound infection and promotion of granulation tissue, 0.90 in graft taking for both treatments and 0.75 for *Lactobacillus plantarum* and 0.77 for SD-Ag in healing. In delayed third-degree burns, values were 0.83 for *Lactobacillus plantarum* and 0.71 for SD-Ag (relative ratio of + 16.90%) for decrease in the bacterial load and providing a granulating tissue wound bed, 0.90 in graft taking for both treatments and 0.75 for *Lactobacillus plantarum* and 0.64 for SD-Ag (relative rate of +17.19%) in healing. Although the number of patients per group was considerably small, those researchers suggested that the treatment with *Lactobacillus plantarum* deserves to be investigated in more details and could be useful as an interesting alternative for the topical treatment of burns.

4.6. Potential Role in Cardiovascular Disease Prevention/Treatment

Cardiovascular diseases have been the *major* health *problem* accounting for up to one-third of deaths worldwide (WHO/FAO, 2003). The risk factors for cardiovascular disease can be grouped into two main categories: unmodifiable factors (such as male gender, and family history of premature heart diseases) and potentially modifiable factors (such as cigarette smoking, high blood pressure, high blood-cholesterol level, physical inactivity, diet rich in fats, salt and refined carbohydrates, diabetes, and obesity). Promising results were achieved with probiotic strain *Lactobacillus plantarum* 299v on the reduction/attenuation of some of these potentially modifiable factors.

Bukowska *et al.* (1997) demonstrated that patients with moderately elevated cholesterol concentrations displayed reduced levels of LDL cholesterol and fibrinogen after supplementation with a functional food product ProViva (Probi AB, Lund, Sweden) containing fruit juice, fermented oat, and *Lactobacillus plantarum* 299v. Intake of this same strain in a roseship drink has been shown to increase the total faecal concentration of carboxylic acids, acetic acid, and propionic acid, indicating increased fermentation in the colon (Johansson *et al.*, 1998). Indeed, beneficial microorganisms can settle in the large intestine and ferment dietary fiber, which main end products include short-chain fatty acids (SCFA), chiefly acetic, propionic, and butyric acids. It has been shown that SCFA, mainly propionic acid, increase glucose tolerance and inhibit cholesterol synthesis in the liver (Venter and Vorster, 1989; Nishima and Freedland, 1990).

Naruszewick *et al.* (2002) performed a controlled, randomized, double-blind trial to determine the influence of strain *Lactobacillus plantarum* 299v on lipid profiles, inflammatory markers, and monocyte function in heavy smokers. Healthy smoker volunteers received daily 400 ml of ProViva (Probi AB) product per day, containing $5x10^7$ CFU/ml of *Lactobacillus plantarum* 299v during six weeks. As a control group, volunteers consumed the same amount of the product without any bacteria. The participants fully agreed in using the product as a dietary supplement and did not change their regular diet or physical activity during the entire study. Among others, the authors observed significant decreases in systolic blood pressure, leptin, and fibrinogen in the experimental group and concluded that administration *Lactobacillus plantarum* 299v could be useful as a protective agent in the primary prevention of artherosclerosis in smokers.

Finally, Karlsson *et al.* (2010) performed a double blind, placebo controlled study to evaluate the microbial change in the intestinal microbiota in patients with cardiovascular disease that daily consumed an oat drink containing *Lactobacillus plantarum* DSM 9843 (ca. 10^{11} CFU of the microorganism ingested per day) (299v; Probi AB) or an unfermented drink. Those researchers evaluated a group of men with atherosclerotic plaque on the carotid wall from whom colonic biopsies were taken before and after four weeks of probiotic treatment and evaluated by Terminal Restriction Fragment Length Polymorphism method. The microbial diversity was calculated and short-chain fatty acids in faeces, and blood markers were also analyzed. Those researchers observed that patients allocated in the probiotic group showed increased intestinal microbial diversity in comparison to subjects from the placebo group and that the fermentation pattern of SCFA in faeces was unaffected for most acids, although the probiotic group displayed decreased concentration of isovaleric and valeric acids. Furthermore, the viable count of lactobacilli increased in the probiotic group but no main changes in blood markers were determined. Karlsson *et al.* (2010) concluded that administration of *Lactobacillus plantarum* DSM 9843 may be an interesting strategy to favour a diverse intestinal microbiota, which play important roles for the health status of the host.

4.7. Cholesterol-Lowering Effects

The reduction in plasma cholesterol levels can lower the risk of coronary events, which can be achieved by appropriate food intake including low-cholesterol and low-fat diets, dietary fiber and yougurt containing specific LAB (Akalin *et al.*, 1997; Schaafsma *et al.*, 1998; Lora *et al.*, 2007; Theuwissen and Mensink, 2008). Concerning the last strategy, Jeun *et al.* (2010) investigated the hypocholesterolemic effects of live and dead cells of *Lactobacillus plantarum* KCTC3928 in C57BL/6 mice fed a high-fat diet. Those authors used double-coated bacteria since they display better acid resistance, bile resistance, and keep stable their physiologic activities, in comparison to non-coated microorganisms. Jeun *et al.* (2010) fed six-week-old C57BL/6 male mice with a commercial chow diet for one week and then the animals were randomly divided into three groups of seven mice each: control group fed with no *Lactobacillus plantarum* KCTC3928, and two treatment groups orally fed with live (PL, 10^9 CFU) or dead (PD, 10^{10} CFU) *Lactobacillus plantarum* KCTC3928 daily for four weeks. According to those authors, at 4-week, oral administration of *Lactobacillus plantarum* KCTC3928 reduced the plasma cholesterol level by 33 % ($P<0.05$), whereas cholesterol levels in PD and control groups did not change (P>0.05). Furthermore, *Lactobacillus plantarum* KCTC3928 feeding resulted in a decreased LDL cholesterol levels in both PL (13 mg/dL, - 42 %) ($P<0.05$) and PD (7 mg/dL, - 20 %) ($P>0.05$) groups. Following *Lactobacillus plantarum* KCTC3928 administration, plasma triacylglycerol levels were decreased by 24 mg/dL (- 32 % reduction) in PL groups assessed at two and four weeks ($P<0.05$). Finally, PD and PL groups showed 45 % and 38 %, respectively, increases in fecal bile acid excretion, compared with the control group ($P>0.05$). The authors concluded that the positive results obtained with double-coated live LAB could be due to the induction of fecal bile acid secretion followed by increased degradation of hepatic cholesterol into bile acids, although further studies in human should be carried out to confirm the observations.

In another study, Kumar *et al.* (2011) studied the anti-hypercholesterolaemic effects of two putative probiotic bile salt hydrolase (Bsh)-producing *Lactobacillus plantarum* strains (Lp19 and Lp21) in rats. Bile salts are secreted as bile into the duodenum in the form of N-acyl compounds conjugated with glycine or taurine (Hofmann and Mysels, 1992), and enhance the emulsification of lipids and aid the absorption of lipid nutrients (Salter *et al.*, 1996) apart from undergoing enterohepatic circulation (Hofmann and Roda, 1984). During this step, deconjugation of bile salts by Bsh liberates the amino acid residue and forms deconjugated bile acids, which are excreted more rapidly from the intestine than the conjugated ones. Once free bile acids are excreted from the body, the synthesis of new bile salts from cholesterol is able to decrease the total cholesterol concentration in the body (Gilliland and Speck, 1977). Kumar *et al.* (2011) evaluated a total of thirty adult male Sprague-Dawley (SD) rats fed with different experimental diets: normal diet, hypercholesterolaemic diet (HD), HD added of *Lactobacillus plantarum* Lp91 (HD91) at ≥ 10^8 UFC/g, HD plus microencapsulated *Lactobacillus plantarum* Lp91 (HDCap91) at ≥ 10^8

UFC/g and HD added of *Lactobacillus plantarum* Lp21 (HD21) at ≥ 10^8 UFC/g during three weeks. Those authors observed positive outcomes after 21 days of treatment, such as a reduction of 23.26 %, 15.71 %, and 15.01 % in total cholesterol determined, a diminution of 21.09 %, 18.77 %, and 18.17 % in triacylglycerol (TAG), a decrement of 38.13 %, 23.22 and 21.42 % in LDL-cholesterol, and an augmentation of 18.94 %, 10.30 %, and 7.78 % in HDL-cholesterol for groups HD91, HDCap91, and HD21, respectively. Plasma HDL-cholesterol and TAG concentrations differed significantly among all groups throughout the experiment ($P<0.05$). Kumar *et al.* (2011) also concluded that the strain *Lactobacillus plantarum* Lp91 has the potential to be explored as a probiotic in the management of hypercholesterolemia by improvement of hydrolysis of bile salts.

Wang *et al.* (2009b) studied the probiotic potential of *Lactobacillus plantarum* MA2, isolated from Chinese traditional Tibet kefir to decrease cholesterol levels *in vivo*. Those authors fed rats on cholesterol-enriched experimental diet, supplemented with *Lactobacillus plantarum* MA2 (lyophilized powder dissolved in physiological saline and administered by gavage with a dose established at 10^{11} microorganisms/day per animal) in comparison to animals that had no supplementation (control). Those researches observed lower serum total cholesterol, low-density lipoprotein cholesterol, and triglycerides levels in the group of rats supplemented with *Lactobacillus plantarum* MA2, while no changes on levels of high-density lipoprotein cholesterol were assessed. Furthermore, liver total cholesterol and triglycerides were significantly increased ($P<0.05$) in probiotic group in comparison with control.

In a study conducted by Nguyen *et al.* (2007), *Lactobacillus plantarum* PH04, a strain isolated from infant feces, was orally administered at doses of 10^7 CFU per day during 14 days to mice previously fed a high-cholesterol diet containing 10 % (w/v) skim milk and 10 % (w/v) cream. As control group, mice receiving a rich-cholesterol experimental diet were supplemented with the LAB. Those authors observed that serum cholesterol and triglycerides were respectively 7 % and 10 % lower in the group fed with *Lactobacillus plantarum* PH04 in comparison with control ($P<0.05$). Interestingly, Nguyen *et al.* (2007) also observed that bacterial translocation was similar in both groups.

Kaushik *et al.* (2009) studied the ability of *Lactobacillus plantarum* Lp9, a strain isolated from buffalo milk, to deconjugate major bile salts such as glycocholate and deoxytaurocholate, and demonstrated its potential to reduce cholesterol levels. Those authors also detected by use of Polimerase Chain Reaction (PCR) method, the presence of a bile salt hydrolase (*bsh*) gene in the LAB. However, *Lactobacillus plantarum* Lp9 was not tested *in vivo* to confirm its probiotic anti-hypercholesterolaemic potential.

4.8. Potential Antiobesity Effects

Once the gut microbiota can be considered an environmental factor involved in the development of obesity (Tilg *et al.*, 2009), probiotics may be useful for the development of foods that aim to prevent and/or improve obesity and its related disorders. In this sense, Takemura *et al.* (2010) investigated the effects of *Lacobacillus plantarum* strain No. 14 (LP14), a bacterial strain previously reported to decrease body fat percentage in healthy volunteers and adipocyte size in mice. Those authors evaluated mice daily fed with normal (NFD) or high-fat (HFD) diet challenged with either $1x10^8$ CFU of LP14 or PBS, for a period of 11 weeks. The researchers verified that in mice given HFD, LP14 administration significantly decreased the mean adipocyte size and showed a tendency of reduction in the weight of white adipose tissue (WAT) and serum total cholesterol and leptin concentrations in comparison with mice supplemented with PBS plus dextrin. The authors concluded that LP14 may be able to exhibit a beneficial effect on the onset of diet-induced obesity by reducing the cell size of WATs, which may involve mechanisms not yet reported in scientific literature.

Lee *et al.* (2007) studied the antiobesity activity of *Lactobacillus plantarum* PL62 in diet-induced obese mice. The strain is able to produce *trans*-10, *cis*-12-conjugated linoleic acid, a compound previously known to exert an antiobesity effect (Park *et al.*, 1997). Lee *et al.* (2007) evaluated four groups of mice including a control group (which received a standard diet) and three groups that were given a high-fat diet and daily supplemented with either skim milk or 10^7 or 10^9 CFU of *Lactobacillus plantarum* PL62. Those researchers observed that although not statistically significant, the weights determined for inguinal, epididymal and perirenal WAT were reduced in the animals supplemented with the LAB. After five weeks of feeding, the weight reduction in mice supplemented with 10^7 or 10^9 of *Lactobacillus plantarum* PL62 was similar. Furthermore, blood levels of total glucose and body weights of mice were lower ($P<0.01$) in the groups treated with PL62 in comparison with the group that received a high-fat diet plus skim milk.

References

Akalin, A. S., Gonc, S. and Duzel, S. (1997). Influence of yogurt and acidophilus yogurt on serum cholesterol levels in mice. *Journal of Dairy Science*, *80*, 2721-2725.

Akın, M. B., Akın, M. S. and Kirmaci, Z. (2007). Effects of inulin and sugar levels on the viability of yoghurt and probiotic bacteria and the physical and sensory characteristics in probiotic ice-cream. *Food Chemistry*, *104*, 93-99.

Anderson, R. C., Cookson, A. L., McNabb, W. C., Kelly, W. J. and Roy, N. C. (2010b). *Lactobacillus plantarum* DSM 2648 is a potential probiotic that enhances intestinal barrier function. *FEMS Microbiology Letters*, *309*, 184-192.

Anderson, R. C., Cookson, A. L., McNabb, W. C., Park, Z., McCann, M. J., Kelly, W. J. and Roy, N. C. (2010a). *Lactobacillus plantarum* MB452 enhances the function of the intestinal barrier by increasing the expression levels of genes involved in tight junction formation. *BMC Microbiology, 10*, 316.

Atiyeh, B. S., Costagliola, M., Hayek, S. N. and Dibo, S. A. (2007). Effect of silver on burn wound infection control and healing: review of the literature. *Burns*, *33*, 139-148.

Badet, C. and Thebaud, N. B. (2008). Ecology of lactobacilli in the oral cavity: a review of literature. *Open Microbiology Journal, 2*, 38-48.

Barbara, G. (2006). Mucosal barrier defects in irritable bowel disease syndrome. Who left the door open? *American Journal of Gastroenterology, 101*, 1295-1298.

Berggren, A., Ahrén, I. L., Larsson, N. and Önning, G. (2011). Randomised, double-blind and placebo-controlled study using a new probiotic lactobacilli for strengthening the body immune defence against viral infections. *European Journal of Nutrition*, *50*, 203-210.

Berthier, F. and Ehrlich, S. D. (1998). Rapid species identification within two groups of closely related lactobacilli using PCR primers that target the 16S/23S rRNA spacer region. *FEMS Microbiology Letters, 161*, 97-106.

Bixquert Jiménez, M. (2009). Treatment of irritable bowel syndrome with probiotics. An etiopathogenic approach at last? *Revista Espanola de Enfermagem y Dignostic*, *101*, 553-564.

Boekhost, J., Siezen, R. J., Zwahlen, M. C., Vilanova, D., Pridmore, R. D., Mercenier, A., Kleerebezem, M., de Voz, W. M., Brüssow, H. and Desiere, F. (2004). The complete genomes of *Lactobacillus plantarum* and *Lactobacillus johnsonii* reveal extensive differences in chromosome organization and gene content. *Microbiology, 150*, 3601-3611.

Bove, P., Capozzi, V., Garofalo, C., Rieu, A., Spano, G. and Fiocco, D. (2011) Inactivation of the *ftsH* gene of *Lactobacillus plantarum* WCFS1: Effects on growth, stress tolerance, cell surface properties and biofilm formation. *Microbiological Research.* doi: 10.1016/j.micres.2011.07.001. (in press)

Bruewer, M., Samarin, S. and Nusrat, A. (2006) Inflammatory bowel disease and the apical junctional complex. *Annals of New York Academy of Science, 1072*, 242-252.

Bujalance, C., Moreno, E., Jimenez-Valera, M. and Ruiz-Bravo, A. A. (2007). probiotic strain of *Lactobacillus plantarum* stimulates lymphocyte responses in immunologically intact and imunocompromised mice. *International Journal of Food Microbiology, 113*, 28-34.

Bukowska, H., Pieczul-Mróz, J., Jastrzêbska, M., Chelstowski, K. and Naruszewicz, M. (1997). Decrease in fibrinogen and LDL-cholesterol levels upon supplementation of diet with *Lactobacillus plantarum* in subjects with moderately elevated cholesterol. *Atherosclerosis, 137*, 437-438.

Cammarota, M., De Rosa, M., Stellavato, A., Lamberti, M., Marzaioli, I. And Giuliano, M. (2009). *In vitro* evaluation of *Lactobacillus plantarum* DSMZ 12028 as a probiotic: Emphasis on innate immunity. *International Journal of Food Microbiology*, *135*, 90-98.

Cebeci, A. and Gürakan, C. (2003). Properties of potential probiotic *Lactobacillus plantarum* strains. *Food Microbiology*, *20*, 511-518.

Chagnaud, G., Machinis, K., Coutte, L., Marecat, A. and Mercenier, A. (2001). Rapid PCR-based procedure to identify lactic acid bacteria: Application to six common *Lactobacillus* spices. *Journal of Microbiological Methods*, *44*, 139-148.

Chao, S. H., Tomii, Y., Sasamoto, M., Fujimoto, J., Tsai, Y. C. and Watanabe, K. (2008). *Lactobacillus capillatus* sp. nov., a motile *Lactobacillus* species isolated from stinky tofu brine. *International Journal of Systematic and Evolutionary Microbiology, 58*, 2555-2559.

Charteris, W. P., Kelly, P. M., Morelli, L. and Collins, J. K. (1998). Antibiotic susceptibility of potentially probiotic *Lactobacillus* species. *Journal of Food Protection*, *61*, 1636-1643.

Choi, S. Y., Chang, E. C., Kim, S. C. and So, J. S. (2003). Antimicrobial susceptibility and strain prevalence of Korean vaginal *Lactobacillus* spp. *Anaerobe*, *9*, 277-280.

Danielsen, M. (2002). Characterization of the tetracycline resistance plasmid pMD5057 from *Lactobacillus plantarum* 5057 reveals a composite structure. *Plasmid*, 48, 98-103.

Danielsen, M. and Wind, A. (2003). Susceptibility of *Lactobacillus* spp. to antimicrobial agents. *International Journal of Food Microbiology*, *82*, 1-11.

Davies, J. (1997). Origins, acquisition and dissemination of antibiotic resistance determinants. *Ciba Foundation Symposium*, *207*, 15-27.

De Bruyne, K., Canu, N., De Vauyst, L. and Vandamme, P. (2009). *Lactobacillus fabifermentans* sp. nov. and *Lactobacillus cacaonum* sp. nov., isolated from Ghanaian cocoa fermentation. *International Journal of Systematic and Evolutionary Microbiology*, *59*, 7-12.

de Carvalho, F. G., Silva, D. S., Hebling, J., Spolidorio, L. C. and Spolidorio, D. M. (2006). Presence of mutans streptococci and *Candida* spp. in dental plaque/dentine of carious teeth and early childhood caries. *Archives in Oral Biology*, *51*, 1024-1028.

Del Piano, M., Carmagnola, S., Anderloni, A., Andorno, S., Ballarè, M., Balzarini, M., Montino, F., Orsello, M., Pagliarulo, M., Sartori, M., Tari, R., Sforza, F. and Capurso, L. (2010). The use of probiotics in healthy volunteers with evacuation disorders and hard stools. *Journal of Clinical Gastroenterology*, *44*, 30-34.

Delgado, S., Flórez, A. B. and Mayo, B. (2007). Antibiotic susceptibility of *Lactobacillus* and *Bifidobacterium* species from the human gastrointestinal tract. *Current Microbiology*, *50*, 202-207.

Dicks, L. M. T. and ten Doeschate, K. (2010). *Enterococcus mundtii* ST4SA and *Lactobacillus plantarum* 423 alleviated symptoms of *Salmonella* infection, as determined in Wistar rats challenged with *Salmonella enterica* serovar Thyphimurium. *Current Microbiology,* 61, 184-189.

World Health Organization (WHO). (2003). Diet, nutrition and the prevention of chronic diseases: report of a joint WHO/FAO expert consultation. Geneva: 81-94.

Duary, R. K., Batish, V. K. and Grover, S. (2010). Expression of the *atpD* gene in probiotic *Lactobacillus plantarum* strains under *in vitro* acidic conditions using RT-qPCR. *Research in Microbiology*, *161*, 399-405.

Edwards-Jones, V., Dawson, M. M. and Childs, C. A. (2003). Survey into TSS in UK burns units. *Burns*, *26*, 323-333.

Edwards-Jones, V. and Greenwood, J. E. (2003). What's new in burn microbiology? James Laing Memorial Prize Essay 2000. *Burns*, *29*, 15-24.

Essid, I., Medini, M. and Hassouna, M. (2009). Technological and safety properties of *Lactobacillus plantarum* strains isolated from a Tunisian traditional salted meat. *Meat Science, 81*, 203-208.

Ewaschuk, J. B., Diaz, H., Meddings, L., Diederichs, B., Dmytrash, A., Backer, J., Looijer-van Langen, M. and Madsen, K. L. (2008). Secreted bioactive factors from *Bifidobacterium infantis* enhance epithelial cell barrier function. *American Journal of Physiology, Gastroenterology and Liver Physiology*, *295*, 1025-1034.

Fayol-Messaoudi, D., Coconiier-Polter, M. H., Lievin-Le Moal, V., Atassi, F., Berger, C. N. and Servin, A. L. (2007). The *Lactobacillus plantarum* strain ACA-DC287 isolated from a Greek cheese demonstrates antagonistic activity *in vitro* and *in vivo* against *Salmonella enterica* serovar Typhimurium. *Journal of Applied Microbiology*, 103, 657-665.

Feld, L., Bielak, E., Hammer, K. and Wilcks, A. (2009). Characterization of a small erythromycin resistance plasmid pLFE1 from the food-isolate *Lactobacillus plantarum* M345. *Plasmid*, *61*, 159-170.

Felske, A., Rheims, H., Wolterink, A., Stackebrandt, E. and Akkermans, A. D. L. (1997). Ribosome analysis reveals prominent activity of an uncultured member of the class Acinetobacteria in grassland soils. *Microbiology, 147*, 2983-2989.

Flórez, A. B., Delgado, S. and Mayo, B. (2007). Antimicrobial susceptibility of lactic acid bacteria isolated from a cheese environment. *Canadian Journal of Microbiology*, *51*, 51-58.

Franz, C. M. A. P., Hummel, A. and Holzapfel, W. H. (2005). Problems related to the safety assessment of lactic acid bacteria starter cultures and probiotics. *Mitt Lebensmittelunters Hygiene, 96*, 39-65.

Fred, E. B., Peterson, W. H. and Anderson, J. A. (1921). The characteristics of certain pentose destroying bacteria, especially as concerns their action on arabinose. *Journal of Biological Chemistry, 48*, 385-412.

Gilliland, S. E. and Speck, M. J. (1977). Deconjugation of bile acids by intestinal lactobacilli. *Applied and Environmental Microbiology*, 33, 15-18.

Gillim-Ross, L. and Subbarao, K. (2006). Emerging respiratory viruses: challenges and vaccine srategies. *Clinical Microbiology Reviews, 19*, 614.

Goldin, B. R. (1998). Health benefits of probiotics. *British Journal of Nutrition, 80*, S203-S207.

Guttman, J. A., Samji, F. N., Li, Y., Vogl, A. W. and Finlay, B. B. (2006). Evidence that tight junctions are disrupted due to intimate bacterial contact and not inflammation during attaching and effacing pathogen infection *in vitro*. *Infections and Immunology, 74*, 6075-6084.

Harrison, A. P. Jr and Hansen, P. A. (1950). A motile *Lactobacillus* from the cecal feces of turkeys. *Journal of Bacteriology, 59*, 444-446.

Hart, A. and Kamm, M. A. (2002). Review article: mechanisms of initiation and perpetuation of gut inflammation by stress. *Alimentation and Pharmacological Therapy, 16*, 2017-2028.

Hasslöf, P., Hedberg, M., Twetman, S. and Stecksén-Blicks, C. (2010). Growth inhibition of oral mutans streptococci and candida by commercial probiotic lactobacilli – an *in vitro* study. *BMC Oral Health*, 10, 8.

Herías, M. V., Hessle, C., Telemo, E., Midtvedt, T., Hanson, L. A. and Wold, A. E. (1999). Immunomodulatory effects of *Lactobacillus plantarum* colonizing the intestine of gnotobiotic rats. *Clinical and Experimental Immunology, 116*, 283-290.

Higashikawa, F., Noda, M., Awaya, T., Nomura, K., Oku, H. and Sugiyama, M. (2010). Improvement of constipation and liver function by plant-derived lactic acid bacteria: a double-blind, randomized trial. *Nutrition, 26*, 367-374.

Hisbergues, M., Magi, M., Rigaux, P., Steuve, J., Garcia, L., Goudercourt, D., Pot, B., Pestel, J. and Jacquet, A. (2007). In vivo and in vitro immunomodulation of Der p 1 allergen-specific response by Lactobacillus plantarum bacteria. *Clinical and Experimental Allergy*, 37, 1286-1295.

Hofmann, A. F. and Mysels, K. J. (1992). Bile acid solubility and precipitation *in vitro* and *in vivo*: the role of conjugation, pH, and Ca^{2+} ions. *Journal of Lipid Research,* 33, 617-626.

Hofmann, A. F. and Roda, A. (1984). Physicochemical properties of bile acids and their relationship to biological properties: an overview of the problem. *Journal of Lipid Research, 24*, 1477-1489.

Holland, D. F. (1920). The families and genera of bacteria. *Journal of Bacteriology*, 5, 191.

Holzapfel, W. H., Haberer, P., Geisen, R., Björkroth, J. and Schillinger, U. (2001). Taxonomy and important features of probiotic microorganisms in food and nutrition. *American Journal of Clinical Nutrition*, 73, 365-373.

Holzapfel, W. H., Haberer, P., Snel, J., Schillinger, U. and Huis In't Veld, J. H. J. (1998). Overview of gut flora and probiotics. *International Journal of Food Microbiology, 41*, 85-101.

Huang, C. H., Chang, M. T., Huang, M. C. and Lee, F. L. (2011). Rapid identification of *Lactobacillus plantarum* group using the SNaPshot minisequencing assay. *Systematic and Applied Microbiology*, *34*, 586-589.

Jacobsen, L., Wilcks, A., Hammer, K., Huys, G., Gevers, D. and Andersen, S. R. (2007). Horizontal transfer of *tet*(M) and *erm*(B) resistance plasmids from food strains of *Lactobacillus plantarum* to *Enterococcus faecalis* JH2-2 in the gastrointestinal tract of gnotobiotic rats. *FEMS Microbiological Ecology*, 59, 158-166.

Jeun, J., Kim, S., Cho, S. Y., Jun, H. J., Park, H. J., Seo, J. G., Chung, M. J. and Lee, S. J. (2010). Hypocholesterolemic effects of *Lactobacillus plantarum* KCTC3928 by increased bile acid excretion in C57BL/6 mice. *Nutrition*, *26*, 321-330.

Johansson, M. L., Nobaek, S., Berggren, A., Nyman, M., Björck, I., Ahrné, S., Jeppsson, B. and Molin, G. (1998). Survival of *Lactobacillus plantarum* DSM 9843 (299v), and effect on the short-chain fatty acid content of faeces after ingestion of a rose-ship drink with fermented oats. *International Journal of Food Microbiology*, *47,* 29-38.

Kailasapathy, K. and Rybka, S. (1997). *L. acidophilus* and *Bifidobacterium* spp.: their therapeutic potential and survival in yogurt. *Australian Journal of Dairy Technology*, *52*, 28-35.

Kandler, O. and Waiss, N. (1986). Genus *Lactobacillus* Beijerinck 1901, 212^{AL}. In Bergey's Manual of Systematic Bacteriology. Edited by Sneath, P. H. A., Mair, N. S., Sharpe, M. E. and Holt, J. G. Baltimore: Williams and Wilkins, *2*, 1209-1234.

Karlsson, C., Ahrné, S., Molin, G., Berggren, A., Palmquist, I., Fredrikson, G. N. and Jeppsson, B. (2010). Probiotic therapy to men with incipient arteriosclerosis initiates increased bacterial diversity in colon: a randomized controlled trial. *Atherosclerosis*, 208, 228-233.

Kaushik, J. K., Kumar, A., Duary, R. K., Mohanty, A. K., Grover, S. and Batish, V. K. (2009). Functional and probiotic attributes of an indigenous isolate of *Lactobacillus plantarum*. *PLoS ONE, 4*, 12.

Klare, I., Konstabel, C., Werner, G., Huys, G., Vankerckhoven, V., Kahlmeter, G., Hildebrandt, B., Müller-Bertling, S., Witte, W. and Goossens, H. (2007). Antimicrobial susceptibilities of *Lactobacillus, Pediococcus* and *Lactococcus* human isolates and cultures intended for probiotic or nutritional use. *Journal of Antimicrobials and Chemotherapy*, *59*, 900-912.

Klarin, B., Wullt, M., Palmquist, I., Molin, G., Larsson, A. and Jeppsson, B. (2008). *Lactobacillus plantarum* 299v reduces colonization of *Clostridium difficile* in critically ill patients treated with antibiotics. *Acta Anaesthesiologica Scandinavica*, *52,* 1096-1102.

Kleerebezem, M., Boekehorst, J., Van Kranenburg, R., Molenaar, D., Kuipers, O. P., Leer, R., Tarchini, R., Peters, S. A., Sandbrink H. M., Fiers, M. W. E. J., Stiekema, W., Lankhorst, R. M. K., Bron, P. A., Hoffer, S. M., Groot, M. N. N., Kerkhoven, R., de Vries, M., Ursing, B., de Vos, W. M. and Siezen, R. J. (2003). Complete genome sequence of *Lactobacillus plantarum* WCFS1. *Proceedings of National Academy of Sciince of the USA*, *100*, 1990-1995.

Kleerebezem, M. and de Vos, W. M. (2011). Lactic acid bacteria: life after genomics. *Microbial Biotechnology*, *4*, 318-322.

Klingberg, T. D., Pedersen, M. H., Cencic, A. and Budde, B. B. (2005). Application of measurements of transepithelial electrical resistance of intestinal epithelial cell monolayers to evaluate probiotic activity. *Applied and Environmental Microbiology*, *71*, 7528-7530.

Ko, J. S., Yang, H. R., Chang, J. Y. and Seo, J. K. (2007). *Lactobacillus plantarum* inhibits epithelial barrier dysfunction and interleukin-8 secretion mediated by tumor necrosis factor-α. *Word Journal of Gastroenterology*, *7*, 1962-1965.

Kõll, P., Mändar, R., Marcotte, H., Leibur, E., Mikelsaar, M. and Hammarström, L. (2008). Characterization of oral lactobacilli as potential probiotics for oral health. *Oral Microbiology and Immunology*, *23*, 139-147.

Kumar, R., Grover, S. and Batish, V. K. (2011). Hypocholesterolaemic effect of dietary inclusion of two putative probiotic bile salt hydrolase-producing *Lactobacillus plantarum* strains in Sprague-Dawley rats. *British Journal of Nutrition*, *105*, 561-573.

Kuratsu, M., Hamano, Y. and Dairi, T. (2010). Analysis of the *Lactobacillus* metabolic pathway. *Applied and Environmental Microbiology*, *76*, 7299-7301.

Lee, K., Paek, K., Lee, H.Y., Park, J. H. and Lee, Y. (2007). Antiobesity effect of *trans*-10, *cis*-12-conjugated linoleic acid-producing *Lactobacillus plantarum* PL62 on diet-induced obese mice. *Journal of Applied Microbiology*, *103*, 1140-1146.

Liu, C. F., Tseng, K. C., Chiang, S. S., Lee, B. H., Hsu, W. H. and Pan, T. M. (2011). Immunomodulatory and antioxidant potential of *Lactobacillus* exopolysaccharides. *Journal of Science and Food Agriculture*, *91,* 2284-2291.

Liu, Z., Li, N. and Neu, J. (2005). Tight junctions, leaky intestines, and pediatric diseases. *Acta Pediatrica*, *94*, 386-393.

Lönnermark, E., Friman, V., Lappas, G., Sandberg, T., Berggren, A. and Adlerberth, I. (2010). Intake of Lactobacillus plantarum reduces certain gastrointestinal symptoms during treatment with antibiotics. *Journal of Clinical Gastroenterology*, *44*, 106-112.

Lora, K. R., Morse, K. L., Gonzalez-Kruger, G. E. and Driskell, J. (2007). A High saturated fat and cholesterol intakes and abnormal plasma lipid concentrations observed in a group of 4- to 8-year-old children of Latino immigrants in rural Nebraska. *Nutrition Research*, *27*, 483-491.

Maeda, N., Nakamura, R., Hirose, Y., Murosaki, S., Yamamoto, Y., Kase, T. and Yoshikai, Y. (2009). Oral administration of heat-killed *Lactobacillus plantarum* L-137 enhances protection against influenza virus infection by stimulation of type I interferon production in mice. *International Immunopharmacology*, *9*, 1122-1115.

Maes, M. (2008). The cytokine hypothesis of depression: inflammation, oxidative and nitrosative stress (IOandNS) and leaky gut as new targets for adjunctive treatments in depression. *Neuroendocrinology Letters*, *29*, 287-291.

Mañé, J., Pedrosa, E., Lorén, V., Gassull, M. A., Espadaler, J., Cuñé, J., Audivert, S., Bonachera, M. A. and Cabré, E. (2011). A mixture of *Lactobacillus plantarum* CECT 7315 and CET 7316 enhances systemic immunity in elderly subjects. A dose-response, double-blind, placebo-controlled, randomized pilot trial. *Nutrition in Hospitals*, *26*, 228-235.

Mangell, P., Nejdfors, P., Wang, M., Ahrné, S., Weström, B., Thorlacius, H. and Jeppsson, B. (2002). *Lactobacillus plantarum* 299v inhibits *Escherichia coli*-induced intestinal permeability. *Digestive Diseases Science*, *47*, 511-516.

Mares, A., Neyts, K. and Debevere, J. (1994). Influence of pH, salt and nitrite on the heme-dependent catalase activity of lactic acid bacteria. *International Journal of Food Microbiology*, *24*, 191-198.

Mattila-Sandholm, T., Matto, J. and Saarela, M. (1992). Lactic acid bacteria with health claims-interations and interference with gastrointestinal flora. *International Dairy Journal*, *9*, 25-35.

McNaught, C. E., Woodcock, N. P., MacFie, J. and Michell, C. J. (2002). A prospective randomized study of the probiotic *Lactobacillus plantarum* 299V on indices of gut barrier function in elective surgical patients. *Gut*, *51*, 827-831.

Mitic, L. L., Van Itallie, C. M. and Anderson, J. M. (2000). Molecular physiology and pathophysiology of tight junctions I. Tight junction structure and function: lessons from mutant animals and proteins. *American Journal of Physiology-Gastrointestinaland Liver Physiology*, *279*, 250-254.

Mullin, J., Valenzano, M., Verrecchio, J. and Kothari, R. (2002). Age- and diet-related increase in transepithelial colon permeability of Fischer 344 rats. *Digestive Diseases Science*, *47*, 2262-2270.

Naruszewicz, M., Johansson, M. L., Zapolska-Downar, D., Bukowska, H. (2002). Effect of *Lactobacillus plantarum* 299v on cardiovascular disease risk factors in smokers. *American Journal of Clinical Nutrition,* 76, 1249-1255.

Nguyen, T. D. T., Kang, J. H. and Lee, M. S. (2007). Characterization of *Lactobacillus plantarum* PH04, a potential probiotic bacterium with cholesterol-lowering effects. *International Journal of Food Microbiology*, *113*, 358-361.

Niedzielin, K., Kordecki, H. and Birkenfeld, B. (2001). A controlled, double-blind, randomized study on the efficacy of *Lactobacillus plantarum* 299v in patients with irritable bowel syndrome. *European Journal of Gastroenterology and Hepatology*, 13, 1143-1147.

Nielsen, D. S., Schillinger, U., Franz, C. M. A. P., Bresciani, J., Amoa-Awua, W., Holzapfel, W. H. and Jakobsen, M. (2007). *Lactobacillus ghanensis* sp. nov., a motile lactic acid bacterium isolate from Ghanaian cocoa fermentations. *International Journal of Systematic and Evolutionary Microbiology*, *57*, 1468-1472.

Nishina, P. M. and Freedland, R. A. (1990). Effects of propionate on lipid biosynthesis in isolated rat hepatocytes. *Journal of Nutrition*, *120*, 668-672.

Omar, N. B., Abriouel, H., Lucas, R., Martínez-Cañamero, M., Guyot, J. P. and Gálvez, A. (2006). Isolation of bacteriocinogenic *Lactobacillus plantarum* strains from ben saalga, a traditional fermented grue from Burkina Faso. *International Journal of Food Microbiology*, *112*, 44-50.

Orla-Jensen, S. (1919). The lactic acid bacteria**.** Copenhagen: Anhr Fred Host and Son, 1-196.

Oudhuis, G. J., Bergmans, D. C., Dormans, T., Zwaveling, J. H., Kessels, A., Prins, M. H., Stobberingh, E. E. and Verbon, A. (2011). Probiotics *versus* antibiotic decontamination of the digestive tract: infection and mortality. *Intensive Care Medicine*, *37*, 110-117.

Ouwehand, A. C., Salvadori, B. B., Fondén, R., Mogensen, G., Salminen, S. and Sellars, R. (2003). Health effects of probiotics and cultures-containing dairy products in humans. *IDF Bulletin*, *380*, 4-19.

Paolillo, R., Romano Carratelli, C., Sorrentino, S., Mazzola, N. And Rizzo, A. (2009). Immunomodulatory effects of *Lactobacillus plantarum* on human colon cancer cells. *International Immunopharmacology*, *9*, 1265-1271.

Park, Y., Albright, K. J., Liu, W., Sorkson, J. M., Cook, M. E. and Pariza, M. W. (1997). Effect of conjugated linoleic acid on body composition in mice. *Lipids*, 32, 853-858.

Pathmakanthan, S., Li, C. K., Cowie, J. and Hawkey, C. J. (2004). *Lactobacillus plantarum* 299: Beneficial *in vitro* immunomodulation in cells extracted from inflamed human colon. *Journal of Gastroenterology and Hepatology*, *19*, 166-173.

Pederson, C. S. (1957). Genus I. *Lacobacillus* Bejerinck. 1901. In: Bergey's Manual of determinative Bacteriology, 7th ed.: Breed, R. S., Murray, E. G. D. and Smith, N. R. (Eds), Williams and Wilkins Co.: Baltimore, MD. 542.

Peral, M. C., Martinez, M. A. and Valdez, J. C. (2009). Bacteriotherapy with *Lactobacillus plantarum* in burns. *International Wound Journal*, *6*, 73-81.

Peral, M. C., Rachid, M. M., Gobbato, N. M., Huaman Martinez, M. A. and Valdez, J. C. (2010). Interleukin-8 production by polymorphonuclear leukocytes from patients with chronic infected leg ulcers treated with Lactobacillus plantarum. *Clinical Microbiology and Infections,* 16, 281-286.

Percival, S. L., Hill, K. E., Malic, S., Thomas, D. W. and Williams, D. W. (2011). Antimicrobial tolerance and the significance of persister cells in recalcitrant chronic wound biofilms. *Wound Repair and Regeneration*, *19,* 1-9.

Puertollano, E., Puertollano, M. A., Cruz-Chamorro, L., Alvarez de Cienfuegos, G., Ruiz-Bravo, A. and de Pablo, M. A. (2008). Orally administered Lactobacillus plantarum reduces pro-inflammatory interleukin secretion in sera from Listeria monocytogenes infected mice. *British Journal of Nutrition*, 99, 819-825.

Putaala, H., Salusjarvi, T., Nordstrom, M., Saarinen, M., Ouwehand, A. C., Bech-Hansen, E. and Rautonen, N. (2008). Effect of four probiotic strains and *Escherichia coli* O157:H7 on tight junction integrity and cyclo-oxigenase expression. *Research in Microbiology*, *195*, 692-698.

Quere, F., Deschamps, A. and Urdaci, M. C. (1987). DNA probe and PCR-specific reaction for *Lactobacillus plantarum*. *Journal of Applied Microbiology*, *82*, 783-790.

Rantsiou, K., Drosinos, E. H., Gialitaki, M., Metaxopoulos, I., Comi, G. and Cocolin, L. (2006). Use of molecular tools to characterize *Lactobacillus* spp. isolated from Greek traditional fermented sausages. *International Journal of Food Microbiology*, *112*, 215-222.

Rojo-Bezares, B., Sáenz, Y., Poeta, P., Zarazaga, M., Ruiz-Larrea, F. and Torres, C. (2006). Assessment of antibiotic susceptibility within lactic acid bacteria strain isolated from wine. *International Journal of Food Microbiology*, *111*, 234-240.

Salter, J., Chaplin, M., Dickerson, J. and Davies, J. (1996). Bile acids and health: is fibre the answer? *Nutrition and Food Science*, *6*, 29-33.

Schaafsma, G., Meuling, W. J., van Dokkum, W. and Bouley, C. (1998). Effects of a milk product, fermented by *Lactobacillus acidophilus* and with fructo-oligosaccharides added, on blood lipids in male volunteers. *European Journal of Clinical Nutrition*, *52*, 436-440.

Sen, S., Mullan, M. M., Parker, T. J., Woolner, J. T., Tarry, S. A. and Hunter, J. O. (2002). Effect of *Lactobacillus plantarum* 299v on colonic fermentation and symptoms of irritable bowel syndrome. *Digestion Diseases Science*, *47*, 2615-2620.

Settanni, L. and Moschetti, G. (2010). Non-starter lactic acid bacteria used to improve cheese quality and provide health benefits. *Food Microbiology*, *27*, 691-697.

Sharpe, M. E. (1981). The genus *Lactobacillus*. In: The Prokaryotes. A handbook on habitats, isolation and identification of bacteria. Starr, M. P., Stolp, H., Truper, H. G., Balows, A. and Schlegel, H.G. (Eds.). Sringer-Verlag: Berlin, p. 1653.

Sherman, P. M., Johnson-Henry, K. C., Yeung, H. P., Ngo, P. S., Goulet, J. and Tompkins, T. A. (2005). Probiotics reduce enterohemorrhagic *Escherichia coli* O157:H7- and enteropathogenic *E. coli* O127:H6- induced changes in polarized T84 epithelial cell monolayers by reducing bacterial adhesion and cytoskeletal rearrangements. *Infection and Immunity*, *73*, 5183-5188.

Siezen, R. J., Enckevort, F. H. J. V., Kleerebezem, M. and Teusink, B. (2004). Genome date mining of lactic acid bacteria: the impact of bioinformatics. *Current Opinions in Biotechnology*, *15*, 105-115.

Simark-Mattsson, C., Emilson, C. G., Hakansson, E. G., Jacobsson, C., Roos, K. and Holm, S. (2007). *Lactobacillus*-mediated interference of mutans streptococci in caries-free vs. caries-active subjects. *European Journal of Oral Science*, *115*, 308-314.

Singh, S., Goswami, P., Singh, R. and Heller, K. J. (2009). Application of molecular identification tools for *Lactobacillus*, with a focus on discrimination between closely related species: A review. *LWT – Food Science and Technology*, *42*, 448-457.

Singh, U. P., Tyagi, P. and Upreti, S. (2007). Manganese complexes as models for manganese-containing pseudocatalase enzymes: Synthesis, structural and catalytic activity studies. *Polyhedron, 26*, 3625-3632.

Sissons, C. H., Anderson, S. A., Wong, L., Coleman, M. J. and White, D. C. (2007). Microbiota of plaque microcosm biofilms: Effect of three times daily sucrose pulses in different simulated oral environments. *Caries Research*, *41*, 413-422.

Snel, J., Vissers, Y. M., Smit, B. A., Jongen, J. M. J., van der Meulen, E. T., Zwijsen, R., Ruinemans-Koerts, J., Jansen, A. P. H., Kleerebezem, M. and Savelkoul, H. F. J. (2011). Strain-specific immunomodulatory effects of *Lactobacillus plantarum* strains on birch-pollen-allergic subjects out of season. *Clinical and Experimental Allergy*, *41*, 232-242.

Takemura, N., Okubo, T. and Sonoyama, K. (2010). *Lactobacillus plantarum* strain No. 14 reduces adipocyte size in mice fed high-fat diet. *Experimental Biological Medicine*, *235*, 849-856.

Tanasupawat, S., Ezaki, T., Suzuki, K., Okada, S., Komagata, K. and Kozaki, M. (1992). Characterization and identification of *Lactobacillus pentosus* and *Lactobacillus plantarum* strains from fermented foods in Thailand. *Journal of General and Applied Microbiology*, *38*, 121-134.

Theuwissen, E. and Mensink, R. P. (2008). Water-soluble dietary fibers and cardiovascular disease. *Physiological Behavier*, *94*, 285-292.

Tilg, H., Moschen, A. and Kaser, A. (2009). Obesity and the microbiota. *Gastroenterology*, *136*, 1476-1483.

Todorov, S. D. and Franco, B. D. G. M. (2010). *Lactobacillus plantarum*: characterization of the species and application in food production. *Food Review International*, *26*, 205-229.

Todorov, S. D., Prévost, H., Lebois, M., Dousset, X., Leblanc, J. G. and Franco, B. D. G. M. (2011). Bacteriocinogenic *Lactobacillus plantarum* ST16Pa isolated from papaya (*Carica papaya*) – From isolation to application: characterization of a bacteriocin. *Food Research International*, *44*, 1351-1363.

Vaarala, O., Atkinson, M. A. and Ney, J. (2008). The "perfect storm" for type 1 diabetes: the complex interplay between intestinal microbiota, gut permeability, and mucosal immunity. *Diabetes*, *57*, 2555-2562.

Valdez, J. C., Peral, M., Rachid, M. Santana, M. and Perdigón, G. (2005). Interference of *Lactobacillus plantarum* on *Pseudomonas aeruginosa in vitro* and in infected burns: the potential use of probiotic in the wound treatment. *Clinical Microbiological Infections, 11*, 472-479.

Vandamme, P., Pot, B., Gillis, M., De Vos, P., Kersters, K. and Swings. J. (1996). Polyphasic taxonomy, a consensus approach to bacterial systematics. *Microbiological Reviews, 60*, 407-438.

Vandeplas, S., Dauphin, R. D., Thiry, C., Beckers, Y., Welling, G. W., Thonart, P. and Théwis, A. (2009). Efficiency of *Lactobacillus plantarum*-xylanase combination on growth performances, microflora populations, and nutrient digestibilities of broilers infected with *Salmonella* Typhimurium. *Poultry Science, 88*, 1643-1654.

Venter, C. S. and Vorster, H. H. (1989). Possible metabolic consequences of fermentation in the colon for human. *Medical Hypotheses, 29*, 161-166.

Wang, B., Huang, Q., Zhang, W., Li, N. and Li, J. (2011). *Lactobacillus plantarum* prevents bacterial translocation in rats following ischemia and reperfusion injury. *Digistive Diseases Science, 56,* 3187-3194.

Wang, B., Li, J., Li, Q., Zhang, H. And Li, N. (2009a). Isolation of adhesive strains and evaluation of the colonization and immune response by *Lactobacillus plantarum* L2 in the rat gastrointestinal tract. *International Journal of Food Microbiology, 132*, 59-66.

Wang, Y., Xu, N., Xi, A., Ahmed, Z., Zhang, B. and Xiaojia, B. (2009b). Effects of *Lactobacillus plantarum* MA2 isolated from Tibet kefir on lipid metabolism and intestinal microflora of rats fed on high-cholesterol diet. *Applied Microbiology and Biotechnology, 84*, 341-347.

Wessels, S., Axelsson, L., Hansen, E. B., De Vuyst, L., Laulund, S., Lähteenmäki, L., Lindgren, S., Mollet, B., Salminen, S. and Von Wright, A. (2004). The lactic acid bacteria, the food chain, and their regulation. *Trends in Food Science and Technology, 15*, 498-505.

Wibowo, D., Esceubruch, R., Davis, C. R., Fleet, G. H. and Lee, T. H. (1985). Occurrence and growth of lactic acid bacteria in wine: a review. *American Journal of Enology and Viticulture, 36*, 302-313.

Wolf, G. and Hammes, W. P. (1988). Effect of hemation on the activity nitrite reductase and catalase in lactobacilli. *Archives in Microbiology, 149*, 220-224.

Yang, Y., Chen, W., Zhang, H. and Tang, J. (2008). Probiotic characteristics of *Lactobacillus plantarum* HO-69 applied in oral cavity. *West China Journal of Stomatology, 26,* 482-485.

Zago, M., Fornasari, M. E., Carminati, D., Burns, P., Suàrez, V., Vinderola, G., Reinheimer, J. and Giraffa, G. (2011). Characterization and probiotic potential of *Lactobacillus plantarum* strains isolated from cheeses. *Food Microbiology, 28*, 1033-1040.

Zhou, J. S., Pillidge, C. J., Gopal, P. K. and Gill, H. S. (2005). Antibiotic susceptibility profiles of new probiotic *Lactobacillus* and *Bifidobacterium* strains. *International Journal of Food Microbiology, 98*, 211-217.

Zhou, Y., Qin, H., Zhang, M., Shen, T., Chen, H., Ma, Y., Chu, Z., Zhang, P. and Liu, Z. (2010). *Lactobacillus plantarum* inhibits intestinal epithelial barrier dysfunction induced by unconjugated bilirubin. *British Journal of Nutrition, 104*, 390-401.

In: Lactobacillus: Classification, Uses and Health Implications ISBN: 978-1-62081-151-1
Editors: Alba I. Perez Campos and Arturo Leon Mena © 2012 Nova Science Publishers, Inc.

Chapter III

Bacteriocin-Producing Lactic Acid Bacteria for Biopreservation: Example of Application in Raw and Processed Salmon

Svetoslav Dimitrov Todorov[1,*], Jean Guy LeBlanc[2], Bernadette Dora Gombossy de Melo Franco[1] and Manuela Vaz-Velho[3]

[1]Universidade de São Paulo, Faculdade de Ciências Farmacêuticas, Departamento de Alimentos e Nutrição Experimental, Laboratório de Microbiologia de Alimentos, São Paulo, SP, Brasil

[2]Centro de Referencia para Lactobacilos (CERELA-CONICET), San Miguel de Tucumán, Tucumán, Argentina

[3]Escola Superior de Tecnologia e Gestão, Instituto Politécnico de Viana do Castelo, Viana do Castelo, Portugal

Abstract

The concept of bio-preservatives has gained significant interest during the last decades based on consumer's requests for more natural and healthier food. Lactic acid bacteria produce different antimicrobial compounds, including bacteriocins, lactic acid, hydrogen peroxide, benzoic acid, fatty acids, diacetyl and other low molecular weight compounds. Bacteriocins produced by lactic acid bacteria have been reported to be active against a wide range of pathogenic microorganisms commonly found in foods. This chapter will discuss the potential of various bacteriocin producing LAB and particularly those from the genera *Lactobacillus* and their potential in the bio-preservation of fish products. These bacteriocins, produced by beneficial microorganisms, can prevent or at least inhibit the growth of pathogens such as *Listeria monocytogenes*, thus increasing the

---

* Corresponding author. Tel.: +55-11-3091 2191; fax: +55-11-3815 4410. E-mail address: slavi310570@abv.bg (SD Todorov).

shelf-life without affecting the physical and sensorial qualities of the end-product. Different methods of application of these LAB strains or of the purified bacteriocins will also be described. Traditional methods for preservation of food products need to be re-evaluated in view on the potential application of bacteriocinogenic LAB.

Introduction

Numerous strains of bacteriocin producing Lactic acid bacteria (LAB) have been isolated from a variety of sources in the last two decades and have been the object of review by Olasupo (1996), Cleveland *et al.* (2001) and Riley and Wertz (2002) amongst others. Bacteriocins are biologically active proteins or protein complexes that display a bactericidal mode of action towards usually closely related species (De Vuyst and Vandamme, 1994). Bacteriocins produced by different LAB have been reported and these have been show to be active against a wide range of microorganisms; however, most bacteriocins have not been fully characterized.

Application of bacteriocins in the food industry is constantly increasing since consumers demand the use of more natural biopreservation methods. The use of chemical preservatives are becoming less and less attractive due to studies showing that some of these additives have negative effects on human health and some are even considered cancerogenic.

The European Food Safety Authority is very strict regarding the application of chemical preservatives and this is one of the reasons for increased searches for more natural methods of food preservation.

Furthermore, the application of conventional antibiotics in several food processes is not plausible due to the increased problems associated to the generation of antibiotic resistant microorganisms. In the last decades, the demand for probiotic products, i.e. that contain microorganisms that confer a health benefit to the host, is increasing and this is another reason for increased interest on LAB that are the most commonly used probiotic bacteria. All of the above mentioned reasons explain why several groups are continuously searching for bacteriocin-producing LAB to improve the safety and quality of a variety of fermented foods.

This chapter will describe the potential of various bacteriocin-producing LAB for the biopreservation of food products with a special emphasis placed on salmon and other fish products.

These bacteriocins, produced by beneficial microorganisms, could prevent or at least retard the growth of pathogens such as *Listeria* (*L.*) *monocytogenes*, thus increasing the shelf-life without affecting the physical and sensorial qualities of the end-product. Besides the obvious economical advantages associated to increased shelf-lives, bacteriocin producing LAB could be one of the factors that could help in reaching the "Zero Tolerance" for *L. monocytogenes* that are required in salmon products and also be helpful in the prevention of other health-threatening pathogens commonly found in all food-derived products.

In this chapter, a brief overview of the classification, characterization, mode of action and production optimization for bacteriocins from LAB in general will be presented. A review of bacteriocins produced by LAB strains isolated from salmon or applied for biopreservation of smock salmon will also be discussed.

Bacteriocins

Lactic acid bacteria are known for their production of antimicrobial compounds, including bacteriocins or bacteriocin-like peptides (De Vuyst and Vandamme, 1994). Bacteriocins of LAB are defined as ribosomally synthesized proteins or protein complexes usually antagonistic to genetically closely related organisms (De Vuyst and Vandamme, 1994). Although new classifications of bacteriocins are constantly emerging (Zouhir *et al.*, 2010), the most accepted classification was proposed by Klaenhammer (1993) and originally defined four distinct classes of bacteriocins produced by LAB.

Most bacteriocins demonstrate a bactericidal mode of action against closely related organisms. These bacteriocins dissipate the proton motive force by disrupting the transmembrane potential and/or the pH gradient of sensitive cells (Montville and Bruno, 1994). Some bacteriocins permit the efflux of relatively large molecules (Chikindas *et al.*, 1993; Maftah *et al.*, 1993; Van Belkum *et al.*, 1991). Two-peptide bacteriocins appear to form relatively specific pores, dissipating the transmembrane potential (Moll *et al.*, 1996). Plantaricin E/F dissipates the pH gradient, causing a pH increase (Moll *et al.*, 1999).

One of the factors that influence the sensitivity of *Listeria* spp. towards certain biopreservatives, including bacteriocins, is the physiological state of the bacteria, i.e., the growth phase at the moment of treatment with the antimicrobial substance (Jydegaard *et al.*, 2000). In general, bacteria are more sensitive to the effects of heat and antimicrobial agents in the exponential growth phase (Hersom and Hulland, 1995). Jydegaard *et al.* (2000) demonstrated that the activity of bacteriocins also depends on the phase of development of the host bacteria. Specifically, nisin and, to a lesser extent, pediocin AcH (PA-1) have shown a higher inhibitory effect on *L. monocytogenes* cells during exponential growth than at the end of the stationary phase.

Stoffels *et al.* (1994) studied the mode of action of carnocin U149, a bacteriocin produced by *Carnobacterium (Carn.) piscicola* isolated from fish. It has previously been observed that nisin-producing *Lactococcus* (*Lc.*) *lactis* subsp. *lactis* strains are at least 10-fold more sensitive to carnocin U149 relative to other LAB.

Addition of carnocin U149 to cells of *Lc. lactis* NZ9700 resulted in a dissipation of the membrane potential and a rapid hydrolysis of internal ATP. These results suggest that carnocin U149 may act on the cytoplasmic membrane. The correlation between production of and/or immunity to nisin and sensitivity to carnocin of *Lc. lactis* strains was further investigated. Several transformed *Lc. lactis* strains carrying a fragment of the nisin gene cluster were tested for their sensitivity to carnocin U149 and their immunity to nisin. The results suggest that NisP, which is one of the membrane-associated proteins involved in the production of nisin acts as receptor for carnocin U149. This may facilitate the binding and/or insertion of carnocin U149 to the cytoplasmic membrane thus increasing its bactericidal activity. Lantibiotic-producing and non-producing mutants of *Staphylococcus (Staph.) epidermidis* and *Lactobacillus* (*Lact.*) *sakei* did not show a difference in sensitivity to carnocin U149. The proposed receptor-mediated action of carnocin U149 at the cytoplasmic membrane therefore seems to be specific for nisin-producing strains (Stoffels *et al.*, 1992; Stoffels *et al.*, 1994).

Interaction of Bacteriocins with the Food Matrix and Influence of Their Activity by Environmental Factors

Practical application of bacteriocin producing organism in food protection is still rather scarce. Indeed, the preconditions for the successful use of these types of cultures are demanding. For example, the organisms have to be metabolically active, have to produce bacteriocin and, as it has been shown by Gänzle *et al.* (1999), the antibacterial activity of the bacteriocin itself depends on the prevailing environmental conditions (Weiss and Hammes, 2006). Weiss and Hammes (2006) showed that when these preconditions were fulfilled, the anti-listerial effect becomes clearly evident. It was observed that the use of the bacteriocinogenic LAB cultures as inoculums is more efficient than the application of a culture supernatant that contains antagonistic activity. Finally, the practical use of the bacteriocinogenic strains can be facilitated by including them in lyophylised preparations that would allow them to maintain their antagonistic potential during prolonged storage (Weiss and Hammes, 2006).

The potential use of bacteriocin-producing LAB as biopreservatives requires good competitive properties of the added culture with the indigenous microflora in the specific product and not cause an increase in the spoilage rate of the product. Furthermore, for their commercial application as protective cultures in vacuum-packed Cold-smoked salmon (CSS), a complete understanding of the influence of the intrinsic and extrinsic conditions of the food matrix on bacteriocin production is essential. In this respect, bacteriocin titers can dramatically change when environmental conditions are altered and optimum production may require certain combinations of factors (Leal-Sánchez *et al.,* 2002). Regarding the complexity of food environments, a better knowledge of the interactions of these factors on bacteriocin production and sustainment of bacteriocin activity is needed. On the other hand, the absence of pathogenicity traits, such as hemolytic activity and antibiotic resistance (Embarek *et al.*,1994), should be demonstrated for cultures suggested for use in foods, especially in the case of antilisterial *Enterococcus* (*Ent.*) spp. in which the horizontal transfer of plasmids may dramatically alter the phenotypes in terms of their pathogenicity as was shown in other microorganisms such as *Bacillus* spp (Helgason *et al*.,2000).

Different aspects of the interaction between *Listeria* spp., bacteriocins and environmental factors were studied by several research groups (Aasen *et al*.,2000; Aasen *et al.*, 2003; Hugas *et al*.,2002; Kumar *et al.*, 1998; Lebois *et al*.,2004; Seaberg *et al*., 2003; Singh *et al*., 2001; Todorov, 2008; Todorov *et al.*, 2007a; Todorov *et al*., 2007b), notwithstanding, the optimal condition for the control of *Listeria* spp. by the bacteriocinogenic culture or the bacteriocins themselves and how they interact in real food systems is still not completetly understood. Most of the research directed at different conditions study some specific aspects of this interaction, but the general conclusion from all these works is that every bacteriocinogenic strain and every bacteriocin are different and no general conclusions can be made for their industrial application.

As an example is the study of the effect of divercin V41 on the growth of *L. monocytogenes* ScottA in BHI broth combined with varying concentrations of NaCl and phenol (liquid smoke) and different pH values. Divercin V41 is a very well studied bacteriocin produced by *Carn. divergence* V41 (Lebois *et al.*, 2004). Lebois *et al.* (2004)

showed that these abiotic factors may have a positive effect on increasing the safety of final product since they could be efficient in delaying the growth of *L. monocytogenes*. In their experimental approach, divercin V41, NaCl content and pH values were the most influential factors on *L. monocytogenes* ScottA growth parameters (lag time and maximum specific growth rate). The term lag time was used to qualify a general situation of delay of *L. monocytogenes* growth which may be the results of different physiological responses including death, injury and repair.

Purified bacteriocins mixed with inhibitory substances produced from traditional spices naturally present in plant tissues have been studied for control of *L. monocytogenes* (Kumar *et al.*, 1998; Seaberg *et al.*, 2003). Garlic has been shown to have a synergetic effect with nisin and sakacin K in inhibiting *L. monocytogenes* (Hugas *et al.*, 2002; Singh *et al.*, 2001).

Zouhir *et al.* (2008) proposed combination of plant extracts and divercin M35 for control of *L. monocytogenes*. The susceptibility of 11 strains of *L. monocytogenes* to divercin M35, a bacteriocin produced by *Carn. divergens* M35, and to aqueous extracts of garlic, onion, oregano, red chili and black pepper at 30°C and 10°C, was evaluated using a microdilution assay, and a divercin-resistant strain was used as a control. Three strains were resistant to divercin M35 (>500 μg/mL) at 30°C, but were more susceptible at 10°C. Garlic was the most inhibitory plant extract, followed by onion, while oregano, red chili and black pepper were less active at both temperatures. Garlic (and other plant) extracts and divercin M35 combined increased the inhibitory activity against the divercin-resistant strains. The garlic/divercin combination was the most effective at inhibiting these strains and was bactericidal at both studied temperatures. It is interesting to point out that log-phase cells were most susceptible to the garlic/divercin combination. Stationary-phase cells were much more resistant at both incubation temperatures. This shows that specific receptors on the surface of *L. monocytogenes* could be present on actively growing cells and probably their configuration slightly changed and became inaccessible during stationary phase. Furthermore, the effect of the garlic/divercin combination inhibition of divercin-resistant *L. monocytogenes* in a food system was also studied using CSS as a food model. Results indicated that this combination could efficiently reduce the viability of *L. monocytogenes* in smoked salmon at 10°C (Zouhir *et al.*, 2008).

Furthermore, it is well known that bacteriocins are amphiphilic peptides susceptible to adsorption to food macromolecules and proteolytic degradation (De Vuyst and Vandamme, 1994). These properties may limit their use as preservation agents. Aasen *et al.* (2003) studied the fate of the bacteriocin sakacin P and nisin in food. Recovery of bacteriocins was studied in homogenates of CSS, chicken cold cuts and raw chicken, with verification of the results in the corresponding food products. More than 80% of the added sakacin P and nisin were quickly adsorbed to proteins present in the food matrix. In foods that had not been heat-treated, proteolytic activity caused a rapid degradation of the bacteriocins, with less than 1% of the total activity left after 1 week in CSS, and even less in raw chicken. In heat-treated foods, the bacteriocin activity was stable for more than 4 weeks. The high fat content in salmon compared to chicken had no adverse effect on bacteriocin recovery or activity. However, mixing of triglyceride oils and bacteriocin solutions caused a considerable loss of activity. No major differences were observed between sakacin P and nisin, but less nisin was adsorbed to muscle proteins at low pH, and the negative effect of oils was less pronounced for nisin. Growth of *L. monocytogenes* was completely inhibited for at least 3 weeks in both chicken cold cuts and cold-smoked salmon by addition of sakacin P (3.5 μg/g), despite the proteolytic degradation in the salmon (Aasen *et al*., 2003). The fat content of the CSS (15-18%) in the study of Katla *et al*. (2001) was higher than what is commonly present (8–14%) in this product (Duffes, 1999). It was shown, using model studies, that increased amounts of soy oil decreased the activity of sakacin P; the potential of sakacin P as an inhibitor substance may thus be greater in products with lower fat contents (Katla *et*., 2001). However, different kinds of fat have been shown to posses varying effects on class IIa bacteriocins (Blom *et al*., 2007).

Food matrices affect the activity of bacteriocins and growth of bacteria in several ways. The activity of sakacin P is reported to increase with increasing salt concentrations up to 7% (Gänzle *et al*., 1999). Katla *et al*. (2001) reported that the salt content in CSS was 2.7 - 3.4%, which is optimal for growth of *L. monocytogenes* at 10°C (Cole *et al*.,1990). Higher salt concentrations are commonly encountered in this product (Hansen *et al*., 1998; Pelroy *et al*., 1994) and would probably contribute both to an increase in the activity of sakacin P (Gänzle *et al*., 1999; Blom *et al*., 1997) and a decrease in the growth of *L. monocytogenes*. Even environmental factors have been shown to affect bacteriocin activities. It was recently reported there is a synergetic effect of carbon dioxide and nisin on *L. monocytogenes* Scott A grown in broth at 4ºC (Nilsson *et al*., 2000).

Very low or even no bacteriocin activity were detected when strains *Ent. faecium* ET12, *Lact. curvatus* ET31, *Lact. fermentum* ET35 and *Ent. faecium* ET88 were grown at high salt concentration and at refrigeration temperatures, despite relatively good cell growth, suggesting that these factors could have a negative effect on the application of these strains as a bio-preservatives in salted products like vacuum-packed CSS fillets (Tome *et al*., 2008a; Tome *et al*., 2008b; Tome *et al*., 2009). Similar detrimental effects of high concentrations of NaCl have been reported by Verluyten *et al*. (2004) Some authors have commented that growth of LAB is inhibited in environments with high salt concentrations (NaCl concentrations higher than 3%) whereas lower amounts (from 1% to 2%) can exert a positive effect (Gänzle *et al*., 1998; Korkeala *et al*., 1992). According to Neysen *et al*. (2003), amylovorin L471 production by *L. amylovorus* DCE 471 was negatively affected by NaCl, because the amount of biomass formed was lower. Most *Ent. faecium* and *Pediococcus* (*Ped.*) species are facultatively anaerobic cocci with the ability to grow in broth containing 6.5% NaCl, at 10°C (Neysen *et al*., 2003). However, among the strains of *Ent. faecium* studied, only *E. faecium* ET05 was able to yield an active bacteriocin at higher salt concentrations, both at 5°C and 10°C. Homofermentative microorganisms like *Lact. delbrueckii* are in general more salt tolerant than heterofermentative LAB. Korkeala *et al*. (1992) have reported enhanced growth of some homofermentative LAB after supplementing media with high salt concentrations. *Lact. curvatus* and *Lact. delbrueckii* cultures have also been isolated from vacuum-packed trout and salmon by González-Rodríguez *et al*. (2002) as well as the former organism from vacuum-packed meat (De Martinis *et al*., 2003). Therefore, the growth of these LABs at salt concentrations up to 5%, at low temperatures both in aerobic and anaerobic atmosphere, is expected. Jeppesen and Huss (1993), while studying the antagonistic activity of LAB isolated from chilled fish products, observed that lowering the temperature from 25°C to 5°C in some cases increased whereas in other cases this decreased the antagonistic effect. This unpredictable and variable effect was explained by the fact that not only the inhibitory strain might be affected, but also the test organism, resulting in altered microbial interactions. The temperature dependency of the pediocin-like bacteriocins is related with the C-terminal disulfide bridge (Fimland *et al*., 2000). In relation to bacteriocin production, most of the literature indicates that the production/activity of class II bacteriocins is pH-dependent as a result of different adsorption characteristics. For instance enterocin production was maximal at neutral pH according to the results of Parente and Ricciardi (1994). The production of enterocin *in vitro* was inhibited below pH 5.5 (Aymerich *et al.*, 2000).

Bacteriocins Produced by Lactic Acid Bacteria Isolated from Salmon and Other Fish or Sea Food

Several studies demonstrated that the microflora present in vacuum-CSS reaches 10^6–10^8 CFU/g after 3 weeks storage at refrigeration temperature (Leroi *et al.*, 1998): 60% of the microflora is represented by the LAB genera *Carnobacterium* spp. and *Lactobacillus* spp., while the remaining 40% are Gram-negative microorganisms such as Enterobacteriaceae, *Shewanella putrefaciens*, *Photobacterium phosphoreum* and *Aeromonas* spp. which are characterized as having a high spoiling potential and are responsible for unpleasant smell and taste (Leroi *et al.*, 2001). Also, members of *Staphylococcus* spp. can be present (Gonzàles-Rodrìguez *et al.*, 2002; Huss *et al.*, 1995). The absence of additives, stabilizers and antioxidants makes smoked salmon a product with high risk of microbial contamination. Moreover, consumer demand for fresher products led to the reduction of salting/drying and smoking of raw material, thus increasing the microbiological risk (Rørvik *et al.*, 1995).

Huss *et al.* (1995) studied the safety hazards related to pathogenic bacteria, parasites and biogenic amines, placing a special emphasis on hazards related to *Clostridium (Cl.) botulinum* and *L. monocytogenes*, and critical control points were identified for all hazards in addition to *L. monocytogenes* growth. For this reason, limiting the shelf life to three weeks at +5°C for cold smoked vacuum-packed salmon that contains at least 3% salt was recommended.

CSS can be contaminated by pathogens, particularly *L. monocytogenes*, an ubiquitary species which could cause serious problems for human health due to its ability to grow at refrigeration temperature, acidic pH and salt concentration up to 10% (Cortesi *et al.*, 1997; Rocourt *et al.*, 2000; Rørvik, 2000; Rørvik *et al.*, 1997). *L. monocytogenes* constitutes the major microbiological risk of CSS, as the frequency of contamination ranges from 10% to 75% (Cortesi *et al.*, 1997; Heinitz and Johnson, 1998). Such a pathogen can be inhibited by sodium nitrite, sodium lactate and different acids, but these additives could affect the sensory qualities of the product (Duffes, 1999). It is important to individuate methods to control the growth of pathogenic and spoilage microflora during the vacuum storage.

Various strategies of biopreservation were proposed to inhibit the growth of *L. monocytogenes* in CSS, such as the use of bacteriocins or bacteriocin-producing strains (Katla *et al.*, 2001; Brillet *et al.*, 2004; Duffes *et al.*, 1999a; Duffes *et al.*,1999b; Richard *et al.*,1994). Some LAB strains, commonly associated with traditional storage technologies, could potentially control *L. monocytogenes* growth without altering the organoleptic characteristics of the product. Promising results were obtained in previous studies concerning the inhibition of such a pathogen by live competitive cultures of *Lactobacillus* and *Carnobacterium* species isolated from smoked salmon (Duffes *et al.*, 1999a; Duffes *et al.*, 1999b; Duffes *et al.*, 2000; Richard *et al.*, 2004; Nilsson *et al.*, 1999; Vaz-Velho *et al.*, 2005).

Bio-preservation methods have gained popularity in the food safety area: e.g., the use of protamines (Johansen *et al.*, 1997), monoglycerides (Wang and Johnson, 1997) and bacteriocins (Todorov *et al.*, 2007; Goff *et al.*, 1996; Leustner *et al.*, 1995; McMiullen and Stiles, 1996; Rodriguez *et al.*, 1997). Wessels and Huss (2006) were the first to propose the use of bacteriocins in smoked fish production, and the interest of biopreservation using LAB to improve the food safety of smoked salmon was evaluated. CSS is a product whose physicochemical characteristics allow the development of *L. monocytogenes* which constitutes the major microbiological risk depending on the consumption of this product. Understanding the biodiversity of the microflora of smoked salmon and its evolution during conservation, the origin of the contamination by *Listeria* and its consequences, and the parameters of the manufacturing process will allow a better control of microbiological quality. The innovative concept of biopreserving smoked salmon consists of an inoculation of salmon by a competitive lactic flora able to inhibit *L. monocytogenes* during storage. The selected strains, such as *Carnobacterium* spp. should neither have any influence on sensory qualities of smoked salmon, nor any effect on the concentration of biogenic amines in the finished product (Weiss and Hammes, 2006).

Bacteriocins Produced by Carnobacterium spp.

According literature (Mora *et al.*, 2003) *Carn. divergens* and *Carn. maltaromaticum* (previously known as *Carn. Piscicola*) are two of the most frequently isolated species showing anti-*Listerial* activity that are used in the biopreservation of salmon (Bhugaloo-Vial *et al.*, 1996; Stoffels *et al.*, 1992; Stoffels *et al.*, 1994; Duffes *et al.*, 2000; Alves *et al.*, 1995; Buchanan *et al.*, 1997; Metivier *et al.*,1998; Nilsson *et al.*, 2002; Pilet *et al.*, 1995; Schöbitz *et al.*, 2003; Schöbitz *et al.*, 2006; Tahiri *et al.*, 2004; Yamazaki *et al.*, 2003).

Pilet *et al.* (1995) screened for bacteriocin producing LAB isolated from fish products (fresh fish, smoked and marinated fish, fish intestinal tract). Twenty-two isolates which were found to produce bacteriocin-like compounds were identified as Carnobacteria, Lactococci and Enterococci. Two Carnobacteria spp. isolates named V1 and V41 were selected for further studies and identified by DNA-DNA hybridization as *Carn. piscicola* and *Carn. divergens*, respectively. Their respective bacteriocins named piscicocin V1 and divercin V41 were heat-resistant and sensitive to various proteolytic enzymes. These bacteriocins were active against *L. monocytogenes* and exhibited a different spectrum of activity against several LAB. Both bacteriocins had a bactericidal and non-bacteriolytic mode of action. Maximum production of piscicocin V1 and divercin V41 in MRS broth occurred at the beginning of the stationary phase and was higher at 20°C than at 30°C. When the cultures were maintained at pH 6.5, bacteriocin production was significantly increased (Pilet *et al.*, 1995). The relationship between the structure and activity of divercin V41 has recently been reviewed in detail (Rihakova *et al.*, 2009).

Production of divercin V41 was successfully heterologous expressed in *Escherichia* (*E.*) *coli* with bacteriocin yields (74 mg DvnRV41 per litre, respectively) among the highest published for heterologous expression systems in shake flasks (Yildirim *et al.*, 2007). Furthermore, the synthetic gene that codes for a mature divercin RV41 peptide production was cloned in the food-grade LAB, *Lc. lactis*. The production of divercin V41 by recombinant *Lc. lactis* was confirmed and quantified by Western blot and ELISA assays. In addition, anti-*Listeria* activity of divercin V41 was determined using an agar diffusion test. Although the levels of divercin V41 produced by recombinant *Lc. lactis* were similar to those produced by the natural host, *Carn. divergens* V41, the specific activities were lower. In conclusion, our data show that the bacteriocin divercin V41 is produced and secreted in an active form by *Lc. lactis* and that this approach may have important applications in the preservation of foods (Bermúdez-Humarán *et al.*, 2007).

The research story of divercin V41 was continued by Connil *et al.* (2002), who studied the potential application of *Carn. divergens* V41 based on the technological properties of this strain, applying a complete factorial design to study some aspects of *Carn. divergens* V41 metabolism (growth, biogenic amine production, and divercin V41 production) in sterile CSS extract at varying temperatures (3°C to 9°C), NaCl levels (2.5% to 6.5%) and glucose concentrations (2 g/L to 6 g/L). The results showed that temperature and NaCl content were the most influential factors on growth parameters in CSS extract. Among the biogenic amines studied, only tyramine was found to be produced by *Carn. divergens* in CSS extract. Furthermore, it was shown that temperature, NaCl, and glucose variations did not greatly affect tyramine and divercin V41 production by the bacteria in the experimental conditions used. Indeed, divercin V41 was produced in CSS extract even under harsh culture conditions. Similarly, tyramine production in CSS extract was delayed at 3°C but reached 35 μg/ml in all experiments after 27 days of storage. However, this tyramine concentration in CSS is low compared with the threshold values of 100 to 800 μg/g reported as the potentially toxic dose in foods. Thus, Connil *et al.* (2002) have found that *Carn. divergens* V41 is a promising strain for the biopreservation of refrigerated cold-smoked salmon (Connil *et al.*, 2002).

The objectives of a study by Vaz-Velho *et al.* (2005) were to compare the efficiencies of *Carn. divergens* V41 and its supernatant V41 to control *L. innocua* 2030c during salmon-trout (*Oncorhynchus mykiss*) cold-smoking processes by treating the raw fish fillets. Fillets were dip-inoculated with 2% (v/v) culture of *L. innocua* 2030c and/or 2% (v/v) culture of *Carn. divergens* V41 culture and 2% or 5% (v/v) of culture supernatant V41. Fillets were dry-salted for 6 h and an average salt content of 5.3% (water phase salt) was obtained. Salted fillets were cold-smoked for 5 h (3 h of drying + 2 h of smoking) and stored at 5°C for 3 weeks in vacuum packs.

All the treatments with *Carn. divergens* V41 or supernatant showed a significant inhibitory effect on *L. innocua* 2030c growth, but the 5% (v/v) *Carn. divergens* supernatant treatment, was by far the most effective against this species. No significant off-odors were found in the products with respect to the tested treatments (Vaz-Velho *et al.*, 2005).

Much less work was performed for the second strain, *Carn. piscicola* V1 originally isolated from Pilet *et al.* (1995) Bhugallo-Vial *et al.* (1996) showed that *Carn. piscicola* V1 produces 2 bacteriocins, named piscicocin V1a and piscicocin V1b that were then purified and characterized. Piscicocin V1a (molecular mass = 4 416 Da) and piscicocin V1b (molecular mass = 4 526 Da) are nonlantibiotic, small, heat-stable antibacterial peptides. Piscicocin V1b was identical to carnobacteriocin BM1, while piscicocin V1a was identified as being a new bacteriocin. Its complete sequence of 44 amino acid residues was determined, and piscicocin V1a was identified as a class IIa bacteriocins due to the presence of the consensus YGNGV motif. These peptides inhibit various gram-positive bacteria, including *L. monocytogenes*. Piscicocin V1a is approximately 100 times more active than piscicocin V1b against indicator strains. However, the antagonistic spectrum is the same for both piscicocins. Comparison of these results with the analysis of the amino acid sequence and secondary structure predictions suggests that (i) the conserved N-terminal conserved domain is involved in the receptor recognition and therefore in an "all-or-none" response against target bacterial cells and (ii) the C-terminal variable and hydrophobic domain determines membrane anchoring and therefore the intensity of the antagonist response (Bhugaloo-Vial *et al.*, 1996).

In a study by Duffes *et al.* (1999a), over one hundred LAB were isolated from commercial vacuum packaged CSS and screened for their antagonistic activity against *L. innocua*. Twenty-two strains were able to produce bacteriocin-like proteinaceous substances. These strains were characterized physiologically and biochemically as *Carnobacterium* strains. One of these strains, *Carn. piscicola* SF668 was selected for further research. This strain was able to reach 10^7 CFU/mL after 21 days in a simulated cold-smoked fish system at 4°C, to produce bacteriocin activity up to 1600 AU/mL and to inhibit *L. monocytogenes* growth. On irradiated vacuum-packed CSS, *Carn. piscicola* SF668 reached 10^7 CFU/g and had a bactericidal effect on *L. monocytogenes* up to the expiry date. Between days 21 and 28, *L. monocytogenes* growth was stopped without emergence of resistant variants (Duffes *et al.*, 2000). In the same study, *Carn. divergens* V41 and *Carn. piscicola* V1 were used as controls since thesehave well known activity against *L. monocytogenes* (Pilet *et al.*, 1995). *Carn. divergens* V41 and *Carn. piscicola* V1 were the most effective strains in co-culture experiments, inhibiting *L. monocytogenes* as early as day 4, whereas *Carn. piscicola* SF668 caused an inhibiting effect at day 13. The potential for using such biopreservation treatments on whole smoked salmon was studied by Duffes *et al.* (1999a).

A different bacteriocin-producing *Carn. piscicola* was isolated from fish (Stoffels *et al.*, 1992a; Stoffels *et al.*, 1992b,). The bacteriocin, termed carnocin UI49, was purified to homogeneity by a four-step purification procedure, including hydrophobic interaction chromatography and reverse-phase chromatography. Carnocin U149 has a bactericidal mode of action, produced during the mid-exponential growth phase at temperatures between 15°C and 34°C. Carnocin UI49 is active against a large number of closely-related LAB including carnobacteria, lactobacilli, pediococci and lactococci. Maximum bactericidal activity was observed at 34°C with a decrease in activity at 15°C where it is completely abolished. It was shown to be heat tolerant and stable between pH 2 and 8. At pH above 8, carnocin U149 was rapidly inactivated. Amino acid analysis revealed a composition of about 35 to 37 amino acids in addition to an unidentified peak which migrates at the position of lanthionine. SDS gel electrophoresis analysis suggests a molecular weight of about 4 500 Da to 5 000 Da. Mass spectrometry gave a molecular weight of 4 635 Da, which is about 1 000 Da larger than that calculated from the amino acid analysis data. Performic acid oxidation of carnocin U149, followed by amino acid hydrolysis, revealed the presence of cysteic acid. The sequence of the first seven amino acid residues was determined to be N-Gly-Ser-Glu-Ile-Gln-Pro-Arg. After the seventh amino acid, carnocin U149 was not available for further Edman degradation. The results suggest that carnocin U149 belongs to the class of bacteriocins termed lantibiotics (Stoffels *et al.*, 1992a; Stoffels *et al.*, 1992b,).

Buchanan and Bagi (1997) described bacteriocins, produced by *Carn. piscicola* LK5 and *Carn. piscicola* 2762. The effects of temperature (4, 12, and 19°C), pH (5, 6, and 7), and NaCl (5, 25, and 45 g/l) on the growth of *L. monocytogenes* Scott A in the presence of either *Carn. piscicola* LK5 or *Carn. piscicola* 2762 were studied quantitatively in BHI broth (Buchanan and Bagi, 1997). Strain LK5 produces a bacteriocin that is released into the environment, whereas strain 2762 appears to produce a bacteriocin that remains cell associated. The primary effect of both *Carn. piscicola* strains was a suppression of the maximum population density attained by *L. monocytogenes*. The extent of this depression was dependent on the three culture variables, and appeared to be a function of their influence on the relative growth rates of the two species. The effects were similar with both strains. However, two bacteriocin-negative strains, *Carn. piscicola* 2305 and *Carn. piscicola* 2818, also suppressed the growth of *L. monocytogenes*. Little of the *Carn. piscicola* isolates' ability to suppress *L. monocytogenes* appears to be attributable to its bacteriocin production. The maximum population density depressing activity of *Carn. piscicola* 2762 could not be attributed to peroxide, pH depression, or oxygen depletion. However, *L. monocytogenes* inhibition may involve nutrient depletion since the extent of maximum population density suppression was decreased in a dose-related manner when the two species were cultured in 3 x and 6 x BHI broth (Buchanan and Bagi, 1997).

Schöbitz *et al.* (2003) studied the antagonism of a bacteriocin-like inhibitory substance produced by *Carn. piscicola* L103 against *L. monocytogenes* strains isolated from salmon and human samples. Results showed an antagonistic effect of the bacteriocin-like inhibitory substance against all the strains of *L. monocytogenes*. Four selected strains presented a higher sensitivity to the bacteriocin in the exponential growth phase and were more resistant in the stationary phase. In Trypton Soy Broth, the inhibitory substance showed a partial bactericidal effect against *L. monocytogenes*. After inactivation of the bacteriocins with a protease, growth of *L. monocytogenes* was observed. The isolate most affected by the action of the bacteriocin was one of salmon origin. From the 86 isolated colonies that grew in the presence of the bacteriocin, 93% showed total resistance and 7% partial resistance, which was maintained through five consecutive culture cycles in the absence of the bacteriocin.

The effectiveness of the antagonistic substance produced by *Carn. piscicola* L103 against *L. monocytogenes* was tested on vacuum packaged salmon. Salmon fillets were inoculated with the bacteriocin containing supernatant (200 AU/ml and 800 AU/ml) and 80 CFU/mL of *L. monocytogenes* and stored at 4°C. The growth of *L. monocytogenes* was determined every 5 days during 15 days. Bacteriocin produced by *Carn. piscicola* L103 had a bacteriostatic effect on *L. monocytogenes* when inoculated on salmon, with counts of 6.0 x 10^3 CFU/mL after 15 days. In the control, where bacteriocin was not added, *L. monocytogenes* counts increased to 1.0 x 10^6 CFU/ml after 15 days of storage (Schöbitz *et al.*, 2006).

Another strain of *Carn. piscicola*, strain CS526 was isolated from frozen surimi and identified as a bacteriocin producer that had strong inhibitory activity against *L. monocytogenes* (Yamazaki *et al.*, 2003). The ability of this bacteriocinogenic strain and nonbacteriocinogenic *Carn. piscicola* JCM5348 to inhibit the growth of *L. monocytogenes* was examined in culture broth incubated at 12°C and CSS stored at 4, 12, and 20°C. *L. monocytogenes* viable counts in the culture broth rapidly declined from 10^6 CFU/mL to less than 10 CFU/mL within 1 day at 12°C in the presence of *Carn. piscicola* CS526. At 4 and 12°C, inhibition of *L. monocytogenes* on salmon depended on the initial inoculum level of *Carn. piscicola* CS526. However, *Carn. piscicola* CS526 showed a bactericidal activity against *L. monocytogenes* within 21 and 12 days at 4 and 12°C in CSS, respectively, even when the initial inoculums levels were low. *Carn. piscicola* CS526 suppressed the maximum cell number of *L. monocytogenes* by two and three log cycles, even at 20°C. However, *Carn. piscicola* JCM5348 did not prevent the growth of the pathogen, except at 4°C. Bacteriocin was detected in the samples coinoculated with *Carn. piscicola* CS526. This study showed that *Carn. piscicola* CS526 might have potential for biopreservation of refrigerated foods against *L. monocytogenes* (Yamazaki *et al.*, 2003).

Surubim (*Pseudoplatystoma* sp.) is a native freshwater fish from Brazil and has been commercially explored. This species is farmed and used for processing. Cold smoking of surubim adds value to the product and gives rise to a novel unique product. The export of this tropical fish has increased in last decade and in order to perform safety and quality assessments of the product, its microbiological status must be determined (Alves *et al.*, 2005). *Carn. piscicola* C2, isolated from Brazilian vacuum-packed cold-smoked surubim and two strains of *Carn. piscicola* isolated from vacuum-packed cold-smoked salmon were capable of limiting or completely inhibiting the growth of *L. monocytogenes* (strain V2) isolated from surubim in fish peptone model system incubated at 10°C. Monocultures of *L. monocytogenes* reached 10^8 CFU/mL, whereas the growth of *L. monocytogenes* was completely inhibited by *Carn. piscicola* C2 strain. The bacteriocinogenic *Carn. piscicola* A9b$^+$ and its nonbacteriocinogenic mutant A9b$^-$ reduced maximum *Listeria* levels by 2 to 3 log units. Both bacteriocinogenic *Carn. piscicola* strains prevented listerial growth in cold-smoked fish juices (surubim and salmon). Although the carnobacteria grew poorly on the cold-smoked surubim at 10°C, the strains were able to reduce maximum *Listeria* counts by 1 to 3 log units in an artificially inoculated product (surubim) (Alves *et al.*, 2005). According Alves *et al.* (2005) *Carn. piscicola* C2 has the potential for use as a biopreservative culture in surubim and other slightly preserved fish.

Nilsson *et al.* (2002) investigated the effect of bacteriocin produced by *Carn. piscicola* A9b, isolated from CSS as an autoinducible, antilisterial class IIa bacteriocin. The specific role of bacteriocin production in the inhibition of *L. monocytogenes* was assessed in other study of Nilsson *et al.* (2004) by comparing the antilisterial effect of the wild-type parent strain with a bacteriocin-negative mutant. These authors suggested that the antilisterial effect of *Carn. piscicola* A9b in both laboratory medium and model fish systems depends on at least two mechanisms of action, one of which one is the production of carnobacteriocin B2. However, the non-bacteriocin-producing variant also produced an important antilisterial effect (Nilsson *et al.*, 2004). *Carn. piscicola* A9b bacteriocin negative mutant did not influence the growth of *L. monocytogenes*, but suppressed growth after 5-8 days incubation (at 5°C), which caused a reduction in final biomass yields in buffered BHI broth, as well in salmon juice and in CSS. The inhibition of *L. monocytogenes* on CSS by the bacteriocin negative variant was, however, not as pronounced as by a nonbateriocin-producing *Carnobacterium* used in other studies (Nilsson *et al.*, 1999). Data shown by Nilsson *et al.* (2004) suggests that *L. monocytogenes* in mixed culture with *Carn. piscicola* A9b enters its stationary phase earlier as compared with monocultures of *L. monocytogenes*. Buchanan and Bagi (1997) also showed that the effect of *Carn. piscicola* on *L. monocytogenes* was largely limited to a suppression of the maximal population density. One explanation for the behavior of *L. monocytogenes* is exhaustion of one or a few essential nutrients and/or accumulation of toxic end products in the growth medium (Finkel *et al.*, 2000; Madigan *et al.*, 1999). The inhibitory effect of a bacteriocin-negative *Carn. piscicola* in sterile-filtered salmon juice was identical to that observed when sterile saline was added in the same volume (Nilsson *et al.*, 2004). Competition for nutrients is considered one of the major evolutionary driving forces in the bacterial world (Christensen *et al.*, 2002) and bacteria with a better utilization of a given energy source may outcompete other bacteria. Several studies have shown that the population of an indigenous microbial flora suppresses the growth of pathogens in food (Babic *et al.*, 1997; Carlin *et al.*, 1996; Mattila-Sandholm *et al.*, 1991; Samelis *et al.*, 1998; Vold *et al.*, 2000) probably due to a competitive advantage for nutrient uptake. It has been suggested that a possible reason for the increasing number of food-borne outbreaks in the US is the low number of a competitive background flora to suppress pathogen survival/invasion and growth in foods (Jay, 1996; Jay, 1997). Although microbial competition is an important safety issue in food production, only limited work have been conducted on the mechanisms of action that are not explainable by the production of antimicrobial compounds. Such studies are needed, since inhibitory effect of bacteriocin-negative mutants *a priori* are not to be expected. Thus, Leisner *et al.* (1996) found that whilst a bacteriocin-producing *Leuconostoc* was strongly inhibitory towards a meat spoiling *Lact. sakei*, its bacteriocin-negative variant did not inhibit growth of the spoilage organisms. Similar results were observed in inhibition of *L. innocua* LMG 13568 grown in salami prepared with starter culture of bacteriocinogenic *Lact. plantarum* 423 and his bacteriocin negative variant (Todorov *et al.*, 2007).

Carn. divergens M35, isolated from a commercial sample of frozen smoked mussels, produces bacteriocin divergicin M35, a class IIa bacteriocin (Tahiri *et al.*, 2004). Divergicin M35 is sensitive to pronase-E, α-chymotrypsin and proteinase K, but not to trypsin and withstands thermal treatments up to 121 °C for 30 min. Divergicin M35 was extracted from the culture supernatant of *Carn. divergens* M35 and purified on a C_{18} SepPack column and further purified by reverse phase-high pressure liquid chromatography. Divergicin M35 had a molecular mass of 4518.75 Da as determined by mass spectrometry, a pI value of 8.3 and positive net charge (+ 3). The amino acid sequence of divergicin M35 was found to consist of 43 amino acid with four cysteine residues (Cys(10), (15, 25, 43)) and showed 80.5% homology with divercin V41 (80.5%) and 80.0% with bavaricin MN. Divergicin M35 showed powerful antilisterial activity, especially against *L. monocytogenes* and was also active against carnobacteria but not against strains of *Lactococcus*, *Lactobacillus*, *Enterococcus*, *Bifidobacteria* and *Escherichia*. Divergicin M35 production began in late exponential phase and reached a maximum activity of 65 000 AU/mL in early stationary phase. Initial pH, Tween 80 and acetate did not affect *Carn. divergens* M35 growth or divergicin production in broth. This bacteriocin may be a potential tool for inhibiting *L. monocytogenes* in seafood products that do not usually undergo an adequate heat treatment (Tahiri *et al.*, 2004). Furthermore Zouhir *et al.* (2008) demonstrated the application of this bacteriocin in combined application with plant extract was effective in inhibiting *L. monocytogenes*.

Bacteriocins Produced by Non Carnobactaerium-Species of LAB Isolated from Salmon and Other Fish or Sea Food

Sixty-one LAB isolates from lightly preserved, chilled fish products were characterized with special emphasis on their antagonistic activity and their possible use as starter cultures (Jeppesen and Huss, 1993a; Jeppesen and Huss, 1993b). The strains were phenotypically identified and characterized with regard to gas production, carbohydrate fermentations, production of off odors, H_2O_2-production, antagonistic activity and growth at 2, 5 and 10°C. Growth at 10°C was detected for all strains within 4 days. At 5°C, 90% of the isolates grew within 3-9 days. At 2°C only 33% of the strains grew within 9 days. Fourteen strains were selected for further testing of antagonistic activity against spoilage- and pathogenic bacteria at different pH. Two strains, a *Leuconostoc* spp. (V6) and a *Lact. plantarum* (LKE5), differed from the other LAB by a very wide inhibitory spectrum for V6 and by distinct inhibition zones for LKE5 (Jeppesen and Huss, 1993a; Jeppesen and Huss, 1993b).

Nine bacteriocin-producing LAB, isolated from vacuum-packaged cold-smoked salmon, were phenotypically and genotypically identified as *Lact. curvatus*, *Lact. delbrueckii, Lact.*

fermentum, Ent. faecium and *Ped. acidilactici*. Their bacteriocins were partially characterised. The antimicrobial spectrum was determined against *L. monocytogenes*, *Ent. faecalis*, *Ent. faecium* and *Staph. aureus*. The molecular size of bacteriocins ranged from 2 800 to 4 500 Da. They were inactivated by treatment with proteolytic enzymes, but neither lipolytic nor glycolytic enzymes. Maximal activity against *L. monocytogenes* ranged between 800 to 10 000 AU/mL at pH 6.5. Most of the bacteriocins maintained full activity in a pH range of 2.0 – 8.0 but were partially or completely inactivated at pH 10.0. Only two bacteriocins from *Lact. curvatus* strains, partially lost activity after heating at 60ºC and 100ºC. All bacteriocins showed a narrow spectrum of activity and a high anti-*Listerial* activity which is characteristic of the class IIa bacteriocins (Tomé *et al*., 2009). The presence of similar LAB species were reported to be present in CSS by several research groups (Gonzàles-Rodrìguez *et al*., 2002), but unfortunately bacteriocin production in these isolates were not evaluated.

The importance of *Ped. acidilactici* strains in the food industry is related to their use as starter cultures in fermented meat and vegetable products (Amézquita and Brashears, 2002). This species, like most other LAB species, is involved in extending the shelf life and improving the hygienic quality of various fermented products, via the production of lactic acid and/or the secretion of bacteriocins (Stiles, 1996). It was very interesting to find that *Ped. acidilactici* is naturally present in CSS and has bacteriocinogenic properties (Tomé *et al*., 2009). The association of pediococci with proteinaceous foods such as fresh and cured meat, and raw sausages has frequently been reported, and particularly for *Ped. acidilactici* in fermented sausages (Parente *et al*., 2001). *Pediococcus* spp. has also been reported in fresh and marinated fish (Paludan-Muller *et al*., 2002).

Isolated bacteriocin-producing LAB (Tomé *et al*., 2009) could be used successfully in the bio-preservation of CSS and development of new potential bio-preservatives in combat against *L. monocytogenes*. Five bacteriocin-producing lactic acid bacteria previously described by Tome *et al.* (2008a; 2008b; 2009) (*Ent. faecium* ET05, *Lact. curvatus* ET06, *Lact. curvatus* ET30, *Lact. delbrueckii* ET32 and *Ped. acidilactici* ET34) were selected by their capacity for growth and producing inhibition in vitro at high salt content, low temperature and anaerobic atmosphere, conditions simulating cold-smoked fish, were inoculated onto salmon fillets, in co-culture with *L. innocua* 2030c, and cold-smoked processed (dry salted for 6 h; dried for 6 h; smoked for 2 h). The finished product was then packed under vacuum and stored at 5°C. Enumeration of LAB and *L. innocua* was performed during storage. Results showed that strain *Ent. faecium* ET05 was the best biopreservative candidate for controlling *L. innocua* growth in vacuum-packaged cold-smoked salmon processed under the salting/drying/smoking parameters referred above. *Lact. curvatus* ET30 and *Lact. delbrueckii* ET32 also showed a good biopreservation potential for CSS although they were less effective than the former. *Lact. curvatus* ET06 and *Ped. acidilactici* ET34 showed a bacteriostatic mode of action against the target bacteria in vitro as well as when inoculated into the salmon fillets (Tome *et al*., 2008a; Tome *et al*., 2008b; Tome *et al*.,2009) Enterococci showed neither haemolytic activity nor vancomycin resistance. The production of histamine was not observed for any of the bacteriocin-producing strains (Tome *et al*., 2008a; Tome *et al*., 2008b; Tome *et al*., 2009).

Lact. sakei (Strain R1333) was isolated from commercially available smoked salmon (Todorov *et al*., 2011). Strain R1333 produces a 3 811 Da class IIa bacteriocin, active against *Ent. faecalis*, *Streptococcus* (*Strep.*) *caprinus*, *Strep. macedonicus*, *Streptococcus* spp., *Lact. sakei*, *Lc. lactis* subsp. *lactis*, *L. innocua*, *L. ivanovii* subsp. *ivanovii* and *L. monocytogenes*.

The mode of activity against *L. innocua* 2030C, *L. ivanovii* subsp. *ivanovii* ATCC19119 and *Ent. faecium* HKLHS was bactericidal, resulting in cell lysis and enzyme- and DNA-leakage. The highest level of activity (1 600 AU/mL) was recorded when cells were grown at 30°C in MRS broth (initial pH 6.5). Only 800 AU/mL was recorded when strain R1333 was grown in MRS without Tween 80. Lower levels of bacteriocin production were recorded when *Lact. sakei* R1333 was grown in MRS at 20°C. Bacteriocin R1333 adsorbs at low levels (200 AU/mL) to producer cells. Bacteriocin R1333 was successfully purified by 60% ammonium sulfate precipitation, followed by separation on a SepPak C_{18} column and reverse phase HPLC on a Nucleosil C_{18} column with a linear gradient from 0.1% TFA to 90% acetonitryl. A molecular mass of 3 811 Da was determined by mass spectrometry. Based on mass spectrometry and sequencing of the PCR amplified fragment targeting the *sak*G gene, *Lact. sakei* R1333 is a potential producer of sakacin G. This was the first report of the identification of sakacin G produced by *Lact. sakei* isolated from smoked salmon (Todorov *et al.*, 2011). A potential application of this bacteriocin and the culture of *Lact. sakei* R1333 in biopreservation of shrimps was recently investigated (personal communication).

The antagonistic activity of LAB against *L. monocytogenes* ScottA and *Yersinia enterocolitica* serotype O:3 was studied by using either a strain of a *Leuconostoc* spp. or *Lact. plantarum* (Jeppesen *et al.*, 1993a). Each of the two potential pathogenic, psychrotrophic bacteria were inoculated with either of the two LAB in sterile shrimp extracts containing citric acid (pH 5.8 and 6.0) and in some experiments 3% (w/v) NaCl was added. An antagonistic effect was obtained with both LAB against both pathogenic organisms, with the effect depending on inoculation levels of the pathogenic organism, the LAB strain and media composition (Jeppesen *et al.*, 1993a).

Ben EmBarek *et al.* (1994) studied nineteen isolates of *Ent. faecium* isolated from vacuum-sealed cooked cod fillets stored during 3-8 weeks at 3°C. Fourteen isolates produced proteinaceous substances that inhibited different strains of *L. monocytogenes*, other pathogenic bacteria and spoilage bacteria. The isolates had similar growth profiles in broth at 0, 3, and 5°C. In the presence of 10^7 CFU/mL *Ent. faecium*, *L. monocytogenes* (10^2 CFU/mL) was strongly inhibited at 3°C and partially inhibited at 5°C and 15°C. When cultivated with 10^4 CFU/mL *Ent. faecium*, *L. monocytogenes* (10^2 CFU/mL) was only slightly inhibited at 15°C and not affected at 3°C or 5°C. Spontaneous resistance phenomena were observed at 15°C after 11 days incubation. The isolates of *Ent. faecium* did not produce off-odours in shrimp extracts or vacuum-sealed cooked fish. The potential use of the isolated strains of *Ent. faecium* as biopreservation was suggested (Ben EmBarek, 1994). Interestingly, a bacteriocin produced by *Ent. faecium* GM-1 was able to inhibit *Vibrio* spp (Kang and Lee, 2005), a serious pathogen of marine fish and invertebrates (Austin and Zhang, 2006).

The biopreservative potential of three antimicrobial-producing LAB strains (*Lact. casei, Lact. plantarum* and *Carn. piscicola*) was evaluated on CSS (Vescovo *et al.*, 2006). *Lact. casei, Lact. plantarum* and *Carn. piscicola* were added individually or in association to CSS which was artificially contaminated with *L. innocua* and stored under vacuum for 30 days at 4°C. All the lactic cultures were able to inhibit *L. innocua* growth, showing a bacteriostatic or bactericidal effect, without negatively affecting the sensory quality of the product. *Lact. casei* was bacteriostatic when inoculated at 6 log CFU/g, and bactericidal at 8 log CFU/g, reducing *L. innocua* 3.3 log CFU/g in comparison with the test at the end of storage. *Lact. plantarum* and *Carn. piscicola* strains, inoculated individually at 6 log CFU/g reduced *L. innocua* counts to 2.8 and 2.7 log CFU/g, respectively. The association *Lact. casei* - *Lact. plantarum* was the most effective among the treatments with 6 log CFU/g inoculum, as *L. innocua* counts decreased 3.2 log CFU/g. The treatment with a *Lact. casei* - *Carn. piscicola* combination was less effective than *Carn. piscicola* alone (Vescovo *et al.*, 2006).

Preliminary trials conducted by Vescovo *et al.* (2006) demonstrated that *Lact. casei*, *Lact. plantarum* and *Carn. piscicola* cultures did not exhibit reliable inhibition on *L. innocua* when inoculated at level lower than 6 log CFU/g. Other authors reported that *L. monocytogenes* growth is suppressed on CSS in the presence of high concentrations of LAB (Brillet *et al.*, 2004; Richard *et al.*, 2004; Vaz-Velho *et al.*, 2005; Rørvik and Yndestad, 1991). Nilsson *et al.* (1999) demonstrated the successful use of a high cell numbers of *C. piscicola* strains (10^6 CFU/mL) to reduce the counts of this pathogen in cold-smoked salmon juice from 10^3 to below 10 CFU/mL after 32 days at 5°C. Moreover, the same strains were able to control *L. monocytogenes* growth in vacuum-packed CSS without causing undesirable sensory changes in the product. The microbiological analysis performed on control samples demonstrated a satisfactory hygienic quality of commercial products at time 0, showing the absence of *Listeria* spp. and total coliforms, low levels of staphylococci (2.2 log CFU/g), enterococci (1.8 log CFU/g) and total aerobic mesophilic microflora (4 log CFU/g); LAB were 4 log CFU/g (Vescovo *et al.*, 2006). The sensory profile, detected on all the samples, did not differ significantly (Vescovo *et al.*, 2006). This study showed that the growth of *L. innocua* VR13560, inoculated in CSS, can be controlled or reduced by the addition of antimicrobial-producing lactic cultures individually or co-culture. With regard to the inhibitory activity, an important role is attributed to the initial inoculum size of the antimicrobial-producing strains and particularly to the contaminant/inhibitor ratio, as previously demonstrated in other food matrices (Scolari *et al.*, 1997; Vescovo *et al.*, 1997). It is also important to observe that such cultures did not have any effect on the quality of the end-products (Vescovo *et al.*, 2006).

Katla *et al.* (2001) investigated the possibility of the application of sakacin P and/or live *Lact. sakei* cultures in the control of *L. monocytogens* in CSS. Even though many bacteriocins have been characterized, only a few have been tested for their potential in the biopreservation of food. One of them, nisin, is successfully used as a food preservative and is approved for human consumption by the United States Food and Drug Administration. However, application of nisin is limited as it has low solubility at a pH above 6 and can be inactivated in certain types of food (Hansen *et al.*, 1998; Jung *et al.*, 1992; Liu and Hansen, 1990; Rose *et al.*, 1999). Other bacteriocins may thus be more suitable than nisin in certain food production processes. Sakacin P produced by *Lact. sakei* (Holck *et al.*, 1994; Tichaczek *et al.*, 1992) is very potent against *L. monocytogenes* (Eijsink *et al.*, 1998; Aasen *et al.*, 2003) and has been extensively studied. The genetics regarding this bacteriocin production have been elucidated (Brurberg *et al.*, 1997 Hühne *et al.*, 1996), the production has been optimized (Aasen *et al.*, 2000; Møretrø *et al.*, 2000) and factors that influence the effect of sakacin P in food system have been analysed (Gänzle *et al.*, 1999; Blom *et al.*, 1997). The possibilities of direct application of sakacin P in food production have been investigated in a study by Katla *et al.* (2001) *L. monocytogenes* was added to vacuum-packed CSS together with known concentrations of the bacteriocin sakacin P or nisin and/or one of two isogenic strains of *Lact. sakei* one of which produced sakacin P whereas the other did not. The vacuum-packaged salmon samples were incubated at 10°C during 4 weeks. O Both sakacin P and nisin had an initial inhibiting effect on growth of *L. monocytogenes*. Addition of the sakacin P-producing or the non-producing *Lact. sakei* had a bacteriostatic effect on *L. monocytogenes* during the complete storage period. When the sakacin P-producing *Lact. sakei* culture was added to vacuum-packed CSS together with sakacin P, a bacteriocidal effect on *L. monocytogenes* was observed (Gänzle *et al.*, 1999). The inhibition effectiveness of a bacteriocin produced by *Lact. curvatus* CWBI-B28 against *L. monocytogenes* was investigated in CSS during storage at 4°C by Ghalfi *et al.* (2006) These researchers applied three bacteriocin-based strategies for the control of *L. monocytogenes* in foods (i.e., producing bacteriocin *in situ*, spraying with partially purified bacteriocin, and packaging in bacteriocin-coated plastic film), plus a newly developed method that uses cell-adsorbed bacteriocin (i.e., a suspension of producer cells on which maximum bacteriocin has been immobilized by pH adjustments), were assessed. Although all the approaches inactivated *L. monocytogenes* in CSS, various efficacy levels were observed. The behavior of *L. monocytogenes* was similar in samples treated with either partially purified bacteriocin or *in situ* bacteriocin production. In both of these cases, the counts of the pathogen declined to below the detectable limit of 0.7 log CFU/mL within the first week, but a 0.95- and 1.3-log increase, respectively, occurred after day 14. The bioactive packaging film resulted in a slower inactivation of the pathogen but prevented any subsequent increase in the CFU counts throughout 22 days of storage at 48°C. Application of the cell-adsorbed bacteriocin was shown to be the most effective means, as it resulted in a complete inactivation of the pathogen within 3 days, and no increase in *Listeria* counts occurred up to 22 days (Ghalfi *et al.*, 2006).

Messi *et al.* (2000) investigated the possibility of applying plantaricin 35d for the biopreservation of salmon. Plantaricin 35d was produced by *Lact. plantarum* isolated from seasoned sausages (Messi *et al.*, 1998). When salmon were artificially contaminated with *L. monocytogenes* and treated at refrigerating temperature with plantaricin 35d, a 3 log reduction of the pathogen was observed, showing the potential of application of *Lact. plantarum* 35d and bacteriocin 35d in the biopreservation of smoked salmon (Messi *et al.*, 2000).

Weiss and Hammes (2006) explored the possibility of applying three bacteriocin-producing strains of *Lact. sakei* (LTH2342, LTH4122 and LTH5754) as individual cultures or in combination to control the growth of *L. innocua* in CSS. Challenge experiments were conducted under practical conditions in a smoke-house. The surface of salmon sides was inoculated with 10^4 CFU/g *L. innocua* and 10^7 CFU/g bacteriocinogenic *Lact. sakei* as well as the bacteriocin non-producing variant. After smoking, the counts of listeria and LAB were determined at days 1 and 14. All bacteriocinogenenic *Lact. sakei* strains reduced the counts of *L. innocua* by more then 2 log units. One of the strains, *Lact. sakei* LTH5754 was an isolate from CSS and achieved a 5 log reduction of *L. innocua* within the storage period. *In vitro* experiments showed that bacteriocinogenic strains of *Lact. sakei* were also effective against three strains of *L. monocytogenes* and one strain of *L. innocua*. The pH as well the sensorial properties of the smoked salmon were not affected by the presence of this tested *Lact. sakei* strain (Weiss and Hammes, 2006).

Combined Application of Different Bacteriocin Preparations for Bio-Preservation of Salmon

The effect of bacteriocin solutions (Microgard™ and Nisin) on reducing total microbial counts, inhibiting *L. monocytogenes*, and prolonging the shelf-life was evaluated by Zuckerman and Avraham (2002). Metabolites of *Propionibacterium shermanii* (Microgard^{tm})

fermented on dextrose containing antimicrobial metabolites were found to be a potential inhibitor for Gram-negative bacteria such as *Pseudomonas*, *Salmonella* and *Yersinia* and some fungi (Al-Zoreky *et al.*, 1991). Among the active compounds were diacetyl as well as lactic, propionic and acetic acids and other undefined low molecular weight inhibitors of about 700 Da in size (Al-Zoreky *et al.*, 1991). Nisin is a polypetide (bacteriocin) produced by certain strains of the food grade *Lc. lactis* subsp. *lactis* and has been used as an antimicrobial in foods since the 1960s. Nisin has a shown a wide range of antimicrobial activity against Gram-positive bacteria (Delves-Broughton *et al.*, 1996).

L. monocytogenes was inoculated onto chilled and on frozen and thawed salmon samples. The combination of Nisin and Microgard^{tm} reduced the total aerobic bacteria populations of fresh chilled salmon by 2 log (P<0.05) and increased its shelf-life, at 6°C, by 3–4 d compared to the control. The above mentioned bacteriocin combination also reduced the growth of inoculated *L. monocytogenes* in frozen–thawed salmon and increased its shelf-life from 5 to 10 d at 6°C. The bacteriocin treatment did not affect surface pH values or colour of the fish (Zuckerman and Avraham, 2002).

Scannell *et al.* (2000) reported that the combination of lactacin (produced by *Lc. lactis*) and Nisin showed inhibiting capabilities against a wide range of Gram-positive bacteria and maintained low microbial population counts during storage. Inhibition by Nisin is based on its adsorption to vegetative cell membranes and disruption of membrane activity via membrane insertion, pore formation and simultaneous depolarization. The increase in membrane permeability disturbs membrane transport of proteins or nucleic acids (Ariyapitipun *et al.*, 2000).

Based on the study of Zuckerman and Avraham (2002) it can be concluded that the combination of the Microgard^{tm} and Nisin at a pH of 4.2 or lower is a feasible mean for increasing the safety and extending the shelf–life of fresh and frozen-thawed salmon. These authors suggested that the cumulative inhibitory action of two bacteriocins is probably due to injury of the bacterial cell membranes and thus causing a reduction in their numbers. This effect depends also on the pH of the treatment solutions due to the increase of undissociated acids which affect cell metabolic activities disturbing the homeostasis of the microorganisms in the salmon (Leistner, 2000). According Zuckerman and Avraham (2002), the antibacterial effect was enhanced with the increased concentrations of the Microgard^{tm} and Nisin.

Szabo and Cahill (1999) studied the effects of nisin and ALTA^{TM} 2341 on the growth of *L. monocytogenes* on smoked salmon packaged under vacuum or 100% CO_2. Smoked salmon slices (pH 6.3) were inoculated with a cocktail of seven *L. monocytogenes* isolates at a level of approximately 2.5 $\log_{10}$ CFU/g. After inoculation, the surface of the smoked salmon slices was treated with either nisin (400 or 1250 IU/g) or ALTA^{TM} 2341 (0.1 or 1%). The smoked salmon was packaged and stored at 4°C (28 d) or 10°C (9 d). On untreated vacuum packaged smoked salmon, *L. monocytogenes* grew 3.8 $\log_{10}$ CFU/g at 4°C and 5.1 $\log_{10}$ CFU/g at 10°C. Growth was reduced on nisin- and ALTA^{TM} 2341-treated vacuum packaged smoked salmon. On the nisin-treated samples, *L. monocytogenes* increased 2.5 (400 IU/g) and 1.5 (1250 IU/g) $\log_{10}$ CFU/g at 4°C, and by 4.3 (400 IU/g) and 2.7 (1250 IU/g) $\log_{10}$ CFU/g at 10°C. With the ALTA^{TM} 2341-treated samples, *L. monocytogenes* increased 2.8 (0.1%) or 1.6 (1.0%) $\log_{10}$ CFU/g at 4°C, and 3.3 (0.1%) or 3.6 (1.0%) $\log_{10}$ CFU/g at 10°C. The growth of *L. monocytogenes* was retarded by packaging the smoked salmon in 100% CO_2. On untreated smoked salmon, only a 0.8 $\log_{10}$ cycle increase was observed at 10°C. Under all the other conditions tested with 100% CO_2, *L. monocytogenes* was detected but growth was prevented

(Szabo *et al*., 1999). Packaging in 100% CO_2 emerged as a clear preservation and safety alternative to bacteriocins in this study, this atmosphere alone effectively suppressing the growth of *L. monocytogenes* at both refrigeration and room temperatures. Presently, this form of packaging does not require regulatory approval for use, at least in some countries such as Australia. The major disadvantage for the manufacturer is in the cost of the equipment and staff training. Before this technology is recommended for commercial use, product quality with respect to chemical and microbial spoilage and consumer acceptability should be identified and evaluated, as should product safety with respect to the growth of other cold-tolerant pathogens (Szabo *et al.*, 1999).

Conclusion

This chapter has shown the potential of various bacteriocin-producing LAB for the biopreservation of vacuum-sealed salmon and other fish products. These bacteriocins, produced by beneficial microorganisms, can prevent or at least retard the growth of pathogens such as *L. monocytogenes*, thus increasing the shelf-life without affecting the physical and sensorial qualities of the end-product. Different methods of application of these LAB strains or of the purified bacteriocins have been described; these should now be applied in the production lines of different fish-based products.

Acknowledgments

Dr. Svetoslav D. Todorov was supported by CAPES and CNPq, Brazilian Government, Brasilia, Brazil. Dr. Jean Guy LeBlanc would like to thank CONICET, ANPCyT and CIUNT, Argentina for their financial support.

References

Aasen, I.M.; Markussen, S.; Møretrø, T.; Katla, T.; Axelsson, L.; Natestad, K.Interactions of the bacteriocins sakacin P and nisin with food constituents. *Int. J. Food Microbiol.* 2003, *87*, 35-43.

Aasen, I.M.; Møretrø, T.; Katla, T.; Axelsson, L.; Storrø, I. Influence of complex nutrients, temperature and pH on bacteriocin production by *Lactobacillus sakei* CCUG 42687. *Appl. Microbiol. Biotechnol.* 2000, *53*, 159-166.

Alves, V.F.; De Martinis, E.C.P.; Destro, M.T.; Vogel, B.F.; Gram, L. Antilisterial activity of a *Carnobacterium piscicola* isolated from Brazilian smoked fish (Surubim [*Pseudoplatystoma* sp.]) and its activity agaist a persistent strain of *Listeria monocytogenes* isolated from surubim. *J. Food Protect*. 2005, *68*, 2068-2077.

Al-Zoreky, N.; Ayres, J.W.; Sandine, W.E. Antimicrobial activity of Microgard against food spoilage and pathogenic microorganisms. *J. Dairy Sci.* 1991, *74*, 758-763.

Amézquita, A.; Brashears, M.M. Competitive inhibition of *Listeria monocytogenes* in ready-to-eat meat products by lactic acid bacteria. *J. Food Protect.* 2002, *65*, 316-325.

Ariyapitipun, T.; Mustapha, A.; Clarke, A.D. Survival of *Listeria monocytogenes* Scott A on vacuum-packaged raw beef treated with polylactic acid, lactic acid, and nisin. *J. Food Protect.* 2000, *63*, 131-136.

Austin, B.; Zhang, X.H. Vibrio harveyi: a significant pathogen of marine vertebrates and invertebrates. *Lett. Appl. Microbiol.* 2006, *43*, 119-124.

Aymerich, T.; Artigas, M.G.; Garriga, M.; Monfort, J.M.; Hugas, M. Effect of sausage ingredients and additives on the production of enterocon A and B by *Enterococcus faecium* CTC492. Optimization of in vitro production and anti-listerial effect in dry fermented sausages. *J. Appl. Microbiol.* 2000, *88*, 866-894.

Babic, I.; Wataba, A.E.; Buta, J.G. Growth of *Listeria monocytogenes* restricted by native microorganisms and other properties of fresh-cut spinach. *J. Food Protect.* 1997, *60*, 912-917.

Ben EmBarek, P.K. Presence, detection and growth of *Listeria monocytogenes* in seafood: a review. *Int. J. Food Microbiol.* 1994, *23*, 17-34.

Bermúdez-Humarán, L.G.; Rihakova, J.; Langella, P.; Memnerova, K.; Nazef, L.; Prevost, H.; Drider, D. Antimicrobial activity of divercin RV41 produced and secreted by *Lactococcus lactis*. *J. Mol. Microbiol. Biotechnol.* 2007, *13*, 259-263.

Bhugaloo-Vial, P.; Dousset, X.; Metivier, A.; Sorokine, O.; Anglade, P.; Boyaval, P.; Marion, D. Purification and amino acid sequence of piscicocin V1a and V1b, two class IIa bacteriocins secreted by *Carnobacterium piscicola* V1 that display significantly different levels of specific inhibitory activity. *Appl. Environ. Microbiol.* 1996, *66*, 4410-4416.

Blom, H.; Katla, T.; Hagen, B.F.; Axelsson, L. A model essay to demonstrate how intrinsic factors affect diffusion of bactericoins. *Int. J. Food Microbiol.* 1997, *38*, 103-109.

Brillet, A.; Pilet, M.-F.; Prevost, H.; Bouttefroy, A.; Leroi, F. Biodiversity of *Listeria monocytogenes* sensitivity to bacteriocin-producing *Carnobacterium* strains and application in sterile cold-smoked salmon. *J. Appl. Microbiol.* 2004, *97*, 1029-1037.

Brurberg, M.B.; Nes, I.F.; Eijsink, V.G.H. Pheromone-induced production of antimicrobial peptides in *Lactobacillus*. *Mol. Microbiol.* 1997, *26*, 347-360.

Buchanan, R.L.; Bagi, L.K. Microbial competition: effect of culture conditions on the suppression of *Listeria monocytogenes* Scott A by *Carnobacterium piscicola*. *J. Food Protect.* 1997, *60*, 254-261.

Carlin, F.; Nguyen, C.; Morris, C.E. Influence of background microflora on *Listeria monocytogenes* on minimally processed fresh broad-leaved endive (*Cichorium endivia* var. *latifolia*). *J. Food Protect.* 1996, *59*, 698-703.

Chikindas, M.L.; García-Garcerá, M.J.; Driessen, A.J.M.; Ledeboer, A.M.; Nissen-Meyer, J.; Nes, I.F.; Abee, T.; Konings, W.N.; Venema, G. Pediocin PA-1, a bacteriocin from *Pediococcus acidilactici* PAC1.0, forms hydrophilic pores in the cytoplasmic membrane of target cells. *Appl. Environ. Microbiol.* 1993, *59*, 3577-3584.

Christensen, B.B.; Haagensen, J.A.J.; Heydorn, A.; Molin, S. Metabolic commensalisms and competition in a two-species microbial consortium. *Appl. Environ. Microbiol.* 2002, *68*, 2495-2505.

Cleveland, J.; Montville, T.J.; Nes, I.F.; Chikindas, M.L. Bacteriocins: safe, natural antimicrobials for food preservation. *Int. J. Food Microbiol.* 2001, *71*, 1–20.

Cole, M.B.; Jones, M.V.; Holyoak, C. The effect of pH, salt concentrations and temperature on the survival and growth *Listeria monocytogenes*. *J. Appl. Bacteriol.* 1990, *69*, 63-72.

Connil, N.; Prévost, H.; Dousset, X. Production of biogenic amines and divercin V41 in coild smoked salmon inoculated with *Carnobacterium divergen* V41, and specific detection of this strain by multiplex-PCR. *J. Appl. Microbiol.* 2002, *92*, 611-617.

Cortesi, M.L.; Sarli, T.; Santoro, N.; Murru, N.; Pepe, T. Distribution and behavior of *Listeria monocytogenes* in three lots of natuarlly-contaminated vacuum-packed smocked salmon stored at 2 and 10°C. *Int. J. Food Microbiol.* 1997, *37*, 209-214.

De Martinis, E.C.P.; Santarosa, P.R.; Freitas, F. Caracterização preliminar de bacteriocinas produzidas por seis cepas de bactérias láticas isoladas de produtos cárneos embalados a vácuo. *Ciência y Tecnología Alimentaria* 2003, *23*, 195-199.

De Vuyst, L.; Vandamme, E.J. Nisin, a lantibiotic produced by *Lactococcus lactis* subsp. *lactis*: Properties, biosynthesis, fermentation and applications. In: De Vuyst, L.; Vandamme, E.J. (eds). Bacteriocins of lactic acid bacteria. *Microbiology, genetics and applications*. Blackie Academic and Professional, London, 1994, 151-221.

Delves-Broughton, J.; Blackburn, P.; Evans, R.J.; Hugenholtz, J. Application of the bacteriocin Nisin. *Antonie van Leeuw.* 1996, *69*, 193-202.

Duffes, F. Improving the control of *Listeria monocytogenes* in cold smoked salmon. *Trends Food Sci. Technol.* 1999, *10*, 211-216.

Duffes, F.; Corre, C.; Leroi, F.; Dousset, X.; Boyaval, P. Inhibition of *Listeria monocytogenes* by in sito produced and semipurified bacteriocins of *Carnobacterium* spp. On vacuum-packed, refrigerated cold-smoked salmon. *J. Food Protect.* 1999a, *62*, 1394-1403.

Duffes, F.; Leroi, F.; Boyaval, P.; Dousset, X. Inhibition of *Listeria monocytogenes* by *Carnobacterium* spp. Strains in a simulated cold smoked fish system stored at 4°C. *Int. J. Food Microbiol.* 1999b, *47*, 33-42.

Duffes, F.; Leroi, F.; Dousset, X.; Boyaval, P. Use of bacteriocin producing *Carnobacterium piscicola* strain, isolated from fish, to control *Listeria monocytogenes* development in vacuum-packed cold-smoked salmon stored at 4°C. *Sci. Alim.* 2000, *20*, 153-158.

Eijsink, V.G.H.; Skeie, M.; Middelhoven, P.H.; Brurberg, M.B.; Nes, I.F. Comparative studies of Class IIa bacteriocins of lactic acid bacteria. *Appl. Environ. Microbiol.* 1998, *64*, 3275-3281.

Embarek, P.K.; Jeppesen, V.F.; Huss, H.H. Antibacteril potential of *Enterococcus faecium* strains to inhibit *Clostridium botulinum* in sous-vide cooked fish filets. *Food Microbiol.* 1994, *11*, 525-536.

Fimland, G.; Johnsen, L.; Axelsson, L.; Brurberg, M.B.; Nes, I.F.; Eijsink, V.G.; Nissen-Meyer, J. A C-terminal disulfide bridge in pediocin-like bactericoins renders bacteriocin activity less temperature dependent and is a mojor determinant of the antimicrobial spectrum. *J. Bacteriol.* 2000, *182*, 2643-2648.

Finkel, S.E.; Zinser, E.R.; Kolter, R. Long-term survival and evolution in the stationary phase. In. Bacterial Stress Responses ed. Storz, G.; Hengge-Aronis, R. Washington, DC. *ASM Press*. 2000, 231-238.

Gänzle, M.G.; Ehrmann, M.; Hammes, W.P. Modeling of growth of *Lactobacillus sanfranciscensis* and *Candida milleri* in response to process parameters of sourdough fermentation. *Appl. Environ. Microbiol.* 1998, *64*, 2616-2623.

Gänzle, M.G.; Weber, S.; Hammes, W.P. Effect of ecological factors on the inhibitory spectrum and activity of bactericoins. *Int. J. Food Microbiol.* 1999, *46*, 207-217.

Ghalfi, H.; Benkerroum, N.; Thonart, P.; Allaoui, A.; Destain, J. Bacteriocin activity by *Lactobacillus curvatus* CWBI-B28 to inactivate *Listeria monocytogenes* in cold-smoked salmon during 4 degrees C storage. J. *Food Protect.* 2006, *69*, 1066-1071.

Goff, J.H.; Bhunia, A.K.; Johnson, M.G. Complete inhibition of low levels of *Listeria monocytogenes* on refrigerated chicken meat with pediocin AcH (PA-1) bound to heat-killed *Pediococcus acidilactici* cells. *J. Food Prot.* 1996, *59*, 1187-1192.

Gonzàles-Rodrìguez, M.N.; Sanz, J.J.; Santos, J.A.; Otero, A.; Garcia-Lopez, M.L. Numbers and types of microorganisms in vacuum-packed cold-smoked freshwater fish at the retail level. *Int. J. Food Microbiol.* 2002, *77*, 161-168.

Hansen, L.T.; Rontved, S.D.; Huss, H.H. Microbiological quality and shelf ife of cold-smoked salmon from three different processing plants. *Food Microbiol.* 1998, *15*, 137-150

Heinitz, M.L.; Johnson, J.M. The incidence of *Listeria* spp., *Salmonella* spp. and *Clostridium botulinum* in smoiked fish and shellfish. *J. Food Protect.* 1998, *61*, 318-323.

Helgason, E.; Andreas, O.; Caugant, A.; Henning, A.; Fouet, A.; Mock, M.; Hegna, I.; Kolsto, A. *Bacillus anthracis*, *Bacillus cereus* and *Bacillus thuringiensis* – one species on the basis of genetic evidences. *Appl. Environ. Microbiol.* 2000, *66*, 2627-2630.

Hersom, A.; Hulland, E. Conservas alimenticas: procesado térmico y microbiología Acribia, Zaragoza. 1995, 451 pp.

Holck, A.L.; Axelsson, L.; Hühne, K.; Kröckel, L. Purification and cloning of sakacin 674, a bacteriocin produced from *Lactobacillus sakei* Lb674. *FEMS Microiol. Lett.* 1994, *115*, 143-150.

Hühne, K.; Axelsson, L.; Holk, A.; Kröckel, L. Analysis of the sakacin P gene cluster from *Lactobacillus sakei* Lb674 and its expression in sakacin-negative *Lb. sakei* starins. *Microbiol.* 1996, *142*, 1437-1448.

Hugas, M.; Garriga, M.; Pascual, M.; Aymerich, M.T.; Monfort, J.M. Enhancement of sakacin K activity against *Listeri monocytogenes* in fermented sausages with pepper or manganese as ingredients. *Food Microbiol.* 2002, *19*, 519-528.

Huss, H.H.; Enembarek, P.K.; Jeppesen, V.F. Control of biological hazards in cold smoked salmon production. *Food Control* 1995, *6*, 335-340.

Jay, J.M. Do background microorganisms play a role in the safety of fresh foods? *Trends Food Sci. Technol.* 1997, *8*, 421-424.

Jay, J.M. Microorganisms in fresh ground meats: the relative safety of products with low versus high number. *Meat Sci.* 1996, *43*, S59-S66.

Jeppesen, V.F.; Huss, H.H. Antagonistic acitivity of 2 strains of lactic acid bacteria against *Listeria monocytogenes* and *Yersinia enetrocolitica* in a model fish production at 5°C. *Int. J. Food Microbiol.* 1993a, *19*, 179-186.

Jeppesen, V.F.; Huss, H.H. Characteristics and antagonistic acidity of lactic acid bacteria isolated from chilled fish products. *Int. J. Food Microbiol.* 1993b, *18*, 306-320.

Johansen, C.; Verheul, A.; Gram, L.; Gill, T.; Abee, T. Protamine-induced permeabilization of cell envelopes of gram-positive and gram-negative bacteria. *Appl. Environ. Microbiol.* 1997, *63*, 1155-1159.

Jung, D.S.; Bodyfelt, F.W.; Daeschel, M.A. Influence of fat and emulsifiers on the efficacity of nisin in inhibiting *Listeria monocytogenes* in fluid milk. *J. Dairy Sci.* 1992, *75*, 387-393.

Jydegaard, A.; Gravesen, A.; Knochel, S. Growth condition-related response of *Listeria monocytogenes* 412 to bacteriocin inactivation. *Lett. Appl. Microbiol.* 2000, *31*, 68-72.

Kang, J.H.; Lee, M.S. Characterization of a bacteriocin produced by Enterococcus faecium GM-1 isolated from an infant. *J. Appl. Microbiol.* 2005, *98*, 1169-1176.

Katla, T.; Møretrø, T.; Aasen, I.M.; Holck, A.; Axelsson, L.; Naterstad, K. Inhibition of *Listeria monocytogenes* in cold smoked salmon by addition of sakacin P and /or live *Lactobacillus sakei* cultures. *Food Microbiol.* 2001, *18*, 431-439.

Klaenhammer, T.R. Genetics of bacteriocins produced by lactic acid bacteria. *FEMS Microbiol. Rev.* 1993, *12*, 39-86.

Korkeala, H.; Alanko, T.; Tiusanen, T. Effect of sodium nitrate and sodium chloride on growth of lactic acid bactéria. *Acta Vet. Scand.* 1992, *33*, 27-32.

Kumar, M.; Berwal, J.S. Sensitivity of food pathogens to garlic (*Allium sativum*). *J. Appl. Microiol.* 1998, *84*, 213-215.

Leal-Sánchez, M.V.; Jiménez-Díaz, R.; Maldonado-Barragán, A.; Garrido-Fernández, A.; Ruiz-Barba, J.L. Optimization of bacteriocin production by batch fermentation of *Lactobacillus plantarum* LPCO10. *Appl. Environ. Microbiol.* 2002, *68*, 4465-4471.

Lebois, M.; Connil, N.; Onno, B.; Prevost, H.; Dousset, X. Effects of divercin V41 combined to NaCl content, phenol (liquid smoke) concentration and pH on *Listeria monocytogenes* Scott A growth in BHI broth by an experimental desigh approach. *J. Appl. Microbiol.* 2004, *96*, 931-937.

Leisner, J.J.; Green, G.G.; Stiles, M.E. Control of beef spoilage by a sulfide-producing *Lactobacilsu sake* strain with bacteriocinogenic *Leuconostoc gelidium* UAL187 during anaerobic strorage at 2°C. *Appl. Environ. Microbiol.* 1996, *62*, 2610-2614.

Leistner, L. Basic aspects of food preservation by hurdle technology. *Int. J. Food Microbiol.* 2000, *55*, 181-186.

Leroi, F.; Joffraud, J.J.; Chevalier, F.; Cardinal, M. Study on the microbial ecology of cold-smoked salmon during storage at 8°C. *Int. J. Food Microbiol.* 1998, *39*, 111-121.

Leustner, J.J.; Green, G.G.; Dilts, B.D.; Stiles, M.E. Effect of growth of selected lactic acid bacteria on storage life of beef stored under vacuum and in air. *Int. J. Food Microbiol.* 1995, *26*, 231-243.

Liu, W.; Hansen, J.N. Some chemical and physical properties of nisin, a small-protein antibiotic produced by *Lactococcus lactis*. *Appl. Environ. Microbiol.* 1990, *56*, 2551-2558.

Madigan, M.T.; Martinko, J.M.; Parker, J. *Brock Biology of Microorganisms.* New-York: Prentice-Hall. 1999.

Maftah, A.; Renault, D.; Vignoles, C.; Héchard, Y.; Bressolier, P.; Ratinaud, M.H.; Cenatiempo, Y.; Julien, R. Membrane permeabilization of *Listeria monocytogenes* and mitochondria by the bacteriocin mesentericin Y105. *J. Bacteriol.* 1993, *175*, 3232-3235.

Mattila-Sandholm, T.; Skytta, E. The effect of spoilage flora on the growth of food pathogens in minced meat stored at chill temperature. *Lebensmittel-Wissenshafl und Technologie* 1991, *24*, 116-120.

McMiullen, L.M.; Stiles, M.E. Potentail for use of bacteriocin-producing lactic acid bacteria in the preservation of meats. *J. Food Pro*tect. 1996, *58*, S64-S71.

Messi, P.; Bondi, M.; Guerrieri, E.; Sabia, C.; Manicardi, G. Plantaricina 35d um antagonista natuare per il controlo di *Listeria monocytogenes* in salmoni affumicati. *Industri. Alim.* 2000, *39*, 343-348.

Messi, P.; Bondi, M.; Sabia, C.; Manicardi, G. Studio di uma nuova batteriocina (plantaricina 35d) attiva nei confronti di patogeni psicrotrofi di interesse alimentare. Atti Del XI Congresso Internazionale dell´Ordine Nazionale dei Biologi, su "*Autocontrollo ed Eco-Audit: concrete opportunita professionali per il Bi*ologo" Napoli 15-18 Ottobre 1998, 1998, *2*, 107-113.

Metivier, A.; Pilet, M.-F.; Dousset, X.; Sorokine, O.; Anglade, P.; Zagorec, M.; Piard, J.C.; Marion, D.; Cenatiempo, Y.; Fremaux, C. Divercin V41, a new bactericoin with two disulphide bounds produced by *Carnobacterium divergens* V41: primary structure and genomic organization. *Microbiol.* 1998, *144*, 2837-2844.

Moll, G.; Ubbink-Kok, T.; Hildeng-Hauge, H.; Nissen-Meyer, J.; Nes, I.F.; Konings, W.N.; Driessen, A.J.M. Lactococcin G is a potassium ion-conducting, two-component bacteriocin. *J. Bacteriol.* 1996, *178*, 600-605.

Moll, G.N.; Konings, W.N.; Driessen, A.J.M. *Bacteriocins: mechanism of membrane insertion and pore formation.* Antonie van Leeuw. 1999, *76*, 185-198.

Montville, T.J.; Bruno, M.E.C. Evidence that dissipation of proton motive force is a common mechanism of action for bacteriocins and other antimicrobial proteins. *Int. J. Food Microbiol.* 1994, *24*, 53-74.

Mora, D.; Scarpellini, M.; Franzetti, L.; Colombo, S.; Galli, A. Reclassification of *Lactobacillus maltaromicus* (Miller et al. 1974) DSM 20342(T) and DSM 20344 and *Carnobacterium piscicola* (Collins et al. 1987) DSM 20730(T) and DSM 20722 as *Carnobacterium maltaromaticum* comb. Nov. *Int. J. Syst. Evol. Microbiol.* 2003, *53*, 675-678.

Møretrø, T.; Aasen, I.M.; Storrø, I.; Axelsson, L. Production of sakacin P by *Lactobacillus sakei* in a completely defined medium. *J. Appl. Microbiol.* 2000, *88*, 536-545.

Neysen, P.; Messens, W.; De Vuyst, L. Effect of sodium chloride on growth and bactericoin production by *Lactobacillus amylovorus* DCE 471. *Int. J. Food Microbiol.* 2003, *88*, 29-39.

Nilsson, L.; Chen, Y.; Chikindas, M.L.; Huss, H.H.; Gram, L.; Montville, T.J. Carbon dioxide and nisin act synergistically on the cytoplasmic membrane of *Listeria monocytogenes. Appl. Environ. Microbiol.* 2000, *66*, 769-774.

Nilsson, L.; Gram, L.; Huss, H.H. Growth control of *Listeria monocytogenes* on cold-smoked salmon using a competitive lactic acid bacteria flora. *J. Food Protect.* 1999, *62*, 336-342.

Nilsson, L.; Ng, Y.Y.; Christiansen, J.N.; Jørgensen, B.L; Grótinum, D.; Gram, L. The contribution of bacteriocin to inhibition of *Listeria monocytogenes* by *Carnobacterium piscicola* strains in cold-smoked salmon system. *J. Appl. Microbiol.* 2004, *96*, 133-143.

Nilsson, L.; Nielsen, M.K.; Ng, Y.; Gram, L. Role of acetate in production of an autoinducible class IIa bacteriocin in *Carnobacteruim piscicola* A9b. *Appl. Environ. Microbiol.* 2002, *68*, 2251-2260.

Olasupo, N.A. Bacteriocins of *Lactobacillus plantarum* strains from fermented foods. Folia Microbiol. 1996, *41*, 130-136.

Paludan-Muller, C.; Valyasevi, R.; Huss, H.H.; Gram, L. Genotypic and phenotypic characterization of garlic-fermenting lactic acid bacteria isolated from som-fak, a Thai low-salt fermented fish product. *J. Appl. Microbiol.* 2002, *92*, 307-314.

Parente, E.; Martuscelli, M.; Gardini, F.; Grieco, S.; Crudele, M.A.; Suzzi, G. Evolution of microbial populations and biogenic amine production in dry sausages produced in Southern Italy. *J. Appl. Mcirobiol.* 2001, *90*, 882-891.

Parente, E.; Ricciardi, A. Influence of pH on the production of enterocin 1146 during batch fermentation. *Lett. Appl. Microbiol.* 1994, *19*, 12-15.

Pelroy, G.A.; Peterson, M.E.; Holland, P.J.; Eklund, M.W. Inhibition of *Listeria monocytogenes* in cold-processed (smoked) salmon by sodium lactate. *J. Food Protect.* 1994, *57*, 108-113.

Pilet, M.-F.; Dousset, X.; Barré, R.; Novel, G.; Desmazeaud, M.; Piard, J.C. Evidence for two bacteriocins produced by *Carnobacterium piscicola* and *Carnobacterium divergens* isolated from fish and active against *Listeria monocytogenes. J. Food Protect.* 1995, *58*, 256-262.

Richard, C. ; Leroi, F.; Brillet, A.; Rachman, C.; Connil, N.; Drider, D.; Pilet, M.-F.; Onno, B.; Dousset, X.; Prevost, H. Maîtrese de développement de *Listeria monocytogenes* dans le saumon fumé: interet de la biopréservation par des acterie lactiques. *Lait* 2004, *84*, 135-144.

Rihakova, J.; Belguesmia, Y.; Petit, V.W.; Pilet, M.F.; Prevost, H.; Dousset, X.; Drider, D. Divercin V41 from gene chracterization to food applications: 1998-2008, a decade of solved and unsolved questions. *Lett. Appl. Microbiol.* 2009, *48*, 1-7.

Riley, M.A.; Wertz, J.E. Bacteriocins: evolution, ecology, and application. *Annu. Rev. Microbiol.* 2002, *56*, 117–137.

Rocourt, J.; Jacquet, C.; Reilly, A. Epidemiology of human listeriosis and seafoods. *Int. J. Food Mirobiol.* 2000, *62*, 197-209.

Rodriguez, E.; Tomillo, J.; Nunez, M.; Medina, M. Combined effect of bacteriocin-producing lactic acid bacteria and lactoperoxidase system activation on *Listeria monocytogenes* in refrigerated raw milk. *J. Appl. Mcirobiol.* 1997, *83*, 389-395.

Rørvik, L.M. *Listeria monocytogenes* in the smoked salmon industry. *Int. J. Food Microbiol.* 2000, *62*, 183-190.

Rørvik, L.M.; Caugant, D.; Yndestad, M. Contamination pattern of *Listeria monocytogens* and other *Listeria* spp. In a salmon slaughterhouse and smoked salmon processing plant. *Int. J. Food Microbiol.* 1995, *25*, 19-27.

Rørvik, L.M.; Skejeve, E.; Knudsen, B.R.; Yndestad, M. Risk factors for contamination of smoked salmon with *Listeria monocytogenes* during processing. *Int. J. Food Microbiol.* 1997, *37*, 215-219.

Rørvik, L.M.; Yndestad, M. *Listeria monocytogenes* in foods in Norway. *Int. J. Food Microbiol.* 1991, *13*, 97-104.

Rose, N.L.; Sporns, P.; Stiles, M.E.; McMullen, L.M. Inactivation of nisin by glutathione in fresh meat. *J. Food Sci.* 1999, *64*, 759-762.

Samelis, J.; Metaxopoulos, J.; Vlassi, M.; Pappa, A. Stability and safety of traditional Greek salami – a microbiological ecology study. *Int. J. Food Microbiol.* 1998, *44*, 69-82.

Scannell, A.G.M.; Paul Ross, R.; Hill, C.; Arendt, E.K. An effective lacticin biopreservative in fresh pork sausage. *Int. J. Food Protect*. 2000, *63*, 370-375.

Schöbitz, R.; Suazo, V.; Costa, M.; Ciampi, L. Effects of a bacteriocin-like inhibitory substance from *Carnobacterium piscicola* against human and salmon isolates of *Listeria monocytogenes. Int. J. Food Microbiol*. 2003, *84*, 237-244.

Schöbitz, R.P.; Borquez, P.A.; Costa, M.E.; Ciampi, L.R.; Brito, C.S. Bacteriocin like substance production by Carnobacterium piscicola in a continuous system with three culture broths. Study of antagonism against *Listeria monocytogenes* on vacuum packaged salmon. *Braz. J. Microbiol.* 2006, *37*, 52-57.

Scolari, G.; Vescovo, M.; Orsi, C.; Sinigaglia, M.; Torriani, S. Lactic acid bacteria na *Aeromonas hydrophila* interactions in ready-to-use vegetable: growth modeling in relation to storage temperature and inoculums size. *Adv. Food Sci.* (CMTL) 1997, *19*, 25-30.

Seaberg, A.C.; Labe, R.G.; Shetty, K. Inhibition of *Listeria monocytogenes* by lite clonal extracts of oregano (*Origanum vulgare*). *Food Biotechnol.* 2003, *17*, 237-244.

Singh, B.; Falahee, M.B.; Adams, M.R. Sinergetic inhibition of *Listeria monocytogenes* by nisin and garlic extract. *Food Microbiol.* 2001, *18*, 133-139.

Stiles, M. *Biopreservation by lactic acid bacteria.* Antonie Van Leeuw. 1996, *70*, 331-343.

Stoffels, G.; Gudmundsdottir, A.; Abee, T. Membrane-associated proteins encoded by nisin gene-cluster may function as a receptor for the lantibiotic carnocin UI49. *Microbiol-UK*, 1994, *140*, 1443-1450.

Stoffels, G.; Nes, I.F.; Guomundsdottir, A. Isolation and properties of a bacteriocin-producing *Carnobacterium piscicola* isolated from fish. *J. Appl. Bacteriol.* 1992a, *73*, 309-316.

Stoffels, G.; Nissen-Meyer, J.; Gudmundsdottir, A.; Sletten, K.; Holo, H.; Nes, I.F. Purification and characterization of a new bactericoin isolated from a *Carnobacterium* sp. *Appl. Environ. Microbiol.* 1992b, *58*, 1417-1422.

Stohr, V.; Joffraud, J.J.; Cardinal, M.; Leroi, F. Spoilage potential and sensory profile associated with bacteria isolated from cold-smoked salmon. *Food Res. Int.* 2001, *34*, 797-806.

Szabo, E.A.; Cahill, M.E. Nisin and ALTA^{tm} 2341 inhibit the growth of *Listeria monocytogenes* on smoked salmon packaged under vacuum or 100% CO_2. *Lett. Appl. Microbiol.* 1999, *28*, 373-377.

Tahiri, I.; Desbiens, M.; Benech, R.; Kheadr, E.; Lacroix, C.; Thibault, S.; Ouellet, D.; Fliss, I. Purification, characterization and amino acid sequencing of divercin M35: A novel class IIa bactericoin produced by *Carnobacterium divergens* M35. *Int. J. Food Microbiol.* 2004, *97*, 123-136.

Tichaczek, P.S.; Nissen-Meyer, J.; Nes, I.F.; Vogel, R.; Hammes, W.P. Characterization of the bacteriocins curvacin A from *Lactobacillus curvatus* LTH1174 and sakacin P from *L. sakei* LTH673. System. *Appl. Microbiol.* 1992, *15*, 460-468.

Todorov, S.D. Bacteriocin production by *Lactobacillus plantarum* AMA-K isolated from Amasi, a Zimbabwean fermented milk product and study of adsorption of bacteriocin AMA-K to *Listeria* spp. *Braz. J. Microbiol.* 2008, *38*, 178-187.

Todorov, S.D.; Koep, K.S.C.; Van Reenen, C.A.; Hoffman, L.C.; Slinde, E.; Dicks, L.M.T. Production of salami from beef, horse, mutton, blesbok (*Damaliscus dorcas phillipsi*) and springbok (*Antidorcas marsupialis*) with bacteriocinogenic strains of *Lactobacillus plantarum* and *Lactobacillus curvatus*. *Meat Sci.* 2007, *77*, 405-412.

Todorov, S.D.; Powell, J.E.; Meincken, M.; Witthuhn, R.C.; Dicks, L.M.T. Factors affecting the adsorption of *Lactobacillus plantarum* bacteriocin bacST8KF to *Enterococcus faecalis* and *Listeria innocua*. *Int. J. Dairy Technol.* 2007, *60*, 221-227.

Todorov, S.D.; Rachman, C.; Fourrier, A.; Dicks, L.M.T.; van Reenen, C.A.; Prévost, H.; Dousset, X. Characterization of bacteriocin produced by *Lactobacillus sakei* R1333 isolated from smoked salmon. *Anaerobe* 2011, *17*, 23-31.

Tomé, E.; Gibbs, P.A.; Teixera, P.C. Growth control of *Listeria innocua* 2030c on vacuum-packed cold-smoked salmon by lactic acid bacteria. *Int. J. Food Microbiol.* 2008a, *121*, 285-294.

Tomé, E.; Pereira, V.L.; Lopes, C.I.; Gibbs, P.A.; Teixera, P.C. *In vitro* tests of suitability of bacteriocin-producing lactic acid bactéria, as potential biopreservation cultures in vacuum-packaged cold-cmoked salmon. *Food Control* 2008b, *19*, 535-543.

Tomé, E.; Todorov, S.D.; Gibbs, P.A.; Teixeira, P.C. Partial characterization of nine bacteriocins produced by Lactic Acid Bacteria isolated from Cold-Smoked Salmon with activity against *Listeria monocytogenes*. *Food Biotechnol.* 2009, *23*, 50-73.

Van Belkum, M.J.; Hayema, B.J.; Jeeninga, R.E.; Kok, J.; Venema, G. Organization and nucleotide sequences of two lactococcal bacteriocin operons. *Appl. Environ. Microbiol.* 1991, *57*, 492-498.

Vaz-Velho, M.; Todorov, S.; Ribeiro, J.; Gibbs, P. Growth control of *Listeria innocua* 2030c during processing and storage of cold-smoked salmon-trout by *Carnobacterium divergens* V41 culture and supernatant. *Food Control* 2005, *16*, 540-548.

Verluyten, J.; Messens, W.; De Vuyst, L. Sodium chloride reduces production of curvacin A, a bacteriocin produced by *Lactobacillus curvatus* strain LTH 1174, originating from fermented sausages. *Appl. Environ. Microbiol.* 2004, *70*, 2271-2278.

Vescovo, M.; Scolari, G.; Orsi, C.; Sinigaglia, M.; Torriani, S. Combined effects of *Lactobacillus sakei* inoculum, modified atmosphere packaging and storage temperature in controlling *Aeromonas hydrophila* in ready-to-use vegetable. *Int. J. Food Sci. Technol.* 1997, *32*, 411-419.

Vescovo, M.; Scolari, G.; Zacconi, C. Inhibition of *Listeria innocua* growth by antimicrobial-producing latic acid cultures in vacuum-packed cold-smoked salmon. *Food Microbiol.* 2006, *23*, 689-693.

Vold, L.; Holck, A.; Wasteson, Y.; Nissen, H. High levels of background flora inhibits of *Escherichia coli* O157:H7 in ground beef. *Int. J. Food Microbiol.* 2000, 56, 219-225.

Wang, L.L.; Johnson, E.A. Control of *Listeria monocytogenes* by monoglycerides in food. *J. Food Protect.* 1997, *60*, 131-138.

Weiss, A.; Hammes, W.P. Lactic acid bacteria as protective cultures against *Listeria* spp. on cold-smoked salmon. *Europ. Food Res. Technol.* 2006, *222*, 343-346.

Yamazaki, K.; Suzuki, M.; Kawai, Y.; Inoue, N.; Montville, T.J. Inhibition of *Listeria monocytogenes* in cold-smoked salmon by *Carnobacterium piscicola* CS526 isolated from frozen surimi. *J. Food Protect.* 2003, *66*, 1420-1425.

Yildirim, S.; Konrad, D.; Calvez, S.; Drider, D.; Prevost, H.; Lacroix, C. Production of recombinant bacteriocin divercin V41 by high cell density *Escherichia coli* batch and fed-batch cultures. *Appl. Microbiol. Biotechnol.* 2007, *77*, 525-531.

Zouhir, A.; Hammami, R.; Fliss, I.; Ben Hamida, J. A New Structure-based Classification of Gram-positive Bacteriocins. *Protein J.* 2010, *29*, 432-439.

Zouhir, A.-M.; Kheadr, E.; Tahiri, I.; Hamida, J.B.; Fliss, I. Combination with plant extracts improves the inhibitory action of divercin M35 agains *Listeria monocytogenes*. *J. Food Qual.* 2008, *31*, 13-33.

Zuckerman, H.; Avraham, R.B. Control of growth of *Listeria monocytogenes* in fresh salmon using Microgard^{tm} and Nisin. Lebensm. *Wiss. Technol.* 2002, *35*, 543-548.

In: Lactobacillus: Classification, Uses and Health Implications ISBN: 978-1-62081-151-1
Editors: Alba I. Perez Campos and Arturo Leon Mena © 2012 Nova Science Publishers, Inc.

Chapter IV

Resistance of Spoilage *Lactobacillus* Spp. to Food Processing Technologies

Fernando Sampedro*

Center for Animal Health and Food Safety, College of Veterinary Medicine,
University of Minnesota, St. Paul, MN, US

Abstract

Food processing preservation requisites are based on achieving an adequate food safety and prolonged shelf-life while maintaining the fresh character and quality properties of food. To preserve food and prolong the shelf-life, thermal pasteurization is the most common processing technology employed, but thermal processing can damage the original characteristics of food. Recently, some emerging non-thermal technologies (high pressure processing, pulsed electric fields, high pressure carbon dioxide, high pressure homogenization) along with new applications of non-conventional thermal technologies such as microwave irradiation have appeared in the market due to the consumers demand for food with fresh-like characteristics. To establish the optimum treatment conditions of a new technology requires not only studies on the influence of the various process parameters on the death of the microorganism and quality factors, but also the choice of a reference microorganism on which to carry out the studies. *Lactobacillus* spp. such as *Lactobacillus plantarum*, *Lactobacillus brevis*, *Lactobacillus sporogenes*, *Lactobacillus hilgardii* and *Lactobacillus viridescens* are considered common spoilage microflora encountered in fruit juices, vegetable-based products, wine and meat. *Lactobacillus* bacteria are capable of growing over a wide pH range and spoiling minimally processed foods owing to its aciduric nature, producing quality detrimental effects. The high resistance to different processing technologies and acidic conditions make *Lactobacillus* genus a target microbial factor to be considered when designing a new processing treatment.

* Center for Animal Health and Food Safety, College of Veterinary Medicine, University of Minnesota, 136G Andrew Boss, 1354 Eckles Avenue, St. Paul, Minnesota 55108. Phone number: (001)-612-624-6949, Fax: 612-624-4906, email: fsampedr@umn.edu.

Introduction

Lactobacillus spp. belong to a group of bacteria-denominated lactic acid bacteria (LAB) characterized to ferment the sugar predominantly to lactic acid (homofermentative) or to lactic acid among other compounds i.e., ethanol, acetate and $C0_2$ (heterofermentative) [1]. They are gram-positive, non-sporulating rods and cocci, grow anerobically but are also aerotolerant. They are involved in the lactic acid fermentation of a variety of vegetable and animal-based products such as sauerkraut, cheeses and fermented olives [2]. Lactic acid fermentation has many benefits on the nutritional value of food and food preservation as a mean for inhibiting pathogen microorganisms due to several factors including acidity, type of organic acid, water activity and redox potential depletion and production of by products (i.e., carbon dioxide, hydrogen peroxide, ethanol, diacetyl) [3]. In addition to that, some bacteriocins-like substances produced by *Lactobacillus* spp. have been proposed as natural biopreservatives [4-6]. Some of *Lactobacillus* spp. (*L. plantarum*, *L. rhamnosus*, *L. gasseri* and *L. acidophilus*) are used as starter cultures for fruits and vegetables fermentation providing consistency and reliability of performance [7].

Fermented foods along with vegetable-based products such as fruit and vegetable juices and their derivatives, are recognized to be microbiologically safe because of their low pH. However, due to the acidic environment (pH values $\leq$ 4.0), some spoilage *Lactobacillus* spp. microflora is able to grow producing a variety of quality detrimental effects such as ''butter'' off-flavor due to the production of diacetyl as a metabolic end product, swelling of packages, increase of the viscosity and the production of hydrogen sulfide and other off-odors [8]. The common *Lactobacillus* strains found in spoiling vegetable-based products are able to survive acidic conditions due to a mechanism that maintains a constant internal pH by regulating the pH gradient between the cytoplasm and the environment through a process called pH homeostasis. That ability is closely related to the activity of the ATPase [9]. Several studies have demonstrated the ability of *Lactobacillus* spp. to survive in acidic environments. *L. plantarum* generally proliferates at the late stage of fermentation of vegetables such as pickles, sauerkraut, processed tomatoes and dressings for salads [10]. The microorganism has also been isolated in a commercial fresh orange juice [11], observing no increase in the population in a reconstituted orange juice (pH 3.9) after 14 d at 4°C [12]. Alwazeer [13] studied the growth of *L. plantarum* in orange juice stored at 15°C and estimated a lag phase time (time to growth) of 29 h and generation time of 5.3 h. After ca.125 h of storage, *L. plantarum* became dominant achieving a population of 8.7 log cycles. Nualkaekul and Charalampopoulos [14] observed good *L. plantarum* survival (less than 0.4 log decrease) during 6 wks of storage of orange, blackcurrant and pineapple juices (pH around 3.8) and about 1.1 log decrease in the case of lemon juice which had the lowest pH (pH 2.5) among all. Champagne [15] inoculated *Lactobacillus rhamnosus* in an apple-pear-raspberry juice blend and observed a decrease of 75% of viable population in juice stored over 5 wks of storage at 7°C in unopened bottles, while in opened bottles, reduction were only between 20% and 40%. Viability losses under anerobic conditions were twice as high at 7°C than at 4°C.

L. plantarum strains have been also found to spoil beer as stated by [16]. Levels of 3% of ethanol were tolerated by the strain and 6% level reduced the growth in 50%, whereas growth inhibition was observed at 10%. This response to ethanol concentration was responsible from the synthesis of ethanol shock proteins. The resistance to hop concentration was also characterized at pH levels of 5.0 and 4.9. Hop extract (200 ppm) inhibited *L. plantarum* growth in 30 and 60% respectively, whereas at pH of 3.5 and 4.0, 10 and 20 ppm respectively, were enough to inhibit growth. HorA activity is related to the capability of LAB to growth in the presence of hop bitter compounds and is crucial for the beer-spoiling bacteria [17].

LAB are also predominant in vacuum-packed meat products and can cause spoilage such as sour flavor, slime formation, off flavor and gas formation. *L. viridescens* is a well-known agent that causes greening of cured meat pigment in processed meat products [18]. Another interesting case is apple cider. The procedure for making cider relies on the spontaneous fermentation of apple juice. Fermentation depends on the presence in the apple juice of indigenous yeasts and LAB. However, after the fermentation process, these bacteria may be a potential source of different cider alterations [19]. One of the most frequent alterations in apple cider is ropiness, caused by the production of exopolysaccharides. Pediococci and lactobacilli (including *Lactobacillus collinoides* and *Lactobacillus diolivorans*) are involved in exopolysaccharide production. The growth of undesirable LAB in apple juice during storage as well as during the fermentation processes for manufacturing beverages may cause significant economic losses to the cider industry [20].

Endospore-forming bacteria are a common problem in fruit and vegetable processing because they are ubiquitous in the starting of raw materials and because endospores can tolerate thermal treatment and adverse conditions and proliferate in packaged fruits and vegetables, causing spoilage. A common type of spoilage in processed vegetables is caused by *Lactobacillus sporogenes*, which is able to germinate and grow at pH values as low as 4 (i.e., tomato products with a pH ranging from 4.1 to 5.0). During storage, *L. sporogenes* can cause flat-sour spoilage from fermentation of the glucose or lactose to lactic acid. For this reason, *L. sporogenes* is the most frequently isolated microorganism from spoiled packaged tomato paste [21].

Conventional thermal pasteurization processes reduce the initial counts of spoilage and pathogenic microorganisms to a safe level, but cause detrimental effects on the overall food quality. New emerging technologies also called "non-thermal" along with "non-conventional" thermal technologies can be an alternative processing strategy to achieve an optimal safety level without affecting the natural properties of food. In the following section, scientific data is provided regarding the effects of different technologies on the survival of *Lactobacillus* spp. in different food matrices.

Thermal Inactivation of *Lactobacillus Spp.*

Thermal process is one of the safest and most frequently used methods for food preservation. In order to validate our process, the food safety of the product needs to be ensured by achieving an adequate inactivation of foodborne pathogens and spoilage microorganisms and an enhanced shelf-life [22]. In thermal pasteurization, several *Lactobacillus* strains have been used to validate the process. The thermal inactivation kinetics of *L. plantarum* have been studied by several authors [23-26]. Several methods have been used to characterize the microbial resistance to thermal treatment such as Thermal Death

Time (TDT) glass tubes, TDT aluminum tubes, glass capillary tubes [25], stainless steel capillary tubes, aluminum TDT disks and glass TDT spheres. Some of these methods (TDT aluminum tubes and disks and TDT glass spheres) have some heating and cooling lag times that may compromise the validity of the analysis inducing bacteria to increase heat resistance by producing heat shock proteins [26].

Generally, the microbial thermal resistance depends on several factors such as type of strain, matrix characteristics (aw, fat content, pH), type of method use for obtaining experimental data and growth phase. Experimental data is usually plotted showing the log numbers of the survivors versus time at different processing temperatures. Plotted curves are usually fitted to a linear regression to obtain the kinetic parameters that characterize the thermal resistance. The decimal reduction time (D) is known as the thermal kinetic parameter and is expressed as the time at certain temperature to obtain 90% of inactivation of microbial population. D values are obtained from the slope of inactivation curves and is considered a primary kinetic parameter. Another parameter that is usually obtained is the thermal coefficient value (z) which is defined as the temperature change needed to reduce 90% D values. The z values are obtained by plotting log numbers of D values against temperature, applied a linear regression and estimate the slope of the inactivation curves. As a secondary parameter, z characterizes the thermal sensitivity of D values. Table 1 shows the D and z values from studies on thermal inactivation of *L. plantarum* in different substrates. Values showed *L. plantarum* thermal resistance varied among substrates with apple juice and milk being the ones with the highest microbial survival. Higher pH value or fat/proteins content could be the main reasons for that behavior. *L. paracasei* was found to be more resistant than *L. plantarum* and able to survive pasteurization processes.

Table 1. Comparison of D and z values of *Lactobacillus* spp. in different matrices

Microorganism	Medium	D values (s)	z value (°C)	Reference
L. plantarum ATCC 8014	0.3% Carboxymethylcellulose (pH=6.95)	76.3-7.3 (52-61°C)	8.9	[26]
L. plantarum DPC 2739	Milk	32.9, 14.7 and 7.1 (60, 72 and 75°C)	9.0	[25]
L. plantarum DPC 1919	MRS broth	42.2-3.0 (50-57.5°C)	6.7	[24]
L. plantarum DPC 2103	MRS broth Milk	10.0-0.66 (48.5-56°C) 3.14 and 1.35 (53.5 and 56.0°C)	6.2	
L. paracasei DPC 2103	MRS broth Milk	14.7-0.52 (60-67.5°C) 22.5-0.32 (60-67.5°C)	5.3	
L. plantarum ATCC 14917	Apple juice (pH=3.4)	52-1.2 (55-80°C)	15.9	[23]

Adapted from [12].

Thermal inactivation of microorganisms is associated with irreversible denaturation of membranes, ribosomes and nucleic acids [22]. However, some other macromolecular changes that occur during thermal death are still unknown. Ananta [27] used flow cytometric analysis to study multiple cellular targets (structural and functional) in the thermal inactivation of *L. rhamnosus*. The results at 60°C showed most of the population was thermally inactivated due to the denaturation of the intracellular esterase without affecting the cytoplasm and absence of membrane degradation. However, at 75°C, a different scenario was observed and most of the cells showed membrane degradation meaning that the target site for cell death is dependent on the temperature used.

Non-Conventional Processing Technologies

Through the years, thermal treatment has been used for producing microbiologically safe food with low enzyme activity and this continues to be the best processing method for the mass production of foods. Nevertheless, in certain products with health beneficial thermolabile bioactive compounds, quality and nutritional content (vitamins, color and flavor, among others) are compromised after thermal processing, even at mild temperatures.

Taking into account the limitations of traditional thermal processes and the trends in consumer demand, scientists have searched for alternative technologies that could ensure the microbial safety while retaining the fresh-like or fresh-prepared quality. These technologies have been called 'emerging preservation technologies' and particularly, the ones based on processing parameters other than temperature, 'non-thermal technologies' [28]. These technologies include pressure-based processing (high pressure processing, high pressure homogenization and high-pressure carbon dioxide) and pulsed electric fields. Other technologies are based on new applications of temperature such as microwave irradiation and ohmic heating called "non-conventional" thermal technologies.

Non-Conventional Thermal Technologies

The fundamentals of microwaves irradiation are based on the interaction of microwaves with water in food products generating heat. Production of heat occurs primarily when ions accelerate and collide or dipoles rotate and line up in the rapidly alternating electric field. Microwave energy has been used to pasteurize or sterilize food at lower temperatures and shorter times than necessary with conventional heating. The mechanism for the lethal effects of microwave radiation on microorganisms is not fully understood. Many studies have shown that heat generated by microwaves in food systems exerts most of the killing effect [29-31]. Shin and Pyun [32] compared three different processes conventional heating (CH), conventional microwave irradiation (CW) and pulsed-microwave irradiation (PW) on the inactivation degree of *L. plantarum*. The temperature-time combination applied was 50°C for 30 min. CH processing had no effects on microorganism survival whereas CW and PW produced 4.0 and 6.0-7.0 log reductions, respectively, being pulsed-microwave the most lethal process. The possible reason argued by the authors about the more efficacy of PW processing is that pulsed irradiation is directly absorbed by molecules, in contrast to CH where molecules are brought to a higher level of energy through collision with other molecules. Consequently, molecules on PW processing would have higher reaction rates than expected based on average product temperatures and thus there would be selective heating on a molecular level [33]. Sublethal injury was observed only after PW treatment. The amount of leakage material (absorbance measured at 260 nm) seemed not to increase during CH whereas microwave processing (CW and PW) showed an increase of absorbance as treatment time increased with PW being the one with the greatest absorbance values indicating a higher membrane damage. The reason for the leakage may reside on the explanation that the energy from the microwave treatment may disrupt the membrane and/or the sub-cellular structure thus liberating lipids and produce the reorientation of long chain molecules (Ke et al., 1978). Cells treated by PW could be recovered from the sublethal injury when incubated at 37°C. Growth rates in MRS and MRS+salt began to increase after 10 h (lag phase), and similar counts were reached after 17 h of incubation on both media indicating the time for the cells to recover. Metabolic activity was shown after 10 and 22 h in cells treated by CH and CW, respectively, whereas PW-treated cells showed a slower rate achieving a lower level of cell mass after 60 h. Lactic acid production was recovered after 10 and 20 h for CH and both types of microwave processing, respectively. Tajchakavit [23] studied the effects of continuous-flow microwave heating system on the inactivation of *L. plantarum* inoculated in apple juice. D values estimated after microwave processing (14.1, 3.8, 0.79, 0.33 s at 57.5, 60.0, 62.5 and 65°C) were lower than those estimated after conventional heating indicating the higher lethality of the former and z value (4.48) much lower showing a more temperature sensitivity of D values in microwave processing. At 60°C, the destruction rate was 6 times faster than conventional heating whereas the relative destruction ratio (microwave/thermal) was 43.

Non-thermal Processing Technologies

High Pressure Processing (HPP)

The potential of using pressure as a food preservation mechanism has been known for over a century [35]. High pressure processing (HPP) is mainly based on the 'isostatic' principle that establishes that pressure distribution is proportional in all food parts independent of its shape and size [36]. HPP technology is a batch system, consisting of pumping a pressurizing fluid (water or oil) from a deposit where it is contained to the treatment vessel. Once the vessel is filled with the liquid, pressure is built-up by a hydraulic pump and is transmitted through the liquid to the vessel until reaching the desired pressure (100-1000 MPa), held during treatment time (1-20 min) and rapidly decompressed. Temperature is increased due to adiabatic heating (3°C/100 MPa) [37].

The mechanism of microbial inactivation by HPP is a combination of different reactions, such as non-covalent bonds breakdown and cell membrane permeability increase [38]. Based on the extensive literature research on microbial inactivation by HPP in different food matrices published in recent years, one can reach a general conclusion that pressure values between 200-600 MPa at room or mild temperatures (35-50°C) for several minutes are usually enough to inactivate the majority of vegetative cells including lactobacilli [35, 38]. Based on this knowledge, industrial HPP treatment is conducted at a pressure level of 600 MPa for a holding time of 1-5 min [39]. HPP technology in industry is growing, including a wide range of products in the meat industry (cured jam, bacon and turkey, pork sausages, foie grass, marinated raw meat and 'ready to eat' meat products), seafood products (oysters, lobster and crab meat) and vegetable products (fruit juices and smoothies, fruit purees, fresh cut fruits, guacamole, jams, deli salads, dressings, sauces and 'ready to eat' vegetable dishes) [40].

Regarding the effects of isostatic pressure on *Lactobacillus* spp., Gervilla [41] studied the effects of pressure treatment on *L. helveticus*, a beneficial strain, used as major starter culture in fermented dairy products and non-starter flavor enhancer (reducing bitterness and accelerating flavor development). That strain could be used in matured cheeses manufactured from milk treated by HPP. *L. helveticus* was reduced to undetectable counts (7.0 log cycles) when inoculated in ovine milk and treated after 450 MPa at 50°C for 5 min with no differences among 10 and 25°C processing temperatures. D values at 450 MPa were 7.1 and 9.1 min at 2 and 25°C, respectively. Authors noticed ovine milk could be protective against pressure enhancing the microbial survival.

In order to elucidate the possible differences in baroresistance among Gram-positive and Gram-negative, Wuytack [42] compared the effectiveness of HPP treatment on several Gram-positive and Gram-negative species. The well-known theory of more resistance of Gram-positive cells due to the thicker cell wall composed by peptideglycan (skeletal structure of bacterial cells) of Gram-positive (40 layers) versus Gram-negative (1-5 layers) is recognized to be the main reason for the observed differences. *L. plantarum* was found to be a moderately baroresistant microorganism with equal resistance of that of *L. innocua.*

LAB are known to maintain a constant internal pH (pH_{in}) even when the pH of the solution is fairly low. However, when the differences between the pH_{in} and the pH of external solution tend to approach to zero, the cell losses its biological activity. Wouters [10] studied the effects of high pressure on the regulation of cytoplasmatic pH (proton efflux) of *L. plantarum*. As commented earlier, ATPase, specifically F_0F_1ATPase, is intimately related to the generation of a proton gradient and effects of 250 MPa produced surprisingly increase in ATPase activity (1.2-1.7 fold), however when higher pressure value was applied (350 MPa), a reduction of 80% was noticed. That phenomenon is not new and some enzymatic activities are enhanced after high pressure treatment [49]. High-pressure treatment also produced a decrease of the cytoplasmatic pH, indicating some damage on the regulation of internal pH occurred mainly due to the decrease on ATPase enzyme activity. No obvious changes in the morphology of pressure-treated cells (250 MPa) could be observed, except some DNA clustering, however, almost 85% of the cells survived the treatment indicating clustering phenomenon could be reversible. Later on, Molina-Gutierrez [43] performed a study to understand better the changes in intracellular pH of *L. plantarum* strain after HPP. It is hypothesized that high pressure treatment may affect the internal pH of bacteria by enhancing the dissociation of acids, increasing the permeability of cell membrane and inactivation of enzymes required for pH homeostasis such as ATPase. The ability of *L. plantarum* to maintain the pH_{in} value was recorded in different solutions at different pH values. *L. plantarum* was found to be moderately baroresistant and treatments at 300 MPa at pH 4.0 and 5.0 for 20 min produced between 4.0-5.0 log reductions whereas at hig, efficacy was reduced and longer times were required (5.0 and 3.0 log reductions after 120 min) were achieved at pH 6.0 and 6.5, respectively. In this case, *L. plantarum* was more sensitive to pressure at acidic pH values. Sublethal injury of 90-99.9% (1-3 log) was estimated after HPP treatment at pH 6.0 and 6.5, however, no injury was produced at pH 4.0 and 5.0 indicating no cells were able to survive the acidic conditions after processing. During compression (pressure build-up), pHin was reduced reversibly (returning to normal values) indicating that both the membrane permeability and proton pump were affected by pressure. Intracellular pH was maintained after pressure treatment at 200 MPa indicating *L. plantarum* was able to maintain a transmembrane proton potential when moderate pressure was applied. However, at 300 MPa, the microorganism lost the ability to maintain the pHin even after pressure release, indicating the irreversibility of the process. That may be due to pressure inactivation of ATPase because of direct denaturation, dissociation of protein quaternary structure or dislocation from the membrane. In consequence, pressure treatment indirectly narrows the pH external range tolerated by the bacteria.

The higher effectiveness of HPP treatment in acidic environments was also evaluated by Mallidis [44] on spoilage microflora encountered in Greek salads. Among all spoilage microbial species, LAB were found to be highly resistant to HPP treatment and no significant effects were found after 415 MPa for 1.5 min. Inactivation kinetics of *L. plantarum* and *L. brevis* isolated from the salad were studied and D and z values estimated in phosphate buffer at pH 7. Increasing the pressure value reduced the D values indicating a higher inactivation rate. At acidic conditions (citrate buffer at pH 4.2), D values were lower in comparison with neutral pH indicating the additional stress produced by the low-pH environment. However, z values were higher at acidic conditions indicating that *Lactobacillus* spp. are less sensitive to pressure changes in that environment. In practical terms, pH had a more significant effect than pressure on reducing the total processing time. Carreño [45] studied the inactivation of *L. plantarum* inoculated in mandarin juice and treated by HPP. No significant effects on the number of survivors were found when pressure was applied for a holding time of 10 s (120 MPa at 45°C or 250 MPa at 30°C), achieving around 1.0 log cycles after 450 MPa. This shows *L. plantarum* has a significant baroresistance. Holding time was increased to 1 min and no significant differences were found between 15 and 30°C achieving 4.4 log cycles at 400 MPa, whereas at 45°C (final temperature 60°C), 6.1 log cycles were achieved indicating synergism between pressure and temperature. Comparing the efficacy of HPP on buffer and food systems, Park [18] studied the inactivation of *L. viridescens* after HPP (600 MPa at 25 and 45°C) on MRS broth and ham samples. Pressures up to 300 MPa seemed not to affect the number of survivors. However, complete inactivation (8.0 log cycles) was achieved after 600 MPa for 5 min. The efficacy of pressure was enhanced by increasing the processing temperature and lethality was increased by 4 log cycles from 20 to 40°C and over 8 cycles at 50°C. The shape of *L. viridescens* cells after HPP treatment at 400 MPa had a nodulation effect, increasing the number of nodes at 600 MPa. Empty cavities between the cytoplasmatic membrane and the outer cell wall were also observed showing enlarged fibrillar regions of denatured DNA at 600 MPa. When the microorganism was inoculated in ham, the inactivation degree was reduced in more than 50% in comparison with MRS broth. Differences in the pressure transfer between the broth and the ham systems where the ham had lower water activity or the protective effect of proteins and other composition constituents of ham could be the main reasons for the reduced efficacy. However, by increasing the temperature of the ham sample to 45°C, the authors achieved a similar inactivation degree than that obtained in the culture broth.

A strategy with promising results to enhance the effectiveness of HPP treatment is the combination of pressure with subzero temperatures. In this sense, Perrier-Cornet [46] studied the inactivation of *L. plantarum* after a combination of pressure (100-300 MPa) and temperatures from -20 to 25°C. Lower temperatures favored the inactivation and at -20°C pressure value of 150 MPa was enough to reduce completely the initial microbial load (8 log cycles). A comparison was made between pressure treatment at ambient temperature (300

MPa at 25°C) and subzero temperature (200 MPa at -10°C) for different holding times (1.5-10 min) showing that increasing holding time favored inactivation reaching 8 log reductions after 8 min and after that inactivation rate grew slowly. Some significant differences were found between the two temperature ranges having the subzero range a faster rate during first 5 min but after that no differences were found. The main reason for the synergy between pressure and subzero temperatures was based on the parallels between water and biological activity of the microorganism because most of the intracellular biochemistry and the cell's metabolism is bound to water and water compression will affect protein denaturation and osmotic equilibrium.

Several studies have tried to shed more light on the mechanisms of pressure inactivation of bacteria. In that sense, two studies performed by the same authors [27,47] used flow cytometry to assess the main cellular sites affected by pressure (metabolism and enzyme activity) of *L. rhamnosus*. Applying pressures up to 400 MPa, no evidence of killing effects were found, however differences were observed in the physiological activity indicating cells were sublethally injured mainly due to perturbation of cellular transport. Loss of cell viability was not correlated with an absence of metabolic activity (mainly from the esterase activity) and loss of membrane integrity whereas non-cultured cells still maintained metabolic activity. The damage and loss of cellular transport system seems to be underlying pressure inactivation. The fact that inactivated bacteria still had metabolic activity and membrane integrity was not fully compromised, may be critical in terms of their potential activity on excreting toxins or food spoiling metabolites in food during shelf-life.

The resistance of *L. plantarum* to different concentrations of ethanol and hop extract that may lead to the beer spoilage has been mentioned earlier in this chapter. Ganzle and Ulmer [16,48] studied the combined effect of ethanol (5 and 10%), hop concentration (50 and 100 ppm) and pressure (300 MPa) on the inactivation of *L. plantarum* strain isolated from beer. Among 90-99.9% (1-3 log) of population was injured. Using a model beer (without ethanol and hop) and pressure treatment (300 MPa for 5 min) 1 log cycle of cells were inactivated and 1 log cycle injured. Storage of pressure-treated cells in the model beer and beer with 5% of ethanol had uninjured cells unaffected but injured cells were inactivated after 48 h in cold storage. No differences in final counts after the storage were observed among the two beers samples indicating ethanol did not affect either viable or injured cells. Hop presence (beer with hop extract) had a more remarkable effect on the survival of pressure-treated cells during storage, reducing 90% of viable cells in 3 h and dropping below detection level after 27 h. Pressure treatment was noticed to reduce the capability of LAB to survive under hop bitter presence. Different actions of mode were observed among ethanol and hop extract compounds. Ethanol was synergetic on the effects of pressure but had no effect during storage, while on the contrary, hop extract had a less pronounced effect during HPP but exerted more lethality during storage. Ethanol is a membrane-active solvent and the microorganism adapts to it by changing membrane fluidity and pressure affects the crystalline phase of membrane. In case of antimicrobial effects of hop compounds is based on the ability to penetrate on cytoplasmatic membrane, pressure has been shown to facilitate the penetration of inhibitors across the cytoplasmatic membrane inactivating the HorA activity so loss of hop resistance is an indirect effect of membrane damage on HorA activity. Authors concluded that HPP treatment involves a two-step process, initial sublethal damage being the cytoplasmatic membrane and membrane-associated transport the main targets, followed by cell death as a result of loss of resistance to adverse environmental conditions.

Regarding the viability, metabolic activity, membrane integrity and functionality of hop resistance mechanisms of *L. plantarum* after HPP treatment, Ulmer [17] observed *L. plantarum* cells underwent sublethal injury after treatment but still conserved almost all of the metabolic activity. Cell death was correlated well with metabolic activity with moderate pressures, however, in treatments able to inactivate 4 log cycles (200 MPa, 60 min; 400 MPa, 1 min; and 500 MPa, 0 min), full metabolic activity was observed indicating that loss of metabolic activity was not the main cause of cell death by HPP. Regarding membrane integrity, treatments of 200, 400 and 500 MPa resulted in 4-6 log cycles inactivation and reducing 50-75% of membrane integrity. In case of 200 MPa, between 50-75% of membrane integrity was still found, however, cells were not damaged or inactivated. That observation indicates that membrane damage may occur prior to cell death. Regarding the HorA activity related to the capability to growth in the presence of hop extract compounds, a pressure treatment of 600 MPA was able to reduce completely the enzyme activity making *L. plantarum* unable to survive in environments with hop presence. In conclusion, the authors stated that pressure treatment was found to involve several steps. Injured cells by mild-pressure values were able to increase the membrane permeability and reduce HorA activity, followed by a loss of cell viability and metabolic activity being membrane damage later than cell death. This may compromise the ability of LAB to survive in stressful environments.

High Pressure Homogenization (HPH)

Homogenization is a well-known fluid mechanical process that involves the subdivision of particles or droplets into micron sizes (0.2-2 μm) by passing the fluid through a minute gap in a homogenization valve, creating a stable dispersion or emulsion [50]. Homogenization at relatively low pressure values (50 MPa) has been extensively used in many industries such as pharmaceutical, cosmetic, chemical and dairy industries (to diminish the fat globule size) [51]. Recently developed robust homogenization equipment has been able to reach pressure values of up to 400 MPa, the so-called dynamic high pressure (DHP) or high pressure homogenization (HPH) [52]. Unlike HPP technology, HPH is a continuous process with treatment time lasting several seconds, which makes it more suitable for liquid food processing.

The microbial inactivation mechanism exerted by HPH is believed to be due to the mechanical destruction of cell integrity, caused by different mechanisms such as spatial pressure and velocity gradients, turbulence, impingement and cavitation, which occur in liquids during high pressure homogenization [53]. The UHPH system consists of two positive displacement high pressure pumps in series with a two stage homogenization valve. A manometer connected to the circuit displays the pressure level [54]. A flow through micrometric gap between the valve and the seat causes the acceleration of the fluid in concurrence with the pressure drop, generating significant mechanical stress, which causes a significant temperature increase depending on the fluid viscosity (estimated as 1.5-2.5°C per 10 MPa) [55].

L. plantarum strain has been used by several authors to assay the effectiveness of pressure homogenization on microbial inactivation in citrus juices [42,45,56,57]. Wuytack [42] studied the effectiveness of HPH treatment on inactivating several Gram-positive (including *L. plantarum*) and Gram-negative strains. A remarkable distinction between Gram-positive and Gram-negative strains resistance was observed with the latter being more susceptible to pressure and thus achieving greater microbial reductions (less than 1.0 log versus 2.0-4.6 log). As stated in HPP treatment, the thicker peptidoglycan cell wall of Gram-positive is recognized to be the main reason for the observed differences. No sublethal injury was observed after the treatment reflecting the 'all or nothing' nature of HPH technology where bacteria are disrupted and dead or not disrupted which are undistinguishable from untreated cells. The phenomenon observed of no injured cells after the homogenization process could mean the inability of HPH technology to be combined with other barriers (low pH, high salt concentration or antimicrobials) as a hurdle approach. Tahiri [56] compared the effects of homogenization pressure (single and several passes) on several spoilage microorganisms (*Leuconostoc mesenteriodes*, *Saccharomyces cerevisiase*, *Pencillium* ssp. and *L. plantarum* ATCC 14917) and *E. coli* O157:H7, inoculated in phosphate buffer and orange juice. *L. plantarum* was found to be the most sensitive strain to HPH treatment in the buffer and 6 log cycles were obtained after 5 passes at 100, 150 and 200 MPa. The differences observed among the microorganisms studied showed the strain-dependent of homogenization pressure lethal effect. The differences on the composition of the cell wall seem to be again the main reason for the observed differences. The effects of HPH on cell morphology of *L. plantarum* was observed by electron microscopy showing severe damage resulting in complete cell rupture and loss of cytoplasmatic contents. An interesting finding was the effect of initial microbial load on the effectiveness of the treatment. Inoculating high initial concentrations of *L. plantarum* (10^7 CFU/mL), the effectiveness of HPH treatment was reduced in comparison with lower concentrations with an explanation of a protective effect against pressure when high-density populations are found in contrast to low-density populations where cells are more exposed to pressure and its dynamic effects. The behavior of *L. plantarum* against homogenization pressure in orange juice was different and a lower inactivation level was achieved in comparison with phosphate buffer (2.3 versus 6.0 log cycles after 5 passes at 200 MPa). The protective effect of some constituents of orange juice (free sugars) or the better acid tolerance of lactic acid bacteria seems to be the main reasons for that behavior. Campos [57] obtained more than 7.0 log cycles of inactivation in a *L. plantarum* strain (CCT 2552) inoculated in orange juice after 250 or 300 MPa achieving a maximum temperature at the homogenizer outlet of 84°C (residence time 0.7 s) before being cooled to 30°C. The combination of high pressure and the inherent high temperature values produced could explain the high microbial inactivation level observed in the study. Pressure values below 200 MPa had a slight effect on *L. plantarum* survival demonstrating the moderate baroresistance. Carreño [45] inoculated *L. plantarum* strain (CECT 220) in mandarin juice and subjected to HPH treatment (pressures up to 150 MPa). Due to the lower pressure values applied in this study, the inactivation level achieved (150 MPa, 30°C, 2.4 log cycles) was less significant than other studies. The authors demonstrated synergetic effect between temperature (30°C inlet, 46°C outlet) and pressure (150 MPa) compared to thermal processing alone (46°C did not produce any effect on microbial reduction) and homogenization pressure alone (only 1 log cycle was reduced after 150 MPa, 15°C inlet-30°C outlet temperature). The authors also applied several predictive models based on the Weibull equation to explain the HPH inactivation knetics [58, 59]. The studies showed a general trend that higher pressure values exert greater inactivation; however, it seems a threshold pressure value needed to be overpassed to induce microbial inactivation.

High-Pressure Carbon Dioxide (HPCD)

The CO_2 inhibitory effect on the growth of microorganisms and its combination with pressure has been known for more than 100 years [60]. However, the application of pressurized CO_2, so-called high-pressure carbon dioxide (HPCD), is relatively new [61]. HPCD processing consists of contacting food with either (pressurized) sub-or supercritical CO_2 for a certain amount of time in a batch (using a pressure vessel), semi-batch or continuous manner. Besides the environmentally benign nature of the HPCD process (CO_2 is non-toxic and easily removed by simple depressurization and degassing), the CO_2 pressures applied for preservation purposes are much lower (generally 20 MPa) as compared to the pressures employed in HPP [62].

Pressure and temperature are the main factors affecting microbial inactivation. The mechanism of microbial inactivation by pressurized CO_2 is exerted by the penetration of CO_2 into the cells, lowering the internal pH and exceeding the buffer capacity of the cytoplasm pool as LAB bacteria are known to have tolerance to acidic environments for long periods. The pH lowering effect cause the inactivation of key enzymes involved in the metabolic activity of the cells [63]. Furthermore, when CO_2 is dissolved in water, carbonic acid is produced, which also increases cellular acidity decreasing the intracellular pH, affecting metabolic activities. Higher temperatures stimulate the diffusivity of CO_2, and can also increase the fluidity of the cell membrane to make penetration easier [64]. However, in contrast to other processing technologies, CO_2 seems not to disrupt the cell membrane but extracting vital hydrophobic compounds and phospholipids from it which cause finally the cell death because of the alteration of the biological system [65].

Regarding the effects of pressurized CO_2 on *L. plantarum* inactivation kinetics using batch systems, Hong [63,65] inoculated a strain isolated from a vegetable fermentation into MRS broth (pH 4.5) and phosphate buffer (pH 7.0) and studied the combined effect of pressure (2-14 MPa) and temperature (25-45°C). Increasing the pressure accelerated the inactivation rate lowering the time needed to achieve 5.0 log reductions from 50-55 to 15-20 min at 7 and 14 MPa, respectively. The main reason of the increasing inactivation rate was the enhancement of solubilization of CO_2 and thus its penetration into the cells. An increasing of temperature also produced greater lethality decreasing the time to achieve 5.0 log reductions from 27 to 5 min at 25 and 45°C, respectively. As stated earlier in this section, an increase of temperature facilitates the CO_2 mass transfer and increase the permeability of cell membrane enhancing the CO_2 lethality effects. Later on, the same authors, Hong [66], performed a study on the effects of pressurized CO_2 on the morphology of *L. plantarum* cells. Microbial cells underwent sublethal injury after the treatment (90% of the population). A good indication of cellular damage is the leakage of cell components into the environment which are UV-absorbing materials. Leakage was analyzed by absorbance measure. The authors found an increase in the absorbance at 260 and 280 nm after the treatment and loss of Mg and K ions indicating cell membrane damage and leakage from cell lysis. The proton efflux was also studied after HPCD treatment by measuring the time to a complete pH equilibration and the difference between cell internal and external pH. Both parameters were modified after CO_2 treatment indicating that the cytoplasm or at least its barrier function was impaired. ATPase activity was not influenced by CO_2, however, some other enzymes were inactivated such as cystine arylamidase, α-galactosidase, β-glucosidase, and N-acetyl-β-glucosaminidase. Authors concluded that membrane damage after CO_2 included loss of salt tolerance, leakage of intracellular substances and collapse of proton gradient. However, glycolysis activity nor ATPase were primary sites for cell death. Cell inactivation probably resulted from damage to the cell membrane and the lowered internal pH because of the penetration of CO_2.

Lately, Yuk [67] studied the inactivation of a *L. plantarum* strain (ATCC 49445) in apple cider using a pilot-plant scale continuous supercritical CO_2 system. The system consisted of a gas-liquid metal contactor. The microbial inactivation was dependent on CO_2 concentration and outlet temperature and an increase on both processing parameters induced greater lethality. For instance, the authors showed an increase from 1.5 to 2.9 log cycles inactivation as CO_2 concentration was increased from 5 to 10% and inactivation increase from 0.7 to 5.2 log cycles when temperature was increased from 34 to 42°C at 8% CO_2. Injured cells among the population were greater as the CO_2 concentration and temperature increased reaching maximum injury of 98%. Scanning electron microscopy (SEM) analysis was applied to pressurized cells to study morphology changes. Authors observed cells with pimples after processing with 5% CO_2 and uneven surfaces and deformed structures after 10 and 12% CO_2. This fact demonstrates *L. plantarum* undergoes changes in cell morphology during HPCD processing. A higher resistance of *L. plantarum* in comparison with *E. coli* regarding the inactivation level achieved and greater morphology changes to the cell membrane was noticed. During storage of processed apple cider (8% CO_2 at 42°C) for 28 d, no significant changes were observed in *L. plantarum* population at 4°C whereas at 8 and 20°C, 6.0 and 6.6 log CFU/mL were reached after 28 and 7 d, respectively. The authors recommended the storage temperature to be set at 4°C to ensure a shelf-life of apple cider of at least 28 d.

High-Voltage Pulsed Electric Fields (PEF)

For many years, high intensity electric fields have been applied to induce electroporation, a phenomenon used to promote bacterial DNA interchange by perforating microbial membranes [68]. Recent advances in the design of the treatment chamber led to the development of a new processing technology known as pulsed electric fields (PEF) [69]. The principle of PEF technology is the treatment of a biological material or food placed between two electrodes installed 0.1-1.0 cm apart in a treatment chamber separated by an insulator with short pulses (1-10 μs) that are generated by a high voltage (5-20 kV) pulse generator. A power source is used to charge a capacitor bank and a switch is used to discharge energy to the treatment chambers [68]. Two principle types of chambers are used in the PEF technology, continuous chambers for pasteurization of liquid food and static chambers more suitable for studies in solid food (extraction of bioactive compounds) [28]. New designs for uniform electric field distribution and geometry such as the co-field continuous chamber, applying the electric field in the same direction as the fluid flow, have made it possible to apply a uniform electric field in large-scale equipment. The main processing parameter is the electric field (kV/cm) which is defined by the relation between the electrical potential difference applied to two electrodes and the distance between them. The main pulse waveform used in a PEF treatment is the square wave which can be applied in monopolar or bipolar mode [70]. Treatment time (μs) is defined by the pulse width (μs), the pulse frequency (Hz), flow rate and chamber dimensions. An oscilloscope is used to monitor voltage, current and pulse width during the treatment. PEF treatment produces a temperature rise due to ohmic heating and a cooling system is used to minimize thermal effects.

Most of the studies on PEF microbial inactivation agree that high voltage treatment produces a series of structural and functional changes in the cellular membrane that lead to microorganism death [71]. The well-known theory exposed by Zimmermann [72] states that when an external electric field is applied to a system, it induces a potential difference across the cellular membrane. The maximum potential difference that the membrane can withstand is known as the 'critical transmembrane potential' or 'critical electric field' (Ec). When the external electric field exceeds Ec, membrane breakdown occurs leading to the formation of pores. The magnitude of Ec principally depends on the membrane characteristics of the microorganism (Gram-positive and Gram-negative), size and shape (smaller cells and rod-shaped cells require higher Ec values) [73-75]. In general, 'reversible' pores are formed under conditions where the electric field is lower than Ec, and the membrane returns to its initial form after the treatment. These pores may cause certain sublethal damage leading to cellular death in stress conditions such as low pH and refrigeration temperature. On the contrary, if the electric field is much higher than Ec, 'irreversible' pores are produced leading to the death of the microorganism. Besides structural changes in the membrane, other changes have been observed inside the cell such as leakage of intracellular material and alterations in cell proteins [76, 77].

Different predictive models [78-81] have been used to characterize the PEF inactivation kinetics of *L. plantarum* by [82-84]. *L. plantarum* CECT 220 showed an inactivation of 2.5 log after 36 kV/cm and 46 μs in orange and carrot juice and orange juice-milk beverage. Different predictive models were used to fit the inactivation data with the Weibull equation [78], the one that produced the most accurate fit and the parameters (a or α) to be an index of the microbial resistance to PEF (the parameter decreased when electric field strength and treatment temperature increased). Exponential and stationary growth stage cells were compared in their resistance to PEF treatment showing that exponentially grown cells were more sensitive to high voltage pulses due to the characteristics of their membranes and critical molecules and structures that seem to be unprotected to different stresses. Gomez [85] used different predictive inactivation models to characterize the PEF inactivation kinetics of *L. plantarum* in a buffer system at different pH values (3.5, 5.0, 6.5 and 7.0). Microbial inactivation was influenced by the pH and higher inactivation was found in more acidic environments. At 22 kV/cm for 400 μs, around 5.0, 2.8, 2.6 and 1.1 log cycles were reduced in the buffer system at pH 3.5, 5.0, 6.5 and 7.0, respectively. The Weibull equation [78] was chosen as the primary model to obtain the kinetic parameters that characterize the behavior of *L. plantarum* against high voltage pulses. After obtaining the primary kinetic parameter (α), some secondary models were applied to establish the relationship between α and electric field and pH. Once that relationship was established, an empirical tertiary model based on the Weibull equation was developed to estimate the final number of microorganisms after PEF treatment changed the pH and electric field strength. The model was validated with data obtained experimentally in orange and apple juices where the model underestimated the inactivation when inactivation degree was higher than 3.0 log cycles. Different predictive models were also used by Selma [86] to characterize the inactivation and growth of *L. plantarum* CECT 220 in different substrates (MRS broth, nutrient broth and orange-carrot juice) after PEF treatment (22-28 kV/cm, 115-70 μs). Both treated and untreated cells were unable to grow at 5 and 8°C in any substrate. The level of inactivation achieved by PEF treatment was 1.3 log cycles but despite the low reduction level achieved, the growth after processing was compromised. Growth rates were not affected by processing but lag times (time to growth) were prolonged indicating some population was injured and longer time required for reparation and adaptation of the cells to the medium. No significant differences were observed among the different PEF treatments in MRS broth at incubation temperatures of 10, 12 or 16°C but lag phase after processing was prolonged from 20 h to 12 d when incubation temperature changed from 16 to 12°C. Cells were unable to grow in the orange-carrot juice at 10°C, indicating the medium does not support growth and lag phase was more prolonged than MRS broth. Authors concluded that the combination of substrate, PEF and low temperature would inhibit *L. plantarum* growth for at least 60 d. The effects of inoculum size (10^3 or 10 CFU/mL) were also studied on growth parameters on both MRS broth and orange-carrot juice. Lag phase was reduced when increasing inoculum size, meaning that population size determined the time needed for the injury to be repaired. However, growth rate was not affected by inoculum size. As stated earlier in this chapter, *L. plantarum* is a spoilage bacteria of beer due to its resistance to high levels of ethanol and hop extract. In this sense, Ulmer [48] studied the effects in cell viability and metabolic changes on *L. plantarum* strain isolated from beer. Electric field values from 10-19 kV/cm did not produce significant effects on cell viability and no sublethal injury was noticed. Increasing electric field values and energy input metabolic activity and membrane integrity were compromised following a linear relationship with cell viability up to 95%. Hop extract (100 ppm at pH 3.6) was combined with PEF treatment (29, 32 and 35 kV/cm) achieving 4.0 log reductions (1.5 log cycles more than the PEF-treated sample) with 99% of injury cells. However, the addition of hop extract did not have any effect on metabolic activity or cell integrity. Authors also performed a combined treatment of PEF and nisin (1 mg/L) increasing the inactivation by 1.5 log cycles achieving 5 log cycles with no sublethal injury. Nisin acted as a pore forming substance integrating it into membrane and dissipating transmembrane gradients. During the storage at 10°C, cells treated by PEF in presence of hop extract and sublethally injured were inactivated within 6 h of storage. This study supported the idea of PEF processing as a "all or nothing" technology with the absence of injury. In contrast, many authors have observed injured cells.

Sampedro [87] assessed the effects of PEF processing parameters (electric field strength, pulse width, energy applied and temperature) on the inactivation of *L. plantarum* in an OJ-milk beverage. For a given quantity of energy applied, the highest inactivation was achieved with high field intensities and short treatment times. An increase in pulse duration did not produce an increase in inactivation independent of temperature and electric field strength applied. When the process temperature was raised to 55°C, the inactivation increased by 0.5 $\log_{10}$, achieving an energy saving of 60%. Wouters [88] studied the membrane permeabilization degree after PEF treatment applying several electric fields (12, 15 and 25 kV/cm) and treatment time varying the energy applied. Higher E values and number of electric pulses induced more permeabilization and inactivation in a linear relationship. After treatment, cells were incubated for 5 h at room temperature or ice water without decreasing the number of permeabilized cells indicating the irreversibility of the membrane damage produced by PEF. The number of permeabilized cells (70%) increased as conductivity of the medium decreased from 1.5 to 0.4 S/m. That observation was true at a constant energy input. However, in a low-conductivity medium, more pulses were needed to be applied to obtain the same energy value so in that case, no differences were observed when treatment time was compared. Low-conductivity substrates are more advantageous to be treated by PEF due allowing more pulses to be applied. Similar membrane permeabilization was observed in mediums with different pH (4.5 and 6.8), however, inactivation was greater in the more acidic environment. Presumably, membrane permeabilization created an influx of H^+ protons and as a result of that, cells in the low-pH environment were more rapidly inactivated. Exponential growth stage cells were more easily inactivated than stationary cells (1.7 vs. 0.5 log) but having the same degree of membrane permeabilization meaning that other phenomenon may be involved in cell inactivation. It is known that cells in a stationary stage adapt better to ensure they survive under stress conditions. Authors also found cell morphology play an important role in membrane permeabilization and inactivation degree. However, treatment time is also related and differences disappeared when treatment times are long enough. No correlation was found between temperature and membrane permeabilization meaning temperatures reached during PEF treatment were not responsible for permeabilization. Finally, *L. plantarum* behavior was compared with a *L. fermentum* strain isolated from tomato paste finding the tomato isolate was more PEF-resistant (0.3 vs. 3.4 log) showing less membrane permeabilization. Unal [89] studied the membrane damage after PEF treatment on *L. leichmannii.* The *Lactobacillus* strains was found to be more resistant than *E. coli* O157:H7 and L. monocytogenes with a reduction of 1.5 log cycles after 20 kV/cm. Membrane damaged measured as an increase of PI fluorescence was correlated well with cell death and an increase in inactivation produce an increase of fluorescence degree. This observation supports the theory of death of cell by PEF is caused by membrane damage.

Abram [90] studied the influence of processing parameters (electric field (E), pulse width and number of pulses) on the inactivation of *L. plantarum.* Authors found that at constant treatment, time and field strength and a longer pulse width produced greater microbial reduction. On the other hand, the effect of the number of pulses seemed to depend on the pulse length, at 1 μs of pulse width increasing number of pulses slightly increased inactivation whereas when 5 μs was applied, no further reduction was achieved. This affirms the idea of including pulse width and number of pulses as parameters when designing a PEF treatment. Higher field strength caused greater inactivation at constant energy input (10 J/mL) and any pulse width. Authors optimized the PEF treatment to obtain the highest lethality with the smallest energy input. 5.8 log cycles were achieved after 5.1 μs of pulse width, E of 35.8 kV/cm and 3.3 pulses or 39.5 kV/cm and 2.6 pulses with an energy input of 23 J/mL. Elez-Martinez [91] inoculated *L. brevis* in orange juice and assessed the effects of PEF processing parameters (electric field (E), treatment time (t), pulse polarity, pulse width and frequency). A maximum reduction of 6 log was achieved after the combination of 1000 μs using 4 μs of pulse width, 35 kV/cm, 200 Hz in bipolar mode with an energy input of 5200 J/mL. The most significant factors on microbial inactivation were E and t whereas pulse polarity was least significant. Pulse polarity effects (monopolar and bipolar) were dependent on E. No statistical differences were found on pulse polarity up to 30 kV/cm, but applying E of 35 kV/cm, bipolar mode was more lethal than monopolar mode. High pulse frequency (50-350 Hz) was found to be detrimental of the microbial reduction being the lower pulse frequency more effective. Increasing the pulse width (1-10 μs) decreased the effectiveness. Shorter pulses may be considered more effective in microbial inactivation. Authors also analyzed morphological changes after PEF treatment and treated cells developed a rough surface and some cells were shrunken. They also exhibited thinner or ruptured cell walls and disorganized cytoplasm with some cytoplasmatic material leaked out of cells.

Other Non-Conventional Technologies

Power ultrasound, also known as high-intensity ultrasound, refers to sound waves with low frequencies (20–100 kHz) and high sound intensity (10–1000 W/cm2) [94]. The use of ultrasonics in industrial processes has two main requirements; a liquid medium and a source of high-energy vibrations (the ultrasound) [95]. During the sonication process, longitudinal waves are created, thereby creating regions of alternating compression and expansion [96]. These regions of pressure change are the main bactericidal effect in ultrasound, creating micromechanical shocks that disrupt cellular, structural and functional components up to the point of cell lysis [97].

Food irradiation is a physical means of food processing that involves exposing the pre-packaged or bulk foodstuffs such as tuber and bulb crops, stored grains, dried ingredients, meat, poultry, fish and fruits to gamma rays, X-rays, or electrons [28]. Foodstuffs are generally irradiated with gamma radiation from a radioisotope source, energetic electrons from particle accelerators and X-rays emitted by high-energy electron beams [98]. The potential application of ionizing radiation in food processing is based mainly on the fact that ionizing radiation damages the DNA very effectively so that living cells become inactivated, therefore microorganisms, insect gametes, and plant meristems are prevented from reproducing, resulting in various preservative effects as a function of the absorbed radiation dose [99]. The main process parameter in the irradiation process is the absorbed dose which is proportional to the ionizing energy absorbed per unit mass of irradiated material measured in gray (Gy) or 1 J/kg [98]. Radiation treatment causes practically no temperature rise in the product and can be used as a terminal treatment in the packaged product [100]. In this sense, Ye [21] performed a combined process of ultrasound and ^{60}Co-γ irradiation on the inactivation of a sporulated microorganism (*L. sporogenes*) in tomato paste. Irradiation dose was the most significant factor on microbial inactivation and thus the inactivation rate increased with increasing irradiation dose. Since irradiation may produce some adverse effects on quality of food, generating off-flavors, it is interesting to reduce its dose to mitigate those effects. Ultrasound power was found to be the second most significant parameter, whereas ultrasonic time had no significant effect. Optimum processing parameters to maximize the microbial reduction and minimize the irradiation dose were 120 W of power, 25 min and 6.5 kGy. However, authors concluded that ultrasound had a negligible effect on reducing the irradiation dose.

Hurdle Approach

All non-conventional technologies cited in this chapter have several drawbacks. In general, high production costs, the unavailability of industrial-scale equipment, the inability to reduce spores and the lack of effective communication to the consumers about the principles and benefits of these technologies are the main limitations. Taking into account all of these limitations, the hurdle approach has been proposed as the most efficient way to improve overall effectiveness. The combination of temperature and/or low pH with other stress factors such as pressure, electric field or radiation has been demonstrated to have a synergetic effect on microbial inactivation in numerous studies. Lowering pH, oxygen concentration and water activity during storage have also been used as a barrier control to the growth of microorganisms and preventing the recovery of any sublethally injured cell during storage. Following this approach, Unal [92] performed a combined treatment of ozone and high-voltage electric fields on the inactivation of *L. leichmannii*. Ozone usually acts rapidly with bacteria producing the oxidation of lipids of the cell envelope, leakage of intracellular contents and finally cell death. However, when used in food, several drawbacks have been detected such as organic components competing with bacteria for ozone and high doses being necessary to achieve a significant microbial reduction. In addition, the ozone doses used in food could cause quality changes and finally altering the sensory properties. Authors found ozone and PEF processing to act synergistically, ozone alone (0.75 and 1.0 μg/mL) produced 2.2 and 3.6 log reductions, respectively, whereas PEF alone (10-30 kV/cm) led to 0.7-7.8 log cycles. A combination of PEF (20 kV/cm) and ozone (0.75 and 1.0 μg/mL) decreased the counts by 7.1-7.2 log cycles. Synergy with ozone was found to be greater at lower than higher doses. This enhancement on the overall effectiveness may result from cell injury during ozone treatment and rapid inactivation of injured cells when treated by PEF. The combined effects of high pressure and CO_2 were studied on *L. plantarum* after inoculation in orange juice [93]. No significant effects were found after the addition of CO_2 (carbonated and non-carbonated samples) in terms of microbial reduction whereas hydrostatic pressure alone was able to reduce 4 log cycles after 455 MPa. However, authors found synergetic effects when CO_2 was combined with pressure achieving more than 5 log reductions. The main reason for that synergism was explained by the fact that CO_2 becomes supercritical when pressurized enhancing its penetration into cells. The subsequent decompression results in a rapid gas expansion physically destroying the cells.

The use of biopreservatives in combination with non-thermal technologies have also been proposed to enhance the effectiveness of these processes against vegetative cells and bacterial spores and help to control microbial growth during storage under chilled conditions. In this sense, bacteriocins and natural antimicrobial compounds such as nisin, lysozyme, lactoferrin, lactoperoxidase system, chitosan and essential oils from diverse plants have been successfully applied. In relation to that, Martinez [20] studied the combination of PEF and enterocin AS-48 (produced from the growth of *Enterococcus faecalis*) on the inactivation of an apple cider spoilage lactobacilli (*L. diolivorans*). Enterocin and PEF acted synergistically and 4.9 log cycles were achieved after the combination of 1000 μs, 35 kV/cm and 2.0 μg/mL of AS-48. Increasing the treatment time increased the log reduction by 1.2 log whereas increasing enterocin from 0.5 to 1.25 μg/mL had very low effect. After treatment, cells were stored at 4 and 22°C for 15 d. At 4°C viable counts of sample with enterocin (2 μg/mL) were reduced by 1.5 log and population increased slowly after storage whereas at 22°C, 3.0 log reductions were observed just after the antimicrobial addition but the surviving fraction reached similar levels than the control sample. Regarding PEF alone (1000 μs, 35 kV/cm), cell reduction accounted for 3.0 log and surviving population did not increase during storage at 4°C whereas at 22°C a slower recovery was observed during storage. Combined PEF and antimicrobial addition led to a reduction of 2.2 log and viable population was completely reduced within the first 24 h with no recovery during 15 d with a similar trend at 22°C observing no recovery for 15 d. The synergism between the enterocin and treatment with high electrical pulses was translated into a reduction of the enterocin concentration or treatment time to achieve a certain inactivation degree. The phenomenon underlying the synergism is explained by the fact that AS-48 compound increases the permeability of cell membrane leading to a complete collapse of motive force and finally, cell death. Bacteriocin could also enhance the pore formation and disruption by PEF by previously sensitization the membrane or preventing the resealing of pores. High-voltage pulses could also facilitate the penetration of antimicrobial compound into the cell.

Conclusion

Because of the limitations of traditional thermal and nonthermal technologies, the future of food preservation is moving towards a-la-carte processing that involves designing a specific treatment for each type of food. The scientific community will continue to search for new developments and applications of non-conventional technologies to fulfill the new challenges of the 21st century with an emphasis on the environmental impact and sustainability of these technologies and the use of non-conventional technologies to produce healthier and convenient foods for high risk population groups such as the elderly, children and the inmunocompromised consumers.

It is the author's belief that in the near future new industrial applications of non-conventional technologies to produce healthy, fresh-like and natural foods in a sustainable manner will become common in the market place.

References

[1] Liu, S.Q. Practical implications of lactate and pyruvate metabolism by lactic acid bacteria in food and beverage fermentations. *Int J Food Micro* 2003; 83 (2):115-131.

[2] Drosinos, E.H., and Paramithiotis. Research trends in lactic acid fermentation, In: *Microbiology Research Trends*, Eds. Palino, MV, 2007, pp. 39-92, Nova Science Publishers, New York, USA.

[3] Adams, M.R., and Nicolaides, L. Review of the sensitivity of different foodborne pathogens to fermentation. *Food Cont* 1997;8 (5/6):227-239.

[4] Jones, R.J., Hussein, H.M., Zagorec, M., Brightwell, G., Tagg, J.R. Isolation of lactic acid bacteria with inhibitory activity against pathogens and spoilage organisms associated with fresh meat. *Food Micro* 2008; 25(2):228-234.

[5] O'Sullivan, L., Ross, R.P. and Hill, C. Potential of bacteriocin-producing lactic acid bacteria for improvements in food safety and quality. *Biochimie* 2008; 84(5-6):593-604.

[6] Trias, R., Bañeras, L., Montesinos, E., Badosa, E. Lactic acid bacteria from fresh fruit and vegetables as biocontrol agents of phytopathogenic bacteria and fungi. *Int Micro* 2008; 11:231-236.

[7] Karovicova, J., Drdak, M., Greif, G., Hybenova, E. The choice of strains of *Lactobacillus* species for the lactic acid fermentation of vegetable juices. *Eur Food Res Technol* 1999; 210(1):53-56.

[8] Basak, S., Ramaswamy, H.S., Piette, J.P.G. High pressure destruction kinetics of Leuconostoc mesenteroides and Saccharomyces cerevisiae in single strength and concentrated orange juice. *Innov Food Sci Emerg Technol* 2002; 3(3):223-231.

[9] Hutkins, R.W., Nannen, N.L. pH homeostasis in lactic-acid bacteria. *J Dairy Sci* 1993; 76(8): 2354-2365.

[10] Wouters, P., Glaasker, E., Smelt, J. Effects of high pressure on inactivation kinetics and events related to proton efflux in Lactobacillus plantarum. *Appl Environ Microbiol* 1998; 64(2):509-514.

[11] Parish, M., and Higgins, D. Isolation and identification of lactic acid bacteria from samples of citrus molasses and unpasteurized orange juice. *J Food Sci* 1988; 53:645-646.

[12] Parish, M.E., Sadler, G.D., and Wicker, L. Viability of *Lactobacillus plantarum* in orange juice under low pH and temperature conditions. *J Food Sci* 1990; 55:1023-1025.

[13] Alwazeer, D., Cachon, R., Divies, C. Behavior of Lactobacillus plantarum and Saccharomyces cerevisiae in fresh and thermally processed orange juice. *J Food Prot* 2002;65(10):1586-1589.

[14] Nualkaekul, S., Charalampopoulos, D. Survival of Lactobacillus plantarum in model solutions and fruit juices. *Int J Food Micro* 2011; 146(2):111-117.

[15] Champagne, C.P., Raymond, Y., and Gagnon, R. Viability of *Lactobacillus rhamnosus* r0011 in an apple-based fruit juice under simulated storage conditions at the consumer level. *J Food Sci* 2008; 73:221-226.

[16] Ganzle, M., Ulmer, H., Vogel, R. High pressure inactivation of Lactobacillus plantarum in a model beer system. *J Food Sci* 2001; 66(8):1174-1181.

[17] Ulmer, H., Ganzle, M., Vogel, R. Effects of high pressure on survival and metabolic activity of Lactobacillus plantarum TMW1.460. *Appl Environ Microbiol* 2000; 66(9):3966-3973.

[18] Park, S., Sohn, K., Shin, J., Lee, H. High hydrostatic pressure inactivation of Lactobacillus viridescens and its effects on ultrastructure of cells. *Int J Food Sci Technol* 2001; 36(7):775-781.

[19] Dueñas, M., Irastorza, A., Fernandez, K., Bilbao, A. Heterofermentative lactobacilli causing ropiness in basque country ciders. *J Food Prot* 1995; 58(1):76-80.

[20] Martinez-Viedma P., Abriouel, H., Sobrino-Lopez, A., Ben, Omar, N, Lucas Lopez, R., Valdivia, E., et al. Effect of enterocin AS-48 in combination with high-intensity pulsed-electric field treatment against the spoilage bacterium Lactobacillus diolivorans in apple juice. *Food Microbiol* 2009; 26(5):491-496.

[21] Ye, S., Qiu, Y., Song, X., Luo, S. Optimization of process parameters for the inactivation of Lactobacillus sporogenes in tomato paste with ultrasound and (60)CO-gamma irradiation using response surface methodology. *Radiat Phys Chem* 2009; 78(3):227-233.

[22] Lee, J., Kaletunc, G. Evaluation of the heat inactivation of Escherichia coli and Lactobacillus plantarum by differential scanning calorimetry. *Appl Environ Microbiol* 2002; 68(11):5379-5386.

[23] Tajchakavit, S., Ramaswamy, H., Fustier, P. Enhanced destruction of spoilage microorganisms in apple juice during continuous flow microwave heating. *Food Res Int* 1998; 31(10):713-722.

[24] Jordan, K., Cogan, T. Heat resistance of Lactobacillus spp. isolated from Cheddar cheese. *Lett Appl Microbiol* 1999; 29(2):136-140.

[25] De Angelis, M., Di Cagno, R., Huet, C., Crecchio, C., Fox, P., Gobbetti, M. Heat shock response in lactobacillus plantarum. *Appl Environ Microbiol* 2004; 70(3):1336-1346.

[26] Augusto, P.E.D., Tribst, A.A.L., Cristianini, M. Thermal Inactivation of Lactobacillus Plantarum in a Model Liquid Food. *J Food Process Eng* 2011; 34(4):1013-1027.

[27] Ananta, E., Knorr, D. Comparison of inactivation pathways of thermal or high pressure inactivated Lactobacillus rhamnosus ATCC 53103 by flow cytometry analysis. *Food Microbiol* 2009; 26(5):542-546.

[28] Barbosa-Cánovas, G.V., Palou, E., Pothakamury, U.R., Swanson, B.G. *Nonthermal preservation of food*, 1998, New York: Marcel Dekker Inc. p. 9-20, 53-110.

[29] Cunningham, F.E. The effect of brief microwave treatment on numbers of bacteria in fresh chicken patties. *Poult. Sci* 1978; 57:296-297.

[30] Dreyfuss, M.S. and Chipley, J.R. Comparison of effects of sublethal microwave radiation and conventional heating on the metabolic activity of Staphylococcus aureus. *Appl. Environ. Microbiol*. 1980; 39: 13-16.

[31] Khalil, H. and Villota, R. The effect of microwave sublethal heating on the ribonucleic acids of Staphylococcus aureus. *J. Food Prot* 1989; 52(8):544-548.

[32] Shin, J.K., Pyun, Y.R. Inactivation of Lactobacillus plantarum by pulsed-microwave irradiation. *J Food Sci* 1997; 62(1):163-166.

[33] Mertens, B., Knorr, D. Developments of non-thermal processes for food preservation. *Food Technol* 1992; 46(5):124.

[34] Ke, P.J. Linke, A.B., and Ackm, R.G. Acceleration of lipid oxidation in frozen mackerel fillet by pretreatment with microwave heating. *J. Food Sci* 1978; 43:38-40.

[35] Wilson, D.R., Dabrowski, L., Stringer, L., Moezelaar, R., Brocklehurst, T.F. High pressure in combination with elevated temperature as a method for the sterilization of food. *Trends Food Sci Tech* 2008; 19:289-299.

[36] Heremans, K. The effects of high pressure on biomaterials. In: Hendrickx, M., Knorr, D., editors. *Ultra High Pressure Treatments of Foods,* 2002, New York: Kluwer Academic. p. 23-25.

[37] Rovere, P. Industrial-scale high pressure processing of foods. In: Hendrickx, M., Knorr, D., editors. *Ultra High Pressure Treatments of Foods* 2002, New York: Kluwer Academic. p. 251-255.

[38] Rastogi, N.K., Raghavarao, K.S.M.S., Balasubramaniam, V.M., Niranjan, K., Knorr, D. Opportunities and challenges in high pressure processing of foods. *Crit Rev Food Sci Nutr* 2007; 47:69-112.

[39] Torres, J.A., Velazquez, G. Commercial opportunities and research challenges in the high pressure processing of foods. *J Food Eng* 2005; 67:95-112.

[40] Tonello, C. "All natural" products through high pressure processing. *Non-thermal Processing Division Newsletter* 2008; 10:5-6.

[41] Gervilla, R., Sendra, E., Ferragut, V., Guamis, B. Sensitivity of Staphylococcus aureus and Lactobacillus helveticus in ovine milk subjected to high hydrostatic pressure. *J Dairy Sci* 1999; 82(6):1099-1107.

[42] Wuytack, E., Diels, A., Michiels, C. Bacterial inactivation by high-pressure homogenization and high hydrostatic pressure *Int J Food Microbiol* 2002; 77(3):205-212.

[43] Molina-Gutierrez, A., Stippl, V., Delgado, A., Ganzle, M., Vogel, R. In situ determination of the intracellular pH of Lactococcus lactis and Lactobacillus plantarum during pressure treatment. *Appl Environ Microbiol* 2002; 68(9):4399-4406.

[44] Mallidis, C., Galiatsatou, P., Taoukis, P., Tassou, C. The kinetic evaluation of the use of high hydrostatic pressure to destroy Lactobacilius plantarum and Lactobacillus brevis. *Int J Food Sci Technol* 2003; 38(5):579-585.

[45] Carreño, J.M., Gurrea, M.C., Sampedro, F., Carbonell, J.V. Effect of high hydrostatic pressure and high-pressure homogenization on Lactobacillus plantarum inactivation kinetics and quality parameters of mandarin juice. *European Food Research and Technology* 2011; 232(2):265-274.

[46] Perrier-Cornet, J., Tapin, S., Gaeta, S., Gervais, P. High-pressure inactivation of Saccharomyces cerevisiae and Lactobacillus plantarum at subzero temperatures. *J Biotechnol* 2005; 115(4):405-412.

[47] Ananta, E., Heinz, V., Knorr, D. Assessment of high pressure induced damage on Lactobacillus rhamnosus GG by flow cytometry. *Food Microbiol* 2004; 21(5):567-577.

[48] Ulmer, H., Heinz, V., Ganzle, M., Knorr, D., Vogel, R. Effects of pulsed electric fields on inactivation and metabolic activity of Lactobacillus plantarum in model beer. *J Appl Microbiol* 2002; 93(2):326-335.

[49] Eisenmenger, M.J., Reyes-De-Corcuera, J.I. High pressure enhancement of enzymes: A review. *Enzyme Microb Technol* 2009; 45(5):331-347.

[50] Diels A.M.J., Michiels, C.W. High-pressure homogenization as a non-thermal technique for the inactivation of microorganisms. *Crit Rev Micro* 2006; 32:201-216.

[51] Diels, A.M.J., De Taeye, J., Michiels, C.W. Sensitization of Escherichia coli to antibacterial peptides and enzymes by high-pressure homogenization. *Int J Food Micro* 2005; 105:165-175.

[52] López-Pedemonte, T., Bríñez, W.J., Roig-Sagués, A.X., Guamis, B. Fate of Staphylococcus aureus in cheese treated by ultrahigh pressure homogenization and high hydrostatic pressure. *J Dairy Sci* 2006; 89:4536-4544.

[53] Donsì, F., Ferrari, G., Lenza, E., Maresca, P. Main factors regulating microbial inactivation by high-pressure homogenization: operating parameters and scale of operation. *Chem Eng Sci* 2009; 64:520-532.

[54] Kumar, S., Thippareddi, H., Subbiah, J., Zivanovic, S., Davidson, P.M., Harte, F. Inactivation of Escherichia coli k-12 in apple juice using combination of high-pressure homogenization and chitosan. *J Food Sci* 2009; 74:8-14.

[55] Popper, L., Knorr, D. 1990. Applications of high-pressure homogenization for food preservation. *Food Tech* 44:84-89.

[56] Tahiri, I., Makhlouf, J., Paquin, P., Fliss, I. Inactivation of food spoilage bacteria and Escherichia coli O157 : H7 in phosphate buffer and orange juice using dynamic high pressure. *Food Res Int* 2006; 39(1):98-105.

[57] Campos, F.P., Cristianini, M. Inactivation of Saccharomyces cerevisiae and Lactobacillus plantarum in orange juice using ultra high-pressure homogenization. *Innovative Food Science & Emerging Technologies* 2007; 8(2):226-229.

[58] Mafart, P., Couvert, O., Gaillard, S., Leguerinel, I. On calculating sterility in thermal preservation methods: application of the Weibull frequency distribution model. *Int J Food Microbiol* 2002;72:107-113.

[59] Coroller, L., Leguerinel, I., Mettler, E., Savy, N., Mafart, P. General model, based on two mixed Weibull distributions of bacterial resistance, for describing various shapes of inactivation curves. *Appl Envion Microbiol* 2006; 72:6493-6502.

[60] Liao, H., Hu, X., Liao, X., Chen, F., Wu, J. Inactivation of Escherichia coli inoculated into cloudy apple juice exposed to dense phase carbon dioxide. *Int J Food Micro* 2007; 118:126-131.

[61] Damar, S., Balaban, M.O. Review of dense phase CO_2 technology: microbial and enzyme inactivation, and effects on food quality. *J Food Sci* 2006; 71:1-11.

[62] García-González, L., Geeraerd, A.H., Elst, K., Van Ginneken, L., Van Impe, J.F., Devlieghere, F. Influence of type of microorganism, food ingredients and food properties on high-pressure carbon dioxide inactivation of microorganisms. *Int J Food Micro* 2009; 129:253-263.

[63] Hong, S., Park, W., Pyun, Y. Non-thermal inactivation of Lactobacillus plantarum as influenced by pressure and temperature of pressurized carbon dioxide. *International Journal of Food Science and Technology* 1999;34(2):125-130.

[64] Hong, S., Pyun, Y. Inactivation kinetics of Lactobacillus plantarum by high pressure carbon dioxide. *J Food Sci* 1999; 64(4):728-733.

[65] Hong S., Park, W., Pyun, Y. Inactivation of Lactobacillus sp. from kimchi by high pressure carbon dioxide. *Food Sci Technol* 1997;30(7):681-685.

[66] Hong, S., Pyun, Y. Membrane damage and enzyme inactivation of Lactobacillus plantarum by high pressure CO2 treatment. *Int J Food Microbiol* 2001; 63(1-2):19-28.

[67] Yuk, H., Geveke, D.J. Non-thermal inactivation and sublethal injury of Lactobacillus plantarum in apple cider by a pilot plant scale continuous supercritical carbon dioxide system. *Food Microbiol* 2011 28(3):377-383.

[68] Zhang, Q.H., Barbosa-Canovas, G.V., Swanson, B.G. Engineering Aspects of Pulsed Electric-Field Pasteurization. *J Food Eng* 1995; 25(2):261-281.

[69] Vega-Mercado, H., Martin-Belloso, O., Qin, B.L., Chang, F.J., Gongora-Nieto, M.M., Barbosa-Canovas, G.V., Swanson, B.G. Non-thermal food preservation: Pulsed electric fields. *Trends Food Sci Technol* 1997; 8(5):151-157.

[70] Ho, S.Y., Mittal, G.S., Cross, J.D., Griffiths, M.W. Inactivation of Pseudomonas fluorescens by high voltage electric pulses. *J Food Sci* 1995; 60(6):1337.

[71] Mañas, P., Pagan, R. Microbial inactivation by new technologies of food preservation. *J Appl Micro* 2005; 98(6):1387-1399.

[72] Qin, B.L., Barbosa-Canovas, G.V., Swanson, B.G., Pedrow, P.D. and Olsen, R.G. Inactivating Microorganisms Using a Pulsed Electric Field Continuous Treatment System. *IEEE Transactions Industr Appl* 1998; 34(1): 43-50.

[73] Heinz, V., Alvarez, I., Angersbach, A., Knorr, D. Preservation of liquid foods by high intensity pulsed electric fields - basic concepts for process design. *Trends Food Sci Technol* 2001; 12 (3-4):103-111.

[74] Toepfl, S., Heinz, V., Knorr, D. High intensity pulsed electric fields applied for food preservation. *Chem Eng Process* 2007; 46(6):537-546.

[75] Wouters, P.C., Alvarez, I., Raso, J. Critical factors determining inactivation kinetics by pulsed electric field food processing. *Trends Food Sci Technol* 2001; 12:112–121.

[76] Aronsson, K., Ronner, U., Borch, E. Inactivation of Escherichia coli, Listeria innocua and Saccharomyces cerevisiae in relation to membrane permeabilization and subsequent leakage of intracellular compounds due to pulsed electric field processing. *Int J Food Micro* 2005; 99(1): 19-32.

[77] Weibull, W. A statistical distribution function of wide applicability. *J Appl Mech* 1951; 51:93-7.

[78] Peleg, M. A model of microbial survival after exposure to pulsed electric fields. *J. Sci. Food Agric* 1995;67:93-99.

[79] Bigelow, W.D. The logarithmic nature of thermal death time curves. *J. Infect. Dis.* 1921; 29:528-536.

[80] Hulsheger, H., Potel, J., Niemann, E.G. Killing of bacteria with electric pulses of high field strength. *Radiat. Environ. Biophys* 1981; 20:53-65.

[81] Rodrigo, D., Martinez, A., Harte, F., Barbosa-Canovas, G., Rodrigo, M. Study of inactivation of Lactobacillus plantarum in orange-carrot juice by means of pulsed electric fields: Comparison of inactivation kinetics models. *J Food Prot* 2001; 64(2):259-263.

[82] Rodrigo, D., Ruiz, P., Barbosa-Canovas, G., Martinez, A., Rodrigo, M. Kinetic model for the inactivation of Lactobacillus plantarum by pulsed electric fields. *Int J Food Microbiol* 2003; 81(3):223-229.

[83] Sampedro, F., Rivas, A., Rodrigo, D., Martinez, A., Rodrigo, M. Effect of temperature and substrate on PEF inactivation of Lactobacillus plantarum in an orange juice-milk beverage. *Eur Food Res Technol* 2006;223(1):30-34.

[84] Gomez, N., Garcia, D., Alvarez, I., Raso, J., Condon, R. A model describing the kinetics of inactivation of Lactobacillus plantarum in a buffer system of different pH and in orange and apple juice. *J Food Eng* 2005; 70(1):7-14.

[85] Selma, M., Salmeron, M., Valero, M., Fernandez, P. Control of Lactobacillus plantarum and Escherichia coli by pulsed electric fields in MRS Broth, Nutrient Broth and orange-carrot juice. *Food Microbiol* 2004; 21(5):519-525.

[86] Sampedro, F., Rivas, A., Rodrigo, D., Martinez, A., Rodrigo, M. Pulsed electric fields inactivation of Lactobacillus plantarum in an orange juice-milk based beverage: Effect of process parameters. *J Food Eng* 2007; 80(3):931-938.

[87] Wouters, P., Bos, A., Ueckert, J. Membrane permeabilization in relation to inactivation kinetics of Lactobacillus species due to pulsed electric fields. *Appl Environ Microbiol* 2001; 67(7):3092-3101.

[88] Unal, R., Yousef, A.E., Dunne, C.P. Spectrofluorimetric assessment of bacterial cell membrane damage by pulsed electric field. *Innov Food Sci Emerg Technol* 2002; 3:247-254.

[89] Abram, F., Smelt, J., Bos, R., Wouters, P. Modeling and optimization of inactivation of Lactobacillus plantarum by pulsed electric field treatment. *J Appl Microbiol* 2003; 94(4):571-579.

[90] Elez-Martinez, P., Escola-Hernandez, J., Soliva-Fortuny, R., Martin-Belloso, O. Inactivation of Lactobacillus brevis in orange juice by high-intensity pulsed electric fields. *Food Microbiol* 2005; 22(4):311-319.

[91] Unal, R., Kim, J., Yousef, A. Inactivation of Escherichia coil O157 : H7, Listeria monocytogenes, and Lactobacillus leichmannii by combinations of ozone and pulsed electric field. *J Food Prot* 2001; 64(6):777-782.

[92] Corwin, H., Shellhammer, T. Combined carbon dioxide and high pressure inactivation of pectin methylesterase, polyphenol oxidase, Lactobacillus plantarum and Escherichia coli. *J Food Sci* 2002 ;67(2):697-701.

[93] Feng, H., Yang, W., Hielscher, T. *Power Ultrasound. Food Sci Tech Int* 2008;14:433-436.

[94] Bates, D., Patist, A. Industrial applications of high power ultrasonics in the food, beverage and wine industry. *Nonthermal Processing Division Newsletter* 2009; 11:2-15.

[95] Sala, F.J., Burgos, J., Condón, S., López, P., Raso, J. Effect of heat and ultrasound on microorganisms and enzymes, In: Gould GW, editor. *New Methods of Food Preservation.* 1995, London: Blackie Academic & Professional. p. 176-204.

[96] Butz, P., Tauscher, B. Emerging technologies: chemical aspects. *Food Res Int* 2002; 35:279-284.

[97] Cleland, M.R. Advances in gamma ray, electron beam, and x-ray technologies for food radiation. In: Sommers CH, Fan X, editors. *Food irradiation research and technology.* 2006, Iowa: Blackwell Publishing. p. 11.

[98] Thayer, D.W. Food irradiation: benefits and concerns. *J Food Quality* 1990; 13:147-169.

[99] Borsa, J. Introduction: Food irradiation moving on. In: Sommers CH, Fan X, editors. *Food irradiation research and technology.* 2006, Iowa: Blackwell Publishing. p. 6-7.

In: Lactobacillus: Classification, Uses and Health Implications ISBN: 978-1-62081-151-1
Editors: Alba I. Perez Campos and Arturo Leon Mena

Chapter V

Probiotic and Health Effects of *Lactobacillus* Strains in Humans

Ana Belén Rey Morán[1], María del Pilar González Abad[2], María Sol Pérez Rodríguez[1], Eva María Martínez Vázquez[1], Ana Isabel Tizón Varela[1] and Nelson Pérez Guerra[*3]

[1]Intensive Care Unit. University Hospital of Ourense, Ramon Puga, Ourense, Spain

[2]Hemodynamic Plant. University Hospital of Ourense, Ramon Puga, Ourense, Spain

[3]Nutrition and Bromatology Group, Department of Analytical and Food Chemistry, Food Science and Technology Faculty, Ourense Campus, University of Vigo, Ourense, Spain

Abstract

The main purpose of this work is to give an overview on the positive and negative effects of some lactobacilli in human health. Firstly, the main characteristics of the most commonly used probiotic strains as well as their use for treatment of human diseases are exposed. Secondly, the manuscript describes the positive results obtained in a number of studies dealing with the evaluation of the potential of different *Lactobacillus* strains in preventing infections by pathogenic bacteria in humans. These positive effects are commonly associated to the capacity of lactobacilli for surviving the stressful environment of the stomach (low pH and bile), for adhering to host epithelial tissue and colonizing the gut, production of ant-adhesion factors, their ability for producing antimicrobial products (lactic acid, hydrogen peroxide, bacteriocins) antagonistic to pathogen growth, and perhaps immune modulation or signalling effects. Evidence of some infections (such as infectious endocarditis, bacteraemia and localized infections) caused by lactobacilli, especially in severely ill or immunocompromised patient, and the

[*] Email: nelsonpg@uvigo.es.

ineffectiveness of lactobacilli treatment for the prevention or treatment of human diseases are also showed.

In addition, examples of different commercial dietary supplements and food products containing probiotic lactobacilli in combination or not with other probiotic bacteria are given, and the major problems limiting the use of these commercial products as probiotic supplements for human and animals are mentioned and discussed.

Keywords: *Lactobacillus* strains, antimicrobial products, health effects, probiotic properties, food containing probiotics

Main Characteristics of Probiotics Strains

There is an increasing scientific and commercial interest in the use of probiotics (live microorganisms which confer a health benefit on the host) for preventing or treating different human diseases (Alvarez and Oberhelman, 2001; Fujiya and Kohgo, 2010), mainly as an alternative to the use of antimicrobials (*e.g.* antibiotics, silver sulphadiazine) that could produce adverse reactions and side effects (Arvola et al., 1999; Trop et al., 2006). These probiotic microorganisms, when administered in adequate amounts, exert health-promoting effects in humans and animals by assisting in the establishment of an intestinal population which is beneficial to the host entity and antagonistic to harmful bacteria (Alvarez and Oberhelman, 2001; Guerra et al., 2007; Fujiya and Kohgo, 2010). For example, in humans, consumption of probiotics can help to maintain intestinal development, nutrition, and treat intestinal inflammations, functional disorders, and other extra-intestinal diseases (Alvarez and Oberhelman, 2001; Fujiya and Kohgo, 2010). In animals, the administration of probiotic bacteria and their metabolites has found to be an effective way to promote body weight gain and feed conversion and to reduce coliform counts in weaned piglets (Guerra et al., 2007).

The most commonly used probiotics are some strains of *Lactobacillus* and *Bifidobacterium* (Tables 1 and 2), which have been formulated into many different types of commercial products such as foods (*e.g.* yogurt drinks, cereals), drugs and dietary supplements in capsules, tablets or sachets. Other microorganisms including the yeast *Saccharomyces cerevisiae* and the bacterium *Bacillus subtilis* have also used as probiotics (Fujiya and Kohgo, 2010). Since the probiotic effects are species and strain specific, after formulating a probiotic product, the probiotic strains must be characterized by using phenotypic and genotypic methods, and the functional and safety aspects of these microorganisms in food must be examined. These probiotic characteristic must be examined by using appropriate target-specific *in vitro* tests (Guerra et al., 2007) that correlate with *in vivo* results (Joint FAO/WHO Working Group, 2002). The main currently *in vitro* tests used to evaluate the probiotic properties of a strain are (Klaver and van der Meer, 1993; Dunne et al., 2001; Joint FAO/WHO Working Group, 2002; Guerra et al., 2007): i) resistance to gastrointestinal transit conditions (gastric acidity and bile toxicity), ii) adhesion to mucus and/or gut epithelial tissue, iii) ability to persist within the gastrointestinal tract, iv) production of antimicrobial substances with activity against pathogenic bacteria, v) ability to reduce pathogen adhesion to surfaces, vi) bile salt hydrolase activity, vii) resistance to spermicides (in case of using probiotics for vaginal use), and viii) ability to influence metabolic activities such as cholesterol assimilation, lactase activity, and vitamin production.

Table 1. Some examples of commercial products containing *Lactobacillus* probiotic strains. *B.*: *Bifidobacterium*, *L.*: *Lactobacillus*, *Lact.*: *Lactococcus*, *Ped.*: *Pediococcus*, *S.*: *Streptococcus*, *Sacc.*: *Saccharomyces*

Manufacturer	Product	Strains
Jarrow Formulas	Jarro-Dophilus EPS®	*B. breve* R0070, *B. longum* BB536, *L. casei* R0215, *L. helveticus* (*L. acidophilus*) R0052, *L. plantarum* R1012, *L. rhamnosus* R0011, *Lact. lactis* ssp. *lactis* R1058, *Ped. acidilactici* R1001
	Jarro-Dophilus® + FOS	*B. lactis* BI-04, *B. longum* BB536, *L. acidophilus* LA-14, *L. casei* R0215, *L. plantarum* R1012, *L. rhamnosus* R0011
	Jarro-Dophilus® Allergen-Free	*B. breve* BR-03, *B. lactis* BS-01, *L. acidophilus* LA-02, *L. paracasei* LPC-00, *L. plantarum* LP-01, *L. rhamnosus* LR-04
	Ultra Jarro-Dophilus®	*B. breve* Bb-03, *B. lactis* BI-04, *B. longum* BB536, *L. acidophilus* La-14, *L. casei* Lc-11, *L. paracasei* Lpc-37, *L. plantarum* Lp-115, *L. rhamnosus* R0011, *L. salivarius* Ls-33, *Lact. lactis* LI-23
NOW Foods: Nutrition for Optimal Wellness	Acidophilus and Bifidus 8 Billion	*B. lactis*, *B. longum*, *L. acidophilus*
	Acidophilus, 3 Billion, Stabilized	*L. acidophilus*
	Acidophilus 4 X 6 Powder	*B. lactis* (*bifidum*), *B. longum*, *L. acidophilus*, *L. bulgaricus*, *L. paracasei* sub-sp., *S. thermophilus*
	BerryDophilus™	*B. lactis, L. acidophilus, L. plantarum, L. salivarius*
	Gr8-Dophilus™	*B. bifidum*, *B. lactis*, *B. longum*, *L. acidophilus*, *L. casei*, *L. rhamnosus*, *L. salivarius*, *S. thermophilus*
	Probiotic-10™ 25 Billion	*B. bifidum*, *B. breve*, *B. longum*, *L. acidophilus*, *L. casei*, *L. paracasei*, *L. plantarum*, *L. rhamnosus*, *L. salivarius*, *S. thermophilus*
EFI: Essential Formulas Incorporated	Dr. Ohhira's Probiotics® Original Formula	*B. breve* ssp. *breve*, *B. infantis* ssp. *infantis*, *B. longum*, *E. faecalis* TH10, *L. acidophilus*, *L. brevis*, *L. bulgaricus*, *L. casei* ssp. *casei*, *L. fermentum*, *L. helveticus* ss. *jugurti*, *L. plantarum*, *S. thermophilus*
Amerifit Nutrition, Inc.	Culturelle® for KIDS	*L. rhamnosus* GG (strain ATCC 53103)
Healthy Origins®	Healthy Origins® Probiotic	*B. breve*, *B. lactis*, *B. longum*, *L cidophilus*, *L. casei*, *L. plantarum*, *L. rhamnosus*, *L. salivarius*
Flora	Udo's Choice Infant's Blend Probiotic	*B. bifidum*, *B. breve*, *B. infantis*, *L. acidophilus*, *L. bulgaricus*, *L. casei*, *S. thermophillus*
	Udo's Choice Children's Probiotic and Udos Choice Adult's Probiotic	*L. casei*, *L. rhamnosus*, *L. acidophilus*, *L. plantarum*, *L. bulgaricus* *S. thermophillus*, *B. breve*, *B. bifidum*

Table 1. (Continued)

Manufacturer	Product	Strains
	Udo's Choice, Super 8 Hi-Potency Probiotic	*L. acidophilus*, *L. rhamnosus*, *L. plantarum*, *L. bulgaricus*, *L. salivarius*, *S. thermophilus*, *B. bifidum*, *B. longum*
	Udo's Choice Advanced Adult's Probiotic	*L. casei*, *L. rhamnosus*, *L. acidophilus*, *L. plantarum*, *L. bulgaricus*, *B. breve*, *B. bifidum*, *B. longum*, *S. thermophillus*
	Udo's Choice Super Bifido Plus Probiotic	*L. rhamnosus*, *L. plantarum*, *L. acidophilus*, *Lact. lactis*, *L. brevis*, *L. bulgaricus*, *L. casei*, *B. breve*, *B. bifidum*, *B. longum*
	Udo's Choice Super 5 Lozenge Probiotic	*L. acidophilus*, *L. bulgaricus*, *L. salivarius*, *B. bifidum*, *S. thermophillus*
Renew Life	Ultimate Flora, Critical Care	*B. bifidum* (HA 132), *B. breve* (HA 129), *B. longum* (HA 135), *L. acidophilus* (HA 122), *L. bulgaricus* (HA 137), *L. casei* (HA 108), *L. lactis* (HA 148), *L. plantarum* (HA 119), *L. rhamnosus* (HA 111), *L. salivarius* (HA 118)
American Health®	Chewable Acidophilus with Bifidus	*L. acidophilus*, *B. bifidum*
Natural Factors	Acidophilus and Bifidus	*L. rhamnosus* HA-111, *L. acidophilus* HA-122, *B. bifidum* HA-132
Nature's Way	Primadophilus® Kids	*L. rhamnosus*, *L. acidophilus*, *B. longum*
	Primadophilus® Bifidus	*L. rhamnosus*, *L. acidophilus*, *B. longum*, *B. breve*
	Primadophilus® Children	*L. rhamnosus*, *L. casei*, *L. acidophilus*, *B. infantis*, *B. bifidum*, *B. breve*, *B. longum*
	Primadophilus® Fortify™	*L. casei*, *L. rhamnosus*, *L. acidophilus*, *B. breve*, *B. bifidum*, *B. longum*
	Primadophilus® Intensive	*L. acidophilus*-122, *L. rhamnosus*-111, *B. breve*-129, *B. longum*-135, *B. bifidum*-132
	Primadophilus® Junior	*L. rhamnosus*, *L. acidophilus*, *B. longum*, *B. breve*, *B. infantis*
	Primadophilus® Kids	*L. rhamnosus*, *L. acidophilus*, *B. longum*
	Primadophilus® Optima	*L. casei*-108, *B. longum*-135, *L. acidophilus*-122, *L. plantarum*-119, *L. rhamnosus*-111, *L. rhamnosus*-114, *B. breve*-129, *B. bifidum*-132, *Lact. lactis*-136, *S. thermophilus*-110, *B. infantis*-116, *L. bulgaricus*-137, *L. salivarius*-118, *L. helveticus*-128
	Primadophilus® Optima Max Bifido	*B. breve* HDS-0129, *B. longum* HDS-0135, *B. bifidum* HDS-0132, *B. infantis* HDS-0116, *L. casei* HDS-0108, *L. paracasei* HDS-0196, *L. rhamnosus* HDS-0111, *L. acidophilus* HDS-0122, *L. rhamnosus* HDS-0114, *L. salivarius* HDS-0118
	Primadophilus® Original	*L. rhamnosus*, *L. acidophilus*

Manufacturer	Product	Strains
Sedona Labs®	iFlora® Multi-Probiotic®	*B. bifidum*, *B. breve*, *B. lactis* (*infantis*), *B. lactis* HN019, *B. longum*, *L. acidophilus*, *L. brevis*, *L. bulgaricus*, *L. casei*, *L. gasseri*, *L. paracasei*, *L. plantarum*, *L. rhamnosus*, *L. salivarius*, *Lact. lactis*, *S. thermophilus*
	iFlora® Kids Multi-Probiotic®	*B. bifidum*, *B. lactis* (*infantis*), *B. lactis*, *B. longum* BB536, *L. acidophilus*, *L. paracasei*, *L. rhamnosus*
	iFlora® Nasal Sinus Support	*B. bifidum*, *L. acidophilus*, *L. rhamnosus*, *S. thermophilus*
	iFlora® Woman Multi-Probiotic®	*B. lactis*, *L. acidophilus*, *L. bulgaricus*, *L. casei*, *L. rhamnosus*, *S. thermophilus*
Garden of life	Primal Defense Ultra	*Sacc. boulardii*, *L. plantarum*, *Bac. subtilis*, *L. paracasei*, *B. longum*, *L. brevis*, *B. bifidum*, *B. breve*, *B. lactis*, *L. acidophilus*, *L. salivarius*, *L. rhamnosus*, *L. casei*
Sigma-Tau Pharmaceuticals	VSL#3	*B. breve*, *B. longum*, *B. infantis*, *L. acidophilus*, *L. plantarum*, *L. paracasei*, *L. delbrueckii* subsp. *bulgaricus*, *S. salivarus* subsp. *thermophilus*
Biogaia	BioGaia Probiotic chewable tablets or drops	*L. reuteri* ATTC 55730
Probiomics	proTract® for Intestinal Health, proTract® IBS, proTract® for atopic dermatitis	*L. fermentum* VRI003 (PCC)
Italfarmaco, S.A.	Prodefen	*L. casei* PXN37, *L. rhamnosus* PXN54, *L. acidophilus* PXN35, *L. bulgaricus* PXN39, *B. infantis* PXN27, *B. breve* PXN25, *S. thermophilus* PXN66
Salvat S.A. Laboratories	Protransitus LP	*L. plantarum* 299v

Table 2. Some examples of currently available food containing *Lactobacillus* strains (World Gastroenterology Organisation Practice Guideline Probiotics and prebiotics, 2009). Notations are as in Table 1

Manufacturer	Brand name	Strains
Danone	Actimel	*L. bulgaricus, S. thermophilus, L. casei* DN-114 001 (*L. casei Immunitas*(s)/Defensis).
	DanActive fermented milk	*L. casei* DN-114 001 (*L. casei Immunitas*)
Arla Foods	Cultura	*L. casei F19*
Yakult	Yakult	*L. casei* Shirota
Nestlé	LC1	*L. johnsonii* La1 (same as NCC533 and formerly *L. acidophilus* La-1)
NextFoods Probi Ferring	GoodBelly, ProViva	*L. plantarum* 299v
Nestlé	*Lactobacillus* fortis	*L. paracasei* St11 (or NCC2461)
Valio	Vifit and others	*L. rhamnosus* ATCC 53013 (LGG)
Norrmejerier	Verum	*L. rhamnosus* LB21, *Lact. lactis* L1A
Danone (Paris, France)	DanActive fermented milk	*L. casei* DN-114 001 (*L. casei Immunitas*)
Probi AB	ProViva (juices and cultured milk)	*L. plantarum* 299v
NextFoods (Boulder, Colorado)	GoodBelly probiotic drinks	*L. plantarum* 299v
Valio Dairy (Helsinki, Finland)	Valio Gefilus®	*L. rhamnosus GG*

Figure 1 shows some of the main desired characteristics of a probiotic strain as well as the strategy of evaluation of these microorganisms before being commercialized as a probiotic product.

Other aspect that must be taken into account is the safety of a probiotic strain in food (Joint FAO/WHO Working Group, 2002).

Thus, although lactobacilli and bifidobacteria associated with food have been considered to be safe, some probiotic bacteria (strains of *Lactobacillus*, *Saccharomyces* and *Bacillus*) have been associated with some clinical infections, mainly in immunosuppressed or critically ill patients (Joint FAO/WHO Working Group, 2002, Slover and Danziger, 2008).

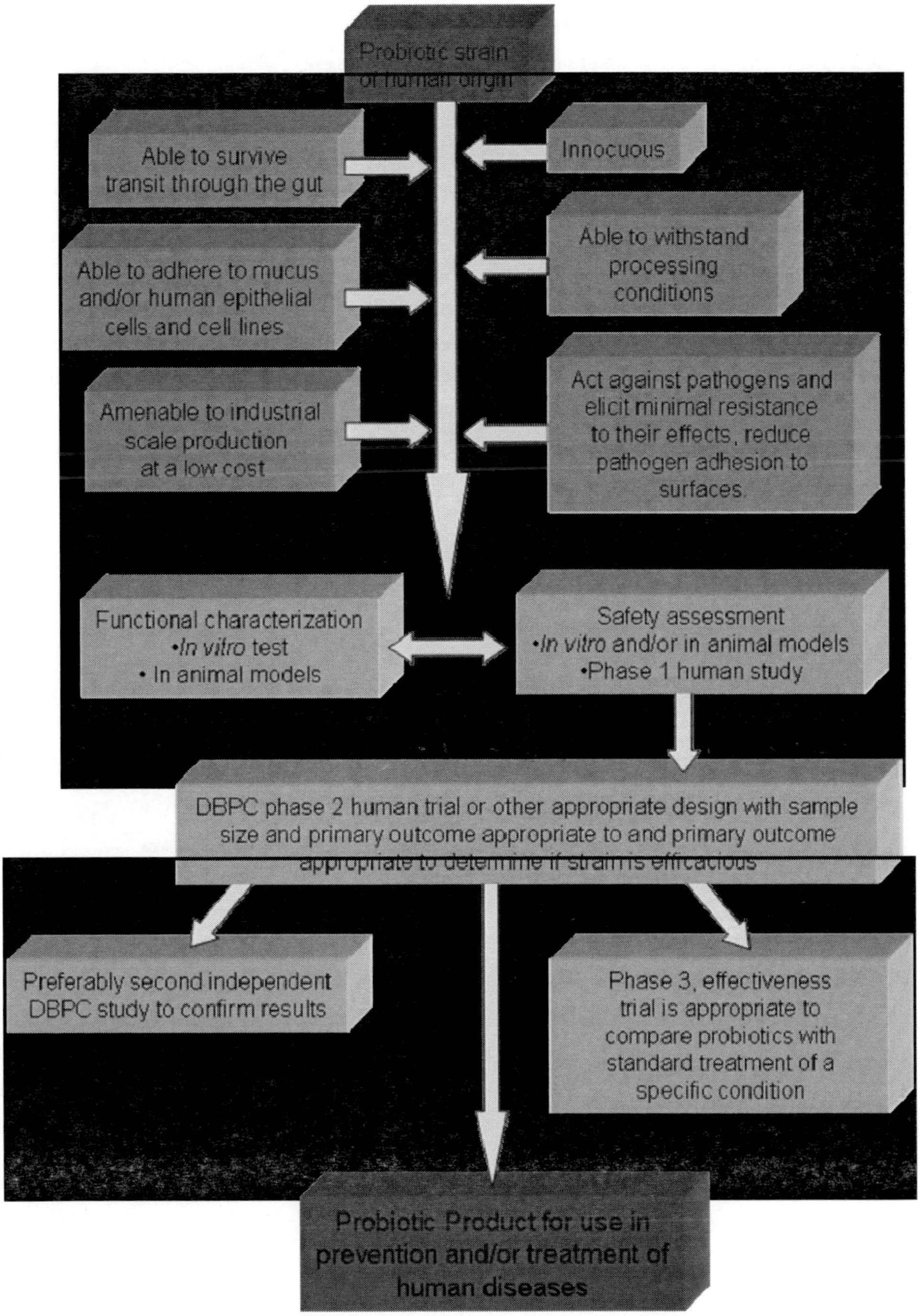

Figure 1. Some of the main desirable characteristics in a probiotic strain (green) and guidelines (blue) for the evaluation of probiotics before being used in food (Dunne et al., 2001; Joint FAO/WHO Working Group, 2002; Alvarez and Oberhelman, 2001; Fajardo et al., 2008). DBPC: Double blind, randomized, placebo-controlled.

Use of Probiotics for Prevention and Treatment of Human Diseases

A number of studies have reported the beneficial effects of the use of probiotics for prevention and treatment of different types of human diseases, including:

i) antibiotic-associated diarrhea in adults, travellers' diarrhea, and diarrheal diseases in young children caused by rotaviruses (Arvola et al., 1999; Szajewska and Mrukowicz, 2001; Kim et al., 2003; Hickson et al., 2007),
ii) lactose intolerance (Kim and Gilliland, 1983; Montes et al., 1995; de Vrese et al., 2001),
iii) allergy caused by food consumption (Majamaa and Isolauri, 1997),
iv) vaginal, oral, gastrointestinal infections, recurrent urinary tract infections (Bruce and Reid, 1988; Hilton et al., 1992; Wagner et al., 1997; Falagas et al., 2006), and
v) cancer (Aso and Akazan, 1992; Aso et al., 1995),
vi) irritable bowel syndrome (Kim et al., 2003; Ait-Belgnaoui et al., 2009; O'Mahony et al., 2005), ulcerative colitis (Kruis et al., 2004), and
vii) dermatitis (Roessler et al., 2007).

Scientific Support for Efficacy of Lactobacilli in Prevention or Treatment of Human Diseases

There are a great number of papers published on the description of clinical studies dealing with the use of bacteriotherapy with different *Lactobacillus* strains for prevention and treatment of different human diseases. Some of these publications describe studies developed with a single *Lactobacillus* strain or with a combination of lactobacilli and other probiotic bacteria.

Although some of these studies have been carried out with rats, or with a reduced amount of patients, or were not double-blind, placebo-controlled trials and this may not guarantee the effectiveness of the treatment with the specific probiotic strains assayed (Falagas et al., 2006), the results obtained suggest that certain strains consumed in adequate amounts could produce health benefits.

Evidence on the Benefits of Some *Lactobacillus* Strains in Human Health

L. Acidophillus DDS-1

This strain has shown to be able of inducing the production of higher levels of immunologic factors (interleukin-1 alpha (IL-1 alpha) and tumor necrosis factor-alpha), with potent cytocidal and cytostatic effects on tumor cells in experimental animals (Rangavajhyala et al., 1997).

L. Acidophillus LA-1

Strain LA-1 was able not only to adhere onto human enterocyte like Caco-2 cells in culture, but also this bacterium also strongly bound to the mucus secreted by the homogeneous cultured human goblet cell line HT29-MTX. LA-1 strain also inhibited, in a dose-dependent manner, the cell association of enterotoxigenic, diffusely adhering and enteropathogenic *Escherichia coli*, and *Salmonella typhimurium* and the cell invasion by enteropathogenic *E. coli*, *Yersinia pseudotuberculosis*, and *Salm. typhimurium* (Bernet et al., 1994).

L. Acidophillus La-5

A probiotic yogurt containing this lactacin B producer strain (Tabasco et al., 2009) and *Bifidobacterium lactis* Bb12 proved to decrease the total cholesterol and low-density lipoprotein cholesterol in people with type 2 diabetes mellitus (Ejtahed et al., 2011). According to the latter researchers, this probiotic yogurt may contribute to the improvement of cardiovascular disease risk factors.

Other researchers (Jain et al., 2004) assayed the effects of a symbiotic preparation (probiotic Trevis^{TM}, Chr Hansen Biosystem, Denmark), that contains 4×10^9 colony forming units of *L. acidophilus* La5, *B. lactis* Bb12, *Streptococcus thermophilus* and *L. bulgaricus*, on gut barrier function and sepsis in critically ill patients. The results showed that the administration of this symbiotic product allows the reduction in the gastric colonisation with multiple organisms and potentially pathogenic bacteria. In contrast, this probiotic preparation did not affect the intestinal permeability, the systemic inflammatory response, septic morbidity, mortality or length of hospital stay.

L. Acidophillus 145

The consumption of fermented milk containing this strain in combination with *Lactobacillus* GG (ATCC 53103), *Bifidobacterium* sp. B420 and *S. thermophilus* led to a significant reduction in the occurrence of potentially pathogenic bacteria in the upper respiratory tract of patients who consumed the probiotic drink (Glück and Gebbers, 2003).

On the other hand, the dietary supplementation with a probiotic yogurt containing the strains *L. acidophilus* 145 and *B. longum* 913 significantly reduced both the genotoxicity of human faecal water and the overall genetic damage in comparison with the supplementation with a yogurt without probiotics (Oberreuther-Moschner et al., 2004).

L. Acidophilus NAS

A study carried out by Metts et al. (2003) demonstrated that vaginal inserts of *L. acidophilus* strain NAS (H_2O_2 positive), with or without oral probiotic capsules proved to be an inexpensive and efficacious approach for treatment of candidal vulvovaginitis in women with recurrent infections.

L. Acidophilus NCFM

This lactacin B producer strain fulfils many of the probiotic criteria, as it is safe, amenable to commercial manipulation, capable to adhere to human fetal intestinal cells and Caco-2 cells and mucus-secreting HT-29 cell culture systems (Kleeman and Klaenhammer, 1982; Greene and Klaenhammer, 1994), survives gastrointestinal tract transit (Crowell, 1998) and was stable during 6 weeks frozen storage at -20ºC (Sanders et. al., 1996). According to Gilliland and Speck (1977), this bacterium produces antibacterial activity against important foodborne disease bacteria including *Staphylococcus aureus*, *Salm. typhimurium*, enteropathogenic *E. coli*, and *Clostridium perfringens*. For these probiotic characteristics this strain has been widely used in commercial foods including milk, yogurt and formulas for children, dietary supplements and in juice (Sanders and Klaenhammert, 2001).

Some studies have showed the potential of this strain for producing favourable effects on health. In this way, consumption of *L. acidophilus* NCFM in adequate amounts: i) reduced incidence of colon cancer in experimental rats (Goldin and Gorbach, 1980), ii) may aid the digestion of lactose in lactose-intolerant subjects (Kim and Gilliland, 1983; Montes et al., 1995), iii) in combination with other probiotic strains (*L. reuteri* and *B. infantis* BBI), decreased the risk of diarrhea in young children (Sanders and Klaenhammert, 2001), iv) decreased significantly the levels of toxic amines in the blood of dialysis patients with small bowel bacterial over-growth (Sanders and Klaenhammert, 2001).

L. Acidophilus ATCC 4356

Although the human diseases (gastritis, gastric and duodenal ulcer and gastric cancer) produced by *Helicobacter pylori* (Uemura et al., 2001) could be effectively treated with the use of antibiotics (Konturek, 2003), *H. pylori* may become resistant to them (Ryan et al., 2001). In addition, there are side effects of prolonged antibiotic use due to the eradication of beneficial bacteria along with the pathogenic bacteria (Cunha, 2001). In this way, *L. acidophilus* ATCC 4356 was found to be able of inhibiting the growth of *H. pylori*, thus indicating that the lactic acid bacterium could be used for treating *H. pylori* infection (Ryan et al., 2008), as an alternative of antibiotic use.

L. Acidophilus LB

Scanning electron microscopy revealed that this strain, isolated from human stool, was able to adhere well to the brush border of differentiated human intestinal epithelial Caco-2 cells in culture (Chauvière, et al., 1992). Due to this ability, strain LB was also capable of inhibiting, depending on the concentration, cell attachment and invasion of Caco-2 cells by the enterovirulent bacteria *Y. pseudotuberculosis* and *Listeria monocytogenes* (Coconnier et al., 1993).

Coconnier et al. (1998) examined the activity of the human *L. acidophilus* strain LB against *H. pylori* 1011 *in vitro* and *in vivo*. The spent culture supernatants of strain LB produced a dramatic decrease in the viability of the pathogen *in vitro* and in parallel a decrease in the adhesion of *H. pylori* to the cultured human mucosecreting HT29-MTX cells was observed. The *in vivo* study in conventional mice showed that the oral administration of strain LB+SCS (spent culture supernatant) inhibited *H. felis* CS1 colonization and decreased urease activity in conventional specific-pathogen-free BALB/c mice (Coconnier et al., 1998).

Coconnier et al. (2000) observed that the treatment with the spent culture supernatant of *L. acidophilus* LB efficiently decreased transcellular passage of *Salm. enterica* serovar *Typhimurium*, inhibited the intracellular growth of the pathogen and also inhibited adhesion-dependent serovar *Typhimurium*-induced interleukin-8 production in Caco-2/TC-7 cells.

Other study showed that *L. acidophilus* LB and its antimicrobial activity produced a protective effect against the brush border lesions promoted by the diarrhoeagenic *E. coli* (Afa/Dr DAEC) strain C1845 in human enterocyte-like cells (Liévin-Le Moal et al., 2002).

An experimental study and a randomized, double-blind, placebo-controlled clinical trial designed to evaluate the antisecretory activity of strain LB against nonrotavirus diarrhoea showed that adding lyophilized, heat-killed *L. acidophilus* LB bacteria plus their culture medium to a solution of oral rehydration solution shortened by 1 day the time until the first normal stool was passed, as compared with children who received placebo oral rehydration solution (Liévin-Le Moal et al., 2007).

L. Bulgaricus Strains

In laboratory investigations, short-lived metabolite mixtures isolated from milk fermented by *L. bulgaricus* plus *S. thermophilus* were more effective in deactivating etiologic risk factors of colon carcinogenesis than those metabolite mixtures isolated from milk fermented by *S. thermophilus* alone (Bodana and Rao, 1990). Some strains of LAB (including *L. bulgaricus*) have demonstrated anti-mutagenic effects probably due to their ability to bind with carcinogenic substances (heterocyclic amines) formed during the cooking of meat (Wollowski et al., 2001).

L. delbrueckii subsp. *bulgaricus* SS1 was found to produce high amounts of angiotensin-I-converting-enzyme (ACE)-inhibitory peptides during fermentation in milk (Gobbetti et al., 2000). Although the inhibition of ACE mainly produces an antihypertensive effect, it may also influence different regulatory systems related with the modulation of blood pressure, immune defense, and nervous system activity (Meisel, 1993).

An *in vitro* study carried out by Bai et al. (2004) showed that *L. bulgaricus* LB10 can suppress interleukin-8 secretion in intestinal epithelia when stimulated by proinflammatory cytokines. This indicates that this strain has a great potential for suppressing inflammatory immune reactions in the intestinal wall that produce tissue damage.

Kitazawa et al. (2003) observed that the DNA from *L. delbrueckii* ssp. *bulgaricus* NIAI B6 (*L. bulgaricus*) can stimulate a significant proliferation of certain immune cells (peyer's patch and splenic B cells) *in vitro*. This result suggests that this probiotic strain could modulate local immunity *in vivo* via the stimulation of lymphoid tissues within the gut.

Other studies related with the consumption of foods containing *L. bulgaricus* have shown that:

- Consumption of fermented *L. bulgaricus* milk improved lactose digestion and reduce symptoms of intolerance in lactose maldigesters (de Vrese et al., 2001). This was related to the high production of microbial β-galactosidase activity by *L. bulgaricus* strain.
- VSL# 3, a composite probiotic containing three *Bifidobacterium* strains (*B. longum*, *B. infantis* and *B. breve*), four *Lactobacillus* strains (*L. acidophilus*, *L. paracasei*, *L. delbrueckii* ssp. *bulgaricus* and *L. plantarum*), and *S. salivarius* ssp. *thermophilus* was found to reduce flatulence scores and retard colonic transit without altering bowel function in patients with irritable bowel syndrome and abdominal bloating (Kim et al., 2003; 2005). Other study showed that this probiotic preparation can regulate the portal vein pressure (De Santis et al., 2000).
- Consumption of a probiotic drink containing *L. bulgaricus* in combination with *L. casei* and *S. thermophilus* was proposed as an effective way for reducing the

incidence of antibiotic associated diarrhoea and *C. difficile* associated diarrhoea (Hickson et al., 2007).

L. Casei Strain Shirota

In humans, a preparation containing this strain has been shown to be safe and effective for preventing recurrence of superficial bladder cancer (Aso and Akazan, 1992; Aso et al., 1995).

Other studies have shown the ability of this strain to inhibit the growth of important pathogens such as *H. pylori*, *L. momnocytogenes*, *Salm. typhimurium* and *E. coli*. In this way, a *L. casei* Shirota strain YIT9029 was able to enhance the host response against oral *L. monocytogenes* infection in rats, probably due to increased cellular-mediated immunity (de Waard et al., 2002).

On the other hand, the regular ingestion of a probiotic beverage containing *L. casei* Shirota was found to produce a beneficial effect on gastrointestinal symptoms of patients with chronic constipation, thus indicating changes in the microflora and the intestinal milieu (Koebnick et al., 2003). In fact, consumption of this probiotic beverage resulted in an improvement of bowel movement and stool consistency.

The use of *L. casei* and yogurt, used as adjuvant in a re-nutrition diet in a non-severe malnutrition experimental model, stimulated the recuperation of the activity of immune cells, thus enhancing the immune response and increasing resistance to certain pathogenic bacteria (*S. typhimurium* and *E. coli*) in the digestive tract (Cano and Perdigon, 2003).

An *in vitro* study with *L. casei* strain Shirota obtained from the fermented milk product Yakult (Yakult Ltd., Tokyo, Japan) showed that this strain significantly inhibited the growth of *H. pylori* SS1 and nine *H. pylori* clinical isolates (Sgouras et al., 2004). The same authors performed *in vivo* experiments to test the inhibitory effect of *L. casei* strain Shirota on *H. pylori* by giving the probiotic bacterium in the normal water supply to SS1 6-week-old C57BL/6 mice previously infected with the pathogen. In this assay, a significant reduction in the levels of *H. pylori* colonization was observed in the antrum and body mucosa, which was accompanied by a significant reduction in the associated chronic and active gastric mucosal inflammation observed at each time point during 9 months (Sgouras et al., 2004).

L. Casei DN-114 001

This strain was found to down-regulate the transcription of a number of genes encoding pro-inflammatory effectors (cytokines and chemokines) and adherence molecules induced by invasive *Shigella flexneri* (Tien et al., 2006). According to these researchers, this fact produced an anti-inflammatory effect that appears to be mediated by the inhibition of the nuclear factor κB (NF-κB) pathway, particularly through stabilization of inhibitory protein I-κBα, which is the specific NF-κB inhibitor.

L. Farciminis

An experiment with rats receiving *L. farciminis* for 15 days before and 4 days after intracolonic administration of trinitrobenzene sulphonic acid (TNBS) or vehicle (Lamine et al., 2004) showed that the probiotic treatment, in absence of TNBS, reduced colonic paracellular permeability and increased the IL-10 level in the colonic wall. The researchers concluded that *L. farciminis* treatment improves TNBS-induced colitis. The results obtained by this research team indicated that, in addition to the nitric oxide released by *L. farciminis*, the anti-inflammatory action of this bacterium involves also normalization of colonic microflora, prevention of bacterial translocation, enhancement of barrier integrity and a decrease in the IL-1β mucosal level (Lamine et al., 2004).

A study with female Wistar rats showed that the treatment with *L. farciminis* CIP 103136 prevents stress induced hypersensitivity to colorectal distension, increase in colonic paracellular permeability, and colonocyte myosin light chain phosphorylation. This positive effect was related with the inhibition of contraction of colonic epithelial cell cytoskeleton and the subsequent tight junction opening. In addition, the researchers suggest that the positive effect produced by *L. farciminis* may also involve direct or indirect effects of nitric oxide produced by this strain (Ait-Belgnaoui et al., 2006).

Treatment with *L. farciminis* CIP 103136 also decreased the stress-induced overexpression of Fos protein (a marker of general neuronal activation, induced by colorectal distension) in spinal and supraspinal sites after colorectal distension in female Wistar rats. According to these researchers, this study supports rationale for selected probiotic strains use in the irritable bowel syndrome management (Ait-Belgnaoui et al., 2009).

L. Johnsonii Strain La1

In experiments with C57BL/6 mice, it was found that administration of *L. johnsonii* La1 (Nestlé, Switzerland) to animals did not affect *H. pylori* colonization, but this treatment reduced *H. pylori*-associated gastritis and reduces levels of proinflammatory chemokines during the early stages of *H. pylori* infection. These results were attributed to a possible reduction in proinflammatory chemotactic signals responsible for the recruitment of lymphocytes and neutrophils in the lamina propria (Sgouras et al., 2005).

L. Kalixensis

This strain, which was isolated from the human stomach (Roos et al., 2005), exhibited a relatively high anti-*H. pylori* activity, which was similar to those of the *L. salivarus* strains UCC119, DSM 20554, DSM 20555, CCUG 27530B, CCUG 44481, JCM 1040, JCM 1046, JCM 1047 and LMG 14476 (Ryan et al., 2008). Therefore, this bacterium may warrant further study related with the treatment of human diseases caused by *H. pylori* (Ryan et al., 2008) and other pathogens.

L. Plantarum ATCC 10241

Pseudomonas aeruginosa is an antibiotic resistant opportunistic pathogen that infects primarily immunocompromised individuals, including patients with cystic fibrosis, cancer or acquired immune deficiency syndrome, or patients with indwelling medical devices or burns (Fergie et al., 1994; Mendelson et al., 1994), or patients having prolonged use of broad-spectrum antibiotics (Brewer et al., 1996).

Valdéz et al. (2005) tested the effect of *in vivo* activity of *L. plantarum* ATCC 10241 against *Pseudomonas aeruginosa* in a burned-mouse model. The burns were infected with the pathogen and then treated with strain ATCC 10241 after 3, 4, 5, 7 and 9 days postinfection. The samples obtained after taking samples from skin, liver and spleen after 5, 10 and 15 days indicated that the lactic acid bacterium not only inhibited the *P. aeruginosa* colonization, but also there was an improvement in tissue repair, enhanced phagocytosis of *P. aeruginosa* by tissue phagocyctes, and a decrease in apoptosis at 10 days. According to Valdéz et al. (2005) and Peral et al. (2009), the cultures of *L. plantarum* ATCC 10241 could be used as a potential therapeutic agent to treat *P. aeruginosa* burn infections. In addition, the bacteriotherapy with strain ATCC 10241 could be used as an alternative to the use of silver sulphadiazine. The latter antibacterial agent is often used to reduce the incidence of burn wound infection, but its application produces adverse reactions and side effects related with the silver deposition on tissues and an increasing resistance to silver sulphadiazine especially in *Pseudomonas* strains (Trop et al., 2006).

However, before establishing the bacteriotherapy treatment with *L. plantarum* ATCC 10241, further studies are required on the treatment and prophylaxis of burn infections with this strain (Valdéz et al., 2005; Peral et al., 2009).

L. Plantarum MB452

A study carried out by Anderson et al. (2010) showed that this strain enhances the function of the intestinal barrier by increasing the expression levels of genes encoding occludin and its associated plaque proteins, ZO-1, ZO-2 and cingulin, which are involved in tight junction formation in health intestinal epithelial cells.

L. Plantarum 299v

Effect of oral administration of *L. plantarum* 299v to children congenitally exposed to human immunodeficiency virus (HIV) was tested by Cunningham et al. (2000). These children had episodes of diarrhoea and frequently experience malabsorption associated with possible bacterial overgrowth. This study showed that the lactic acid bacterium was able to colonize children with HIV and to elicit specific systemic immune response after oral supplementation. The results obtained suggest that the treatment with *L. plantarum* 299v has the potential to promote growth and development in immunocompromised hosts.

In a study with 60 patients with irritable bowel syndrome (IBS) and a normal colonoscopy or barium enema (Nobaek et al., 2000), administration with *L. plantarum* 299v decreased pain and flatulence in these patients. In the same way, a controlled, double-blind, randomized study (Niedzielin et al., 2001) with 40 patients also showed a beneficial effect of *L. plantarum* 299v in patients with IBS receiving the strain over a period of 4 weeks.

Other study showed that when this lactic acid bacterium was administered to Sprague-Dawley rats in the drinking water, the *E. coli*-induced increase in intestinal permeability was suppressed, thus showing that *L. plantarum* 299v may exert beneficial effects in the gastrointestinal tract (Mangell et al., 2002).

On the other hand, a small but well designed randomized clinical trial in patients with acute pancreatitis receiving a freeze-dried preparation containing live *L. plantarum* 299v together with a substrate of oat fiber revealed that the effectiveness of this treatment to reduce pancreatic sepsis and the number of surgical interventions related to pancreatic damage (Oláh et al., 2002). However, the researchers also consider that the probiotic regimen may need to be given for a longer duration to exert a maximum effect, because treatment failure often occurs in the post-administration period. In fact, experience from a controlled trial in liver transplant recipients indicates that early enteral nutrition with fiber-containing solutions and living *L. plantarum* 299v decreased markedly the rate of postoperative infections (Rayes et al., 2002a). A similar study in patients with major abdominal surgery (gastric and pancreas resections) showed that patients with early enteral nutrition supplemented with *L. plantarum*

299v and fiber significantly decreased postoperative infections (Rayes et al., 2002b). In this way, pneumonia seemed to be prevented and no catheter-related sepsis occurred.

L. Reuteri ATCC 55730

This strain was effective as a therapeutic agent in treatment of acute rotavirus diarrhoea in children between 6 and 36 months of age hospitalized with acute diarrhoea (75% rotavirus), because this probiotic bacterium reduced significantly duration of watery diarrhoea in comparison to the group of patients receiving placebo (Shornikova et al., 1997). Similarly, in a double-blind, placebo-controlled, randomized trial, infants fed a formula supplemented with *L. reuteri* ATCC 55730 had a significant decrease of number of days with fever, clinic visits, child care absences, and antibiotic prescriptions in comparison with the children receiving *Bifidobacterium lactis* BB-12 and the control group, which received a formula without probiotics (Weizman et al., 2005).

L. Rhamnosus GG

A great number of researches have proven the efficacy of *L. rhamnosus* strain GG (LGG) for the treatment of infectous and allergic diseases. Clinical studies found that strain GG given to infants and children reduced the risk or duration of diarrhoea caused by *Cl. difficile* and other pathogens (Mattar et al., 2001; Szajewska and Mrukowicz, 2001; Basu et al., 2007; Alvarez and Oberhelman, 2011).

Glück and Gebbers (2003) observed a significant reduction in the occurrence of nasal colonization by potentially pathogenic bacteria, including *Staph. aureus*, *S. pneumoniae*, β-hemolytic streptococci, and *Haemophilus influenzae*, in patients who consumed a fermented milk product containing LGG in combination with other probiotic bacteria (*Bifidobacterium* sp. strain B420, *L. acidophilus* 145, and *S. thermophilus*). The researchers attributed this result to the possible stimulation of the B lymphocytes of the the gut-associated lymphoid tissue and their migration to the upper respiratory immune system. This led to the production of more effective secretory immunoglobulin A2, which probably contributed to the elimination of the potentially pathogenic bacteria.

L. rhamnosus GG was also found to be effective in prevention of atopic diseases in newborns (Reid and Kirjaivanen, 2005) and in children at high risk (Kalliomäki et al., 2001). In addition, Vendt et al. (2006) observed that newborns receiving LGG-supplemented formula grew better than those fed with the regular formula.

Zhang et al. (2005) observed that the exposure of the intestinal epithelium to high doses of live *L. rhamnosus* GG decreased tumor necrosis factor-α–induced Interleukin-8 production in Caco-2 cells and may effectively ameliorate inflammation. Other study with rats showed that preventive administration of LGG to infants may lead to enhanced resistance to neonatal bacterial sepsis and meningitis due to suppression of pathogen translocation across the gut barrier (Huang et al., 2009).

L. Rhamnosus GR-1

This strain in combination with *L. fermentum* RC-14 produced a significant reduction in colonization of the vagina by potential pathogenic bacteria and yeast in healthy women given daily oral freeze-dried capsules of these strains for 60 days (Reid et al., 2003).

Other study showed that the oral supplementation with *L. rhamnosus* GR-1 and *L. reuteri* RC-14 was an effective way for the treatment of vaginosis and bacterial vaginitis, and in relapse prevention (Cianci et al., 2008). The positive effect of these combinations of *Lactobacillus* has been related to the ability of these strains for colonizing the vagina and re-establishing the vaginal ecosystem remarkably (Reid et al., 2003; Cianci et al., 2008).

In addition, a recent case report has revealed that the combination of strains *L. rhamnosus* GR-1 and *L. reuteri* RC-14 could have applicability to the care of males with urinary tract infection (Anukam et al., 2009).

L. Salivarius Strains UCC118 and UCC119

Studies with *L. salivarius* UCC118, a probiotic strain of human origin, have shown that this strain is able to inhibit the growth of the invasive foodborne pathogen *L. monocytogenes* mainly due to the production of the Abp118 bacteriocin (Corr et al., 2007). However, strain UCC118 did not shown antibacterial activity against *H. pylori* (Ryan et al., 2008). In contrast, other strains of *L. salivarius* (UCC119, DSM 20554, DSM 20555, CCUG 27530B, CCUG 44481, JCM 1040, JCM 1046, JCM 1047 and LMG 14476) exhibited a potent anti-*H. pylori* activity, thus suggesting that these lactic acid bacteria have potential for the treatment of human diseases produced by *H. pylori*. Particularly, *L. salivarus* UCC119 showed ability to inhibit the growth of six antibiotic-resistant clinical isolates of *H. pylori*, thus suggesting that strain UCC119 has a great potential to be used as an adjunct in the treatment of human diseases produced by the antibiotic-resistant pathogens (Ryan et al., 2008).

L. Salivarius WB1004

In an *in vitro* study, Aiba et al. (1998) observed that *L. salivarius* WB1004 was able to produce high amounts of lactic acid that completely suppressed the growth of *H. pylori* in a mixed culture.

The results obtained by the same researchers, in an *in vivo* experiment with the use of a *H. pylori*-infected gnotobiotic murine model showed that *L. salivarus* produced lactic acid concentrations between 2.3 and 18.9 μmol/g in the gastric contents. These levels of lactic acid inhibited almost completely the *H. pylori* viability and completely abolished urease activity *in vitro*. This indicated that administration of *L. salivarius* to mice can produce an efficient regression of inflammation in the animals. Therefore, Aiba et al. (1998) concluded that the strain WB1004 has potential to be used as a probiotic strain in the treatment of *H. pylori*-related diseases, mainly because the lactic acid bacterium can bind tightly to gastric epithelial cells, grow well, and thus produce a high amount of lactic acid in the stomach.

L. Sporogenes

A study on the effect of oral administration of *L. sporogenes* (*Bacillus coagulans*) to 17 patients with type II hyperlipidemia showed that this bacterium was able to reduce LDL cholesterol levels associated with a small but significant increase in HDL cholesterol (Mohan et al., 1990). However, larger trials are necessary to confirm the use of *L. sporogenes* as a side-effect free alternative to drug therapy in the treatment of high cholesterol (Mohan et al., 1990).

To determine the efficacy of a fructo-oligosaccharides-*Lactobacillus sporogenes* preparation in the prevention of diarrhoea, a multi-centre double-blind versus placebo controlled trial was carried out with 120 children with active infections requiring antibiotics (La Rosa et al., 2003). The children were treated for 10 days either with *L. sporogenes* or with placebo. The results showed that, out of 98 evaluable patients, 71% in the *L. sporogenes*-treated group had no diarrhoea versus 38% in the placebo group. Moreover, duration of diarrhoea in the probiotic-treated group was lower (0.7 days) than that observed in the placebo group (1.6 days).

In addition to its ability for lowering cholesterol, other studies reported that *L. sporogenes* has been used in the treatment of aphthous stomatitis (Sharma et al., 1980), gut dysbiosis (Voĭchishina et al., 1991) and vaginal diseases (Kale et al. 2005).

Infections Caused by Lactobacilli

Despite that many *Lactobacillus* strains have found to produce beneficial health effects, some of them have shown to be associated with various types of infections, especially in severely ill or immunocompromised patient (Rautio et al., 1999; Cannon et al., 2005; Falagas et al., 2006; Slover and Danziger, 2008). These infections include infectious endocarditis, bacteraemia and localized infections (pneumonia, peritonitis, abdominal and liver abscesses, meningitis, endophthalmitis, esophageal infection, erysipeloid infection, throat infection, meningitis, wound infection, vascular graft infection, and fistula). Although a number of lactobacilli species (*acidophilus*, *brevis*, *confuses*, *curvatus*, *delbrueckii*, *fermentum*, *gasseri*, *jensenii*, *lactis*, *leichmannii*, *paracasei* and *salivarus*) appeared to be associated with these infections, the most common species causing these diseases were *L. casei*, *L. rhamnosus* and *L. plantarum* (Cannon et al., 2005).

However, only in a few cases, it was possible to associate lactobacilli consumed with those isolated from infectious sites (Rautio et al., 1999; Mackay et al., 1999; Falagas et al., 2006). In fact, the numbers of infections associated with probiotics are scarce in comparison with the considerable and gradually increasing consumption of probiotics (Salminen et al., 2002; Falagas et al., 2006).

Ineffectiveness of Lactobacilli Treatment for the Prevention or Treatment of Human Diseases

Saltzman et al. (1999) investigated the strategy of using organisms with avid intestinal adherence properties combined with ß-galactosidase activity to colonize the upper gastrointestinal tract and to potentially convert a lactose-intolerant subject with hypochlorhydria to a lactose-tolerant one. The researchers observed that ingestion of *L. acidophilus* BG2FO4 for 7 d in subjects with lactose maldigestion with and without hypochlorhydria failed to result in any significant overall improvement in hydrogen production after lactose ingestion. For this reason, the researchers concluded that the strategy of changing the enteric flora in the upper gastrointestinal tract was not an effective way for changing an upper intestinal condition such as lactase deficiency.

Helin et al. (2002) studied the effect of oral treatment with *L. rhamnosus* GG in 18 patients (teenagers and young adults), who were allergic to birch pollen and apple food and had intermittent symptoms of atopic allergy and/or mild asthma by using a double-blind, placebo-controlled study. The results obtained showed that the treatment with the probiotic bacterium neither alleviate the symptoms nor reduce the use of the medications of the patients during the birch-pollen season or the subsequent 2 months (Helin et al., 2002). In addition, in a clinical human trial, consumption of the same strain (two capsules of LGG) did not show a significant beneficial effect for patients with rheumatoid arthritis in comparison with the patients receiving a placebo twice daily in double-blind fashion for 12 months (Hatakka et al., 2003). The oral administration of *Lactobacillus* GG neither prevented endoscopic recurrence at one year nor reduced the severity of recurrent lesions in patients operated on for complicated Crohn's disease (Prantera et al., 2002). In addition, according with the report of Schultz et al. (2004), the treatment with *Lactobacillus* GG was also ineffective in inducing and maintaining remission of Crohn's disease. In a randomized, double-blind, placebo-controlled clinical trial, 230 male infants and young children aged 4 to 24 months with acute watery diarrhoea of <2 days' duration were administered with *L. paracasei* strain ST11 (10^{10} colony forming units) or placebo, daily for 5 days. Although strain ST11 feeding significantly reduced the cumulative stool output, stool frequency and oral rehydration solution intake in children with less-severe nonrotavirus diarrhoea, administration with this strain had no overall effect on severe rotavirus diarrhoea (Sarker et al., 2005). A randomized controlled trial showed that supplementation with *L. acidophilus* LAVRI-A1 not only failed to reduce the risk of atopic dermatitis in high-risk infants but also it was associated with increased allergen sensitization in infants receiving the probiotic (Taylor et al., 2007).

Kukkonen et al. (2007) studied the effects of probiotics and prebiotics combination on allergy prevention in 1223 pregnant women carrying high risk children. The mothers were randomized to receive either placebo or a probiotic preparation (1 capsule containing *L. rhamnosus* GG, *L. rhamnosus* LC705, *B. breve* Bb99 and *Propionibacterium freudenreichii* ssp. *shermanii* JS) for 2 to 4 weeks prior to delivery allergy-prone infants. The newborns infants were then supplemented with either the same probiotic preparation in combination with galacto-oligosaccharides (461 babies) or placebo (464 babies) for six months after birth. These researchers did not observe a significant difference between the groups receiving the probiotic preparation and the placebo for allergic diseases (food allergy, eczema, asthma, and allergic rhinitis). However, probiotic treatment significantly reduced eczema and especially atopic eczema (Kukkonen et al., 2007). According to Delcenserie et al. (2008), the results obtained with the probiotic treatments for chronic inflammatory bowel diseases have been disappointing. In addition, when a beneficial effect of the probiotic consumption was observed, the positive effect was not necessarily an improvement over the use of previously prescribed drugs.

Commercial Dietary Supplements and Food Products Containing Probiotic Lactobacilli

The probiotic products are commonly commercialized as either functional foods or supplements. Production of these products has increased during last years in the first world countries, as a response to the growing consumer´s demand for more attractive and healthy foods and due to the increasing knowledge of consumers about nutrition and resulting implications to health (Jones, 2002; Mollet and Rowland, 2002). The purported health benefits commonly are reflected on the label of the product package. Thus, according to some manufacturers (Table 1), consumption of these products can:

- help to maintain healthy intestinal flora.
- enhance the body's natural defences.
- assist intestinal functioning.
- adhere and colonize intestinal tract and stimulate intestinal immune response.
- help to support the body's natural resistance.
- improve digestion of dairy products.
- support oral and throat health, healthy gums and nasal sinuses, good breath.
- enhance absorption of nutrients from food and other supplements.
- help to reduce the length and severity of antibiotic-associated diarrhoea.
- help to break up mucus to flush out irritants.
- promote optimal vaginal and urinary tract health.

In tables 1 and 2 are shown some commercial dietary supplements and food products containing probiotic lactobacilli, as reported by the manufacturers. However, the purpose of these tables is to provide examples of probiotic products commercially available, not provide recommendations for their use.

Major Problems Limiting the Use of Commercial Products as a Probiotic Supplement for Human and Animals

According to the recommendations of the Working Group of the FAO/WHO and the International Scientific Association for Probiotics and Prebiotics (ISAPP), the suitable description of a probiotic product as reflected on the label should include:

- Genus and species identification, with nomenclature consistent with current scientifically recognized names.
- Strain designation.
- Minimum viable count of each strain at the end of shelf-life.
- Recommended storage conditions.
- Safety in the conditions of recommended use.
- Recommended dose, which should be based on induction of the physiological effect.
- Proper storage conditions.
- An accurate description of the physiological effect, as far as is allowable by law.
- Corporate contact information for post-market surveillance.

However, some major factors restricted the use of many of these preparations.

These include low numbers of viable cells (zero in some cases), contamination or incorrect labelling of these products, mainly because there is no indication that the strains used possess any of the recommended probiotic properties (Weese, 2002; Gardiner et al., 2004).

Conclusion

Mechanistic studies in laboratory assays or in animal models have shown that antibacterial substances (bacteriocins, hydrogen peroxide, organic acids: lactic acid and acetic acid) produced by some lactobacilli inhibit the growth and survival of different pathogens (*E. coli*, *Salm. typhimurium*, *Sh. sonnei*, *H. pylori*, *Campylobacter jejuni*, *Camp. coli*, *Cl. difficile* and *Candida albicans*). In addition, a great number of the studies described in this work indicated a beneficial health effect of bacteriotherapy with *Lactobacillus* strains in humans.

However, despite the promising results of some of these studies, a particularly significant clinical effect of bacteriotherapy with lactobacilli remains to be clearly established. This is mainly because most of the trials either included a small number of patients or were not placebo-controlled and consequently, the efficacy of these probiotic strains was not well established.

Therefore, further research is needed to prove the effectiveness of lactobacilli in preventing or treating different human diseases before being marketed. These studies must be carried out with a larger sample size (a high number of patients) and using more randomized, double-blind, placebo-controlled trials.

References

Aiba, Y., Suzuki, N., Kabir, A.M., Takagi, A., Koga, Y. (1998). Lactic acid-mediated suppression of *Helicobacter pylori* by the oral administration of *Lactobacillus salivarius* as a probiotic in a gnotobiotic murine model. *Am. J. Gastroenterol.* 93, 2097-2101.

Ait-Belgnaoui, A., Eutamene, H., Houdeau, E., Bueno, L., Fioramonti, J., Theodorou, V. (2009). *Lactobacillus farciminis* treatment attenuates stress-induced overexpression of Fos protein in spinal and supraspinal sites after colorectal distension in rats. *Neurogastroenterol. Motil.* 21, 567-573.

Ait-Belgnaoui, A., Han, W., Lamine, F., Eutamene, H., Fioramonti, J., Bueno, L., Theodorou, V. (2006). *Lactobacillus farciminis* treatment suppresses stress induced visceral hypersensitivity: a possible action through interaction with epithelial cell cytoskeleton contraction. *Gut* 55, 1090-1094.

Alvarez, M.I., Oberhelman, R.A. (2001). Probiotic agents and infectious diseases: a modern perspective on a traditional therapy. *Clin. Infect. Dis.* 32, 1567-1576.

Anderson, R.C., Cookson, A.L., McNabb, W.C., Park, Z., McCann, M.J., Kelly, W.J., Roy, N.C. (2010). *Lactobacillus plantarum* MB452 enhances the function of the intestinal barrier by increasing the expression levels of genes involved in tight junction formation. *BMC Microbiol.* 2010, 10, 316.

Anukam, K.C., Hayes, K., Summers, K., Reid, G. (2009). Probiotic *Lactobacillus rhamnosus* GR-1 and *Lactobacillus reuteri* RC-14 may help downregulate TNF-Alpha, IL-6, IL-8, IL-10 and IL-12 (p70) in the neurogenic bladder of spinal cord injured patient with urinary tract infections: A two-case study. Adv. Urol. 2009. http://www.ncbi.nlm.nih.gov/pmc/articles/PMC2694319/.

Arvola, T., Laiho, K., Torkkeli, S., Mykkänen, H., Salminen, S., Maunula, L., Isolauri, E. (1999). Prophylactic *Lactobacillus* GG reduces antibiotic-associated diarrhea in children with respiratory infections: a randomized study. *Pediatr.* 104, 1-4.

Aso, Y., Akaza, H., Kotake, T., Tsukamoto, T., Imai, K., Naito, S. (1995). Preventive effect of a *Lactobacillus casei* preparation on the recurrence of superficial bladder cancer in a double-blind trial. The BLP study group. *Eur. Urol.* 27, 104-109.

Aso, Y., Akazan, H. (1992). Prophylactic effect of a *Lactobacillus casei* preparation on the recurrence of superficial bladder cancer. *Urol Int.* 49, 125-129.

Bai, A.P., Ouyang, Q., Zhang, W., Wang, C.H., Li, S.F. (2004). Probiotics inhibit TNF-α-induced interleukin-8 secretion of HT29 cells. *World J. Gastroenterol.* 10, 455-457.

Basu, S., Chatterjee, M., Ganguly, S., Chandra, P.K. (2007). Effect of *Lactobacillus rhamnosus* GG in persistent diarrhea in Indian children: a randomized controlled trial. *J. Clin. Gastroentero1.* 41, 756-760.

Bernet, M.F., Brassart, D., Neeser, J.R. Servin, A.L. (1994). *Lactobacillus acidophilus* LA 1 binds to cultured human intestinal cell lines and inhibits cell attachment and cell invasion by enterovirulent bacteria. *Gut* 35, 483-489.

Bodana, A.R., Rao, D.R. (1990). Antimutagenic activity of milk fermented by *Streptococcus thermophilus* and *Lactobacillus bulgaricus*. *J. Dairy Sci.* 73, 3379-3384.

Brewer, S.C., Wunderink, R.G., Jones, C.B., Leeper K.V. (1996). Ventilator-associated pneumonia due to *Pseudomonas aeruginosa. Chest* 109, 1019-1029.

Bruce, A.W., Reid, G. (1988). Intravaginal instillation of lactobacilli for prevention of recurrent urinary tract infections. *Can. J. Microbiol.* 34, 339-343.

Cannon, J.P, Lee, T.A., Bolanos, J.T., Danziger, L.H. (2005). Pathogenic relevance of *Lactobacillus*: a retrospective review of over 200 cases. Eur. J. Clin. Microbiol. Infect. Dis. 24, 31-40.

Cano, P.G., Perdigon, G. (2003). Probiotics induce resistance to enteropathogens in a renourished mouse model. *J. Dairy Res.* 70, 433-440.

Chauvière, G., Coconnier, M.H., Kerméis, S., Fourniat, J., Servin, A.L. (1992). Adhesion of human *Lactobacillus acidophilus* strain LB to human enterocyte-like Caco-2 cells. *J. Gen. Microbiol.* 138, 1689-1696.

Cianci, A., Giordano, R., Delia, A., Grasso, E., Amodeo, A., De Leo, V., Caccamo, F. (2008). Efficacy of *Lactobacillus rhamnosus* GR-1 and of *Lactobacillus reuteri* RC-14 in the treatment and prevention of vaginosis and bacterial vaginitis relapses. *Minerva Ginecol.* 60, 369-376.

Coconnier, M.H., Bernet, M.F., Kernéis, S., Chauvière, G., Fourniat, J., Servin, A.L. (1993). Inhibition of adhesion of enteroinvasive pathogens to human intestinal Caco-2 cells by *Lactobacillus acidophilus* strain LB decreases bacterial invasion. *FEMS Microbiol. Lett.* 110, 299-305.

Coconnier, M.H., Lievin, V., Hemery, E., Servin, A.L. (1998). Antagonistic activity against *Helicobacter* infection *in vitro* and *in vivo* by the human *Lactobacillus acidophilus* strain LB. *Appl. Environ. Microbiol.* 64, 4573-4580.

Coconnier, M.H., Liévin, V., Lorrot, M., Servin, A.L. (2000). Antagonistic activity of *Lactobacillus acidophilus* LB against intracellular *Salmonella enterica* serovar

Typhimurium infecting human enterocyte-like Caco-2/TC-7 cells. *Appl. Environ. Microbiol.* 66, 1152-1157.

Corr, S.C., Li, Y., Riedel, C.U., O'Toole, P.W., Hill, C., Gahan, C.G.M. (2007). Bacteriocin production as a mechanism for the antiinfective activity of *Lactobacillus salivarius* UCC118. *Proc. Natl. Acad. Sci. USA.* 104, 7617-7621.

Crowell, D.C. (1998). Microbial analysis of human intestinal flora after feeding *Lactobacillus acidophilus*. M.S. Thesis, North Carolina State Univ., Raleigh.

Cunha, B.A. (2001). Antibiotic side effects. *Med. Clin. North Am.* 85, 149-185.

Cunningham, S., Ahrné, S., Bengmark, S., Johann-Liang, R., Marshall, F., Metakis, L., Califano, C., Dunn, A.M., Grassey, C., Hinds, G., Cervia, J. (2000) Probiotics and immune response. *Am. J. Gastroenterol.* 95, S22-S25.

De Santis, A., Famularo, G., De Simone, C. (2000). Probiotics for the hemodynamic alterations of patients with liver cirrhosis. *Am. J. Gastroenterol.* 95, 323-324.

de Vrese, M., Stegelmann, A., Richter, B., Fenselau, S., Laue, C., Schrezenmeir, J. (2001). Probiotics-compensation for lactase insufficiency. *Am. J. Clin. Nutr.* 73, 421S-429S.

de Waard, R., Garssen, J., Bokken, G.C.A.M., Vos, J.G. (2002). Antagonistic activity of *Lactobacillus casei* strain Shirota against gastrointestinal *Listeria monocytogenes* infection in rats. *Int. J. Food Microbiol.* 73, 93-100.

Delcenserie, V., Martel, D., Lamoureux, M., Amiot, J., Boutin, Y., Roy, D. (2008). Immunomodulatory effects of probiotics in the intestinal tract. *Curr. Issues Mol. Biol.* 10, 37-54.

Dunne, C., O'Mahony, L., Murphy, L., Thornton, G., Morrissey, D., O'Halloran, S., Feeney, M., Flynn, S., Fitzgerald, G., Daly, C., Kiely, B., O'Sullivan, G.C., Shanahan, F., Collins J.K. (2001). In vitro selection criteria for probiotic bacteria of human origin: correlation with in vivo findings. *Am. J. Clin. Nutr.* 73, 386S-392S.

Ejtahed, H.S., Mohtadi-Nia, J., Homayouni-Rad, A., Niafar, M., Asghari-Jafarabadi, M., Mofid, V., Akbarian-Moghari, A. (2011). Effect of probiotic yogurt containing *Lactobacillus acidophilus* and *Bifidobacterium lactis* on lipid profile in individuals with type 2 diabetes mellitus. *J. Dairy Sci.* 94, 3288-3294.

Fajardo, P., Fuciños, C., Méndez, J., Pastrana, L., Pérez-Guerra, N. (2008). Performance and intestinal coliform counts in weaned piglets fed a probiotic culture (*Lactobacillus casei* ssp. *casei* CECT 4043) or an antibiotic. *J. Food Prot.* 71, 1797-1805.

Falagas, M.E., Betsi, G.I., Athanasiou, S. (2006). Probiotics for prevention of recurrent vulvovaginal candidiasis: a review. *J. Antimicrob. Chemother.* 58, 266-272.

Fergie, J.E., Shema, S.J., Lott, L., Crawford, R., Patrick, C.C. (1994). *Pseudomonas aeruginosa* bacteremia in immunocompromised children: analysis of factors associated with a poor outcome. *Clin. Infect. Dis.* 18, 390-394.

Forsum, U., Holst, E., Larsson, P. G., Vasquez, A., Jakobsson, T. and Mattsby-Baltzer, I. (2005). Bacterial vaginosis – a microbiological and immunological enigma. *Acta Pathol. Microbiol. Immunol. Scand.* 113, 81–90.

Fujiya, M., Kohgo, Y. (2010). Novel perspectives in probiotic treatment: the efficacy and unveiled mechanisms of the physiological functions. *Clin. J. Gastroenterol.* 3, 117-127.

Gardiner, G.E., Casey, P.G., Casey, G., Lynch, P.B., Lawlor, P.G., Hill, C., Fitzgerald, G.F., Stanton, C., Ross, R.P. (2004). Relative ability of orally administered *Lactobacillus murinus* to predominate and persist in the porcine gastrointestinal tract. *Appl. Environ. Microbiol.* 70, 1895-1906.

Gilliland, S.E., Speck, M.L. (1977). Antagonistic action of *Lactobacillus acidophilus* toward intestinal and foodborne pathogens in associative cultures. *J. Food Prot.* 40, 820-823.

Glück, U., Gebbers, J.O. (2003). Ingested probiotics reduce nasal colonization with pathogenic bacteria (*Staphylococcus aureus*, *Streptococcus pneumoniae*, and β-hemolytic streptococci). *Am. J. Clin. Nutr.* 77, 517-520.

Gobbetti1, M., Ferranti, P., Smacchi, E., Goffredi, F., Addeo, F. (2000). Production of angiotensin-I-converting-enzyme-inhibitory peptides in fermented milks started by *Lactobacillus delbrueckii* subsp. *bulgaricus* SS1 and *Lactococcus lactis* subsp. *cremoris* FT4. *Appl. Environ. Microbiol.* 66, 3898-3904.

Goldin, B.R., Gorbach, S.L. (1980). Effect of *Lactobacillus acidophilus* dietary supplements on 1,2-dimethylhydrazine dihydrochloride-induced intestinal cancer in rats. *J. Natl. Cancer Inst.* 64, 263-265.

Greene, J.D., Klaenhammer, T.R. (1994). Factors involved in adherence of lactobacilli to human Caco-2 cells. *Appl. Environ. Microbiol.* 60, 4487–4494.

Guerra, N. P., Fajardo, P., Méndez, J., Cachaldora, P., Pastrana, L. (2007). Production of four potentially probiotic lactic acid bacteria and their evaluation as feed additives for weaned piglets. *Anim. Feed Sci. Technol.* 134, 89-107.

Hatakka, K., Martio, J., Korpela, M., Herranen, M., Poussa, T., Laasanen, T., Saxelin, M., Vapaatalo, H., Moilanen, E., Korpela, R. (2003). Effects of probiotic therapy on the activity and activation of mild rheumatoid arthritis--a pilot study. *Scand. J. Rheumatol.* 32, 211-215.

Helin, T., Haahtela, S., and Haahtela, T. (2002). No effect of oral treatment with an intestinal bacterial strain, *Lactobacillus rhamnosus* (ATCC 53103), on birch-pollen allergy: a placebo-controlled double-blind study. *Allergy* 57, 243-246.

Hickson, M., D'Souza, A.L., Muthu, N., Rogers, T.R., Want, S., Rajkumar, C., Bulpitt, C.J. (2007). Use of probiotic *Lactobacillus* preparation to prevent diarrhoea associated with antibiotics: randomised double blind placebo controlled trial. *B.M.J.* 335, 80.

Hilton, E., Isenberg, H.D., Alperstein, P., France, K., Borenstein, M.T. (1992). Ingestion of yogurt containing *Lactobacillus acidophilus* as prophylaxis for candidal vaginitis. *Ann. Intern. Med.* 116, 353-357.

Huang, S.H., He, L., Zhou, Y., Wu, C.H., Jong, A. (2009). *Lactobacillus rhamnosus* GG suppresses meningitic *E. coli* K1 penetration across human intestinal epithelial cells in vitro and protects neonatal rats against experimental hematogenous meningitis. *Int. J. Microbiol.* 2009: http://downloads.hindawi.com/journals/ijmb/2009/647862.pdf.

Jain, P.K., McNaught, C.E., Anderson, A.D.G., MacFie, J., Mitchell, C.J. (2004). Influence of synbiotic containing *Lactobacillus acidophilus* La5, *Bifidobacterium lactis* Bb 12, *Streptococcus thermophilus*, *Lactobacillus bulgaricus* and oligofructose on gut barrier function and sepsis in critically ill patients: a randomised controlled trial. *Clin. Nut.* 23, 467-475.

Joint FAO/WHO Working Group (2002). Guidelines for the evaluation of probiotics in food: report of a Joint FAO/WHO Working Group on Drafting Guidelines for the Evaluation of Probiotics in Food, London, Ontario, Canada. Available at: http://www.who.int/food safety/fs_management/en/probiotic_guidelines.pdf on Nov. 7, 2011.

Jones, P.J. (2002). Clinical nutrition: 7. Functional foods - more than just nutrition. *C.M.A.J.* 166, 1555-1563.

Kale, V.V., Trivedi, R.V., Wate, S.P., Bhusari, K.P. (2005). Development and evaluation of a suppository formulation containing *Lactobacillus* and its application in vaginal diseases. *Ann. N. Y. Acad. Sci.* 1056, 359-365.

Kalliomäki, M., Salminen, S., Arvilommi, H., Kero, P., Koskinen, P., Isolauri, E. (2001). Probiotics in primary prevention of atopic disease: a randomized placebo-controlled trial. *Lancet* 357, 1076-1079.

Kim, G.S., Gilliland, S.E. (1983). *Lactobacillus acidophilus* as a dietary adjunct for milk to aid lactose digestion in humans. *J. Dairy Sci.* 66, 959-966.

Kim, H.J., Camilleri, M., McKinzie, S., Lempke, M.B., Burton, D.D., Thomforde, G.M., Zinsmeister, A.R. (2003). A randomized controlled trial of a probiotic, VSL#3, on gut transit and symptoms in diarrhoea-predominant irritable bowel syndrome. *Aliment. Pharmacol. Ther.* 17, 895-904.

Kim, H.J., Vazquez, M.I., Camilleri, M., Stephens, D., Burton, D.D., Baxter, K., Thomforde, G., Zinsmeister, A.R. (2005). A randomized controlled trial of a probiotic combination VSL# 3 and placebo in irritable bowel syndrome with bloating. *Neurogastroenterol. Motil.* 17, 687-696.

Kitazawa, H., Watanabe, H., Shimosato, T., Kawai, Y., Itoh, T., Saito, T. (2003). Immunostimulatory oligonucleotide, CpG-like motif exists in *Lactobacillus delbrueckii* ssp. *bulgaricus* NIAI B6. *Int. J. Food Microbiol.* 85, 11-21.

Klaver, F.A., van der Meer, R. (1993). The assumed assimilation of cholesterol by lactobacilli and *Bifidobacterium bifidum* is due to their bile salt-deconjugating activity. *Appl. Environ. Microbiol.* 59, 1120-1124.

Kleeman, E.G., Klaenhammer, T.R. (1982). Adherence of *Lactobacillus* species to human fetal intestinal cells. *J. Dairy Sci.* 65, 2063-2069.

Koebnick, C., Wagner, I., Leitzmann, P., Stern, U., Zunft, H.J. (2003). Probiotic beverage containing *Lactobacillus casei* Shirota improves gastrointestinal symptoms in patients with chronic constipation. *Can. J. Gastroenterol.* 17, 655-659.

Konturek, J.W. (2003). Discovery by Jaworski of *Helicobacter pylori* and its pathogenetic role in peptic ulcer, gastritis and gastric cancer. *J. Physiol. Pharmacol.* 54, 23-41.

Kruis, W., Fric, P., Pokrotnieks, J., Lukás, M., Fixa, B., Kascák, M., Kamm, M.A., Weismueller, J., Beglinger, C., Stolte, M., Wolff, C., Schulze, J. (2004). Maintaining remission of ulcerative colitis with the probiotic *Escherichia coli* Nissle 1917 is as effective as with standard mesalazine. *Gut* 53, 1617-1623.

Kukkonen, K., Savilahti, E., Haahtela, T., Juntunen-Backman, K., Korpela, R., Poussa, T., Tuure, T., Kuitunen, M. (2007). Probiotics and prebiotic galactooligosaccharides in the prevention of allergic diseases: a randomized, double-blind, placebo-controlled trial. *J. Allergy Clin. Immunol.* 119, 192-198.

La Rosa, M., Bottaro, G., Gulino, N., Gambuzza, F., Di Forti, F., Ini, G., Tornambe, E. (2003). Prevention of antibiotic-associated diarrhea with *Lactobacillus sporogens* and fructo-oligosaccharides in children. A multicentric double-blind vs placebo study. *Minerva Pediatr.* 55, 447-452.

Lamine, F., Eutamène, H., Fioramonti, J., Buéno, L., Théodorou, V. (2004). Colonic responses to *Lactobacillus farciminis* treatment in trinitrobenzene sulphonic acid-induced colitis in rats. Scand. *J. Gastroenterol.* 39, 1250-1258.

Liévin-Le Moal, Amsellem, V.R., Servin, A.L., Coconnier, M.H. (2002). *Lactobacillus acidophilus* (strain LB) from the resident adult human gastrointestinal microflora exerts activity against brush border damage promoted by a diarrhoeagenic *Escherichia coli* in human enterocyte-like cells. *Gut* 50, 803-811.

Liévin-Le Moal, V., Sarrazin-Davila, L.E., Servin, A.L. (2007). An experimental study and a randomized, double-blind, placebo-controlled clinical trial to evaluate the antisecretory activity of *Lactobacillus acidophilus* strain LB against nonrotavirus diarrhea. *Pediatr.* 120, 795-803.

Mackay, A.D., Taylor, M.B., Kibbler, C.C., Hamilton-Miller, J.M. (1999). *Lactobacillus endocarditis* caused by a probiotic organism. *Clin. Microbiol. Infect.* 5, 290-292.

Majamaa H, Isolauri E. (1997). Probiotics: a novel approach in the management of food allergy. *J. Allergy Clin. Immunol.* 99, 179-185.

Mangell, P., Nejdfors, P., Wang, M., Ahrné, S., Weström, B., Thorlacius, H., Jeppsson, B. (2002). *Lactobacillus plantarum* 299v inhibits *Escherichia coli*-induced intestinal permeability. *Dig. Dis. Sci.* 47, 511-516.

Mattar, A.F., Drongowski, R.A., Coran, A.G., Harmon, C.M. (2001). Effect of probiotics on enterocyte bacterial translocation in vitro. Pediatr. Surg. Int. 17, 265-268.

Meisel, H. (1993). Casokinins as inhibitors of angiotensin-converting-enzyme, p. 153-159. In G. Sawatzki and B. Renner (ed.), New perspectives in infant nutrition. Thieme, Stuttgart, Germany.

Mendelson, M.H., Gurtman, A., Szabo, S., Neibart, E., Meyers, B.R., Policar, M., Cheung, T.W., Lillienfeld, D., Hammer, G., Reddy, S., Choi, K., Hirschman, S.Z. (1994). *Pseudomonas aeruginosa* bacteremia in patients with AIDS. *Clin. Infect. Dis.* 18, 886-895.

Metts, J., Famula, T.R., Trenev, N., Clemens, R.A. (2003). *Lactobacillus acidophilus*, strain NAS (H_2O_2 positive), in reduction of recurrent candidal vulvovaginitis. *J. Appl. Res.* 3, 340-348.

Mohan, J.C., Arora, R., Khalilullah, M. (1990). Short term hypolipidemic effects of oral *Lactobacillus sporogenes* therapy in patients with primary dyslipidemias. *Indian Heart J.* 42, 361-364.

Mollet, B., Rowland, I. (2002). Functional foods: at the frontier between food and pharma. *Curr. Opin. Biotechnol.* 13, 483-485.

Montes, R.G., Bayless, T.M., Saavedra, J.M., Perman, J.A. (1995). Effect of milks inoculated with *Lactobacillus acidophilus* or a yogurt starter culture in lactose-maldigesting children. *J. Dairy Sci.* 78, 1657-1664.

No author. (2009). World Gastroenterology Organisation practice guideline: Probiotics and prebiotics. *Arab. J. Gastroenterol.* 10, 33-42.

Nobaek, S., Johansson, M.L., Molin, G., Ahrné, S., Jeppsson, B. (2000). Alteration of intestinal microflora is associated with reduction in abdominal bloating and pain in patients with irritable bowel syndrome. Am. *J. Gastroenterol.* 95, 1231-1238.

Niedzielin, K., Kordecki, H., Birkenfeld, B. (2001). A controlled, double-blind, randomized study on the efficacy of Lactobacillus plantarum 299V in patients with irritable bowel syndrome. *Eur. J. Gastroenterol. Hepatol.* 13, 1143-1147.

Oberreuther-Moschner, D.L., Jahreis, G., Rechkemmer, G., Pool-Zobel, B.L. (2004). Dietary intervention with the probiotics *Lactobacillus acidophilus* 145 and *Bifidobacterium longum* 913 modulates the potential of human faecal water to induce damage in HT29clone19A cells. *Br. J. Nutr.* 91, 925-932.

Oláh, A., Belágyi, T. Issekutz, Á., Gamal, M.E., Benkmark, S. (2002). Randomized clinical trial of specific *Lactobacillus* and fibre supplement to early enteral nutrition in patients with acute pancreatitis. *Br. J. Surg.* 89, 1103-1107.

O'Mahony, L., McCarthy, J., Kelly, P., Hurley, G., Luo, F., Chen, K., O'Sullivan, G.C., Kiely, B., Collins, J.K., Shanahan, F., Quigley, E.M. (2005). *Lactobacillus* and *Bifidobacterium* in irritable bowel syndrome: symptom responses and relationship to cytokine profiles. *Gastroenterol.* 128, 541-551.

Peral, M.C., Martinez, M.A., Valdez, J.C. (2009). Bacteriotherapy with *Lactobacillus plantarum* in burns. *Int. Wound J.* 6, 73-81.

Prantera, C., Scribano, M.L., Falasco, G., Andreoli, A., Luzi, C. (2002). Ineffectiveness of probiotics in preventing recurrence after curative resection for Crohn's disease: a randomised controlled trial with *Lactobacillus* GG. *Gut* 51, 405-409.

Rangavajhyala, N., Shahani, K.M., Sridevi, G., Srikumaran, S. (1997). Nonlipopolysaccharide component(s) of *Lactobacillus acidophilus* stimulate(s) the production of interleukin-1 alpha and tumor necrosis factor-alpha by murine macrophages. *Nut. Cancer* 28, 130-134.

Rautio, M., Jousimies-Somer, H., Kauma, H., Pietarinen, I., Saxelin, M., Tynkkynen, S., Koskela, M. (1999). Liver abscess due to a *Lactobacillus rhamnosus* strain indistinguishable from *L. rhamnosus* strain GG. *Clin. Infect. Dis.* 28, 1159-1160.

Rayes, N., Seehofer, D., Hansen, S., Boucsein, K., Müller, A.R., Serke, S., Bengmark, S., Neuhaus, P. (2002a). Early enteral supply of *Lactobacillus* and fiber versus selective bowel decontamination: A controlled trial in liver transplant recipients. *Transplant.* 74, 123-128.

Rayes, N., Hansen, S., Seehofer, D., Müller, A.R., Serke, S., Bengmark, S., Neuhaus, P. (2002b). Early enteral supply of fiber and lactobacilli versus conventional nutrition: a controlled trial in patients with major abdominal surgery. *Nutr.* 18, 609-615.

Reid, G., Charbonneau, D., Erb, J., Kochanowski, B., Beuerman, D., Poehner, R., Bruce, A.W. (2003). Oral use of *Lactobacillus rhamnosus* GR-1 and *L. fermentum* RC-14 significantly alters vaginal flora: randomized, placebo-controlled trial in 64 healthy women. *FEMS Immunol. Med. Microbiol.* 35, 131-134.

Reid, G., Kirjaivanen, P. (2005). Taking probiotics during pregnancy. Are they useful therapy for mothers and newborns? *Can. Fam. Physician.* 51, 1477-1479.

Roessler, A., Friedrich, U., Vogelsang, H., Bauer, A., Kaatz, M., Hipler, U.C., Schmidt, I., Jahreis, G. (2007). The immune system in healthy adults and patients with atopic dermatitis seems to be affected differently by a probiotic intervention. *Clin. Exp. Allergy* 38, 93-102.

Roos, S., Engstrand, L., Jonsson, H. (2005). *Lactobacillus gastricus* sp. nov., *Lactobacillus antri* sp. nov., *Lactobacillus kalixensis* sp. nov. and *Lactobacillus ultunensis* sp. nov., isolated from human stomach mucosa. *Int. J. Syst. Evol. Microbiol.* 55, 77-82.

Ryan, K.A., Daly, P., Li, Y., Hooton, C., O'Toole, P.W. (2008). Strain-specific inhibition of *Helicobacter pylori* by *Lactobacillus salivarius* and other lactobacilli. *J. Antimicrob. Chemother.* 61, 831-834.

Ryan, K.A., van Doorn, L.J., Moran, A.P., Glennon, M., Smith, T., Maher, M. (2001). Evaluation of clarithromycin resistance and cagA and vacA genotyping of *Helicobacter pylori* strains from the west of Ireland using line probe assays. *J. Clin. Microbiol.* 39, 1978-1980.

Salminen, M.K., Tynkkynen, S., Rautelin, H., Saxelin, M., Vaara, M., Ruutu, P., Sarna, S., Valtonen, V., Järvinen, A. (2002). *Lactobacillus* bacteremia during a rapid increase in probiotic use of *Lactobacillus rhamnosus* GG in Finland. *Clin. Infect. Dis.* 35, 1155-1160.

Saltzman, J.R., Russell, R.M., Golner, B., Barakat, S., Dallal, G.E., Goldin, B.R. (1999). A randomized trial of *Lactobacillus acidophilus* BG2FO4 to treat lactose intolerance. *Am. J. Clin. Nutr.* 69, 140-146.

Sanders, M.E., Walker, D.C., Walker, K.M., Aoyama, K., Klaenhammer, T.R. (1996). Performance of commercial cultures in fluid milk applications. *J. Dairy Sci.* 79, 943-955.

Sanders, M.E., Klaenhammer, T.R. (2001). The scientific basis of *Lactobacillus acidophilus* NCFM functionality as a probiotic. *J. Dairy Sci.* 84, 319-331.

Sarker, S.A., Sultana, S., Fuchs, G.J., Alam, N.H., Azim, T., Brüssow, H., Hammarström, L. (2005). *Lactobacillus paracasei* strain ST11 has no effect on rotavirus but ameliorates the outcome of nonrotavirus diarrhea in children from Bangladesh. *Pediatr.* 116, 221 228.

Schultz, M., Timmer, A., Herfarth, H.H., Sartor, R.B., Vanderhoof, J.A., Rath, H.C. (2004). *Lactobacillus* GG in inducing and maintaining remission of Crohn's disease. BMC Gastroenterol. 4, http://www.biomedcentral.com/content/pdf/1471-230X-4-5.pdf.

Sgouras, D., Maragkoudakis, P., Petraki, K., Martinez-Gonzalez, B., Eriotou, E., Michopoulos, S., Kalantzopoulos, G., Tsakalidou, E., Mentis, A. (2004). In vitro and in vivo inhibition of *Helicobacter pylori* by *Lactobacillus casei* strain Shirota. *Appl. Environ. Microbiol.* 70, 518-526.

Sgouras, D.N., Panayotopoulou, E.G., Martinez-Gonzalez, B., Petraki, K., Michopoulos, S., Mentis, A. (2005). *Lactobacillus johnsonii* La1 attenuates *Helicobacter pylori*-associated gastritis and reduces levels of proinflammatory chemokines in C57BL/6 mice. *Clin. Diagn. Lab. Immunol.* 12, 1378-1386.

Sharma, J.K., Kapoor, K.K., Mukhija, R.D. (1980). Clinical trial of Sporlac in the treatment of recurrent aphthous ulceration. *Uttar Pradesh State Dent. J.* 11, 7-12.

Shornikova, A., Casas, I.A., Isolauri, E., Mykkänen, N., Vesikari, T. (1997). *Lactobacillus reuteri* as a therapeutic agent in acute diarrhea in young children. *J. Pediatr. Gastroenterol.* Nutr. 24, 399-404.

Slover, C.M., Danziger, L. (2008). *Lactobacillus*: a Review. *Clin. Microbiol. Newslett.* 30, 23-27.

Szajewska, H., Mrukowicz, J.Z. (2001). Probiotics in the treatment and prevention of acute infectious diarrhea in infants and children: a systematic review of published randomized, double-blind, placebo-controlled trials. *J. Pediatr. Gastroenterol. Nutr.* 33, S17-S25.

Tabasco, R., García-Cayuela, T., Peláez, C., Requena, T. (2009). *Lactobacillus acidophilus* La-5 increases lactacin B production when it senses live target bacteria. *Int. J. Food Microbiol.* 132, 109-116.

Taylor, A.L., Dunstan, J.A., and Prescott, S.L. (2007). Probiotic supplementation for the first 6 months of life fails to reduce the risk of atopic dermatitis and increases the risk of allergen sensitization in high-risk children: a randomized controlled trial. *J. Allergy Clin. Immunol.* 119, 184-191.

Tien, M.T., Girardin, S.E., Regnault, B., Le Bourhis, L., Dillies, M.A., Coppee, J.Y., Bourdet-Sicard, R., Sansonetti, P.J., Pedron, T. (2006). Antiinflammatory effect of *Lactobacillus casei* on *Shigella*-infected human intestinal epithelial cells. *J. Immunol.* 176, 1228-1237.

Trop, M., Novak, M., Rodl, S., Hellbom, B., Kroell, W., Goessler, W. (2006). Silver-coated dressing acticoat caused raised liver enzymes and argyria-like symptoms in burn patient. *J. Trauma* 60, 648-652.

Uemura, N., Okamoto, S., Yamamoto, S., Matsumura N, Yamaguchi, S., Yamakido, M., Taniyama, K., Sasaki, N., Schlemper, R.J. (2001). *Helicobacter pylori* infection and the development of gastric cancer. *N. Engl. J. Med.* 345, 784-789.

Valdéz, J.C., Peral, M.C., Rachid, M., Santana, M., Perdigón G. (2005). Interference of *Lactobacillus plantarum* with *Pseudomonas aeruginosa in vitro* and in infected burns: the potential use of probiotics in wound treatment. *Clin. Microbiol. Infect.* 11, 472-479.

Vendt, N., Grünberg, H., Tuure, T., Malminiemi, O., Wuolijoki, E., Tillmann, V., Sepp, E., Korpela. R. (2006). Growth during the first 6 months of life in infants using formula enriched with *Lactobacillus rhamnosus* GG: double-blind, randomized trial. *J. Hum. Nutr. Diet.* 19, 51-58.

Voĭchishina, L.G., Chaplinskiĭ, V., V'iunitskaia, V.A. (1991). The use of sporulating bacteria in treating patients with dysbacteriosis. *Vrach. Delo.* 12, 73-75.

Wagner, R.D., Pierson, C., Warner, T., Dohnalek, M., Farmer, J., Roberts, L., Hilty, M., Balish, E. (1997). Biotherapeutic effects of probiotic bacteria on candidiasis in immunodeficient mice. *Infect. Immun.* 65, 4165-4172.

Weese, J.S. (2002). Microbiologic evaluation of commercial probiotics. *J. Am. Vet. Med. Assoc.* 220, 794-797.

Weizman, Z., Asli, G., Alsheikh, A. (2005). Effect of a probiotic infant formula on infections in child care centers: comparison of two probiotic agents. *Pediatrics* 115, 5-9.

Wollowski, I., Rechkemmer, G., Pool-Zobel, B.L. (2001). Protective role of probiotics and prebiotics in colon cancer. *Am. J. Clin. Nutr.* 73, 451S-455S.

Zhang, L., Li, N., Caicedo, R., Neu, J. (2005). Alive and dead *Lactobacillus rhamnosus* GG decrease tumor necrosis factor-alpha-induced interleukin-8 production in Caco-2 cells. *J. Nutr.* 135, 1752-1756.

In: Lactobacillus: Classification, Uses and Health Implications ISBN: 978-1-62081-151-1
Editors: Alba I. Perez Campos and Arturo Leon Mena

Chapter VI

Lactic Acid Bacteria in Meat and Fish: New Approaches for Traditional Applications

Silvina Fadda[1,*], Lucila Saavedra[1], Carolina Belfiore[2] and Graciela Vignolo[1]
[1]Centro de Referencia para Lactobacilos, CERELA-CONICET, Tucumán, Argentina
[2]PROIMI – CONICET, Tucumán, Argentina

Abstract

Lactic acid bacteria (LAB) are the main characters during fermentation of muscle-based foods, guaranteeing safety and sensorial quality of the products. The diversity of LAB resulting from the prevailing environmental conditions in processed meat and fish as well as the results in which LAB are involved as functional cultures are discussed. Particularly, the role of bacteriocinogenic *Enterococcus* during meat fermentation and the effect of curing agents on the antimicrobial activity is here described. In addition, LAB adaption to stressful conditions is analyzed by post-genomic technologies.

Introduction

Lactic Acid Bacteria (LAB) constitute the most important group of microorganisms involved in food fermentations in which they are either present as contaminants or deliberately added as starter cultures. They contribute to the flavor and texture characteristics of fermented products and inhibit food spoilage by producing large amounts of lactic acid and growth-inhibiting compounds. LAB are involved during yogurt, cheese, cultured butter, sour cream, sausages, cucumber pickles, olives and sauerkraut production among others, while some species may act as beer, wine and processed meat spoilers. Moreover, LAB are also

* Corresponding author: email: sfadda@cerela.org. ar.

considered the most important microorganisms responsible for the health-promoting effects of fermented foods, such as immunomodulation, intestinal integrity and pathogen resistance [124]. Indeed, strains of some species have traditionally been used as probiotics and added as functional bacteria in various food commodities [74]. Due to the huge economic significance of industrial applications of LAB as starters, biopreservatives and probiotics, a research emphasis on their metabolism, genetic and applications has been placed in the last 25 years [48].

Meat Fermentation

Fermentation and drying can be considered as the oldest way to preserve raw materials. Although the historical origin of fermented meat products remains unknown, fragmentary bibliographical research has traced it back more than 2500 years in China. Many of these products have been known in Europe since the 13-14th century after being introduced by Marco Polo. Preservative and palatability effects must have been experienced after mixing comminute fresh meat with salt contaminated with nitrate, spices or herbs, stuffed into animal intestines and then dried. Proof of sausage production was first documented in ancient Greece, where it may have been propitiated by the existing climate conditions [72]. This tradition was inherited by the Romans and from then on fermented sausages spread to Central, Eastern and Northern European countries as well as to America and Australia [32,40]. Despite the widespread production of fermented sausages, Europe is the major producer and consumer of these meat products; production and per capita consumption figures being the highest in Germany, Italy, Spain and France [75,33]. The remarkable technological advances and significant improvements in meat hygiene that occurred about fifty years ago have been capitalized for the development of a range of meat fermented products in which differences among countries and regions are the result from meat species availability, environmental conditions and traditions. Nevertheless, the stability of fermented meat products is mainly determined by a combination of acidification brought about by lactic acid bacteria (LAB) and lowering of water activity (a_w) during curing and drying. In addition, biochemical and physicochemical changes occur as a result of the interactions between microorganisms-meat and fat-processing technology that originates the wide range of available fermented sausages [130].

Fermented Sausages

Fermented sausages can be defined as a meat product made of a mixture of mainly beef and pork meat and less often of poultry, lamb, goat, horse, ostrich and game meat [132]; pork fat, salt, curing agents, sugar, spices and in many cases starter cultures are added. The mix, including as little oxygen as possible is placed into steam-permeable casings and subjected to fermentation and drying process.

Curing

When applying to meat, this term refers to shelf life improvement by preserving and avoiding spoilage with the help of salt, acid and drying and/or smoking. Originally meat curing was the addition of rock salt, sea salt, or mined salt to an unheated piece of meat or small cuts of meat to lower water activity, prevent microbial growth and chemical spoilage, tenderize the meat and add flavour to the product. Although there was difference on salts effectiveness, some were better than others at enhancing and stabilizing the produt's red color. Saltpeter (KNO_3) was recognized as the contaminant of these "better" salts. The preserving and preventing rancidity remained a secret until Polenske (1891) [90]. Then, Lehmann (1899) [68] confirmed that nitrite was the agent producing the red meat color and heat stability. Currently it is well known the mechanism by which nitrate is responsible for the stable colour of meat. Briefly, nitrite is formed from nitrate during meat fermentation by the action of nitrate-reducing bacteria (*Micrococcaceae*) Nitrite or nitrous acid formed, when reduced by microbial action, is able to sequester oxygen in the meat batter, chemically reacting in the inverse direction to form nitrate, thus preventing oxidative (rancidity) processes. Furthermore, nitrite undergoes chemical reductions to NO at pH 5.4-5.5, this binding to meat myoglobin to form the heat stable NO-myoglobin responsible for the typical bright red colour of fermented sausages [57]. Additionally, NO_2^- and NO_3^- anions bind Fe^{2+}/Fe^{3+}, which prevents the growth of certain microorganisms such as *Clostridium botulinum* [76]. NaCl is normally added at levels ranging from 2 to 4% depending on technology and market demands. Salt performs many functions including microbial growth suppression, a_w reduction, salt soluble proteins release and prooxidant effect. Nitrite and/or nitrate are added to the meat batter at a level of 150-250 ppm depending on the meat product and country regulations [57].

Other Ingredients

The use of ascorbates has become a common practice, the main objective being the improvement of the stability of the red nitrosylated pigment and the prevention of lipid oxidation [9]. Sugars are also added to fermented sausages, among which the most often used sugars are dextrose, glucose, sucrose and lactose as well as corn syrup and different starches [103]. The main role of sugars in fermented meat products is to act as substrates for LAB to produce lactic acid, the type of sugar influencing the rate of pH decline. Dextrose and glucose promote a more rapid acidification rate compared to the disaccharides lactose and sucrose [32]. Short-processed fermented products are usually supplemented with 0.5-0.7% glucose or sucrose or 1% lactose while for long-ripening dry sausages, common levels are around 0.3% glucose or sucrose or 0.5% lactose [101]. Spices are mostly which differentiate fermented sausages. Ground pepper (0.2-0.3%) is usually present in all types of sausages and, particularly in Mediterranean fermented sausages; they also may contain high levels (1-3%) of paprika and/or garlic. Whole mustard seed, coriander, ginger, cardamom, nutmeg and cloves, among others spices are also used in semidry fermented sausages formulation [26]. Spices have also proved to act as effective antioxidants, to stimulate LAB activity by supplying Mn and to be inhibitory against undesirable organisms [1,26]. During industrially produced fermented sausages, a variable number of other additives are also included, among which natural (cochineal and paprika extracts) and artificial colorants are added to improve cured-red pigment stability [100]. Variable amounts (0.5-3%) of exogenous proteins to assure protein gellation and phosphates to act as thickeners, humectants and gelling agents are often used. To achieve consumer demands for extremely savoury products, flavour enhancers (glutamic and guanylic acids) and flavouring agents (protein hydrolysates, herbs and smoke extracts) are also added [100]. For a rapid pH drop chemical acidulants such as lactic or citric acid or glucono-delta lactone may be used although different flavours in final products may be obtained [103].

Starter Cultures

The importance of lactic acid bacteria in the production of dry fermented sausages was described for the first time in 1940. These bacteria reduce the fermentation time, minimize manufacturing defects, improve sensory properties and inhibit the development of pathogenic and spoilage microbiota. The use of bacteria has progressed continuously and still today, the inoculation with starters guarantees the hygienic and sensory quality of fermented products such as dry sausages [43]. Bacteria used as starters are inoculated at a rate of 10^6 viable germs/g of meat mixture. In Europe, bacterial starters are made up of a balance between lactic acid bacteria (*Lactobacillus*) and cocci Gram positive bacteria (*Staphylococcus*, *Kocuria*). In Mediterranean countries, yeasts (*Debaryomyces*, *Candida*) and moulds (*Penicillium*) are also inoculated onto the surface in order to make up the covering flora of the dry sausage [119].

Lab Biodiversity in Fermented Meats

Food ecosystems constitute a large source of microbial diversity in which a number of bacterial communities coexist. In particular, meat ecosystems are able to support the growth of highly specific microbial associations, the presence of them depending on factors that persist during processing, transportation and storage. The ecological strategies adopted by microorganisms to grow in meat ecosystems are the consequence of the prevailing environmental conditions, in which the intrinsic and extrinsic factors governing microbial growth will determine the type and number of bacteria present in meat. Among these factors, those physicochemical (concentration and availability of nutrients, pH, redox potential, buffering capacity, a_w, meat structure) and related to storage and processing conditions (temperature and oxygen availability) are the most prevalent [129].

Table 1. Lactic acid bacteria species isolated from traditional fermented sausages

LAB	Source
Lb. sakei/curvatus/plantarum/bavaricus	Ferm* sausages (Spain)
Lb. sakei/curvatus/plantarum; *Carnobacterium sp.; Enterococcus sp.*	Salami (Greece)
Lb. sakei/curvatus; Pediococcus sp.	*Chorizo* (Spain)
Lb. sakei/curvatus/bavaricus;	Naples-type salami (Italy)
Lb. sakei/curvatus/plantarum/casei/brevis, alimentarius	Ferm sausages (Italy)

34LAB	Source
Lb. sakei/plantarum/curvatus; *Leuc. carnosum/gelidum/pseudomesenteroides* *P. pentosaceus*	*Salsiccia* and *Soppressata* (Italy)
Lb. sakei/curvatus/plantarum/paracasei	*Soppressata* (Italy)
Lb. sakei/curvatus/plantarum	Salami Friulano (Italy)
Lb. sakei/curvatus/plantarum/buchneri/ paracasei; E. faecium; *Pediococcus sp.* *Leuconostoc sp.*	Ferm sausages (Greece)
Lb. sakei/curvatus/plantarum; E. faecium	*Fuet* and *Chorizo* (Spain)
Lb. sakei/curvatus/casei; E. Casseliflavus, Leuc. mesenteroides; Lac. lactis	Fresh sausages (Italy)
Lb.sakei/curvatus/plantarum/brevis/paraplantarum; Lac. lactis; E. pseudoavium *Leuc. citreum/mesenteroides; W. hellenica*	Salami Friulano (Italy)
Lb. sakei/curvatus/plantarum/casei/ paraplantarum; Lac. lactis; Leuc. citreum/mesenteroides; E. faecium/pseudoavium Weissella sp.	Ferm sausages (Hungary, Greece and Italy)
Lb. sakei/curvatus/paracasei; Lac. garviae	Salami Fruilano (Italy)
Lb. sakei/curvatus/plantarum; Enterococcus	Ferm sausages (Argentina)
Lb. sakei/plantarum/paracasei/pentosus/ brevis/rhamnosus; Lac. lactis; E. faecium Leuconostoc sp.	Ferm sausages (Greece)
Lb. sakei/curvatus; Leuc. mesenteroides	*Chorizo*, *fuet* and *Salchichón* (Spain)
Lb. curvatus/plantarum/brevis; P. pentosaceus/acidilactici; Leuc. mesenteroides	*Salchichón* and *Chorizo* (Spain)
Lb. sakei/curvatus	*Soppressata* (Italy)
Lb. sakei/curvatus/plantarum; Lac. lactis	*Ciauscolo* salami (Italy)
Lb. sakei/plantarum/brevis; P. pentosaceus; Leuc. carnosum/mesenteroides	Ferm sausages (Italy)
Lb. sakei/curvatus/plantarum/pentosus/ rhamnosus/brevis/paracasei/alimentarius/ fermentum/bavaricus; *Lac. lactis;* *E. faecium; P. pentosaceus/acidilactici; Leuc. mesenteroides*	Ferm sausages (South-east Europe)
Lb. sakei/plantarum/paraplantarum/brevis/ zeae/rhamnosus/paracasei; E. faecium; Leuc. mesenteroides; W. cibaria/viridescens; P. acidilactici/pentosaceus	*Alheira* Ferm sausage (Portugal)

* Ferm: Fermented, Lb: Lactocacillus; Lac: Lactococcus; C: Carnobacterium; Leuc: Leuconostoc; W: Weissella; P: Pediococcus; E:Enterococcus.

Adapted from Fontana et al. (2011).

In traditional fermented sausages, *Lactobacillus sakei*, *Lactobacillus curvatus*, *Lactobacillus plantarum* and *Pediococcus* spp. are by far the most often isolated species among LAB, which must have adapted to the existing stringent conditions during meat fermentation. As a versatile bacterium, *Lb. plantarum* has been identified as part of the fermented sausages microbiota, however this species lacks the meat specialization found in *Lb. sakei* [22-23,59]. Less information is available on the physiology and genetics of *Lb. curvatus* other than the production of antibacterial substances. Although pediococci do not compose a relevant part of the microbial community of European fermented sausages, they are used as a starter culture in North American sausage style. Moreover, during spontaneous meat fermentation, different *Enterococcus* species can be found in relatively high numbers, contributing together with lactobacilli to meat fermentation [60].

In the last decades, microbial ecology has undergone profound changes as the result of the huge impact of emerging technologies applied to the analysis of natural communities. New molecular and genomic approaches are revealing a more dynamic and diverse view of the microbial world than originally anticipated. Culture-dependent and -independent molecular methods showed to provide a better knowledge of the microbial biodiversity in a wide range of environments establishing a link between function and complex relationships among the members of natural microbial consortia [129]. Moreover, the genome sequencing of *Lb. plantarum* [64], *Lb. sakei* [22] as well as *P. pentosaceus* [81] represented a great help in the design of appropriate primers and probes for the molecular identification, control and monitor of LAB species in fermented sausage ecosystem. From the picture of fermented sausage microbiota the predominance of *Lb. sakei* and *Lb. curvatus* emerges (Table 1). The isolation frequency of *Lb. plantarum* was not as high as that of the other two lactobacilli, although it has also been identified as an important species participating in sausage fermentation. In a significantly lower numbers, other *Lactobacillus* species were also reported such as *Lactobacillus bavaricus, Lactobacillus casei, Lactobacillus paracasei, Lactobacillus brevis, Lactobacillus paraplantarum, Lactobacillus alimentarius, Lactobacillus rhamnosus, Lactobacillus buchneri, Lactobacillus pentosus* and *Lactobacillus fermentum* [129]. As regard as pediococci, *P. pentosaceus* and *P. acidilactici* were often isolated during European sausages fermentation [12,14,66,2]. As shown in Table 1, species from the genera *Leuconostoc*, *Carnobacterium* and *Weissella* were sporadically isolated, underlining possible pitfalls in their identification or simply their low incidence in this ecological niche, these LAB genera being considered as undesirable spoilage during meat fermentation. More recently, *Lb. sakei* and *Lb. curvatus* were reported as the most frequently isolated lactobacilli in traditional fermented sausages produced in the North of Italy [27,29] while a vast number of LAB were isolated from a short ripened fermented product from Portugal [3]. Since the main species found in the sausages studied so far were the same among the different products, it can be suggested that plant-specific biotypes develop within the species [91]. This microbiota, usually referred to as "house flora" and resulting from the selective action of processing conditions, will predominate during fermentation and characterize the final product of a specific plant. These well adapted biotypes are promising strains for autochthonous starter cultures development that will enable sausages to be produced with both high higienic and sensory quality [19,118].

Ocurrence of Enterococci in Meat and Fermented Meat Products

In food environments, enterococci have always been associated with dairy products. Particularly in areas of southern Europe, they are recognized to play a role in the development of sensory characteristics during ripening of many cheeses, such as artisanal cheeses from goat, ewe, or bovine milk in which they occur as nonstarter lactic acid bacteria (NSLAB) [50-51]. Some strains have been proposed in combination with both mesophilic and thermophilic LAB species as part of defined starter cultures for different Italian semi-cooked cheeses

[21,50-51,35]. Although the role of this LAB group in meat fermented products has not been fully elucidated.

In raw meat, they often appear as contaminant microorganisms and they are mainly transfer from the intestinal environment to food during slaughtering and conditioning processes. In a study conducted by Sustackova [116], *Enterococcus faecalis* (67%) and *Enterococcus faecium* (13.7%) were most frequently isolated species among contaminate chilled and frozen processed meat, heat-treated meat products and unheated fermented dry salami. In addition, from a total of 120 food samples analyzed in Brazil, a high prevalence of enterococci (52.5%) was reported, meat and cheese being the most contaminated [53]. In meat fermented products, a quite different scenery with significant differences in the presence and persistence of enterococci was observed [88,28,42,44,92,89,66,2]. From these studies, *E. faecium* was reported as the predominant species, followed by *E. faecalis, E. casseliflavus, E. gallinarum, E. pseudoavium and E. mundtii.* In spite of the controversial presence of *Enterococcus* in foods, *E. faecium* from meat, showed to have a much lower pathogenicity potential than clinical strains and indeed some of them are used as starter cultures due to their competitiveness in meat environment and their contribution to final flavour and safety [60,46]. Since the dominant microbiota of fermented sausages is mainly constituted by several species of *Lactobacillus* and coagulase-negative staphylococci (CNS), it seems that these microorganisms create an ecological niche for the optimal development of *Enterococcus* [60]. Moreover, due to enterococci are poor acidifiers in meat [8] their growth might become more difficult in sausages which contain a high content of fermentable sugars in the formulation. However, in artisanal sausages with high pH, they may find better conditions for survival and growth [60]. For example, in Iberian dry-cured sausages, a low-acid fermented meat products, *E. faecium* and *E. faecalis* represent one of the LAB species that are found in relatively high numbers [102]. Similarly, enterococci were reported to steadily increase their numbers from 10^2 to 10^5 CFU g^{-1} after 14 days of ripening of traditional Argentinean sausages manufactured without starter cultures [42].

Enterococci as Non-traditional Starter Culture for Meat Fermentation. *E. Mundtii* crl35, a Case Study

E. mundtii CRL35, a bacteriocinogenic strain isolated from artisanal cheeses produced in Argentina, was genetically and biochemically characterized in our laboratory [104,102,107,105,106].

In addition, this strain was found to lack intrinsic virulence factors and antibiotic resistance traits. Although the dairy origin of *E. mundtii* CRL35, and considering its safety characteristics, the potential to be used as a co-culture for cured fermented sausages, was analyzed. The competitiveness of *E. mundtii* CRL35 was evaluated in a meat fermentation model (MFM) added with the curing agents (CA) used for sausage production (3% NaCl, 100ppm $NaNO_2$ and 75ppm ascorbic acid).

Results showed that *E. mundtii* CRL35 was able to grow at 25°C in the MFM at a similar growth rate than the obtained in laboratory culture media.

The maximal bacterial growth was attained at 24 h in the absence of CA mixture, even with an initial slower growth, similar cell numbers at 72 h was obtained in the presence of curing additives (Figure 1A). These results are in agreement with the enterococcci competitiveness under meat fermented conditions as previously reported by others [18,61,5,45].

In contrast to the known low acidifying potential reported for enterococci [8,110], acidification rates achieved by *E. mundtii* CRL35 during growth in the MFM added with CA (4.6 at 72 h) were as expected for meat fermentation, this constituting a key advantage to be used as starter culture or co-culture to drive fermentation.

Competitiveness of *E. Mundtii* crl35 during Meat Fermentation

Bacteriocin Production

In the last thirty years, many studies have focused on bacteriocins from LAB due to their potential biotechnological applications. Many of them have shown both, *in vitro* and *in vivo* effectiveness against food-borne, human pathogens and spoilage microorganisms [107,41,54,20]. Since antibiotics are not allowed in food, the use of these natural antagonistics with preservative properties has become a trademark approach in food safety; bacteriocin-producing LAB or their purified bacteriocins used as extra hurdles for preservation is highly beneficial, preventing the outgrowth not only of food-borne pathogen but also other spoilage LAB [20].

However, some food additives or complex food environments were reported to affect bacteriocin production. Ingredients such as spices commonly applied in fermented sausage manufacture, nitrite and salt concentration were found to affect either the growth of bacteriocinogenic strain and/or the bacteriocin production [126-128]. The production of enterocin CRL35 by *E. mundtii* CRL35, an antilisterial class IIa bacteriocin [104,107,105] was investigated in MFM in the presence or absence of the curing mixture. In contrast to the findings by others, the maximal bacteriocin activity in MFM was observed in the presence of the complete CA mixture (data not shown).

Since food additives might produce inhibition to target organisms by themselves, *Listeria* inhibition experiments were performed adding individual additives or their combination to sterile MFM. Results showed a lack of inhibition, indicating no inhibitory effect of additives towards *Listeria* cells. When time-kill curves were performed with a synthetic enterocin CRL35 alone or in combination with the CA, a rapid decrease of *Listeria* counts occurred in the presence of CA, this evidencing an enhancing effect of curing additives on enterocin CRL35 activity (Figure 1B). These results are in disagreement with the negative effect of curing salts on LAB biopreservative potential reported by others [71,113]. Particularly, it has been shown that *E. faecium* CTC492 was able to lead dry sausage fermentation but not to produce enterocins A and B at an active antilisteria level because of the inhibitory effect of salt and pepper and the low pH [8].

Conversely, *E. faecium* CCM 4231 and *E. faecium* RZS were found to be only partially competitive during meat fermentation but strongly inhibitory against *Listeria* and improved organoleptic properties of sausages [18].

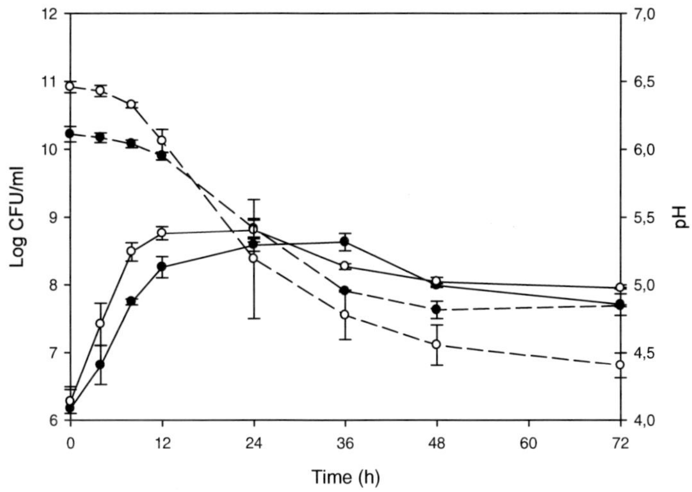

Figure 1A. Growth (solid line) and pH (dashed line) evolution of *Enterococcus mundtii* CRL35 at 25ºC in Meat Fermentation Model (MFM) during 72 h. MFM without curing additives, -CA (○), MFM with curing additives, +CA (●).Data correspond to three independent experiments.

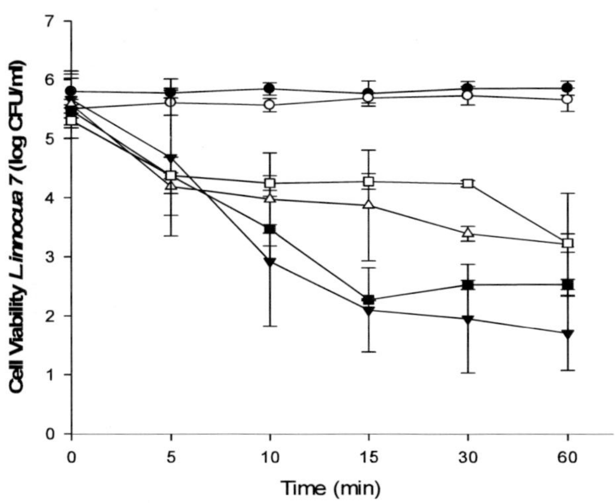

Figure 1B. Time–kill kinetics of S-Ent CRL35 (synthetic enterocin CRL35) and cell free meat supernatant (CFMS) with and without CA mixture against *L. innocua* 7. Cell viability was expressed as $\log_{10}$ CFU/ml. Control *Listeria* cells with (●) and without (○) CA mixture; S-Ent CRL35 with (▼) and without (Δ) CA mixture, CFMS with (■) and without (□) CA mixture. The error bars represent the standard deviations. Data correspond to three independent experiments.

Proteolytic Activity

It has been shown that during cured meat processing sarcoplasmic and myofibrillar proteins undergo proteolysis, which has a marked effect on product texture and flavor [108-109]. Muscle proteinases play a key role in the initial breakdown of meat proteins during the first days of fermentation [63, 121]. Although the role of microbial enzymes in this event has been accepted with great reluctance, LAB play a major role in secondary proteolysis producing small peptides and free amino acids, mainly by their intracellular amino, di and tripeptidase activity, contributing to sensory charcateristics of sausages as direct flavor enhancers or as precursors of other flavor compounds [85, 38-39, 37]. In this regard, enterococci are known as low proteolytic organisms with the exception of *E. faecalis* strains [6,7,117]. Most of the studies on enterococci proteolytic activity were performed using milk proteins as substrate [117,110]. Since up to date, there was no information about proteolytic activity of this LAB group on meat proteins, the hydrolytic ability of *E. mundtii* CRL35 was analyzed by SDS-PAGE. When sarcoplasmic proteins were used as substrate, the hydrolysis of higher molecular weight bands already at 24 h was observed, while in the presence of CA combination (NaCl, $NaNO_2$ and ascorbic acid) a faster and more intense proteolysis was carried out by *E. mundtii* CRL35 (Figures 2A,B). In addition, an increase of free amino acids was found at 72 h in the presence of ascorbic acid suggesting aminopeptidase activity from *E. mundtii* CRL35 (data not shown).

The competitiveness of *E. mundtii* CRL35 in a cured meat environment was demonstrated. Some key features exhibited by this strain point out its suitability to be used as starter or adjunct culture in meat fermentation: the lack of intrinsic virulence factors and antibiotic resistance traits, the rapid meat colonization, the improvement of organoleptic characteristics as well as the biopreservative potential.

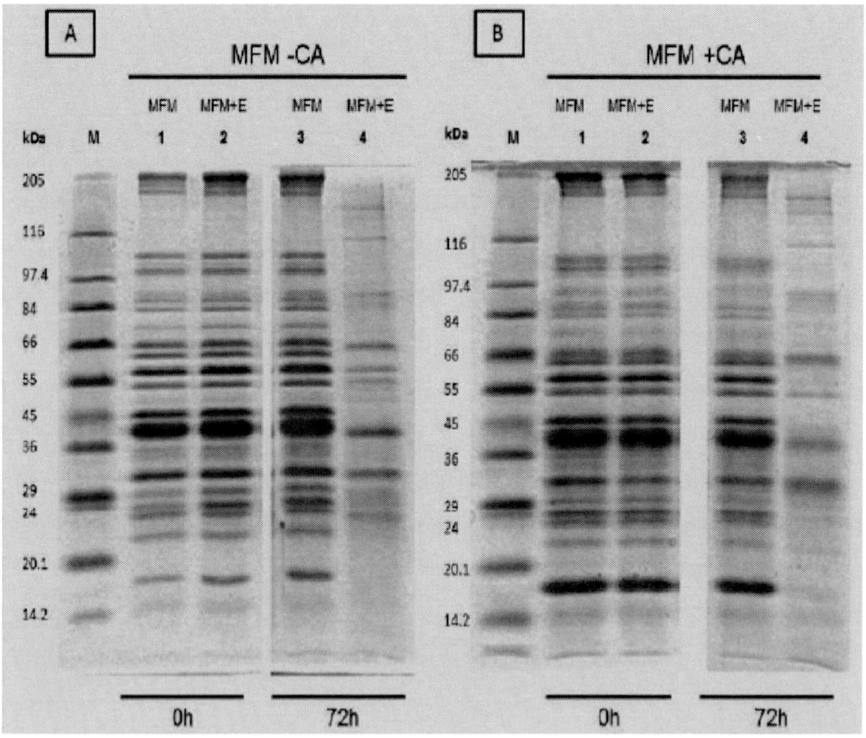

Figure 2AB. SDS-PAGE profiles of sarcoplasmic proteins from MFM inoculated with *E. mundtii* CRL35 during 72 h at 25°C. A MFM without additives (MFM - CA): lane M: Molecular weight marker; lane 1: MFM (non-innoculated control at 0 h); lane 2: MFM + *E. mundtii* CRL35 (0 h); lane 3: MFM (non-innoculated control at 72 h); lane 4: MFM + *E. mundtii* CRL35 (72 h).B. MFM supplemented with the curing additives mixture (MFM + CA). Lane M: Molecular weight marker; lane 1: MFM (non-innoculated control at 0 h); lane 2: MFM + *E. mundtii* CRL35 (0 h); lane 3: MFM (non-innoculated control at 72 h); lane 4: MFM + *E. mundtii* CRL35 (72h).

Fish Ecosystems

The range of fish products is very large including products prepared using a broad spectrum of both traditional and modern food technologies. Gram and Huss (1996)[55] reported fish ecosystems to fall into three categories: (i) lightly preserved fish products to which low level of salt (< 6%) and other preservatives are applied constituting a group of high-value delicatessen products (cold-smoked, pickled and marinated fish) that are consumed as ready-to-eat foods; (ii) salt cured and fermented fish added with high salt level (15-20%), involving dry-salted and wet-salted fish products and (iii) heat treated fish products. In addition, vacuum packaging is used to extend the shelf life and to keep the quality of fresh and processed fish foods, in which strict temperature control and storage at refrigeration is required to ensure food safety and to guarantee high product quality [78]. Today, salting is still an important preservation technique used to produce specialty products, in which the ripening process imparts specific desirable flavors [134].

Wet-salted fish constitute one of the oldest intermediate moisture foods, these products being available in markets around the world. In European countries this products are produced using herring (*Clupea harengus*), anchovy (*Engraulis encrasicholus*) and sprat (*C. sprattus*); in South America, anchovy (*E. anchoita and E. ringens*) is used in Argentina, Uruguay, Brazil and Peru while sardine (*Sardinella aurita*) is mostly used in Brazil. Traditional lightly preserved (salted) fermented fish are widespread in Africa and South-East of Asia. In Thailand, India and Nepal they are typically composed of freshwater or saltwater fish species, salt (2–7%), a carbohydrate source (boiled/roasted rice, millet, sugar or fruit) and highly spiced with garlic, ginger, chilli and pepper [87,120,65,4,67].

In particular, the biomass of anchovy (*Engraulis anchoita*), the most abundant fish species in the Southwest Atlantic Ocean, is considered to be around 4,300,000 tons, with a maximum sustainable yield of 1,140,000 tons [17], being at present under exploited at 27,800 tons/year (SAGPyA, 2008).

Recently, as part of a FAO project which addresses the use of wild fish to feed cultured fish and its implications to food security and poverty alleviation, a study on the use of small pelagic fish for new products developments, has been carried out [80]. Anchovy is processed for human consumption only in Argentina, where a diversified production (whole frozen, salt-cured, marinated and tinned anchovy fillets) targets mainly the external market. *E. anchoita* is described as a species with high-lipid content which undergoes significant variations according to the time of the year [137]. Due to lipid oxidation, anchovy is highly perishable; however high-polyunsaturated fatty acids (omega-3) content [82] makes anchovy a healthy raw material; its dietary intake is in fact inversely associated with cardiovascular diseases [25].

Salted Anchovy

Among pelagic species, *E. anchovy* is the most suitable to apply technologies of salted with maturation [13]. This traditional process in Mediterranean regions has been introduced in Argentina in the Atlantic coast.

Figure 3. Flow diagram of the processing of salted anchovies.

Table 2. Identified *Lb. sakei* CRL1756 proteins significantly modified during its growth in the presence of 10%NaCl

Spot Nº	Protein	Accession Nº	Function	Gene	Mass (Da)	*Volume/spot* Salt /	Control	Comments
1	Hsp20	LSA0050	Folding	*lsa50*	16844	8968.52	ND	D
2	ClpB	LSA1040	Folding	*clpB*	96937	2422.06	ND	D
11	Extracellular protein	LSA0647	Unknown function protein	*lsa647*	64436	1724.56	ND	D
12	Adenylosuccinate synthetase	LSA0063	Nucleotides and nucleic acids metabolism	*purA*	46823	1360.92	ND	D
14	Ferritin-type putative protein	LSA1782	Atypical conditions Adaptation	*lsa1782*	18040	1713.18	ND	D
9	Malate dehydrogenase	LSA0441	Carbohydrates metabolism and related molecules	*mleS*	59331	3195.89	ND	D
15	Pyruvate oxidase	LSA1830	Carbohydrates metabolism and related molecules	*pox2*	66727	627.71	ND	D
3	Hypothetical protein	LSA0964	Unknown function protein	*lsa964*	30400	ND	103.29	R
7	ClpL ATPase protein	LSA0207	Folding	*clpL*	78400	2047.18	5971.17	OE
19	Chaperone GrpE	LSA1237	Folding	*grpE*	22520	1556.82	3125.40	OE
22	Dyp type peroxidase	LSA1831	Inorganic ions transport and detoxification processes	*lsa1831*	35719	1111.47	2050.91	OE
86	Chaperone DnaK	LSA1236	Folding	*dnaK*	66250	17140.57	25162.44	OE

Table 2. (Continued)

Spot Nº	Protein	Accession Nº	Function	Gene	Mass (Da)	*Volume/spot* Salt /	Control	Comments
5	tRNA synthetase	LSA1421	Aminoacyl-tRNA synthetase	*argS*	62933	5948.03	1864.20	UE
6	Triosephosphate isomerase	LSA0606	Glycolysis	*pgk*	27480	1976.59	651.32	UE
8	Phosphoglycerate mutase	LSA1855	Glycolysis	*gpm5*	25836	6845.34	2369.12	UE
16	Phosphoglycerate kinase	LSA0605	Glycolysis	*pgk*	42700	5761.91	2654.37	UE
17	Fructose bisphosphate aldolase	LSA1527	Glycolysis	*fba*	30979	14742.98	7239.99	UE
18	Enoyl-(acyl carrier protein) reductase	LSA0823	Lipid metabolism	*fabl*	26786	2182.23	1076.07	UE

D: exclusively detected, ND: non detected, R: repressed, OE: overexpressed and UE: underexpressed in the presence of NaCl.

The methodology was adapted from that of salted and ripened Italian and Spanish sardines. Salting process for anchovies can be divided into two stages: diffusion of salt into the fish flesh and elimination of water and ripening, longer and slower than the first, yields a product with tender consistency and a characteristic pleasant aroma and taste. The major preservative effect of salt is the removal of water by osmosis. The ripening of salted anchovy is a biochemical process causing chemical and physicochemical changes in fish tissues. It is recognized that the ripening of salted anchovy takes place via enzymatic pathways. The relative importance attributed to tissue enzymes versus microbial enzymes is controversial. Endogenous fish enzymes play an essential role in anchovy preparation and protein hydrolysis probably due to trypsin-like enzymes [56]. It is a relatively slow process, requiring several months (6-9) after which a final gastronomically appreciated and palatable product is obtained. The decrease of a_w during salting-in process largely regulates microbial growth and the more time-consuming maturation is the result of fish meat and bacterial enzyme activities. Spread of tissue and digestive enzymes from the pyloric tract to the muscle has a key role during product ripening, this occurring during the hand heading and gutting in which pyloric appendages as reservoirs of digestive enzymes are preserved. The action of fermentative microorganisms also plays an important role although it has not been yet elucidated. In Figure 3 the technological process is summarized. Briefly, after anchovies capture and selection, they are peeled and brined during 12 h. This first salting step is followed by manual headed/gutted and a final washing in a brine solution, anchovies are then placed in barrels alternating with layers of dry salt forming a fish crown. When filling of barrels is completed anchovies are pressed with a weighty-fitting lid on the top of the mass exerting a decreasing pressure from approximate 60 to 10 gr/cm^2 and ripened in a closed room at 21ºC for up to 8 months. This process allows tissue water output and salt intake by avoiding air inlet causing rancidity. Anchovy fillets are manually done removing the remaining skin and bones, then packaged in cans or glass jars. In Argentina most of salted anchovy produced is exported, thus the ripening process is known to take place during shipping in barrels, the final removal of salt and fish filleted in the country of destination.

Lab Associated with Fish and Fish Products

Different genera of LAB have been isolated from intestinal contents and gills of healthy marine and freshwater fish and their surrounding environment [96,93,16,112]. Although a few LAB species from the genera *Streptococcus, Lactococcus, Vagococcus* and *Weissella* have been described as pathogens for fish [125,84,73,69]. In marine farmed fish, some epizootics linked to streptococci were reported [36], these LAB being then reclassified as *Lactococcus garvieae* as responsible for septicemias, opthalmias and hemorrhages. *Carnobacterium maltaromaticum* has also been isolated from different diseased fish and its virulence has been clearly established in rainbow trout and striped bass experimentally infected [122]. Recently, a novel *Weissella* species has been described as an opportunistic pathogen for rainbow trout [73].

However, most indigenous LAB are harmless and may be considered as potential functional cultures in the production of fermented fish products or as probiotics to improve fish health. The balance of this microbiota in fish is influenced by a variety of factors including food, animal physiology and immunological factors. The setting up of a normal gut flora may be regarded as complementary to the establishment of digestive enzymes, and under normal conditions it serves as a barrier against invading pathogens. Lactobacilli, notably *Lb. plantarum,* have been found in Atlantic salmon, pollock, Arctic char and cod [95,94,111,115]. *Carnobacteria*, including *C. divergens*, *C. gallinarum* and *C. inhibens* have been isolated from all these fish species, as well as from rainbow trout [62,98,58]. *Carnobacteria* have even been quoted as being the dominant genus in the gastro-intestinal tract of juvenile Atlantic salmon and cod [95,112]. The presence of *Leuconostoc mesenteroides*, *Lactococcus piscium*, *Vagococcus salmoninarum*, *Lactobacillus fuchuensis*, *Streptococcus* spp., and *Weissella* spp. have also been reported [133,135,97,94,73,83]. Factors such as water salinity or stress were observed to affect the presence of LAB in fish [97]. Indeed, lactobacilli numbers in the gastro-intestinal tract of Arctic char were reported to be lower in those fish reared in sea water than in fresh water, while any change in the number of *Leuconostoc* and *Enterococcus* was found [97]. In addition, Atlantic salmon stressed by daily manipulations were shown to experience a drop in Gram-negative numbers in their intestinal microbiota and an increase of carnobacteria counts [94].

LAB growth on processed fish products may be stimulated by different technological processes such as drying, salting, marinating and vacuum-packaging in which they constitute the major microbiota contributing to the quality of final products [78,120,65,4]. The type of LAB present in fish ecosystems is highly variable [49,69]. LAB population in traditional smoked and sun-dried fish products from the Eastern Himalayan regions of Nepal and India were reported to range between 4.7–8.3 log CFU/g [120] while a LAB load of 4.5 log CFU/g was found in fresh anchovies [11]. Although many studies have described the natural and variable occurrence of LAB in fish, there are only a few studies reporting the presence of LAB in anchovies (*Engraulis anchoita*). *Tetragenococcus halophilus* (previously known as *Pediococcus halophilus*) has been described as the main species in salted anchovy [131], while recently *Leuconostoc mesenteroides* and *Leuconostoc carnosum, Lb sakei* and *Lb. curvatus* were found to be the predominant lactic acid bacteria in fresh anchovy [11]. Similarly, Najjari [86] identified *Lb. sakei* as the predominant LAB from oil salted anchovies from Morocco.

On the other hand, a great LAB diversity has been reported from different vacuum-packaged (VP) fish products, this usually considered as a spoilage microbiota producing gas, slime and sourness in the product. LAB dominating in spoiled VP cold-smoked fish products includes the genera of *Lactobacillus, Leuconostoc, Lactococcus* and *Carnobacterium.* In VP cold-smoked salmon, herring and rainbow trout, *Lb. curvatus* has been found in majority together with *Lb. sakei, Lb. plantarum, Lactobacillus* farciminis, *Lactococcus* spp., *Lc. mesenteroides, Leuconostoc citreum* and *Carnobacteria piscicola* [123,47,70,87,77]. However, *Lb. sakei, Lb. curvatus* and *C. piscicola* were the three main species identified from sugar-salted (gravad) fish, a traditionally manufactured fish product in the Nordic countries while *Lactobacillus alimentarius,* was found to be the specific spoilage organisms in marinated herrings [79].

Physiological Adaption of Lab to Salted Environments. *Lb. Sakei* CRL1756, a Case Study

It is known that to promote their survival, bacteria have evolved quick and complex responses under restrictive growth conditions such as starvation, high temperatures, salinity and changes in pH [10,31].

Because the osmolarity of the environmental medium is one of the most variable parameters during food processing, much attention has been paid in recent years to understand the adaption mechanisms of bacteria to this stressful condition.

A sudden increase in the osmolarity of the environment results in the movement of water from the cell to the outside, which causes a detrimental loss of cell turgor pressure changing the intracellular solute concentration and cell volume.

Osmoprotection

One of the most powerful adaptive strategies bacterial cells have developed for their acclimation to osmotic stress involve intracellular accumulation of the so-called compatible solutes which equilibrate cellular osmotic pressure [30,52,15,136]. These osmolytes such as glycine-betaine (GB), proline, trehalose, and glycerol may also protect native proteins from denaturation and favor the formation of native protein oligomers, some of these behaving as "chemical chaperones" by promoting the correct refolding of unfolded proteins in vitro and in the cell [34]. Such compatible solutes occur widely in food and other environments where they may protect bacterial cells from deleterious effects of hyper-osmolarity.

In LAB, GB was reported to serve as the major effective accumulating osmoprotectant from an exogenous supply, its transport appearing to be activated by exposure of the cells to high osmotic pressure [99]. In agreement with the accumulation of osmoprotectants, salt-stressed bacteria may alter their gene expression, which is revealed by up- and down-regulation of protein synthesis as adaptive response.

The integration of these stress responses is accomplished by networks of regulators which allow the cell to react to various and complex environmental shifts [136,31]. Recently, the role of different osmoprotectants on *Lb. sakei* CRL1756 growth adaption in the presence of 10% NaCl was investigated.

When the anchovy isolate, was cultured at 30ºC in MRS broth containing 0 and 10% NaCl, growth rates were significantly affected by NaCl (Figure 4). Even when bacterial cells were viable in the presence of 10% NaCl, the pre-adaption to increasing salt concentrations and the effect of different compatible solutes were assayed to enhance its osmotolerance. The addition of GB showed to reduce more efficiently the lag phase of *Lb. sakei* CRL1756 than proline and carnitine.

The long lag phase observed for GB pre-adapted cells, in presence of 10% NaCl, may suggest the induction of transport systems for compatible solutes (Figure 4).

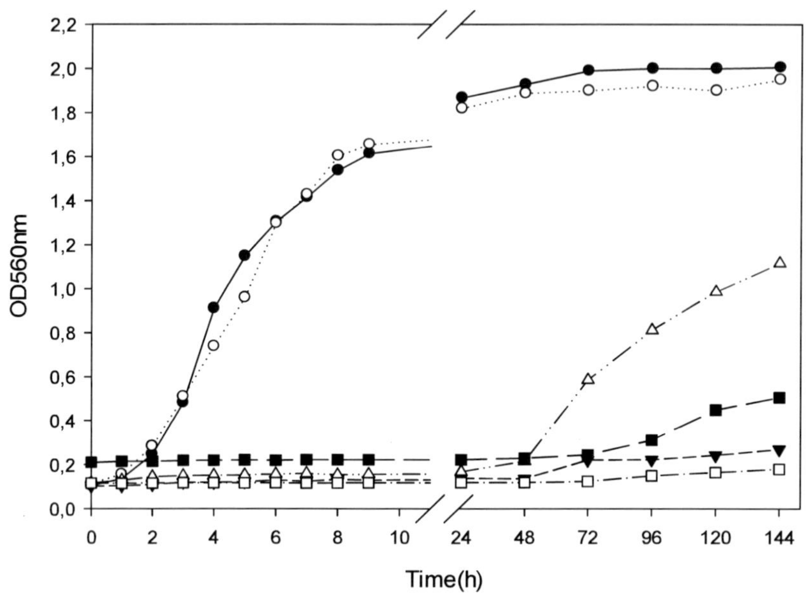

Figure 4. Growth of *Lb. sakei* CRL1756 (pre-adapted with the putative osmoprotectant at a final concentration of 3 mM) at 30°C in MRS (•); MRS + 10% NaCl (□); MRS + GB (○); MRS + GB +10% NaCl (Δ), MRS + proline + 10% NaCl (▼) and MRS + carnitine + 10% NaCl (■).

Proteins Involved in *Lb. Sakei* CRL1756 Adaption to Salt

Nowadays proteomics together with bioinformatics provide new approaches for bacterial functional genomics [24]. In particular, LAB adaption to different stress challenges (oxidative, acid, cold/heat, osmotic, bile, ethanol) has been illuminated by this approach [31,114]. To gain insight in the osmoadaptive strategy of *Lb. sakei* CRL1756, the response to the presence of 10% NaCl was investigated by a comparative proteomic analysis. After pre-adaption in the presence of GB, cells grown in MRS + GB with 0 and 10% NaCl were harvested at mid-log phase growth (Figure 4) and submitted to two-dimension electrophoresis (2DE). Intracellular bacterial proteins putatively involved in salt response were determined by comparison of protein patterns obtained in the absence or presence of 10% NaCl. Comparison of protein profiles revealed a total of 50 spots differentially expressed under the assayed conditions. Eighteen out of 21 statistically significant spots were identified by Maldi Tof-MS (Figure 5). The identified spots were assigned to different functional categories, namely general stress, energy, translation, lipid and nucleotide metabolism, adaption to atypical conditions as well as inorganic ions transport and detoxification as shown in Table 2. Results indicated that the presence of 10% NaCl would represent a major stress condition for the growth of *Lb. sakei* CRL1756. However, the addition of GB as compatible solute was observed to be crucial for *Lb. sakei* cells to overcome osmotic constraint. The development of tolerance mechanisms to high salt concentration was represented by the induction and overexpression of proteins related to stress response (ClpL ATPase, putative specificity factor for CLpP protease, spot 7; GrpE, co-chaperone, spot 19 and DnaK, chaperone, spot 86), adaption to atypical conditions (putative ferritin-like DNA-binding protein, LSA1782), transport of inorganic ions and detoxification mechanisms (Dyp-type peroxidase, LSA1831).

Moreover, to counteract the decreased synthesis of glycolytic enzymes under hypertonic stress (triosephosphate isomerase, spot 6; phosphoglycerate mutase, spot 8; phosphoglycerate kinase, spot 16 and fructose bisphosphate aldolase, spot 17), *L. sake*i CRL1756 was able to induce the enzymes malate dehydrogenase (spot 9) and pyruvate oxidase (spot 15) to be used as an alternative energetic pathway for cellular metabolism maintenance. The knowledge of *Lb. sakei* CRL1756 response to high NaCl concentrations is essential for the selection of appropriate functional starter cultures for the development of new salted anchovy-based products.

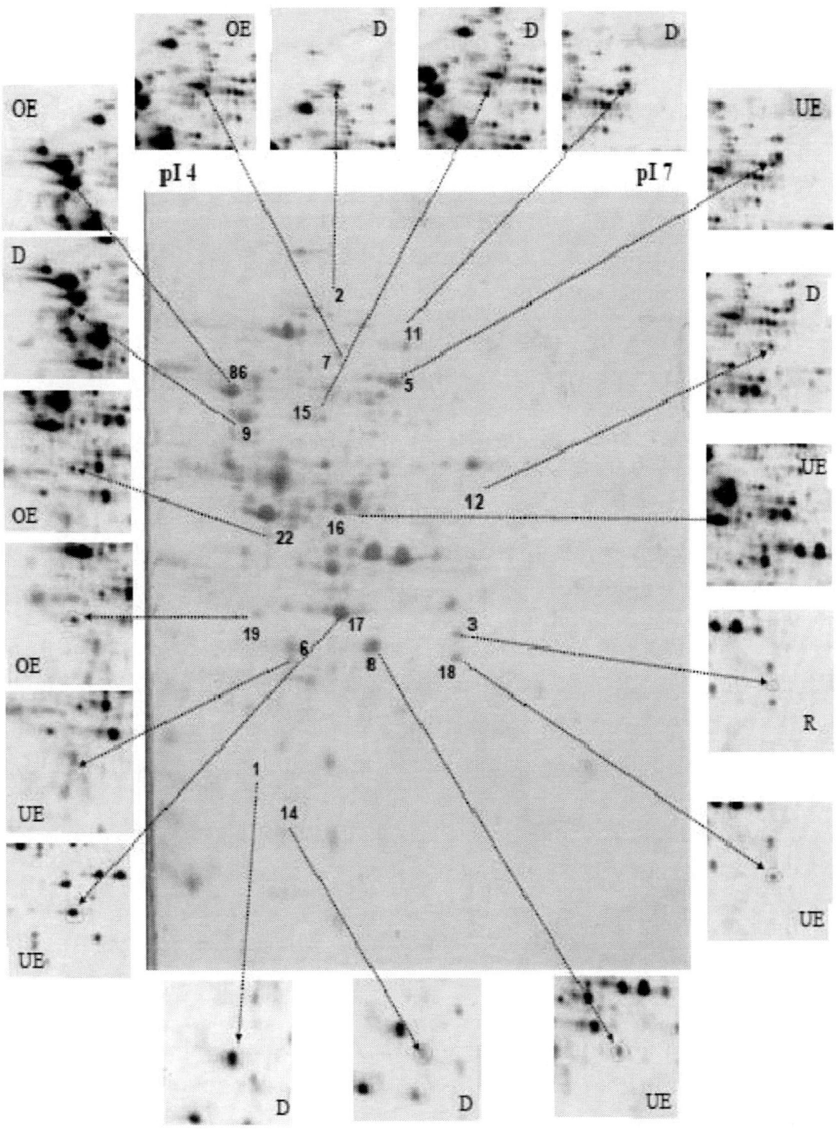

Figure 5. 2DE images of *Lb. sakei* CRL1756 cytosolic proteins expressed during growth in absence (central gel) and presence of 10% NaCl (portions of gels). Arrows indicate spots differentially expressed under osmotic stress. D: exclusively detected; R: repressed; OE: overexpressed and UE: underexpressed in the presence of 10% NaCl.

Conclusion

LAB plays a key role in industrial and artisan meat and fish fermentations. They are known to contribute to sensorial and hygienic characteristics of raw materials and processed foods by means of acid production from carbohydrates, muscle protein degradation as well as bacteriocins production. The ecological strategies adopted by LAB to grow in meat/fish ecosystems are the consequence of the prevailing environmental conditions during processing, transportation and storage that will determine the final type and number of bacteria. The topics covered in this chapter not only confirms existent knowledge about the importance of LAB in foods, but also calls into relevance the potential role of Enterococci as complementary starters for fermented meats. Based on the competitiveness during meat fermentation as well as the sensorial and safety features exhibited this strain may find a practical application as a novel functional starter culture. On the other hand, by means of the proteomic approach described herein, it was possible to gain insight on the adaptive response of the anchovy isolate *Lb. sakei* CRL1756 to high salt concentrations, this being of paramount importance in the selection of appropriate functional starter cultures. Thus, using novel approaches, a better understanding of LAB metabolism can be obtained which certainly will impact in a greater diversification of muscle-based foods.

References

[1] Aguirrezábal, M., J. Mateo, M. Domínguez, & J. Zumalacárregui. (2000). The effect of paprika, garlic and salt on rancidity in dry sausages. *Meat Sci,* 54, 77-81.

[2] Albano, H., I. Henriques, A. Correia, T. Hogg, & P. Teixeira. (2008). Characterization of microbial population of "Alheira" (a traditional Portuguese fermented sausage) by PCR and traditional cultural microbiological methods. *J Appl Microbiol,* 105, 2187–2194.

[3] Albano, H., C. A. van Reenen, S. D. Todorov, D. Cruz, L. Fraga, T. Hogg, L.M.T. Dicks, P. Teixeira. (2009). Phenotypic and genetic heterogeneity of lactic acid bacteria isolated from ''Alheira", a traditional fermented sausage produced in Portugal. *Meat Sci,* 82, 389–398.

[4] An, C., H. Takahashi, B. Kimura, & T. Kuda. (2010). Comparison of PCR-DGGE and PCR-SSCP analysis for bacterial flora of Japanese traditional fermented fish products, aji-narezushi and iwashi-nukazuke. *J Sci Food Agric,* 90, 1796-1801.

[5] Ananou, S., M. Garriga, M. Hugas, M. Maqueda, M. Martinez-Bueno, A. Galvez, & E. Valdivia. (2005). Control of *Listeria monocytogenes* in model sausages by enterocin AS-48. *Int J Food Microbiol,* 103, 179-90.

[6] Andrighetto, C., E. Knijff, A. Lombardi, S. Torriani, M. Vancanneyt, K. Kersters, J. Swings, & F. Dellaglio. (2001). Phenotypic and genetic diversity of enterococci isolated from Italian cheeses. *J Dairy Res,* 68, 303-316.

[7] Arizcun, C., Y. Barcina, & P. Torre. (1997). Identification and characterization of proteolytic activity of *Enterococcus* spp. isolated from milk and Roncal and Idiazabal cheese. *Int J Food Microbiol,* 38, 17-24.

[8] Aymerich, T., M.G. Artigas, M. Garriga, J.M. Monfort, & M. Hugas. (2000). Effect of sausage ingredients and additives on the production of enterocin A and B by *Enterococcus faecium* CTC492. Optimization of in vitro production and anti-listerial effect in dry fermented sausages. *J Appl Microbiol,* 88, 686-94.

[9] Balev, D., T. Vulkova, S. Dragoev, M. Zlatanov, & S. Bahtchevanska. (2005). A comparative study on the effect of some antioxidants on the lipid and pigment oxidation in dry-fermented sausages. *Int J Food Sci Technol,* 40, 977-983.

[10] Baliarda, A., H. Roberta, M. Jebbarc, M. Blanco,A. Deschamps, & C. Le Marrec. (2003). Potential osmoprotectants for the lactic acid bacteria *Pediococcus pentosaceus* and *Tetragenococcus halophila. Int J Food Microbiol,* 84, 13-20.

[11] Belfiore, C., J. Björkroth, E. Vihavainen, R. Raya, & G. Vignolo. (2010). Characterization of *Leuconostoc* strains isolated from fresh 1 anchovy (Engraulis anchoita). *J Gen Appl MIcrobiol,* 55, 175-180.

[12] Benito, M., A. Martin, E. Aranda, E. Pérez-Nevado, S. Ruiz-Moyano, M. Córdoba. (2007). Characterization and selection of autochthonous lactic acid bacteria isolated from traditional Iberian dry-fermented salchichón and chorizo sausages. *J Food Sci,* 72, M193–M201.

[13] Bertullo, E. (2006). *Validación tecnológica del proceso productivo de anchoita (Engraulis anchoita) y de otras especies pelágicas pequeñas, saladas y maduradas.* <http://csic1.csic.edu.uy/directorio/html/ SP/Modalidad2/resumen-2006/Bertullo- Resumen-M2-2006.html >.

[14] Bonomo, M., A. Ricciardi, T. Zotta, E. Parente, & G. Salzano. (2008). Molecular and technological characterization of lactic acid bacteria from traditional fermented sausages of Basilicata region (Southern Italy). *Meat Sci,* 80, 1238–1248.

[15] Bremer, E., & R. Krämer. Coping with osmotic challenges: osmoregulation through accumulation and release of compatible solutes in bacteria. In: G. Stortz, R. Hengge-Aronis (Eds.) *Bacterial stress responses.* ASM Press, Washington DC, (2000). pp. 79-97.

[16] Bucio, A., R. Hartemink, J.W. Schrama, J. Verreth, & F.M. Rombouts. (2006). Presence of lactobacilli in the intestinal content of freshwater fish from a river and from a farm with a recirculation system. *Food Microbiol, 2*3, 476-482.

[17] Cabrer, A. I., M. R Casales, & M.I. Yeannes. (2002). Physical and chemical changes in anchovy (*Engraulis anchoita*) flesh during marination. *J Aquat Food Product Technol,* 11, 19-30.

[18] Callewaert, R., M. Hugas, & L.D. Vuyst. (2000). Competitiveness and bacteriocin production of Enterococci in the production of Spanish-style dry fermented sausages. *Int J Food Microb,* 57, 33-42.

[19] Casaburi, A., M.-C. Aristoy, S. Cavella, R. Di Monaco, D. Ercolini, F. Toldrá & F. Villani. (2007). Biochemical and sensory characteristics of traditional fermented sausages of Vallo di Diano (Southern Italy) as affected by the use of starter cultures. *Meat Sci,* 76, 295–307.

[20] Castellano, P., C. Belfiore, S. Fadda, & G. Vignolo. (2008). A review of bacteriocinogenic lactic acid bacteria used as bioprotective cultures in fresh meat produced in Argentina. *Meat Sci,* 79, 483-99.

[21] Centeno, J.A., S. Menendez, M. Hermida, & J.L. Rodriguez-Otero. (1999). Effects of the addition of *Enterococcus faecalis* in Cebreiro cheese manufacture. *Int J Food Microbiol,* 48, 97-111.

[22] Chaillou, S., M. C Champomier-Vergès, M. Cornet, A. M. Crutz-Le Coq, A. M. Dudez, V. Martin, S. Beaufils, E. Darbon-Rongere, R. Bossy, V. Loux, & M. Zagorec. (2005). The complete genome sequence of the meat-borne lactic acid bacterium *Lactobacillus sakei* 23 K. *Nature Biotechnol,* 23, 1527–1533.

[23] Chaillou, S., M. Daty, F. Baraige, A.M. Dudez, P. Anglade, R. Jones, C.A. Alpert, M.C. Champomier-Vergès, & M. Zagorec. (2009). Intraspecies genomic diversity and natural population structure of the meat-borne lactic acid bacterium *Lactobacillus sakei. Appl Environ Microbiol,* 75, 970–980.

[24] Champomier-Verges, M-C., M. Zagorec, & S. Fadda. Proteomics: a tool for understanding lactic acid bacteria adaptation to stressful environments. In: F. Mozzi, R. Raya, G. Vignolo, (Eds.) Biotechnology of Lactic Acid Bacteria. *Novel applications.* Wiley-Blackwell, Ames, Iowa, (2010). pp. 57-72.

[25] Cicero, A. F., S. Ertek, & C. Borghi. (2009). Omega-3 polyunsaturated fatty acids: their potential role in blood pressure prevention and management. *Curr Vasc Pharmacol.,* 7, 330-337.

[26] Chi, S-P. & Y-C. Wu. Spices and Seasonings. In: F. Toldrá (Ed.) *Handbook of fermented meat and poultry*, Blackwell Publishing, Iowa, (2007). pp. 87-100.

[27] Cocolin, L., P. Dolci, K. Rantsiou, R. Urso, C. Cantoni, G. Comi. (2009). Lactic acid bacteria ecology of three traditional fermented sausages produced in the North of Italy as determined by molecular methods. *Meat Sci,* 82, 125–132.

[28] Comi, G., R. Urso, L. Iacumin, K. Rantsiou, P. Cattaneo, C. Cantoni, and L. Cocolin. (2005). Characterization of naturally fermented sausages produced in North East of Italy. *Meat Sci,* 69,381–392.

[29] Coppola, S., G. Mauriello, M. Aponte, G. Moschetti, & F. Villani. (2000). Microbial succession during ripening of Naples-type salami, a southern Italian fermented sausage. *Meat Sci,* 56,321–329.

[30] Csonka, L. (1989). Physiological and genetic responses of bacteria to osmotic stress. *Microbiol Rev,* 53, 121-147.

[31] De Angelis, M., & M. Gobbetti. (2004). Environmental stress responses in *Lactobacillus*: a review. *Proteomics* 4, 106-122.

[32] Demeyer, D. Meat Fermentation: Principles and Applications. In: H. Hui, L. Meunier-Goddik, A. Hansen, J. Josephsen, W-K. Nip, P. Stanfield, and F. Toldrá (Eds.) *Handbook of food and beverage fermentation technology,* Marcel Deker Inc., New York, (2004). pp. 353-368.

[33] Di Cagno, R., C. Cháves López, R. Tofalo, G. Gallo, M. De Angelis, A. Paparella, W. Hammes, & M. Gobbetti. (2008). Comparison of the compositional, microbiological, biochemical and volatile profile characteristics of three Italian PDO fermented sausages. *Meat Sci,* 79, 224-235.

[34] Diamant, S., N. Eliahu, D. Rosenthal, & P. Goloubinoff. (2001). Chemical chaperones regulate molecular chaperones in vitro and in cells under combined salt and heat stresses. *J Biol Chem,* 276, 39586-39591.

[35] Dimov, S.G., S. Peykov, D. Raykova, P. Ivanova, N. Kirilov, M. Dalgalarrondo, J.M. Chobert, T. Haertle, & I. Ivanova. (2010). A newly discovered bacteriocin produced by *Enterococcus faecalis* 3915. *Benef Microbes,* 1, 43-51.

[36] Eldar, A., C. Ghittino, L. Asanta, E. Bozzeta, M. Goria, M. Prearo, & H. Bercovier. (1996). *Enterococcus seriolicida* is a junior synonym of *Lactococcus garvieae,* a causative agent of septicemia and meningoencephalitis in fish. *Curr Microbiol,* 32,85–88.

[37] Fadda, S., C. Lopez, & G. Vignolo. (2010). Role of lactic acid bacteria during meat conditioning and fermentation: peptides generated as sensorial and hygienic biomarkers. *Meat Sci,* 86, 66-79.

[38] Fadda, S., Y. Sanz, G. Vignolo, M. Aristoy, G. Oliver, & F. Toldra. (1999). Characterization of muscle sarcoplasmic and myofibrillar protein hydrolysis caused by *Lactobacillus plantarum. Appl Environ Microbiol,* 65, 3540-6.

[39] Fadda, S., Y. Sanz, G. Vignolo, M. Aristoy, G. Oliver, & F. Toldra. (1999). Hydrolysis of pork muscle sarcoplasmic proteins *by Lactobacillus curvatus* and *Lactobacillus sake. Appl Environ Microbiol,* 65, 578-84.

[40] Fadda S, & G. Vignolo. Central and South American Products. In: F. Toldrá and R. Talon (Eds.) Handbook of Fermented Meat and Poultry. Part 9. *Ripened meat products.* Blackwell Publishing Inc, Malden, MA, USA, (2007). pp. 387-391.

[41] Fimland, G., L. Johnsen, B. Dalhus, & J. Nissen-Meyer. (2005). Pediocin-like antimicrobial peptides (class IIa bacteriocins) and their immunity proteins: biosynthesis, structure, and mode of action. *J Pept Sci,* 11, 688-96.

[42] Fontana, C., P. Cocconcelli, & G. Vignolo. (2005). Monitoring the bacterial population dynamics during fermentation of artisanal Argentinean sausages. *Int J Food Microbiol,* 103,131–142.

[43] Fontana, C., S. Fadda, P.S. Cocconcelli, & G. Vignolo. Lactic Acid Bacteria in Meat Fermentations. In: S. Salminen (Ed.). *Lactic Acid Bacteria: Microbiological and Functional Aspects,* Fourth Edition. Taylor and Francis Group, LLC. Oxford, (2011). pp 247-263.

[44] Fontana, C., S. Gazzola, P.S. Cocconcelli, & G. Vignolo. (2009). Population structure and safety aspects of *Enterococcus* strains isolated from artisanal dry fermented sausages produced in Argentina. *Lett Appl Microbiol,* 49, 411-414.

[45] Foulquie Moreno, M.R., P. Sarantinopoulos, E. Tsakalidou, & L. De Vuyst. (2006). The role and application of enterococci in food and health. *Int J Food Microbiol,* 106, 1-24.

[46] Franz, C.M., G-S. Cho, W.H. Holzapfel, & A. Gálvez. Safety of lactic acid bacteria. In: F. Mozzi, R. Raya, G. Vignolo (Eds.) *Biotechnology of Lactic Acid Bacteria. Novel Applications,* Wiley-Blackwell, Ames, Iowa, (2010). pp. 341–359.

[47] Gancel F., F. Dzierszinski, & R. Tailliez. (1999). Identification and characterization of *lactobacillus* species isolated from fillets of vacuum-packed smoked and salted herring (*Clupea harengus*). *J Appl Microbiol.* 82, 722.

[48] Gasson, M.J. & W. De Vos. Genetics and biotechnology of Lactic Acid Bacteria. *Ed Scientific book or proceedings*. Springer. London, Blackie (2004).

[49] Gatesoupe, F.J. (2008). Updating the importance of lactic acid bacteria in fish farming natural occurrence and probiotic treatment. *Biotechnol,* 14, 107–114.

[50] Giraffa, G. (2002). Enterococci from foods. *FEMS Microbiol Rev,* 26, 163-71.

[51] Giraffa, G. (2003). Functionality of enterococci in dairy products. *Int J Food Microbiol,* 88, 215-22.
[52] Glaasker, E., W.N. Konings, & B. Poolman. (1996). Osmotic regulation of intracellular solute pools in *Lactobacillus plantarum. J Bacteriol,* 178, 575–582.
[53] Gomes, B.C., C.T. Esteves, I.C. Palazzo, A.L. Darini, G.E. Felis, L.A. Sechi, B.D. Franco, & E.C. De Martinis. (2008). Prevalence and characterization of *Enterococcus* spp. isolated from Brazilian foods. *Food Microbiol,* 25, 668-75.
[54] Guinane, C.M., P.D. Cotter, C. Hill, & R.P. Ross. (2005). Microbial solutions to microbial problems; lactococcal bacteriocins for the control of undesirable biota in food. *J Appl Microbiol,* 98, 1316-25.
[55] Gram L. & H.H. Huss. (1996). Microbiological spoilage of fish and fish products. Int J Food Microbiol, 33, 121-137.
[56] Hernández-Herrero M., X. Arthur, E. Roig-Sagués, J.J. López-Sabater, M. Rodríguez-Jerez, T. Mora-Ventura. (2000). SDS-PAGE of salted anchovies (*Engraulis encrasicholus* L) during the ripening process. *Eur Food Res Technol,* 212, 26–30.
[57] Honikel, K.-O. (2008). The use and control of nitrate and nitrite for the processing of meat products *Meat Sci,* 7, 68–76.
[58] Huber, I., B. Spanggaard, K.F. Appel, L. Rossen, T. Nielsen, & L. Gram. (2004). Phylogenetic analysis and in situ identification of the intestinal microbial community of rainbow trout (*Oncorhynchus mykiss, Walbaum*). *J Appl Microbiol,* 96, 117-132.
[59] Hüfner, E., T. Markieton, S. Chaillou, A.M. Crutz-Le Coq, M. Zagorec, & C. Hertel. (2007). Identification of *Lactobacillus sakei* genes induced during meat fermentation and their role in survival and growth. *Appl Environ Microbiol,* 73, 2522–2531.
[60] Hugas, M., M. Garriga, & T. Aymerich. (2003). Functionality of enterococci in meat products. *Int J Food Microbiol,* 88, 223-233.
[61] Hugas, M., M. Garriga, T. Aymerich, &. J. Monfort. Bacterial cultures and metabolites for the enhancement of safety and quality in meat products. In: F. Toldrá (Ed.). *Research Advances in the Quality of Meat and Meat Products.* Research Signpost, Kerala, India, (2002), pp 225-247.
[62] Jöborn, A., M. Dorsch, J.C. Olsson, A. Westerdahl, & S. Kjelleberg. (1999). *Carnobacterium inhibens* sp. nov., isolated from the intestine of Atlantic salmon (Salmo salar). *Int J Syst Bacteriol,* 49, 1891-1898.
[63] Kato, T., T. Matsuda, T. Tahara, M. Sugimoto, R. Sato, and R. Nakamura. (1994). Effects of meat conditioning and lactic fermentation on pork muscle protein degradation. *Biosci Biotechnol Biochem,* 58, 408-410.
[64] Kleerebezem, M., J. Boekhorst, , R. van Kranenburg, D. Molenaar, O.P. Kuipers, R. Leer, R. Tarchini, S. A. Peters, H. M. Sandbrink, M.W. Fiers, W. R. M. Stiekema, K. Lankhorst, P. A. Bron, S. M. Hoffer, M. N. N. Groot, R. Kerkhoven, M. de Vries, B. Ursing, W. de Vos, & R. Siezen. (2003). Complete genome sequence of *Lactobacillus plantarum* WCFS1. *PNAS,* 100, 1990-1995.
[65] Kopermsub, P. & S. Yunchalard. (2010). Identification of lactic acid bacteria associated with the production of plaa-som, a traditional fermented fish product of Thailand. *Int J Food Mirobiol,* 138, 200-204.
[66] Kozačinski, L., E. Drosinos, F. Čaklovica, L. Cocolin, J. Gsparik-Reichardt, & S. Vesković. (2008). Investigation of microbial association of traditionally fermented sausages. *Food Technol Biotechnol,* 46, 93–106.

[67] Lakshmanan R, R. Jeya Shakila, & G. Jeyasekaran. (2002). Changes in the halophilic amine bacterial flora during salt-drying of sardines (*Sardinella gibbosa*). *Food Res Intern,* 35, 541-546.

[68] Lehmann, K. B. (1899). Über das Haemorrhodin. Ein neues weit verbreitetes Blutfarbstoffderivat. *Über Physical Medicine. Gesellschaft Würzburg,* 4, 57–61.

[69] Leroi, F. (2010). Occurrence and role of lactic acid bacteria in seafood. *Food Microbiol,* 27, 698-709.

[70] Leroi, F., J.J. Joffraud, F. Chevalier, M. Cardinal. (1998). Study of the microbial ecology of cold smoked salmon during storage at 8ºC. *Int J Food Microbiol,* 39, 111-121.

[71] Leroy, F. & L. de Vuyst. (1999). Temperature and pH conditions that prevail during fermentation of sausages are optimal for production of the antilisterial bacteriocin sakacin K. *Appl Environ Microbiol,* 65, 974-81.

[72] Liepe, H.U. Food and feed products with microorganism. In: H. J. Rehm & G. Reeds (Eds.) *Biotechnology, Pohleim, Verlag Chemia GMBH,* Weinheim. Germany, (1983). pp. 399-424.

[73] Liu, J.Y., A.H Li, C. Ji & W.M. Yang. (2009). First description of a novel *Weissella* species as an opportunistic pathogen for rainbow trout Oncorhynchus mykiss (Walbaum) in China. *Vet Microbiol, 136*, 314-320.

[74] Ljungh A., & T. Wadstrom. (2006). Lactic acid bacteria as probiotics. *Curr Iss Intest Microbiol,* 7, 73–89.

[75] Lücke, F-K. (1998). Fermented meat products. *Food Res International,* 27, 199-307.

[76] Lücke F-K. (2000). Utilization of microbes to process and preserve meat. *Meat Sci,* 56, 105-115.

[77] Lyhs U., J. Björkroth, & H. Korkeala. (1999). Characterization of lactic acid bacteria from spoiled, vacuum-packaged, cold-smoked rainbow trout using ribotyping. *Int J Food Microbiol,* 52, 77-84.

[78] Lyhs, U., J. Björkroth, & H. Korkeala. (2002). Characterisation of lactic acid bacteria from spoiled, vacuum-packaged, 'gravad' rainbow trout using ribotyping. *Int J Food Microbiol,* 72, 147–153.

[79] Lyhs U., H. Korkeala, P. Vandamme & J. Björkroth. (2001). *Lactobacillus alimentarius*: a specific spoilage organism in marinated herring. *Int J Food Microbiol, 64*, 355-360.

[80] Madureira, L., I. Queiroz, M.E. Santo, P. Abdallah, J. Castello, J. Hansen, C. Prentice-Hernández, S. Amorin, M. Bertolotti, M.I Yeannes, E. Manca, N. Avdalov, & W. Ruiz. (2007). Case study: South American anchovy, (*Engraulis anchoita*). Towards sustainable aquaculture: selected issues and guidelines. Component 4: Use of wild fish and/or other aquatic species to feed cultured fish and the implications to food security and poverty alleviation. In *FAO Fisheries Technical Paper.* Rome, pp. 1-26.

[81] Makarova, K., A. Slesarev, Y. Wolf, A. Sorokin, & B. Mirkin, E. Koonin, A. Pavlov, N. Pavlova, V. Karamychev, N. Polouchine, V. Shakhova, I. Grigoriev, Y. Lou, D. Rohksar, S. Lucas, K. Huang, D. M. Goodstein, T. Hawkins, V. Plengvidhya, D. Welker, J. Hughes, Y. Goh, A. Benson, K. Baldwin, J.H. Lee,I. Díaz-Muñiz, B. Dosti, V. Smeianov, W. Wechter, R. Barabote, G. Lorca, E. Altermann, R. Barrangou, B. Ganesan, Y. Xie, H. Rawsthorne, D. Tamir, C. Parker, F. Breidt, J. Broadbent, R. Hutkins, D. O'Sullivan, J. Steele, G. Unlu, M. Saier, T. Klaenhammer, P. Richardson,

S. Kozyavkin, B. Weimer, & D Mills. (2006). Comparative genomics of the lactic acid bacteria. *PNAS,* 103, 15611–15616.

[82] Massa, A. E., M. I. Yeannes, & E. A. Manca. (2007). Ácidos grasos poliinsaturados de la serie Omega-3 en ejemplares bonaerenses y patagónicos de anchoita argentina. *Rev Aceites y Grasas,* 66 (1), 568- 572.

[83] Matamoros, S., F. Leroi, M. Cardinal, F. Gigout, F. Kasbi Chadli, J. Cornet, F. Prevost, & M.F. Pilet (2009). Psychrotrophic lactic acid bacteria used to improve the safety and quality of vacuum-packaged cooked and peeled tropical shrimp and cold smoked salmon. *J Food Prot,* 72, 365-374.

[84] Michel, C., C. Pelletier, M. Boussaha, D.G Douet, A. Lautraite, & P. Tailliez. (2007). Diversity of lactic acid bacteria associated with fish and the fish farm environment, established by amplified rRNA gene restriction analysis. *Appl Environ Microbiol,* p. 2947–2955.

[85] Molly, K., D. Demeyer, G. Johansson, M. Raemaekers, M. Ghistelinck&, & I. Geenen. (1997). The importance of meat enzymes in ripening and flavour generation in dry fermented sausages. First results of a European project. *Food Chem,* 59, 539-545.

[86] Najjari, A., H. Ouzari, A. Boudanous, and M. Zagorec. (2008). Method for reliable isolation of *Lactobacillus sakei* strains originating from Tunisian seafood and meat products. *Int J Food Microbiol,* 121, 342-351.

[87] Paludan-Müller, C., H. Huss, L. Gram. (1999). Characterization of lactic acid bacteria isolated from a Thai low-salt fermented fish product and the role of garlic as substrate for fermentation. *Int J Food Microbiol,* 46, 219-229.

[88] Papamanoli, E., N. Tzanetakis, P. Litopoulou-Tzanetaki, & P. Kotzekidou. (2003). Characterization of lactic acid bacteria isolated from Greek dry-fermented sausage in respect of their technological and probiotic properties. *Meat Sci,* 65,859–867.

[89] Paramithiotis, S., D-M. Kagkli, V. Blana, G-J.Nychas, & E. Drosinos. (2008). Identification and characterization of *Enterococcus* spp. in Greek spontaneous sausage fermentation. *J Food Prot,* 71, 1244–1247.

[90] Polenske, E. (1891). Uber den Verlust, welchen Rindfleisch and Nahrwert durch das Pokeln erleidet sowie uber die Veranderungen salpeterhaltiger Pokellaken. *Arbeiten aus dem kaiserlichen Gesundheitsamt,* 7, 471–474.

[91] Rantsiou, K., E. Drosinos, M. Gialitaki, R. Urso, J. Krommer, J. Gasparik-Reichardt, S. Tóth, I. Metaxopoulos, G. Comi, & L. Cocolin. (2005). Molecular characterization of *Lactobacillus* species isolated from naturally fermented sausages produced in Greece, Hungary and Italy. *Food Microbiol,* 22, 19–28.

[92] Rantsiou, K., R. Urso, L. Iacumin, C. Cantoni, P. Cattaneo, G. Comi, & L. Cocolin, (2005). Culture dependent and -independent methods to investigate the microbial ecology of Italian fermented sausages. *Appl Environ Microbiol,* 71, 1977–1986.

[93] Ringø, E. Lactic acid bacteria in fish farming. In: Salminen, S., wright, A., Ouwehand, A. (Eds.). *Lactic Acid Bacteria: Microbiological and Functional Aspects,* third ed. CRC press, New York, (2004). pp. 581-610.

[94] Ringø E., & F.J. Gatesoupe (1998) *Lactic acid bacteria in fish: a review. Aquaculture* 160, 177-203.

[95] Ringø, E., R.E. Olsen, Ø. Øverli, & F. Løvik. (1997). Effect of dominance hierarchy formation on aerobic microbiota associated with the epithelial mucosa of subordinate

and dominant individuals of Arctic charr, Sallelinus alpinus. *L Aquacult Res,* 28, 901–904.

[96] Ringø, E., M. Seppola, A. Berg, R.E. Olsen, U. Schillinger, & W. Holzapfel. (2002). Characterization of *Carnobacterium divergens* strain 6251 isolated from intestine of Arctic charr (*Salvelinus alpinus L*). *Syst Appl Microbiol,* 25, 120-129.

[97] Ringø, E., & E. Strom. (1994). Microflora of Arctic char, *Salvelinus alpinus* (L.); gastrointestinal microflora of free-living fish, and effect of diet and salinity on the intestinal microflora. *Aquacult Fish Manage,* 25, 623-629.

[98] Ringø, E., M.S. Wesmajervi, H.R. Bendiksen, A. Berg, , R.E. Olsen, T. Johnsen, H. Mikkelsen, M. Seppola, E. Strom, & W.H. Holzapfel. (2001). Identification and-characterization of carnobacteria isolated from fish intestine. *Syst Appl Microbiol,* 24, 183-191.

[99] Robert, H., C. Le Marrec, C. Blanco, & M. Jebbar. (2000). Glycine betaine, carnitine and choline enhance salinity tolerance and prevent the accumulation of sodium to a level inhibiting growth to *Tetragenococcus halophila. Appl Environ Microbiol,* 66, 509-517.

[100] Roncalés, P. Additives. In: F. Toldrá (Ed.) *Handbook of fermented meat and poultry,* Blackwell Publishing Iowa, (2007). pp. 77-86.

[101] Ruiz, J. Ingredients. In: F. Toldrá (Ed.) *Handbook of fermented meat and poultry,* Blackwell Publishing Iowa, (2007). pp. 59-76.

[102] Ruiz-Moyano, S., A. Martin, M.J. Benito, F.P. Nevado, & M. de Guia Cordoba. (2008). Screening of lactic acid bacteria and bifidobacteria for potential probiotic use in Iberian dry fermented sausages. *Meat Sci,* 80, 715-21.

[103] Rust, R. 2007. "U.S. Products". In: F. Toldrá (Ed.) *Handbook of fermented meat and poultry,* Blackwell Publishing Iowa, pp. 303-306.

[104] Saavedra, L., C. Minahk, A.P. de Ruiz Holgado, & F. Sesma. (2004). Enhancement of the enterocin CRL35 activity by a synthetic peptide derived from the NH2-terminal sequence. *Antimicrob Agents Chemother,* 48, 2778-81.

[105] Salvucci, E., E.M. Hebert, F. Sesma, & L. Saavedra. (2010). Combined effect of synthetic enterocin CRL35 with cell wall, membrane-acting antibiotics and muranolytic enzymes against Listeria cells. *Lett Appl Microbiol,* 51, 191-5.

[106] Salvucci, E., L. Saavedra, E.M. Hebert, C. Haro, & F. Sesma. (2012). Enterocin CRL35 inhibits *Listeria monocytogenes* in a murine model. *Foodborne Pathog Dis,* 9, 68-74.

[107] Salvucci, E., L. Saavedra, & F. Sesma. (2007). Short peptides derived from the NH2-terminus of subclass IIa bacteriocin enterocin CRL35 show antimicrobial activity. *J Antimicrob Chemother,* 59, 1102-8.

[108] Sanz, Y., S. Fadda, G. Vignolo, M.C. Aristoy, G. Oliver, & F. Toldra. (1999). Hydrolysis of muscle myofibrillar proteins by *Lactobacillus curvatus* and *Lactobacillus sake. Int J Food Microbiol,* 53, 115-25.

[109] Sanz, Y., S. Fadda, G. Vignolo, M.C. Aristoy, G. Oliver, & F. Toldra. (1999). Hydrolytic action of *Lactobacillus casei* CRL 705 on pork muscle sarcoplasmic and myofibrillar proteins. *J Agric Food Chem,* 47, 3441-8.

[110] Sarantinopoulos, P., G. Kalantzopoulos, & E. Tsakalidou.(2001). Citrate metabolism by *Enterococcus faecalis* FAIR-E 229. *Appl Environ Microbiol,* 67, 5482-5487.

[111] Schroder, K., E. Clausen, A.M. Sandberg, & J. Raa. (1980). Psychrotrophic *Lactobacillus plantarum* from Fish and its ability to produce antibiotic substances. In: J.J. Connell (Ed.) Fishing News Books, London. *Adv. Fish. Sci. Technol.*, pp. 480-483.

[112] Seppola, M., E.R. Olsen, E. Sandaker, P. Kanapathippillai, W. Holzapfel, and E. Ringø. (2006). Random amplification of polymorphic DNA (RAPD) typing of carnobacteria isolated from hindgut chamber and large intestine of Atlantic cod (*Gadus morhua I.*). *Syst Appl Microbiol,* 29, 131-137.

[113] Settanni, L., & A. Corsetti. (2008). Application of bacteriocins in vegetable food biopreservation. *Int J Food Microbiol,* 121, 123-38.

[114] Spano, G., & S. Massa. (2006). Environmental stress response in wine lactic acid bacteria: beyond *Bacillus subtilis*. *Crit Rev Microbiol,* 32, 77-86.

[115] Strom, E., & J.A. Olafsen. The indigenous microflora of wild-captured juvenile cod in net-pen rearing. In: Lèsel, R. (Ed.), *Microbiology in Poecilotherms.* Elsevier, Amsterdam, (1990). pp. 181-185.

[116] Sustackova, A., E. Napravnikova, & J. Schlegelova. (2004). Antimicrobial resistance of *Enterococcus* spp. isolates from raw beef and meat products. *Folia Microbiol* (Praha), 49, 411-417.

[117] Suzzi, G., M. Caruso, F. Gardini, A. Lombardi, L. Vannini, M.E. Guerzoni, C. Andrighetto, & M.T. Lanorte. (2000). A survey of the enterococci isolated from an artisanal Italian goat's cheese (semicotto caprino). *J Appl Microbiol,* 89, 267-274.

[118] Talon, R., S. Leroy, I. Lebert, , P. Giammarinaro, J. Chacornac, M. Latorre-Moratalla, C. Vidal-Carou, E. Zanardi, M. Conter, & A. Lebecque. (2008). Safety improvement and preservation of typical sensory qualities of traditional dry fermented sausages using autochthonous starter cultures. *Int J Food Microbiol,* 126,227–234.

[119] Talon, R., S. Leroy-Sétrin, & S. Fadda. Bacterial starters involved in the quality of fermented meat products. In: F. Toldrá (Ed.) *Research Advances in the Quality of Meat and Meat Products. Research Signpost,* Kerala, India, (2002). pp. 175-191.

[120] Thapa, N., J. Pal, & J. P. Tamang. (2006). Phenotypic identification and technological properties of lactic acid bacteria isolated from traditionally processed fish products of the Eastern Himalayas. *Int J Food Microbiol,* 107, 33-38.

[121] Toldrá, F. (1998). Proteolysis and lipolysis in flavour development of dry-cured meat products. *Meat Sci,* 67, 711−719.

[122] Toranzo, A. E., J. L. Romalde, S. Nuñez, A. Figueras, & J. L. Barja. (1993). An epizootic in farmed, market-size rainbow trout in Spain caused by a strain of *Carnobacterium piscicola* of unusual virulence. *Dis. Aquat Organ,* 17,87–99.

[123] Truelstrup H. L. (1995). *Quality of chilled vacuum-packed cold-smoked salmon.* Danish Institute for Fisheries Research and The Royal Veterinary and Agricultural. University of Copenhagen, Denmark, Ph.D.

[124] Vaughan, E.E., H.G. Heilig, K. Ben-Amor, & W. de Vos. (2005). Diversity, vitality and activities of intestinal lactic acid bacteria and bifidobacteria assessed by molecular approaches. *FEMS Microbiol Rev,* 29, 477–490.

[125] Vendrell, D., J.L. Balcázar, I. Ruiz-Zarzuela, I. de Blas, O. Gironés, & J.L. Múzquiz. (2006). *Lactococcus garviae* in fish: a review. *Comp Immun Microbiol Infect Dis, 29,* 177-198.

[126] Verluyten, J., F. Leroy, & L. De Vuyst. (2004). Effects of different spices used in production of fermented sausages on growth of and curvacin A production by *Lactobacillus curvatus* LTH 1174. *Appl Environ Microbiol,* 70, 4807-13.

[127] Verluyten, J., F. Leroy, & L. De Vuyst. (2004). Influence of complex nutrient source on growth of and curvacin a production by sausage isolate *Lactobacillus curvatus* LTH 1174. *Appl Environ Microbiol,* 70, 5081-5088.

[128] Verluyten, J., W. Messens, & L. De Vuyst. (2004). Sodium chloride reduces production of curvacin A, a bacteriocin produced by *Lactobacillus curvatus* strain LTH 1174, originating from fermented sausage. *Appl Environ Microbiol,* 70, 2271-2278.

[129] Vignolo, G., C. Fontana, & P.S. Cocconcelli. New approaches for the study of lactic acid bacteria biodiversity: A focus on meat ecosystems. In: F. Mozzi, R. Raya, G. Vignolo (Eds.) B*iotechnology of Lactic Acid Bacteria. Novel Applications*, Wiley-Blackwell, Ames, Iowa, (2010). pp. 251–271.

[130] Vignolo, G., C. Fontana, & S. Fadda. Semidry and dry fermented sausages. In: F. Toldrá (Ed.), *Handbook of Meat Processing, Blackwell Publishing*, Ames, Iowa, (2010) pp. 379–398.

[131] Villar, M., A. Pesce de Ruiz Holgado, J.J. Sanchez, R. Trucco, & G. Oliver. (1985). Isolation and characterization of *Pediococcus halophilus* from salted anchovies (Engraulis anchoita*). Appl Environ Microbiol,* 49, 664-666.

[132] Vural, H., & E. Özvural. Fermented sausages from other meats. In: F. Toldrá (Ed.) *Handbook of fermented meat and poultry, Blackwell Publishing Ames,* Iowa, (2007). pp. 369-374.

[133] Wallbanks, S., A.J. Martinez.Murcia, J.L. Fryer, B.A. Phillips, & M.D. Collins. (1990). 16S rRNA sequence determination for members of the genus Carnobacterium and related lactic acid bacteria and description of *Vagococcus salmoninarum* sp. nov. *Int J Syst Bacteriol,* 40, 224-230.

[134] Wheaton F.W, & T.B Lawson. Other preservation methods. In: F.W Wheaton, T.B Lawson (Eds.) *Processing aquatic food products.* Wiley, New York, (1985). pp 273–328.

[135] Williams, A.M., J.L. Fryer, & M.D. Collins. (1990). *Lactococcus piscium* sp. nov. a new Lactococcus species from salmonid fish. *FEMS Microbiol Lett,* 68, 109-114.

[136] Wood, J.M., E. Bremer, L. Csonka, R. Krämer, B. Poolman, T. van der Heide, & L.T. Smith. (2001). Osmosensing and osmoregulatory compatible solute accumulation by bacteria. *Comp Biochem Physiol A Mol Integr Physiol,* 130, 437-460.

[137] Yeannes, M. I. & M. R. Casales. (1995) *Estudio de las variables de proceso de marinados de anchoita* (E. anchoita). Alimentaria, Mayo, 87-91.

In: Lactobacillus: Classification, Uses and Health Implications ISBN: 978-1-62081-151-1
Editors: Alba I. Perez Campos and Arturo Leon Mena

Chapter VII

Strategies for Low-Cost Production and Modeling of Highly Concentrated Cultures of *Lactobacillus Casei* CECT 4043

Nelson P. Guerra**, *Paula Fajardo Bernárdez* and *Lorenzo Pastrana Castro
Nutrition and Bromatology Group, Department of Analytical and Food Chemistry, Food Science and Technology Faculty, Ourense Campus, University of Vigo, Ourense, Spain

Abstract

Two re-alkalized fed-batch cultures with the potential probiotic strain *Lactobacillus casei* subsp. *casei* CECT 4043 were carried out in whey by using different mixtures of feeding substrates (concentrated whey plus a concentrated solution of lactose and mussel processing wastes plus a concentrated solution of glucose). Mass balance equations were carried out in the two cultures to calculate the volumes of feeding substrates, the concentrations of nutrients (lactose, glucose, proteins, nitrogen and phosphorous) added to the fermentation medium, as well as the concentrations of biomass and products in the fermentor justly after each feeding.

The data obtained was used to develop mathematical models to describe the growth and product formation by the CECT 4043 strain. The growth models were based on Monod kinetics and related to the phases of nitrogen consumption observed in each culture. The antibacterial activity produced in these fermentations was described by using a modified form of the Luedeking and Piret model that includes a term for the specific effect of pH drop on the antibacterial activity synthesis. Production of lactic acid and acetic acid was modeled by using the Luedeking and Piret model without modifications. A reasonable description of the growth and product formation over the time by the

* Author whom correspondence should be addressed. Tel. +34-988-387062. Fax +34-988-387001. E-mail address: nelsonpg@uvigo.es.

Lactobacillus casei strain was provided by the developed models, which appear to be useful as a control tool in the large scale production of a potentially probiotic *L. casei* CECT 4043 culture.

Keywords: Health effects, *Lactobacillus casei*, mathematical modeling, mussel processing wastes, probiotic culture, re-alkalized fed-batch fermentation, whey

1. Introduction

Growth-promoting antibiotics have been widely used in feeds as therapeutic agents for the prevention or treatment of diseases produced by pathogenic microorganisms (Kyriakis et al., 1997), and as prophylactic agents to improve the production results (Guerra et al., 2007a). However, extensive use of antibiotics in animal feed produces several serious side effects including allergic reactions and the presence of antibiotic residues in edible animal products. Consumption of these products with antibiotic residues may contribute to the development of antibiotic-resistant disease-causing bacteria in humans (Abe et al., 1995; Pascual et al., 1999; Thorlacius and Jeppsson, 2003; Verdon, 2010).

Therefore, the interest has been focussed on using some probiotic microorganisms (mainly lactic acid bacteria (LAB) belonging to the species of *Lactobacillus*, *Pediococcus*, *Streptococcus* and *Enterococcus*) and organic acids as an alternative to the use of growth-promoting antibiotics in feeds (Guerra et al., 2007a). Some others probiotics used in animal feed are bacteria belonging to the genus *Bacillus* (*B. cereus* var. *toyoi*, *B. licheniformis*, *B. subtilis*) and strains of yeast including *Saccharomyces cerevisiae* and *Kluyveromyces* species (Anadón et al., 2006).

Probiotics are viable microorganisms that once ingested in sufficient amounts, exert health benefits by assisting in the establishment of an intestinal population which is beneficial to the host entity and antagonistic to harmful bacteria (Schrezenmeir and de Vrese, 2001; Matijašić et al., 2004). With regard to this, LAB have shown to produce a number of well-established benefits. Thus, it has been reported that consumption of these bacteria can: i) improve lactose digestion (Martini et al., 1991), ii) aid to prevent and treat diarrhea (Arvola et al., 1999; Szajewska and Mrukowicz, 2001; Kim et al., 2003; Hickson et al., 2007) and iii) stimulate the immune activity in the intestinal mucosa, which helps the body resist and fight off infection (Perdigón et al., 1993).

A number of *Lactobacillus* strains have been associated with these health benefits. Thus, *L. casei* strain Shirota has been shown to be safe and effective for preventing or treating different human diseases. This is basically because consumption of this strain stimulated the recuperation of the activity of immune cells, thus enhancing the immune response and increasing resistance to certain pathogenic bacteria including *Listeria momnocytogenes*, *Helicobacter pylori*, *Salmonella typhimurium* and *Escherichia coli* (Aso and Akazan, 1992; Aso et al., 1995; de Waard et al., 2002; Cano and Perdigon, 2003; Koebnick et al., 2003; Sgouras et al., 2004). In the same way, *L. casei* DN-114 001 was found to down-regulate the transcription of a number of genes encoding pro-inflammatory effectors (cytokines and chemokines) and adherence molecules induced by invasive *Shigella flexneri* (Tien et al., 2006).

During the last decades, a variety of commercial animal probiotic preparations have been developed and marketed for veterinary administration in European countries (Fasoli et al., 2003). However, taking into account the relatively high prices of these commercial probiotic preparations, their use as additives in feed for farm animals could be cost-prohibitive. To solve this problem, it could be more adequate to select firstly an adequate probiotic strain and secondly to produce a highly concentrated culture of this strain in a cheaper culture medium (Guerra et al., 2007a). However, for the large scale production of these probiotic cultures, it is necessary to design an appropriate fermentation procedure and an adequate process control through the use of mathematical models that accurately describe the fermentation kinetic (Guerra et al., 2007c; 2008).

In a previous study developed in our laboratory (Guerra and Pastrana, 2002a), a total of 50 LAB strains were screened for producers of growth inhibitory substances in batch cultures in de Man Rogosa and Sharpe (MRS) broth. The major producer of lactic acid (21.7 g/L) was *L. casei* subsp. *casei* CECT 4043, a strain isolated from a Spanish traditional variety (Majorero) of semi-hard goats' cheese (Requena et al., 1991). In addition, this bacterium also fulfils many of the probiotic criteria, because it showed to be: i) non-pathogenic, ii) bile and acid tolerant, iii) able to survive during storage with skim milk at -20ºC for 3 months and in a piglet feed at room temperature for 8 days (Guerra et al., 2007a).

Considering the substantial availability of whey and mussel processing wastes at very low prices by local dairy and mussel processing plants of Galicia (Spain), the use of these effluents as culture media for production of probiotic products could provide a profitable substrate for a low cost production. In this way, culture media prepared with both wastes were successfully used to produce high amounts of biomass and antibacterial extracellular products (AECPs) by *Lactococcus lactis* spp. *lactis* CECT 539 (Guerra and Pastrana, 2003a) and *Pediococcus acidilactici* NRRL-B 5624 (Guerra et al., 2005b; 2007b).

Therefore, in the present study, we described the production of a highly-concentrated probiotic culture of *L. casei* subsp. *casei* CECT 4043 in a whey-based medium by using a fermentation technique based on successive re-alkalizations and feeding of the culture medium.

Then, the growth of the lactic acid bacterium was modeled by using different modeling strategies to select the most adequate model describing biomass production. Subsequently, the Luedeking and Piret model (Luedeking and Piret, 1959) or a modification of it was used to fit the experimental concentration data corresponding to the products (lactic acid, acetic acid and antibacterial activity).

2. Materials and Methods

2.1. Bacterial Strains

L. casei subsp. *casei* CECT 4043 (a high lactic acid-producing strain) and *Carnobacterium piscicola* CECT 4020 (the target organism in the antibacterial activity assay) were obtained from the Spanish Type Culture Collection (CECT). Stock cultures of both strains were maintained at -40°C in Nutrient broth (Cultimed Panreac Química S.A., Barcelona, Spain) with 15% (v/v) added glycerol (Panreac Química S.A., Barcelona, Spain).

Working cultures maintained at 4°C on MRS agar (Cultimed Panreac Química S.A., Barcelona, Spain) were prepared monthly from frozen stock cultures.

2.2. Culture Medium, Fermentation Conditions and Inoculation

A whey-based medium, which was obtained from a local dairy plant as diluted whey (DW: concentrated whey (CW) mixed with wash waters) was used as the fermentation medium.

Re-alkalized fed-batch cultures were carried out at a controlled temperature of 30 °C in a 6 L bench top fermentor (New Brunswick Scientific, Edison, New Jersey, U.S.A.) with a 4 L working volume of DW medium adjusted at an initial pH of 7.0. A controlled constant agitation speed of 200 rpm and a controlled constant aeration flow rate of 0.5 L/h were used during the fermentations.

In the first re-alkalized fed-batch culture (fermentation I), two fresh substrates were used to feed the fermentor: a medium prepared with CW (the liquid remaining after the first cheese pressing) and a 400 g/L concentrated lactose. In the second fed-batch culture (fermentation II), the feeding substrate consisted of a mixture of a concentrated mussel processing waste (CMPW) medium and a 310 g/L concentrated glucose.

The DW, CW and CMPW media were prepared as previously described (Guerra et al., 2001; Guerra and Pastrana, 2002b). The mean composition of these culture media is shown in Table 1.

Table 1. Nutrients concentration (g/L) in the diluted whey (DW) used as fermentation medium and in the feeding media: concentrated whey (CW) and concentrated mussel processing wastes (CMPW)

	DW	CW	CMPW
Lactose	20.27 ± 0.27	51.02 ± 2.88	-
Glucose	-	-	102.00 ± 1.67
Proteins	2.01 ± 0.04	5.05 ± 0.05	3.51 ± 0.04
Total nitrogen	0.50 ± 0.06	1.02 ± 0.07	0.78 ± 0.06
Total phosphorous	0.24 ± 0.03	0.43 ± 0.02	0.31 ± 0.08

Fed-batch fermentations were carried out as a batch fermentation without pH control during the first 12 h, when the culture reached the stationary phase (Fajardo et al., 2008). At this time, a sample of 100 mL was taken from the fermentation medium to perform analytical determinations. After determining the total sugars concentration in the withdrawn sample, the amounts of sugars consumed by *L. casei* CECT 4043 were calculated. The fermentor was fed with the same volume (100 mL) of the feeding substrate to restore the initial total sugars concentration in the fermentation medium. Immediately after feeding, the medium was re-alkalized up to pH 7.0 with 4M NaOH. These sampling, feeding and re-alkalization strategies were repeated every 12 h until the CECT 4043 strain was unable to bring about the decrease of pH.

The two fed-batch cultures were inoculated with 2% (v/v) of a 12 h culture of *L. casei* CECT 4043 in DW medium.

2.3. Analytical Methods

Growth was monitored by absorbance at 700 nm and converted into dry cell weight from a standard curve (Guerra et al., 2001). Cells were harvested by centrifugation (12,000 × *g* for 15 min at 4°C) of culture samples and washed twice with saline (0.8% NaCl). The culture supernatants were used for analytical determinations. Total sugars (*TS*), nitrogen (*TN*), phosphorous (*TP*) and proteins (*Prot*) were determined by methods described in a previous work (Guerra and Pastrana, 2003a).

The concentrations of lactose (*Lact*), glucose (*Gluc*), lactic acid (*L*A) and acetic acid (*AA*) were measured by HPLC using an ION-300 Organic Acids column (length 300 mm, internal diameter 7.8 mm) with a pre-column IONGUARD^{TM} (polymeric guard column), both obtained from Tecknokroma S. Coop. C. Ltda, Barcelona, Spain. The mobile phase consisted of 3 mM H_2SO_4 at a flow rate of 0.4 mL/min at 60-65 °C and the refractive index of the peaks was measured by a refractometer with a refractive-index detector (Guerra and Pastrana, 2003a).

2.4. Preparation of Cell-Free Supernatants (CFS) of *L. Casei* and Quantification of the Antagonistic Activity

Samples of *L. casei* cultures were acidified to pH 3.5 with 5 M HCl, heated at 80 °C for 3 min. and centrifuged at 27,200 × *g* for 15 min. at 4 °C. The supernatants containing total antibacterial activity (*AbA*) were frozen until needed.

Using *C. piscicola* CECT 4020 as the indicator, the antibacterial activities of the CFS were assayed by using a photometric assay (Cabo et al., 1999). The CFS were serially diluted in distilled sterile water (this step eliminated the need to correct the pH of samples). A 2.5 mL portion of the diluted CFS was added in sterile culture tubes. Each tube was inoculated with 2.5 mL of an overnight culture of the indicator strain, which was diluted to an absorbance of 0.2 with sterile buffered MRS broth (pH 6.3) just before being used. Controls consisted of three culture tubes in which the diluted CFS were replaced by distilled sterile water. All tubes were then incubated at 30 °C for 6 h. Growth inhibition was measured spectrophotometrically at 700 nm. Dose-response curves were obtained from this data. Antibacterial activity was calculated as antibacterial activity units (AU). It was defined as the dilution which produced a 50% growth inhibition (ID_{50}) compared to control tubes. The reciprocal of the dilution gave the titer of antibacterial activity in AU per mL cell-free supernatant (Murado et al., 2002).

2.5. Mass Balance Equations in the Re-alkalized Fed-Batch Fermentations

In this work, although the feeding volumes (*VF*) were matched with the sampling volume (*VS*), the volume of the fermentation medium (*V*) in the two re-alkalized fed-batch fermentations were not maintained constant due to the addition of the necessary volumes of 4M NaOH (*VNaOH*) for re-alkalizing the medium up to the initial pH value of 7.0.

Then, the initial fermentation volume at the beginning of each re-alkalization cycle in the two re-alkalized fed-batch fermentations was calculated as:

$$V_{t_n} = V_{t_{n-1}} - VS_{t_{n-1}} + VNaOH_{t_n} + VF_{t_n} \quad [I.1]$$

Since $VF_{tn} = VS_{tn\text{-}1}$ in the two re-alkalized fed-batch fermentations, equation [I.1] takes the form:

$$V_{t_n} = V_{t_{n-1}} + VNaOH_{t_n} \quad [I.2]$$

Thus, the sum of the volumes of feeding substrates that must be added to the culture for restoring the initial total sugars (*TS*) in the fermentation medium are (Guerra et al., 2010):

$$VS_{t_{n-1}} = VF_{t_n} = \left(VCL_{t_n} + VCW_{t_n}\right), \text{ for the first fermentation} \quad [I.3]$$

$$VS_{t_{n-1}} = VF_{t_n} = \left(VCG_{t_n} + VCMPW_{t_n}\right), \text{ for the second fermentation} \quad [I.4]$$

where *VCL*, *VCW*, *VCG* and *VCMPW* are respectively, the volumes (in L) of concentrated lactose, CW medium, concentrated glucose and CMPW medium added to the fermentor at the beginning of each feeding cycle.

Then, both the *VCW* and *VCMPW* were respectively calculated as:

$$VCW_{t_n} = VS_{t_{n-1}} - VCL_{t_n} \quad [I.5]$$

$$VCMPW_{t_n} = VS_{t_{n-1}} - VCG_{t_n} \quad [I.6]$$

The reduction in the amounts (in grams) of *TS* in the medium due to the joint effect of the extraction of samples and the consumption of *TS* by the growing strain (TS_{C+E}) in both fermentations was calculated by applying a mass balance equation for the total sugars:

$$TS_{C+E} = V_{t_{n-1}} \cdot [TS_{t_{n-1}}] - \left(V_{t_{n-1}} - VS_{t_n}\right) \cdot [TS_{t_n}] \quad [I.7]$$

where $[TS_{tn\text{-}1}]$ and $[TS_{tn}]$ are respectively, the concentrations of total sugars (in g/L) in the fermentation medium at the beginning (20 g/L) and at the end of each feeding cycle. The difference ($V_{tn\text{-}1}$ - VS_{tn}) represents the remaining volume (in L) in the fermentor after the extraction of samples.

Therefore, the amounts of *TS* that must be added to the fermentor to restore the initial *TS* concentration in the culture were calculated by the following expressions:

$$VCW \cdot \left[TS_{CW}\right] + VCL \cdot \left[TS_{CL}\right] = TS_{C+E}, \text{ for the first fermentation} \quad [I.8]$$

$$VCMPW \cdot [TS_{CMPW}] + VCG \cdot [TS_{CG}] = TS_{C+E} \text{, for the second fermentation} \quad [I.9]$$

where $[TS_{CW}]$, $[TS_{CMPW}]$, $[TS_{CL}]$ and $[TS_{CG}]$ are the concentrations of total sugars in the CW (51 g/L) and CMPW (102 g/L) media and in the concentrated solutions of lactose (400 g/L) and glucose (310 g/L), respectively.

Substituting respectively equations [I.5] and [I.6] into equations [I.8] and [I.9] gives:

$$\left(VS_{t_{n-1}} - VCL_{t_n}\right) \cdot [TS_{CW}] + VCL_{t_n} \cdot [TS_{CL}] = TS_{C+E} \text{, for the first fermentation} \quad [I.10]$$

$$\left(VS_{t_{n-1}} - VCG_{t_n}\right) \cdot [TS_{CMPW}] + VCG_{t_n} \cdot [TS_{CG}] = TS_{C+E} \text{, for the second fermentation} \quad [I.11]$$

Thus, the VCL and VCG values can be calculated as:

$$VCL_{t_n} = \frac{\left(TS_{C+E} - VS_{t_{n-1}} \cdot [TS_{CW}]\right)}{\left([TS_{CL}] - [TS_{CW}]\right)} \text{, for the first fermentation} \quad [I.12]$$

$$VCG_{t_n} = \frac{\left(TS_{C+E} - VS_{t_{n-1}} \cdot [TS_{CMPW}]\right)}{\left([TS_{CG}] - [TS_{CMPW}]\right)} \text{, for the second fermentation} \quad [I.13]$$

Now, the VCW and $VCMPW$ values were obtained by introducing respectively, the values of VCL and VCG into equations [I.5] and [I.6].

The concentrations of biomass (X) and products (AbA, LA, AA) in both fermentations, at the beginning of each re-alkalization cycle, justly after the addition of the feeding substrates, were calculated by using the following equation:

$$[P]_{beg} = \frac{\left(V_{end} - VS_{end}\right) \cdot [P]_{end}}{V_{beg}} \text{ being} \qquad V_{beg} = V_{end} + VF + VNaOH \quad [I.14]$$

where $[P]_{end}$ and $[P]_{beg}$ are respectively, the biomass or product concentrations in the fermentor at the end of a re-alkalization and feeding cycle and at the beginning of the following cycle. The difference (V_{end} - VS_{end}) represents the remaining volume of fermentation medium at the end of each re-alkalization and feeding cycle, justly after sampling. V_{beg} is the volume of fermentation medium at the beginning of each feeding cycle, justly after feeding and re-alkalizing the medium with 4M NaOH.

The concentrations of nutrients ($Lact$, $Gluc$, TN, TP and $Prot$) at the beginning of each re-alkalization cycle in the fermentations (I) and (II), were respectively calculated by using the following equations:

$$[S]_{beg} = \frac{\left(V_{end} - VS_{end}\right) \cdot [S]_{end} + VCW_{beg} \cdot [S_{CW}] + VCL_{beg} \cdot [S_{CL}]}{V_{beg}} \quad [I.15]$$

$$[S]_{beg} = \frac{\left(V_{end} - VS_{end}\right) \cdot [S]_{end} + VCMPW_{beg} \cdot [S_{CMPW}] + VCG_{beg} \cdot [S_{CG}]}{V_{beg}} \quad [I.16]$$

where $[S]_{end}$ and $[S]_{beg}$ are respectively, the nutrient concentrations in the fermentor at the end of a re-alkalization and feeding cycle and at the beginning of the following cycle. $[S_{CW}]$, $[S_{CMPW}]$, $[S_{CL}]$ and $[S_{CG}]$ are the nutrient concentrations in the feeding substrates (CW, CMPW, CL and CG).

2.6. Model Parameters Determination and Model Evaluation

The model parameters and their confidence intervals were obtained by using the non-linear curve-fitting software of SigmaPlot (version 9.0, Systat Software, Inc., 2004), which minimized the deviations between model predictions and experimental data according to the sum of squares of errors (*SSE*) of the model fit:

$$SSE = \sum_{i=1}^{n}\sum_{j=1}^{m}\Delta_{i,j}^{2} = \sum_{i=1}^{n}\sum_{j=1}^{m}\left(v_{exp} - v_{pred}\right)^{2} \qquad [I.17]$$

where $\Delta_{i,j}$ represents the difference between the experimental data values (v_{exp}) and the values predicted by the corresponding model (v_{pred}) and, n and m represent the number of experimental data points and the number of variables, respectively.

The coefficients of the models with P values lower than 0.05 were considered statistically significant. Parameters were removed from the models when their asymptotic interval of confidence included zero.

The criteria used to evaluate the goodness-of-fit of each model were the determination coefficient (R^2) and the mean relative percentage deviation modulus (*RPDM*):

$$RPDM = \frac{100}{N}\sum_{i=1}^{N}\frac{\left|X_{exp} - X_{pred}\right|}{X_{exp}} \qquad [I.18]$$

where X_{exp} is the experimental value, X_{pred} is the calculated value by the corresponding model, and N is the number of experimental data. The *RPDM* parameter is widely used to determine the quality of the fit, being a value of *RPDM* below 10% indicative of a good fit for practical purposes (Lomauro et al., 1985; Aguerre et al., 1989; Guerra et al., 2007c).

2.7. Statistical Analyses

Individual experiments were performed in duplicate and the analytical analyses were performed in triplicate as stated in the figure legends. Comparisons were made between the results (concentration of biomass, organic acids and antibacterial activity) of the two cultures. Statistical analysis was performed with SPSS 8.0 for Windows (Statsoft 2001, Tulsa, OK, U.S.A.). Differences were considered as statistically significant, when the P-value was less than 0.05.

3. Results and Discussion

3.1. Realkalized Fed-Batch Fermentations

The cumulative concentrations of biomass, nutrients (total nitrogen, phosphorous and sugars) and products (lactic acid, acetic acid and antibacterial activity) obtained in the two fed-batch cultures were used in a previous work, not only for discussing the fed-batch growth kinetics of *L. casei* CECT 4043, but also for calculating the fermentation yield coefficients and the efficiencies in nutrient consumption (Fajardo et al., 2008).

However, in the present study, the concentrations of biomass, nutrients and products inside the fermentor at each sampling time and after each re-alkalization and feeding cycle (calculated with equations [I.14], [I.15] or [I.16]) were used to develop the corresponding mathematical models.

On the basis of batch fermentation results (Fajardo et al., 2008), a first re-alkalized fed-batch fermentation was carried out in the DW medium adjusted at an initial pH of 7.0. The fermentor was fed with a mixture of CW medium and a concentrated lactose (400 g/L) each 12 h, when the culture reached the stationary phase. The feeding substrates were fed into the fermentation medium to replenish the levels of total sugars consumed and other nutrients in each feeding cycle. The culture was re-alkalized up to a pH of 7.0 immediately after each feeding to facilitate the nutrient consumption by *L. casei* CECT 4043. This approach was used to avoid or reduce substrate-associated growth inhibition observed in CW medium by appropriately controlling the carbon source concentration in the fermentor (Fajardo et al., 2008).

The time-course of the concentrations of biomass, nutrients and products inside the fermentor in the first re-alkalized fed-batch culture of *L. casei* CECT 4043 in DW medium is shown in Figure 1. Three important observations can be made from the results obtained in this culture. Firstly, the maximum concentrations of biomass (~ 0.8 g/L), AbA (10.0 AU/mL), lactic acid (8.4 g/L) and acetic acid (0.49 g/L) inside the fermentor obtained in this first re-alkalized fed-batch fermentation were significantly higher ($P < 0.05$) than those observed by Fajardo et al., (2008) in the homolactic batch fermentation on DW (X = 0.3 g/L, AbA = 3.4 AU/mL, LA = 2.2 g/L).

Secondly, a typically homolactic fermentation was observed during the first 60 h (only lactic acid was produced). However, the heterolactic fermentation began with the production of acetic acid from the 60 h of incubation. This metabolic shift from homolactic to heterolactic fermentation was observed before for other LAB in cultures under limitation of glucose (Thomas et al., 1979; Borch et al., 1991; Cocaign-Bousquet et al., 1996) or nitrogen (Thomas et al., 1979) and under carbon excess conditions with lactose (Garrigues et al., 1997; 2001; Torino et al., 2001), galactose (Thomas et al., 1980; Garrigues et al., 1997) and maltose (Snoep et al., 1992; Hofvendahl et al., 1999). This metabolic pathway shift has been associated to a modification of pyruvate metabolism with a decreased activity of lactate dehydrogenase (Lopez de Felipe et al., 1998; Zhu and Yang, 2006).

Thirdly, the growth stopped and the lactic acid-producing strain completely lost its ability to recover the acidic pH after 156 h of incubation (Figure 1), although the nutrients (*Lact*, *TN*, *TP* and *Prot*) were not completely consumed.

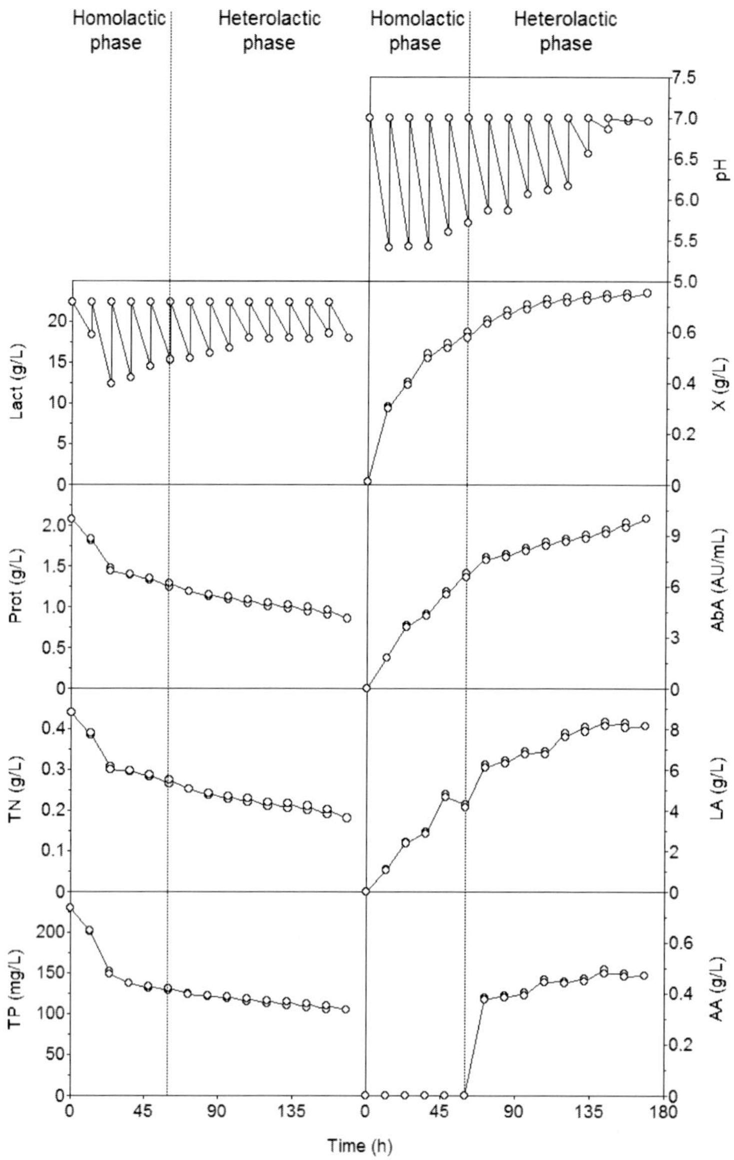

Figure 1. Time course of the re-alkalized fed-batch fermentation of *L. casei* CECT 4043 on diluted whey medium with feeding with concentrated whey medium and a 400 g/L concentrated lactose solution. Lact, lactose; Prot, proteins; TN, total nitrogen; TP, total phosphorus; X, biomass; AbA, antibacterial activity; AU, activity units; LA, lactic acid; AA, acetic acid. Results shown are means of two independent experiments with three replicates each.

Then, the cessation of growth could be related to the exhaustion of some essential micronutrients (vitamins or minerals) or amino acids in the medium (van Niel and Hahn-Hägerdal, 1999, Fajardo et al., 2008), that could not be replenished with the volumes of CW medium added to the fermentor.

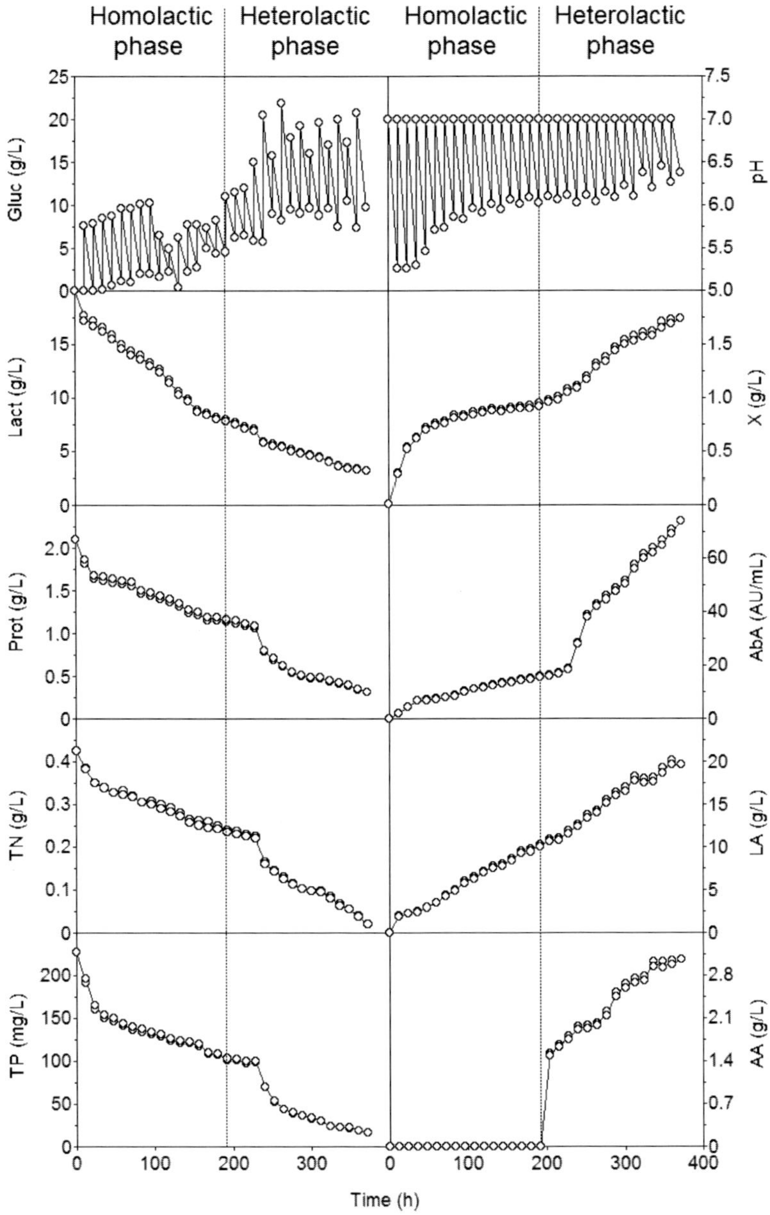

Figure 2. Time course of re-alkalized fed-batch culture of *L. casei* CECT 4043 on diluted whey medium with feeding with concentrated mussel processing waste medium and a 310 g/L concentrated glucose solution. Gluc, glucose. Other abbreviations are the same as in Figure 1. Results shown are means of two independent experiments with three replicates each.

Since the feeding with glucose enhanced both the biomass and AECPs production by *Pediococcus acidilactici* NRRL B-5627 in comparison with the feeding with lactose in fed-batch cultures in whey (Guerra et al., 2007b), a second re-alkalized fed-batch culture with *L. casei* CECT 4043 was performed in DW medium (pH 7.0), using a mixture of substrates composed of CMPW medium and a 310 g/L concentrated glucose.

In comparison with the first culture (Figure 1), five important differences could be noted in the second fed-batch fermentation (Figure 2):

1. The cells had an active period higher than that observed in the first re-alkalized fed-batch culture, because the producer strain was able to bring about the decrease of pH during the whole fermentation period. Consequently, the productions of biomass (1.7 g/L), AbA (74.2 AU/mL), lactic acid (19.6 g/L) and acetic acid (3.1 g/L) inside the fermentor were higher ($P < 0.05$) than those obtained in the first fed-batch culture (Figure 1).
2. Although the shift from homolactic to heterolactic fermentation was also observed in this second fed-batch culture, the heterolactic phase began later (after 192 h) than that of the first fermentation, probably as a consequence of the nutrient feeding with CMPW medium.
3. The culture displayed two growth phases similar to that of a diauxic growth pattern. The first growth pulse corresponded to the homolactic phase and the second growth pulse corresponded to the heterolactic phase of fermentation. Each growth pulse was characterized by an exponential growth phase followed by a stationary phase. These two growth phases had been observed before in other re-alkalized fed-batch cultures with different LAB (Guerra and Pastrana, 2003a; Guerra et al., 2005b; 2007b; 2010). This behavior was associated with a biphasic nitrogen consumption (Niven et al., 1998). Thus, the free amino acids and utilizable oligopeptides originally present in the medium are reported to be the main nitrogen source for growth during the first growth phase, whereas the caseins are the source of amino acids in the second growth phase (Flambard et al., 1997; Juillard et al., 1995; 1998; Letort et al., 2002).
4. The nutrients (nitrogen, proteins and phosphorous) were almost completely consumed at the end of the fermentation. These results suggest that the glucose and the proteins contained in the CMPW medium can be more efficiently utilized as carbon and nitrogen sources by *L. casei* CECT 4043 than the lactose and the proteins present in the CW medium. Other lactobacilli also showed a higher preference by glucose in comparison to lactose. Thus, glucose was a better carbon source than lactose for biomass and exopolysaccharide production by *L. casei* CG11 (Cerning et al., 1994) and for lactic acid production by *L. helveticus* ATCC 15807 (Torino et al., 2005). Lactose utilization by *L. casei* ATCC 393 was strongly repressed by the presence of glucose, since this strain displayed a likely diauxie when grown in a culture medium containing both carbon sources (Veyrat et al., 1994). The efficiency in lactose consumption in the second fed-batch culture (E_L = 48%) decreased in comparison with the first fed-batch culture (E_L = 74%), probably due to the preferential utilization of glucose, which was consumed at a high efficiency (E_G = 89%) by *L. casei* CECT 4043 (Fajardo et al., 2008). According to some researchers (Veyrat et al., 1994; Gosalbes et al., 2000; Yebra et al., 2000), this difference could be related with the negative control that the glucose exerts upon structural organization, regulation and expression of the *lac* genes in *L. casei* strains.

In addition, the hypothesis that proteins of CMPW medium were more efficiently consumed than those of the DW medium is supported by the fact that the efficiency in protein consumption was higher in the second culture (62.0 %) than that (53.2 %) obtained in the first culture (Fajardo et al., 2008).

The stimulation that glucose produced on the growth of *L. casei* CECT 4043 in the second fed-batch culture, also led to an increase in the efficiencies in nitrogen and phosphorous (E_{TN} and E_{TP}) consumption (Fajardo et al., 2008). As it was observed by other researchers (Cabo et al., 2001; Guerra and Pastrana, 2003a; Guerra et al., 2005a, 2005b), the increase in efficiencies of nutrient consumption demonstrated that the system evolved in a more balanced way with respect to the nutrients consumed and added.

5. The $Y_{X/TSc}$ yield was similar in both re-alkalized fed-batch fermentations, but the $Y_{AbA/TSc}$ yield in the second culture was higher than in the first culture. However, both yields were lower than those obtained the non-re-alkalized batch culture (Fajardo et al., 2008), probably due to an overconsumption of the carbon sources (glucose and lactose) by the growing cells in the fed-batch cultures to compensate for the stress generated as a consequence of both the inappropriate final pH values for growth and the rapid increase of pH at the beginning of each re-alkalization period. The same trend was observed in other re-alkalized cultures with two *Lactococcus* strains (Cabo et al., 2001; Guerra and Pastrana, 2003a) and with *Pcd. acidilactici* NRRL B-5627 (Guerra et al., 2005a; 2005b).

In contrast, the yields $Y_{X/TNc}$ and $Y_{AbA/TNc}$ were higher in the second fed-batch fermentation, indicating a best efficiency in the utilization of the nitrogen source for biomass and antibacterial activity production (Fajardo et al., 2008). The increase in $Y_{AbA/X}$ observed in the second fed-batch culture indicated that in this case, the *L. casei* CECT 4043 cells were more productive (Cabo et al., 2001; Guerra and Pastrana, 2003a).

The results obtained indicate that the fed-batch culture performed in DW medium with feeding with CMPW medium and concentrated glucose allows the production of a highly concentrated culture of *L. casei* CECT 4043 in whey at a low cost, with the additional advantage of reducing the initial organic load of the waste, and consequently its contaminant effect. This culture, containing both the cells and AECPs, was successfully used as a probiotic additive in piglet feeds. In fact, a significant increase in body weight gain and final body weight and a significant decrease in the viable coliform counts were observed in weaned piglets fed a diet supplemented with this lactic acid bacterium as compared with piglets (controls) fed the non-supplemented feed (Guerra et al., 2007a).

3.2. Mathematical Modeling

To model the biomass production by *L. casei* CECT 4043 in the two re-alkalized fed-batch cultures, different growth models were evaluated. Therefore, the experimental biomass concentration data were firstly modeled with the Logistic equation [1]:

$$X = \frac{K_1}{1+e^{(c_1 - b_1 \cdot t)}}, \text{ being } c_1 = \ln\left(\frac{K_1}{X_0} - 1\right) \qquad [1]$$

where X_0 is the initial biomass concentration (g/L), K_1 is the maximum biomass concentration (g/L), b_1 is a logistic rate parameter (h^{-1}) and t is the time (h).

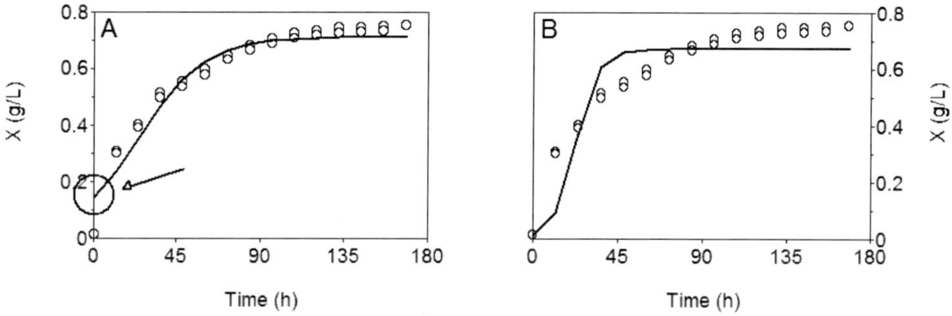

Figure 3. Modeling the biomass production obtained in the first re-alkalized fed-batch fermentation of *L. casei* CECT 4043. The solid lines drawn through the experimental biomass data (symbols) were obtained according to the logistic model [1] without restrictions (A) or with the logistic model [1] subjected to the restriction that the adjusted X_0 = experimental X_0 (B).

The use of model [1] to describe the growth of *L. casei* CECT 4043 in the two re-alkalized fed-batch cultures was unsatisfactory (Figure 3A and 4A). Thus, initial biomass concentration (X_0) was considerably overestimated with the use of model [1] in the two cultures (see circles and arrows in Figures 3A and 4A, Table 2). In an attempt to solve this problem, the experimental growth data was remodeled by using model [1] subjected to the restriction that adjusted X_0 = experimental X_0. However, a clear lack of fit was obtained with this approach (Figure 3B and dashed lines in Figure 4A). These results indicated that the Logistic model [1] can not be used to describe the growth of *L. casei* CECT 4043 in the two re-alkalized fed-batch cultures, basically because the second fed-batch culture displayed two growth phases (Figure 2).

Table 2. Statistically significant ($P < 0.05$) coefficients (means ± standard errors) of the modified logistic model [1] used for predicting the growth of *L. casei* CECT 4043 in DW medium in the re-alkalized fed-batch fermentations I and II

	Re-alkalized fed-batch fermentation I	Re-alkalized fed-batch fermentation II
Parameters	Logistic model	Logistic model
K_1	0.712 ± 0.0194	3.308 ± 1.6409*
b_1	1.379 ± 0.2696	1.806 ± 0.5008
c_1	0.056 ± 0.0093	0.005 ± 0.0013
$[X_0]_1$* (g/L)	0.1431	0.4669
R^2	0.9511	0.9171
RPDM	72.74	115.07

$[X_0]$* Mean initial biomass concentration calculated according to the model [1]. The mean experimental $[X_0]$ value in each culture was 0.013 g/L.
* No significant ($P > 0.05$).

To investigate this latter hypothesis, the time-series data in the second fed-batch fermentation was split in two by considering that the duration *of* the first and second growth phases corresponded to those of the homolactic (0-192 h) and heterolactic (192-372 h) phases, respectively. Subsequently, the two growth phases were modelled separately by using equation [1] to model the first growth phase and equation [2] to model the second growth phase:

$$X = \frac{K_2}{1 + e^{(c_2 - b_2 \cdot t)}}, \text{ being: } c_2 = \ln\left(\frac{K_2}{K_1} - 1\right) \quad [2]$$

where K_2 is the maximum biomass concentration (g/L) in the second growth phase and b_2 is the logistic rate parameter (h^{-1}). Other terms are as previously described.

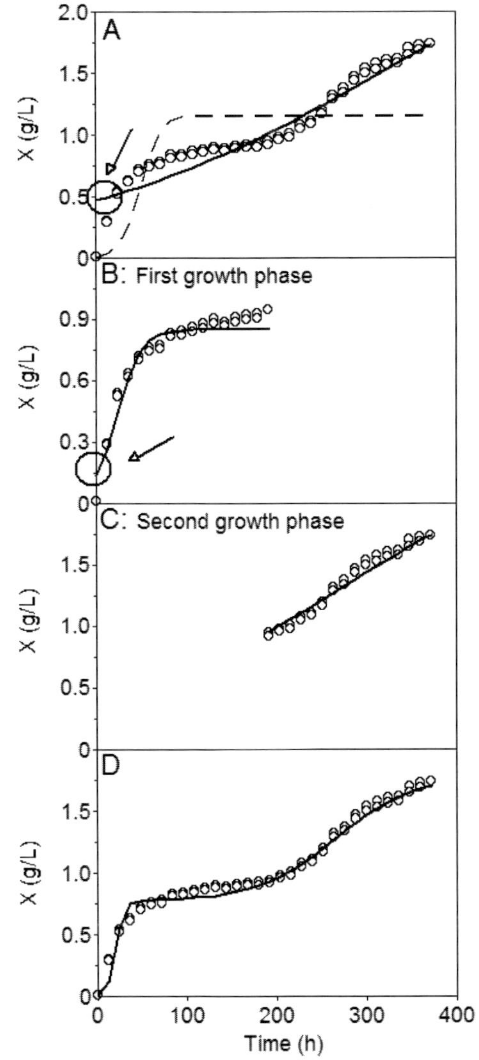

Figure 4. Modeling the biomass production obtained in the second re-alkalized fed-batch fermentation of *L. casei* CECT 4043. The solid lines drawn through the experimental biomass data (symbols) were obtained according to the logistic models [1] (A and B), or [2] (C) without restrictions, or with the Bi-Logistic model [3] (D). The dashed line drawn through the experimental biomass data in Figure 4A was obtained according to the logistic model [1] subjected to the restriction that the adjusted X_0 = experimental X_0.

The results obtained are shown in Figures 4 B and C and in Table 3. After analyzing the results, it can be noted that the use of models [1] and [2] to describe respectively, the first and second growth phases of *L. casei* CECT 4043 in the re-alkalized fed-batch culture II had some drawbacks. On the one hand, the use of model [1] to describe the biomass production in the first growth pulse lead again, to a considerable overestimation of the initial biomass concentration (see circle and arrow in Figure 4B and Table 3), and although the R^2 value (0.9588) was acceptable, the value of *RPDM* (62.10%) was higher than 10% (Table 3). However, the fitting improved in the second growth pulse with the use of model [2], because a high R^2 value (0.9835) and a low *RPDM* value (2.51%) were obtained. These results indicate that modeling the two growth pulses separately was not an adequate alternative to describe accurately the behavior of this culture that exhibited a diauxic growth.

Therefore, the following strategy to model biomass production in the second fed-batch culture was focused on the use of a Bi-logistic model comprised of the sum of two 3-parameter logistic growth pulses (subjected to the restrictions that adjusted X_0 = experimental X_0, as indicated below):

$$X = \frac{K_3}{1 + e^{(c_3 - b_3 \cdot t)}} + \frac{K_4}{1 + e^{(c_4 - b_4 \cdot t)}} \qquad [3]$$

where K_3 *and* K_4 are the maximum biomass concentration (g/L) in the first and second growth phases respectively, b_3 and b_4 are the logistic rate parameters (h^{-1}) in each growth phase. Other terms were as previously defined.

With this procedure, the results obtained (Figure 4D, Table 4) were satisfactory because model [3] fitted well the experimental data ($R^2 > 0.98$ and $RPDM < 10\%$) obtained in the second re-alkalized fed-batch culture and the initial biomass was not overestimated in this case.

Table 3. Statistically significant ($P < 0.05$) coefficients (means ± standard errors) of the modified logistic models [1] and [2] used to describe the two growth cycles in the second re-alkalized fed-batch fermentation of *L. casei* CECT 4043 in DW medium

Parameters	First growth phase (0-180 h)	Second growth phase (180-372 h)
K_1	0.853 ± 0.0185	2.125 ± 0.1750
b_1	1.657 ± 0.2986	2.207 ± 0.1791
c_1	0.072 ± 0.0113	0.010 ± 0.0015
$[X_0]_1$* (g/L)	0.1362	-
R^2	0.9588	0.9835
RPDM	62.10	2.51

$[X_0]$* Mean initial biomass concentration calculated according to the model [1]. The mean experimental $[X_0]$ value in each culture was 0.013 g/L.

However, since the diauxic growth observed in the second fermentation seemed to be associated to a biphasic nitrogen consumption, it could be more adequate to develop a model that relates the growth rates with the nitrogen consumption. Thus, the Monod model (in case of fed-batch fermentation I) or a diauxic Monod-type equation (in case of fed-batch fermentation II), that relates the growth with the concentration of the growth controlling (limiting) substrate could be used for this purpose (Guerra et al., 2007c; 2008). Since the sugars consumed and extracted from the fermentation medium at each sampling time were replenished in each feeding cycle, the nitrogen source was the growth limiting substrate in both fed-batch cultures (Figures 1 and 2).

Table 4. Statistically significant ($P < 0.05$) coefficients (means ± standard errors) of the Bi-logistic model [3] used for predicting the biphasic growth of *L. casei* CECT 4043 in DW medium in the second realkalized fed-batch fermentation

Parameters	Mean values ± standard error
K_1	1.041 ± 0.0683
b_1	5.782 ± 0.4748
c_1	0.022 ± 0.0023
$[X_0]_1$* (g/L)	0.0134
K_2	0.767 ± 0.0100
b_2	4.240 ± 1.0331
c_2	0.213 ± 0.0482
R^2	0.9859
RPDM	5.23

$[X_0]$* Mean initial biomass concentration calculated according to the model [3]. The mean experimental $[X_0]$ value in the culture was 0.013 g/L.

Therefore, the specific growth rate (μ in h^{-1}) in the first fed-batch culture was expressed as a function of the nitrogen source (Monod model) multiplied by a dimensionless inhibition function (δ_i) for accounting the inhibition by product formation (Callewaert and De Vuyst, 2000; Guerra et al., 2007c; 2008):

$$\mu \cdot \delta_i = \left(\frac{\mu_{max1} \cdot (N - N_{min1})}{K_{N1} + N - N_{min1}} \right) \cdot \delta_i \qquad [4]$$

where μ_{max1} is the maximum specific growth rate (h^{-1}), N is the initial nitrogen concentration (g/L) in each feeding and re-alkalization cycle justly after feeding the fermentor. K_{N1} is the nitrogen affinity constant (g/L) and N_{min1} is the minimum nitrogen concentration (g/L) that supports the growth.

In the second fed-batch culture, the specific growth rate was expressed as the sum of the specific growth rates (μ_1 and μ_2) on each nitrogen consumption phase multiplied by the inhibition function δ_i (Guerra et al., 2007c; 2008):

$$\mu \cdot \delta_i = (\mu_1 + \mu_2) \cdot \delta_i = \left(\frac{\mu_{max1} \cdot (N - N_{min1})}{K_{N1} + N - N_{min1}} + \frac{\mu_{max2} \cdot (N - N_{min2})}{K_{N2} + N - N_{min2}} \right) \cdot \delta_i \qquad [5]$$

where μ_{max1} and μ_{max2} are the specific growth rates (h^{-1}) of the exponential growth phases 1 and 2, respectively. K_{N1} and K_{N2} are the nitrogen affinity constants (g/L) in each phase of nitrogen consumption. N_{min1} and N_{min2} are the minimum nitrogen concentrations (g/L) that support the growth in each phase. Other terms are as previously described.

The inhibition function δ_i accounts for the self-inhibition of the cells due to the production of antimicrobial products (Callewaert and De Vuyst, 2000; Åkerberg et al., 1998), so that:

$$\delta_i = \delta_{LA} \cdot \delta_{AA} = \left(\frac{K_{iLA}}{K_{iLA} + LA} \right) \cdot \left(\frac{K_{iAA}}{K_{iAA} + AA} \right) \qquad [6]$$

where δ_{LA} and δ_{AA} are individual inhibition functions which account for the inhibition of cell growth by the increase in the concentrations (g/L) of lactic acid (LA) and acetic acid (AA). K_{iLA} and K_{iAA} are the inhibition constants (g/L) for the lactic acid and acetic acid, respectively.

In order to take into account the effect of dilution, which occurred due to the sample extraction and the feeding of nutrients, mass balance was carried out across the fermentor and the biomass accumulation rate (r_X) was expressed as:

$$rX = \frac{dX}{dt} = \frac{\left(X(t_n) - X(t_{n-1}) \cdot \left(1 - \frac{V_S}{V_F} \right) \right)}{dt} = \mu \cdot \delta_i \cdot X \qquad [7]$$

where X is the biomass concentration (g/L), V_S and V_F are the volumes (in litres) of sampling and fermentation medium, respectively. $X(t_{n-1})$ and $X(t_n)$ are the biomass concentrations (g/L) inside the fermentor at the beginning and at the end of each re-alkalization and feeding cycle, respectively. Other terms are as previously defined.

Table 5. Statistically significant ($P < 0.05$) coefficients (means ± standard errors) of the model [7] used for predicting the monophasic (in case of fermentation I) and biphasic (in case of fermentation II) growth data of *L. casei* CECT 4043

Parameter	Fed-batch fermentation I (Monod model)	Fed-batch fermentation II (Double-Monod model)
μ_{max1} (h^{-1})	0.209 ± 0.0321	0.177 ± 0.0073
K_{N1} (g/L)	0.090 ± 0.0148	0.055 ± 0.0015
N_{min1} (g/L)	0.204 ± 0.0213	0.228 ± 0.0020
μ_{max2} (h^{-1})	-	0.025 ± 0.0072
K_{N2} (g/L)	-	0.183 ± 0.0769
N_{min2} (g/L)	-	0.023 ± 0.0016
K_{iLA} (g/L)	3.652 ± 0.1984	19.453 ± 7.9976
K_{iAA} (g/L)	0.306 ± 0.0974	0.155 ± 0.0681
R^2	0.9974	0.9867
RPDM	1.64	4.94

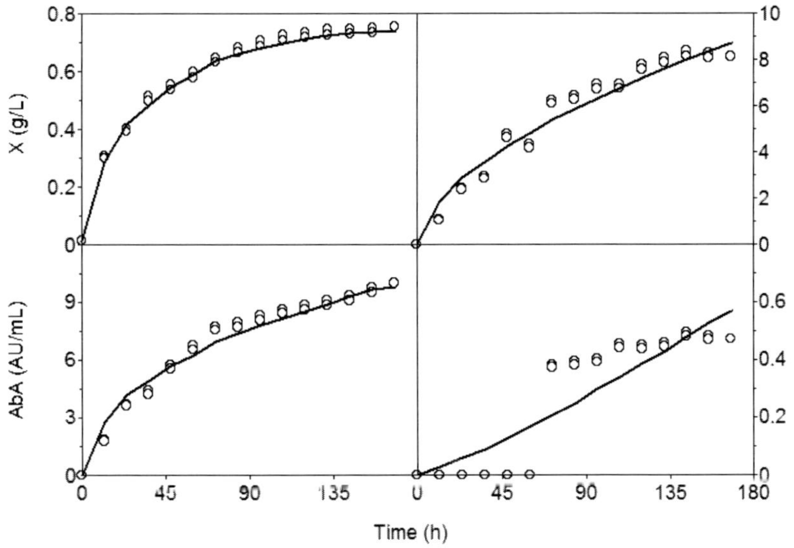

Figure 5. Modeling the biomass production and product (LA, AA and AbA) formation obtained in the first re-alkalized fed-batch fermentation of *L. casei* CECT 4043. The solid lines drawn through the experimental biomass data (symbols) were obtained according to the models [7] (in case of biomass), [8] (in case of lactic acid and acetic acid) and [9] (in case of antibacterial activity).

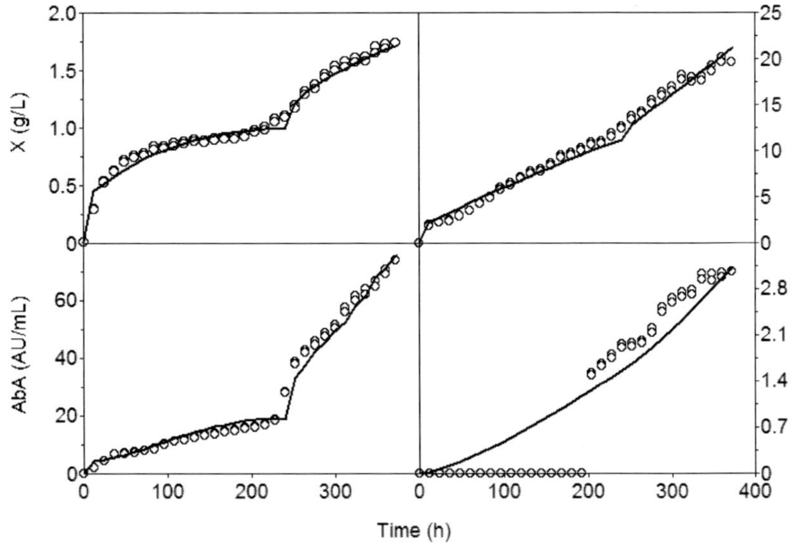

Figure 6. Modeling the biomass production and product (LA, AA and AbA) formation obtained in the second re-alkalized fed-batch fermentation of *L. casei* CECT 4043. The solid lines drawn through the experimental biomass data (symbols) were obtained according to the models [7] (in case of biomass), [8] (in case of lactic acid and acetic acid) and [9] (in case of antibacterial activity).

The use of model [7] to describe the experimental growth data of *L. casei* CECT 4043 was satisfactory, because an excellent agreement was found between model predictions and experimental results for biomass production in both fed-batch fermentations (left higher part of Figures 5 and 6). In addition, considerably higher R^2 values and low *RPDM* values were obtained (Table 5). In addition, the use of both the Monod model and the double Monod model resulted in improved predictions of the experimental data over the previously mentioned Logistic and Bi-logistic models in the two re-alkalized fed-batch cultures.

These results strengthened the usefulness of the models based on Monod kinetics for describing the growth of the strain CECT 4043 in the two re-alkalized fed-batch fermentations in whey.

After selecting the most adequate growth model, the products (lactic acid, acetic acid and antibacterial activity) formation was modeled by using the Luedeking and Piret model or a modified form of it.

Accumulation rates (*rP*) of lactic acid and acetic acid were represented to contain growth (α) and non-growth (β) associated constants (Luedeking and Piret, 1959) as follows:

$$rP = \frac{\left(P(t_n) - P(t_{n-1})\cdot\left(1 - \frac{V_S}{V_F}\right)\right)}{dt} = \alpha \cdot rX + \beta \cdot X \qquad [8]$$

where $P(t_{n-1})$ and $P(t_n)$ are the lactic acid or acetic acid concentrations (g/L) inside the fermentor at the beginning and at the end of each re-alkalization and feeding cycle, respectively. Other terms are as previously defined.

Antibacterial activity rate (*rAbA*) was modeled by using a modified form of the Luedeking and Piret model that includes a term for the effect of pH drop (*rpH*) on *AbA* production (Guerra and Pastrana, 2003b):

$$rAbA = \frac{\left(AbA(t_n) - AbA(t_{n-1})\cdot\left(1 - \frac{V_S}{V_F}\right)\right)}{dt} = \left(\alpha \cdot rX + \beta \cdot X\right)\cdot\left(1 + e^{-\kappa \cdot \left(r_{pHmax} - r_{pH}\right)}\right) \qquad [9]$$

where $AbA(t_{n-1})$ and $AbA(t_n)$ are the antibacterial activity concentrations (AU/mL) inside the fermentor at the beginning and at the end of each re-alkalization and feeding cycle, respectively. κ is a constant of proportionality (h) and *rpHmax* is the maximum pH drop (h^{-1}). Other terms are as previously defined.

The models [8] and [9] described adequately the productions of lactic acid and antibacterial activity, respectively (Figures 5 and 6), and provided relatively higher values for the R^2 coefficient and *RPDM* values lower or slightly higher than 10% (Table 6). From the results obtained with both models (Table 6), it can be concluded that lactic acid was produced as a primary metabolite ($\alpha >> \beta$) and the antibacterial activity was synthesized as a primary metabolite ($\alpha >> \beta$) dependent on the pH drop ($\kappa \neq 0$). Thus, the model [9] could be used to describe satisfactorily the synthesis of antibacterial activity that is influenced by the pH drop rate, which is usually difficult to describe.

Modelling the production of acetic acid in the two realkalized fed-batch cultures with the use of equation [8] was unsatisfactory (Figures 5 and 6, Table 6). This was probably due to the fact that this product was not synthesized from the beginning of the cultures. With regard to this, the appearance of mixed acid products has been usually related with a nutrient (carbon or nitrogen source) limitation (De Vuyst, 1995; Kim et al., 1997). Then, it is reasonable to suppose that the production of acetic acid started when the culture medium reached a nitrogen concentration that limits the growth of *L. casei* CECT 4043. This suggests that the production of acetic acid was inhibited, in the homolactic fermentation phase, by nitrogen concentrations higher than the limiting level (substrate inhibition). In the same way, it could be considered that nitrogen concentrations lower than the limiting concentration, favored the production of this metabolite.

Table 6. Statistically significant ($P < 0.05$) coefficients (means ± standard errors) for products (lactic acid and acetic acid, antibacterial activity) in the two re-alkalized fed-batch fermentations of *L. casei* CECT 4043 in DW medium

Parameters	Fed-batch fermentation I	Fed-batch fermentation II
Lactic acid		
α_{LA} (dimensionless)	5.593 ± 2.2702	4.080 ± 0.8655
β_{LA} (h^{-1})	0.038 ± 0.0138	0.028 ± 0.0063
R_{LA}^2	0.9650	0.9856
$RPDM_{LA}$	13.07	7.87
Acetic acid		
α_{AA} (dimensionless)	0.046 ± 0.3446*	-0.261 ± 0.5828*
β_{AA} (h^{-1})	0.004 ± 0.0040*	0.087 ± 0.0043*
R_{AA}^2	0.7976	0.8837
$RPDM_{AA}$	45.15	55.46
Antibacterial activity		
α_{AbA} (AU/mg)	4.700 ± 1.3012	4.853 ± 0.036
β_{AbA} (AU/mg/h)	0.002 ± 0.0040	0.001 ± 0.0001
κ (h)	14.530 ± 0.7215	35.547 ± 0.1110
rpH_{max} (h^{-1})	0.145 ± 0.0063	0.146 ± 0.0003
R_{AbA}^2	0.9861	0.9849
$RPDM_{AbA}$	6.36	13.93

* No significant ($P > 0.05$).

From the results obtained, it can be pointed out that the models [4], [5], [7], [8] and [9] successfully described the productions of biomass, lactic acid and antibacterial activity by *L. casei* CECT 4043 in the two re-alkalized fed-batch cultures in whey.

Conclusion

The use of the models [4], [5], [7], [8] and [9] may lead to the development of better strategies for the optimization of the production of a potentially probiotic culture of *L. casei* CECT 4043 to ensure the economic viability of the fermentation process. These models could be used as a tool that has potential in optimization studies of industrial reactors and in the development of new bioproductions with the same strain.

Acknowledgments

The research presented in this paper was financially supported by the Instituto Nacional de Investigación y Tecnología Agraria y Alimentaria (INIA), Spain (project CAL01-045-C2-2) and The Xunta de Galicia, Spain (project PGIDT00BIO1E). We thank COREN, S.C.L. for their collaboration in the elaboration of this work.

References

Abe, F., Ishibashi, N., Shimamura, S. (1995). Effect of administration of bifidobacteria and lactic acid bacteria to newborn calves and piglets. *J. Dairy Sci.* 78, 2838-2846.

Aguerre, R. J., Suarez, C., Viollaz, P. E. (1989). New BET type multilayer sorption isotherms. Part II. Modeling water sorption in foods. Lebensm. Wiss. *Technol. – Food Sci. Technol.* 22, 192-195.

Åkerberg, C., Hofvendahl, K., Hahn-Hägerdal, B., Zacchi, G. (1998). Modeling the influence of pH, temperature, glucose and lactic acid concentrations on the kinetics of lactic acid production by *Lactococcus lactis* ssp. *lactis* ATCC 19435 in whole-wheat flour. *Appl. Microbiol. Biotechnol.* 49, 682-690.

Anadón, A., Martínez-Larrañaga, M.R., Aranzazu, M. (2006). Probiotics for animal nutrition in the European Union. Regulation and safety assessment. *Regul. Toxicol. Pharmacol.* 45, 91-95.

Arvola, T., Laiho, K., Torkkeli, S., Mykkänen, H., Salminen, S., Maunula, L., Isolauri, E. (1999). Prophylactic *Lactobacillus* GG reduces antibiotic-associated diarrhea in children with respiratory infections: a randomized study. *Pediatr.* 104, 1-4.

Aso, Y., Akazan, H. (1992). Prophylactic effect of a *Lactobacillus casei* preparation on the recurrence of superficial bladder cancer. *Urol Int.* 49, 125-129.

Aso, Y., Akazan, H., Kotake, T., Tsukamoto, T., Imai, K., Naito, S. (1995). Preventive effect of a *Lactobacillus casei* preparation on the recurrence of superficial bladder cancer in a double-blind trial. The BLP study group. *Eur. Urol.* 27, 104-109.

Borch, E., Berg, H., Holst, O. (1991). Heterolactic fermentation by a homofermentative *Lactobacillus* sp. during glucose limitation in anaerobic continuous culture with complete cell recycle. *J. Appl. Microbiol.* 71, 265-269.

Cabo, M.L., Murado, M.A., González, Mª.P., Pastoriza, L. (1999). A method for bacteriocin quantification. *J. Appl. Microbiol.* 87, 907-914.

Cabo, M.L., Murado, M.A., González, Mª.P., Pastoriza, L. (2001). Effects of aeration and pH gradient on nisin production. A mathematical model. *Enzyme Microb. Technol.* 29, 264-273.

Callewaert, R., De Vuyst, L. (2000). Bacteriocin production with *Lactobacillus amylovorus* DCE 471 is improved and stabilized by fed-batch fermentation. *Appl. Environ. Microbiol.* 66, 606-613.

Cano, P.G., Perdigon, G. (2003). Probiotics induce resistance to enteropathogens in a re-nourished mouse model. *J. Dairy Res.* 70, 433-440.

Cerning, J., Renard, C.M.G.C., Thibault, J.F., Bouillanne, C., Landon, M., Desmazeaud, M., Topisirovic, L. (1994). Carbon source requirements for exopolysaccharide production by

Lactobacillus casei CG11 and partial structure analysis of the polymer. *Appl. Environ. Microbiol.* 60, 3914-3919.

Cocaign-Bousquet, M., Garrigues, C., Lubiere, P., Lindley, N. (1996). Physiology of piruvate metabolism in *Lactococcus lactis*. *Ant. van Leeuw.* 70, 253-267.

De Vuyst, L. (1995). Nutritional factors affecting nisin production by *Lactococcus lactis* subsp. *lactis* NIZO 22186 in a synthetic medium. *J. Appl. Microbiol.* 78, 28-33.

de Waard, R., Garssen, J., Bokken, G.C.A.M., Vos, J.G. (2002). Antagonistic activity of *Lactobacillus casei* strain Shirota against gastrointestinal *Listeria monocytogenes* infection in rats. *Int. J. Food Microbiol.* 73, 93-100.

Fajardo, P., Rodríguez, I., Pastrana, L., Pérez-Guerra, N. (2008). Production of a potentially probiotic culture of *Lactobacillus casei* subsp. *casei* CECT 4043 in whey. *Int. Dairy J.*, 18, 1057-1065.

Fasoli, S., Marzotto, M., Rizzotti, L., Rossi, F., Dellaglio, F., Torriani, S. (2003). Bacterial composition of comercial probiotic products as evaluated by PCR-DGGE analysis. *Int. J. Food Microbiol.* 82, 59-70.

Flambard, B., Richard, J., Juillard, V. (1997). Interaction between proteolytic strains of *Lactococcus lactis* influenced by different types of proteinase during growth in milk. *Appl. Environ. Microbiol.* 63, 2131-2135.

Garrigues, C., Loubiere, P., Lindley, N.D., Cocaign-Bousquet, M. (1997). Control of the shift from homolactic acid to mixed-acid fermentation in *Lactococcus lactis*: predominant role of the NADH/NAD^+ ratio. J. *Bacteriol.* 179, 5282-5287.

Garrigues, C., Mercade, M., Cocaign-Bousquet, M., Lindley, N.D., Loubiere, P. (2001). Regulation of pyruvate metabolism in *Lactococcus lactis* depends on the imbalance between catabolism and anabolism. *Biotechnol. Bioeng.* 74, 108-115.

Gosalbes, M.J., Esteban, C.D., Galán, J.L., Pérez-Martínez, G. (2000). Integrative food-grade expression system based on the lactose regulon of *Lactobacillus casei*. *Appl. Environ. Microbiol.* 66, 4822-4828.

Guerra, N.P., Rua, M.L., Pastrana, L. (2001). Nutritional factors affecting the production of two bacteriocins from lactic acid bacteria on whey. *Int. J. Food Microbiol.* 70, 267-281.

Guerra, N.P., Pastrana, L. (2002a). Modeling the influence of pH on the kinetics of both nisin and pediocin production and characterization of their functional properties. *Proc. Biochem.* 37, 1005-1015.

Guerra, N.P., Pastrana, L. (2002b). Production of bacteriocins from *Lactococcus lactis* subsp. *lactis* CECT 539 and *Pediococcus acidilactici* NRRL B-5627 using mussel-processing wastes. *Biotechnol. Appl. Biochem.* 36, 119-125.

Guerra, N.P., Pastrana, L. (2003a). Enhancement of nisin production by *Lactococcus lactis* in periodically re-alkalized cultures. *Biotechnol. Appl. Biochem.* 38, 157-167.

Guerra, N.P., Pastrana, L. (2003b). Influence of pH drop on both nisin and pediocin production by *Lactococcus lactis* and *Pediococcus acidilactici*. *Lett. Appl. Microbiol.* 37, 51-55.

Guerra, N.P., Fajardo, P., Torrado A., López, C., Pastrana, L. (2005a). Fed-batch pediocin production by *Pediococcus acidilactici* NRRL B-5627 on whey. *Biotechnol. Appl. Biochem.* 42, 17-23.

Guerra, N.P., Torrado, A., López, C., Pastrana, L. (2005b). Modeling the fed-batch production of pediocin using mussel-processing wastes. *Proc. Biochem.* 40, 1071-1083.

Guerra, N.P., Fajardo, P., Méndez, J., Cachaldora, P., Pastrana, L. (2007a). Production of four potentially probiotic lactic acid bacteria and their evaluation as feed additives for weaned piglets. *Anim. Feed Sci. Technol.* 134, 89-107.

Guerra, N.P, Fajardo, P., Pastrana, L. (2007b). Fed-batch pediocin production on whey using different feeding media. *Enzyme Microb. Technol.* 41, 397-406.

Guerra, N.P., Agrasar, A.T., Macías, C.L., Fajardo, P., Castro, L.P. (2007c). Dynamic mathematical models to describe the growth and nisin production by *Lactococcus lactis* subsp. *lactis* CECT 539 in both batch and re-alkalized fed-batch cultures, *J. Food Eng.* 82, 103-113.

Guerra, N.P., Fajardo, P., Castro, L.P. (2008). Modeling the stress-inducing biphasic growth and pediocin production by *Pediococcus acidilactici* NRRL B-5627 in re-alkalized fed-batch cultures. *Biochem. Eng. J.* 40, 465-472.

Guerra, N.P., Fajardo, P., Fuciños, C., Rodríguez, I., Alonso, E., Torrado, A., Pastrana, L. (2010). Modeling the biphasic growth and product formation by *Enterococcus faecium* CECT 410 in realkalized fed-batch fermentations in whey. *J. Biomed. Biotechnol.* 2010, 1-16.

Hickson, M., D'Souza, A.L., Muthu, N., Rogers, T.R., Want, S., Rajkumar, C., Bulpitt, C.J. (2007). Use of probiotic *Lactobacillus* preparation to prevent diarrhea associated with antibiotics: randomized double-blind placebo controlled trial. *B.M.J.* 335, 80.

Hofvendahl, K., van Niel, E.W.J., Hahn-Hägerdal, B. (1999). Effect of temperature and pH on growth and product formation of *Lactococcus lactis* ssp. *lactis* ATCC 19435 growing on maltose. *Appl. Microbiol. Biotechnol.* 51, 669-672.

Juillard, V., Le Bars, D., Kunji, E. R. S., Konings, W. N., Gripon, J. C., Richard, J. (1995). Oligopeptides are the main source of nitrogen for *Lactococcus lactis* during growth in milk. *Appl. Environ. Microbiol.* 61, 3024-3030.

Juillard, V., Helinck, S., Flambard, B., Foucaud, C., Richard, J. (1998). Amino acid supply of *Lactococcus lactis* during growth in milk. *Recent Res. Dev. Microbiol.* 2, 233-252.

Kim, H.J., Camilleri, M., McKinzie, S., Lempke, M.B., Burton, D.D., Thomforde, G.M., Zinsmeister, A.R. (2003). A randomized controlled trial of a probiotic, VSL#3, on gut transit and symptoms in diarrhea-predominant irritable bowel syndrome. *Aliment. Pharmacol. Ther.* 17, 895-904.

Kim, W.S., Hall, R.J., Dunn, N.W. (1997). The effect of nisin concentration and nutrient depletion on nisin production of *Lactococcus lactis*. *Appl. Microbiol. Biotechnol.* 48, 449-453.

Koebnick, C., Wagner, I., Leitzmann, P., Stern, U., Zunft, H.J. (2003). Probiotic beverage containing *Lactobacillus casei* Shirota improves gastrointestinal symptoms in patients with chronic constipation. *Can. J. Gastroenterol.* 17, 655-659.

Kyriakis, S.C., Tsiloyiannis, V.K., Lekkas, S., Petridou, E., Vlemmas, J., Sarris, K. (1997). The efficacy of enrofloxacin in-feed medication, by applying different programmes for the control of post-weaning diarrhea syndrome of piglets. *J. Vet. Med.* B 44, 513-521.

Letort, C., Nardi, M., Garault, P., Monnet, V., Juillard, V. (2002). Casein utilization by *Streptococcus thermophilus* results in a diauxic growth in milk. *Appl. Environ. Microbiol.* 68, 3162-3165.

Lomauro, C.J., Bakshi, A.S., Labuza, T.P. (1985). Evaluation of food moisture isotherm equations. Part I. Fruit, vegetable and meat products. *Lebensm. Wiss. Technol. – Food Sci. Technol.* 18, 111-117.

Lopez de Felipe, F., Kleerebezem, M., de Vos, W.M., Hugenholtz, J. (1998). Co-factor engineering: a novel approach to metabolic engineering in *Lactococcus lactis* by controlled expression of NADH-oxidase. *J. Bacteriol.* 180, 3804-3808.

Luedeking, R., Piret, E.L. (1959). A kinetic study of the lactic acid fermentation. Batch process at controlled pH. *J. Biochem. Microbiol. Technol. Eng.* 1, 393-412.

Martini M.C., Kukielka, D., Savaiano, D. (1991). Lactose digestion from yogurt: influence of a meal and additional lactose. *Am. J. Clin. Nutr.* 53, 1253-1258.

Matijašić, B.B., Stojkovic, S., Salobir, J., Malovrh, S., Rogelj, I. (2004). Evaluation of the *Lactobacillus gasseri* K7 and LF221 strains in weaned piglets for their possible probiotic use and their detection in the faeces. *Anim. Res.* 53, 35-44.

Murado, M.A., González, Mª.P., Vázquez, J.A. (2002). Dose-response relationships: an overview, a generative model and its application to the verification of descriptive models. *Enzyme Microb. Technol.* 31, 439-455.

Niven, G.W., Knight, D.J., Mulholland, F. (1998). Changes in the concentrations of free amino acids in milk during growth of *Lactococcus lactis* indicate biphasic nitrogen metabolism. *J. Dairy Res.* 65, 101-107.

Pascual, M., Hugas, M., Badiola, J.I., Monfort, J.M., Garriga, M. (1999). *Lactobacillus salivarius* CTC2197 prevents *Salmonella enteritidis* colonization in chickens *Appl. Environ. Microbiol.* 65, 4981-4986.

Perdigon, G., Medici, M., Bibas Bonet de Jorrat, M.E., Valverde De Budeguer, M., Pesce De Ruiz Holgado, A. (1993). Immunomodulating effects of lactic acid bacteria on mucosal and tumoral immunity. *Int. J. Immunother.* 9, 29-52.

Requena, T., Pelaez, C., Desmazeaud, M.J. (1991). Characterization of lactococci and lactobacilli isolated from semi-hard goats' cheese. *J. Dairy Res.* 58, 137-145.

Schrezenmeir, J., de Vrese, M. (2001). Probiotics, prebiotics and synbiotics – approaching a definition. *Am. J. Clin. Nutr.* 73, 361S-364S.

Sgouras, D., Maragkoudakis, P., Petraki, K., Martinez-Gonzalez, B., Eriotou, E., Michopoulos, S., Kalantzopoulos, G., Tsakalidou, E., Mentis, A. (2004). In vitro and in vivo inhibition of *Helicobacter pylori* by *Lactobacillus casei* strain Shirota. *Appl. Environ. Microbiol.* 70, 518-526.

Snoep, J.L., Teixeira de Mattos, M.J., Starrenburg, M., Hugenholtz, J. (1992). Isolation, characterization, and physiological role of the pyruvate dehydrogenase complex and alpha-acetolactate synthase of *Lactococcus lactis* subsp. *lactis* bv. *diacetylactis*. *J. Bacteriol.* 174, 4838-4841.

Szajewska, H., Mrukowicz, J.Z. (2001). Probiotics in the treatment and prevention of acute infectious diarrhea in infants and children: a systematic review of published randomized, double-blind, placebo-controlled trials. *J. Pediatr. Gastroenterol. Nutr.* 33, S17-S25.

Thomas, T.D., Ellwood, D.C., Longyear, M.C. (1979). Change from homo- to heterolactic fermentation by *Streptococcus lactis* resulting from glucose limitation in anaerobic chemostat cultures. *J. Bacteriol.* 138, 109-117.

Thomas, T.D., Turner, K.W., Crow, V.L. (1980). Galactose fermentation by *Streptococcus lactis* and *Streptococcus cremoris*: Pathways, products, and regulation. *J. Bacteriol.* 144, 672-682.

Thorlacius, H., Jeppsson, B. (2003). Anti-infective effects of lactobacilli in the gastrointestinal tract. *Curr. Med. Chem. Anti-Infective Agents*, 2, 263-267.

Tien, M.T., Girardin, S.E., Regnault, B., Le Bourhis, L., Dillies, M.A., Coppee, J.Y., Bourdet-Sicard, R., Sansonetti, P.J., Pedron, T. (2006). Anti-inflammatory effect of *Lactobacillus casei* on *Shigella*-infected human intestinal epithelial cells. *J. Immunol.* 176, 1228-1237.

Torino, M.I., Taranto, M.P., Sesma, F., Font de Valdez, G. (2001). Heterofermentative pattern and exopolysaccharide production by *Lactobacillus helveticus* ATCC 15807 in response to environmental pH. *J. Appl. Microbiol.* 91, 846-852.

Torino, M.I., Hébert, E.M., Mozzi, F., Font de Valdez, G. (2005). Growth and exopolysaccharide production by *Lactobacillus helveticus* ATCC 15807 in an adenine-supplemented chemically defined medium. *J. Appl. Microbiol.* 99, 1123-1129.

van Niel, E.W.J., Hahn-Hägerdal, B. (1999). Nutrient requirements of lactococci in defined growth media. *Appl. Microbiol. Biotechnol.* 52, 617-627.

Verdon, E. (2010). Antibiotic residues in muscle tissues of edible animal products. In Nollet, L.M.L., Toldrá, F. (Eds.), Safety analysis of foods of animal origin. (Chapter 10, pp. 249 -347). Boca Raton, USA: CRC Press.

Veyrat, A., Monedero, V., Pérez-Martínez, G. (1994). Glucose transport by the phosphoenolpyruvate: mannose phosphotransferase system in *Lactobacillus casei* ATCC 393 and its role in carbon catabolite repression. *Microbiol.* 140, 1141-1149.

Yebra, M.J., Veyrat, A., Santos, M.A., Pérez-Martínez, G. (2000). Genetics of L-sorbose transport and metabolism in *Lactobacillus casei*. *J. Bacteriol.* 182, 155-163.

Zhu, Y., Yang, S.T. (2006). Effect of pH on metabolic pathway shift in fermentation of xylose by *Clostridium tyrobutyricum*. *J. Biotechnol.* 110, 143-157.

In: Lactobacillus: Classification, Uses and Health Implications ISBN: 978-1-62081-151-1
Editors: Alba I. Perez Campos and Arturo Leon Mena

Chapter VIII

Unraveling Genomics of Lactic Acid Bacteria and Flavor Formation in Dairy Products

Ana Paula do Carmo[1], Arnaldo Chaer Borges[1], Célia Alencar de Moraes[1] and Antônio Fernandes de Carvalho[2*]

[1]Instituto de Biotecnologia Aplicada à Agropecuária (BIOAGRO), Departamento de Microbiologia, Universidade Federal de Viçosa, Viçosa, Minas Gerais, Brazil
[2]Departamento de Tecnologia de Alimentos, Universidade Federal de Viçosa, Viçosa, Minas Gerais, Brazil

Abstract

Lactic acid bacteria (LAB) are industrially important organisms and have been used as starter cultures in various food-fermentation processes. The improvement of gene annotation has led to a better prediction of flavor producing pathways in specific lineages. LAB of the genus *Lactobacillus* are widely used in the dairy industry. The most recent phylogenetic classification of this genus includes 20 completely sequenced genomes. Thus, it was possible to distinguish three groups: NCFM, GG and WCFS; this nomenclature is derived from the reference strains of each group, *Lactobacillus acidophilus* NCFM, *Lactobacillus rhamnosus* GG and *Lactobacillus plantarum* WCFS1, respectively. *Lactobacillus delbrueckii* species can be represented by three main subspecies, *delbrueckii*, *bulgaricus* and *lactis*. These subspecies are evolutionary distinct, colonizing different habitats and also presenting distinct fermentation patterns. While the subspecies *bulgaricus* and *lactis* are almost exclusively associated with milk and dairy products, the subspecies *delbrueckii* colonizes plant sources. Flavor compounds can be produced from the degradation of three principal constituents: lactose (glycolysis and Leloir pathway), lipids (lipolysis) and protein (proteolysis), the latter being the most

* Corresponding author: Departamento de Tecnologia de Alimentos, Universidade Federal de Viçosa, Viçosa - Minas Gerais, Brazil. Email: antoniofernandes@ufv.br.

complex and important process for the formation of flavor compounds in fermented foods. The proteolytic systems of LAB can be divided into specific components that degrade proteins into amino acids by sequence reactions involving cell-wall proteinases, peptide transport systems and different enzymes. These include intracellular proteinases (endopeptidases and aminopeptidases), whose degradation reactions usually involve release of flavor compounds. The introduction of probiotic LAB in starter cultures for manufacturing fermented products, in addition to meeting current demands of consumers who seek innovative and differentiated products while at the same time providing health benefits to the organism, has provided significant economic return, driving a multimillion-dollar market worldwide. *Lactobacillus delbrueckii* UFV H2b20 was isolated from faecal material of a breast fed newborn child in the first week of life, and since then has been the focus of numerous studies to demonstrate promising characteristics for its use as a probiotic. The operon that encodes ribosomal rRNAs in *L. delbrueckii* UFV H2b20 was partially characterized. Data suggests that *L. delbrueckii* UFV H2b20 is more closely related to *L. delbrueckii* subsp. *delbrueckii* than *L. delbrueckii* subsp. *lactis*. On the other hand, this bacterium can be compared to *L. delbrueckii* subsp. *lactis*, *L. delbrueckii* subsp. *bulgaricus* and *L. acidophilus* by a phenotypical aspect, since both of them are able to degrade lactose and therefore can grow on milk. In previous studies, viable or inactivated cells of *L. delbrueckii* UFV H2b20 were concentrated in stabilized Minas Frescal cheese whey, from which it was possible to manufacture dairy products containing this bacterium. Subsequently, an efficient method for dehydration and maintenance of cell viability in the dehydrated state was evaluated. Thus, a probiotic starter culture with promising characteristics for use in dairy products was produced at the laboratory scale. Recent studies have demonstrated that *L. delbrueckii* UFV H2b20 possesses several proteins needed for an efficient proteolytic system that supports growth on milk. Based on the genes identified so far, it was possible to propose a schematic model to explain how these proteolytic enzymes are involved in flavor formation. Moreover, this strain has some unexpected metabolic characteristics which suggest that *L. delbrueckii* UFV H2b20 might be considered as a new and unusual subspecies among *L. delbrueckii* subspecies, the first one identified as a probiotic. Since *L. delbrueckii* UFV H2b20 has been extensively characterized and because it is a probiotic, this review includes some more detailed information about this bacterium. Additionally, current scientific researches about genomics of LAB and flavor formation in dairy products are thoroughly discussed.

1.1. Genomics of Lactic Acid Bacteria

Lactic acid bacteria (LAB) are low-GC-content, Gram-positive bacteria which are found in nutrient-rich environments such as milk, meat and the mammalian gastrointestinal tract (GIT). LAB undergo rapid fermentative growth on the sugars found in these environments, with lactic acid as the main fermentation product. LAB can be classified as homofermentative, facultative heterofermentative or heterofermentative, based on their glucose fermentation features (Klaenhammer *et al.*, 2005; Claesson *et al.*, 2007). These microorganisms are important industrially and have been used as starter cultures in various food-fermentation processes.

In general, the selection criteria for starter cultures are their rate of acidification and their flavor-producing characteristics (Teusink and Smid, 2006). Flavor formation is an important feature for industrial applications of LAB.

The genus *Lactobacillus* is the most representative among LAB, and currently, various *Lactobacillus* species are decribed in the NCBI database. For instance, several genome projects for *Lactobacillus* and *Bifidobacteria* genera are underway (http://www.ncbi.nlm.nih.gov/genomes/lproks.cgi/). Thousands of probiotic *Lactobacillus* strains have already undergone complete genome sequencing. (*L. acidophilus*, *L. casei*, *L. amylovorus*, *L. crispatus*, *L. paracasei*, *L. fermentum*, *L. gasseri*, *L. johnsonii*, *L. plantarum*, *L. rhamnosus*, *L. reuteri*, and *L. salivarius* strains representatives). While *Bifidobacterium* genomes available are characterized by a high degree of syntheny along their entire genomes, the *Lactobacillus* genomes in turn reflect a phylogenetic, phenotypical and ecological diversity inherent to this genus (Ventura *et al.*, 2007).

Before the genome sequencing era, the Lactobacillales classification presented an unsolved question, mainly because the classification was based on fermentation patterns. These fermentation patterns usually went unconfirmed by rRNA phylogeny (Vaughan *et al.*, 2005). Taxonomists have always known that the genus *Lactobacillus* could be subdivided into groups, and extensive analysis revealed that the evolutionary mechanisms of LAB genomes are driven by genome decay and lateral gene transfer, which expands the possibilities of selected gene families (Makarova *et al.*, 2006; Makarova and Koonin, 2007). Initially, three groups were proposed based on type of fermentation: the *Lactobacillus casei/Pediococcus* group, the *Leuconostoc* group and the *L. delbrueckii* group, later renamed the *L. acidophilus* group.

The *L. casei* group was split into further subgroups and a new genus (Felis and Dellaglio, 2007). The overall phylogenetic structure of the rRNA tree in the genus *Lactobacillus* is quite complicated, and led to the suggestion of eight groups.

Another reclassification was proposed for lactobacilli. This classification distinguished four subgeneric groups (Claesson *et al.*, 2008). Recently, a study encompassing 20 complete LAB genomes (Kant *et al.*, 2010) were organized them into three main groups, based on the phylogeny of the proteins encoded by *Lactobacillus* core genome. These groups are NCFM, GG, and WCFS.

The nomenclature is derived from the reference strains *Lactobacillus acidophilus* NCFM, *Lactobacillus rhamnosus* GG, and *Lactobacillus plantarum* WCFS1, respectively. In this classification, *L. acidophilus* NCFM, *Lactobacillus helveticus* DPC 4571, *Lactobacillus delbrueckii* subsp. *bulgaricus* ATCC 11842, *L. delbrueckii* subsp. *bulgaricus* ATCC BAA-365, *Lactobacillus johnsonii* NCC 533, *Lactobacillus johnsonii* FI 9785, *Lactobacillus crispatus* ST 1, and *Lactobacillus gasseri* ATCC 33323 are representatives of the NCFM group.

These strains share 119 signaturre genes, the highest identity ratio found when comparing values discovered for the other two groups (Kant *et al.*, 2010). Table 1 summarizes the evolution of classification of the Lactobacillae, Leuconostocaceae, and Streptococcaceae families based on comparative genomic studies published so far.

The initial LAB genome sequencing primarily involved *Streptococcus* genomes, which are major representatives of pathogenic LAB (Stiles and Holzapfel, 1997).

However, the non-pathogenic LAB genome sequencing project has been conducted by the Lactic Acid Bacteria Genome Sequencing Consortium since 2002 (Klaenhammer *et al.*, 2002).

Table 1. Evolution of genome classification of Lactobacillae, Leuconostocaceae, and Streptococcaceae, with focus on *Lactobacillus* strains

Microrganism	Makarova et al (2006)[a,b]	Claesson et al (2008)[b]	Kant et al (2010)[b]
Lactobacillae family			
Lactobacillus plantarum WCFS1	Group 1	Group B	WCFS Group
Lactobacillus plantarum JDM 1	-	-	WCFS Group
Lactobacillus johnsonii NCC 533	Group 1	Group A	NCFM Group
Lactobacillus johnsonii FI 9785	-	-	NCFM Group
Lactobacillus acidophilus NCFM	Group 1	Group A	NCFM Group
Lactobacillus salivarius subsp. *salivarius* UCC 118	Group 1	Group B	WCFS Group
Lactobacillus sakei subsp. *sakei* 23K	Group 1	Group C	GG Group
Lactobacillus delbrueckii subsp. *bulgaricus* ATCC 11842	Group 1	Group A	NCFM Group
Lactobacillus delbrueckii subsp. *bulgaricus* ATCC BAA-365	-	Group A	NCFM Group
Lactobacillus gasseri ATCC 33323	Group 1	Group A	NCFM Group
Lactobacillus brevis ATCC 3567	Group 1	Group B	WCFS Group
Lactobacillus casei ATCC 334	Group 2	Group C	GG Group
Lactobacillus casei BL23	-	-	GG Group
Lactobacillus crispatus ST1	-	-	NCFM Group
Lactobacillus reuteri F275	-	Group B	-
Lactobacillus reuteri BSM 20016	-	-	WCFS Group
Lactobacillus reuteri JCM 1112	-	-	WCFS Group
Lactobacillus helveticus DPC 4571	-	Group A	NCFM Group
Lactobacillus rhamnosus GG	-	-	GG Group
Lactobacillus rhamnosus Lc 705	-	-	GG Group
Lactobacillus fermentum IFO 3956		-	WCFS Group
Leuconostocaceae family			
Oenococcus oeni PSU-1	Group 1	Group D	-
Leuconostoc mesenteroides subsp. *mesenteroides* ATCC 8293	Group 2	Group D	-
Streptococcaceae family			
Pediococcus pentosaceus ATCC 25745	Group 1	Group B	-
Lactococcus lactis subsp. *lactis* Il1403	Group 2	-	-
Lactococcus lactis subsp. *cremoris* MG1363	Group 2	-	-
Streptococcus thermophilus CNRZ 1066	Group 2	Outgroup	-
Streptococcus thermophilus LMG 18311	Group 2	-	-
Streptococcus thermophilus LMD-9	Group 2	-	-

[a] – Data adaptated from Makarova et al (2006) and Makarova and Koonin (2007).

[b] – The genomes signed with a minus signal (-) were not used in the study or were not yet available at NCBI database until publication the mentioned work. The classification proposed by Kant et al (2010) only refers to species of the *Lactobacillus* genus.

The Lactobacillales genomes are relatively small compared to non-obligately parasites or symbiont bacteria. They number about 2 Mb with 2000 genes, although the number of genes in different species may vary from 1600 to 3000. Because of this variation, it can be inferred that LAB genome evolution involved an active process of gene loss, duplication and acquisition of genes (Makarova and Koonin, 2007). There is a high degree of amino acid auxotrophies, as revealed through *Lactobacillus* genome sequences. Lactobacilli can compensate for these auxotrophies by synthesizing several oligopeptide transport systems. When comparing lactobacilli genomes from different origins (gut, plants or from dairy products), it can be noted that certain functional groups match their specific niche adaptation (Makarova *et al.*, 2006; Makarova and Koonin, 2007; Cai *et al.*, 2009; Kant *et al.*, 2010). *L. delbrueckii* subsp. *bulgaricus* and *L. helveticus* genomes contain various pseudogenes for carbohydrate utilization, and they are well adapted to grow on lactose (Ventura *et al.*, 2007; Cai *et al.*, 2009). Moreover, it was observed that their evolutionary genome history involved expressive gene loss (van de Guchte *et al.*, 2006; Callanan *et al.*, 2008).

The acidification capacity and the enzymatic processes that normally occur during growth of LAB may exert beneficial effects on fermented foods, mainly dairy products. These features contribute to the development of their typical flavors and textures as well as to the microbiological quality of the fermented foods (Klaenhammer *et al.*, 2005).

LAB are widely used in industrial applications because of the physiological characteristics which are useful in the dairy industry, such as substrate utilization, stress response, microbial community interactions and probiotic properties. However, the inherent mechanisms which lead to these features are very complex. Current functional genomics studies, based on *high-throughput* approaches, combined with experimental and computational methodologies, are being used to discover new genes, signal pathways, metabolic routes and regulatory circuits as well.

1.2. Microbial Ecology of Probiotic Lactobacilli

The term microbiome was first suggested by Joshua Lederberg to describe the collective genome of human indigenous microbiota (Hooper and Gordon, 2001). It is estimated that the gut microbiome has approximately 10 times more genes than the human genome (Turnbaugh *et al.*, 2007). This genetic apparatus may synthesize biochemical pathways otherwise absent in human genome and can considerably contribute to human health.

The small and large intestines harbor numerous microbial ecosystems throughout their length with approximately 10^{14} cells Among these cells anaerobic bacteria and bifidobacteria are predominant (Eckburg *et al.*, 2005). Some LAB, notably lactobacilli, occupy important niches in both human and animal GIT, and have been associated with beneficial effects on individuals' health. For this reason, they have been designated as "probiotics", meaning live microorganisms that, when administrated in suitable concentrations, exert health benefits on the host (Saxelin *et al.*, 2005).

The scientific research regarding the probiotic potential of lactobacilli species was initiated by Elie Metchnikoff, who devoted the last decade of his life to studying the link between lactobacilli and human health (Kleerebezem and Vaughan, 2009). More recent studies have linked probiotics to other beneficial effects in the host, including symptom alleviation of diarrhea and inflammatory bowel disease, as well as showing beneficial effects over the common cold, lactose digestion, lipid blood modulation, and so on (Lenoir-Wijnkoop *et al.*, 2007).

The global market for probiotic foods is constantly growing. The worldwide market generated US$ 16 billion in 2008 and that figure is estimated to reach US$ 19.6 billion by 2013. The market for functional foods varies greatly around the world, with Asia/Australia being the clearly dominant region, since Japan is regarded as the birthplace of functional foods. Eastern Europe and Africa remain underdeveloped in this area. Germany, France, the United Kingdom, and the Netherlands are the countries that most significantly contribute to the European functional food market. In Brazil, although the probiotic market is underdeveloped, it is beginning to grow due to increasing numbers of industries that have created and improved marketing and advertising for new functional foods..65 % of all Brazilian functional foods are probiotic products. In North America, the functional foods market is expected to reach US$ 16.7 billion in 2013 (Granato *et al.*, 2010).

Yogurt containing *Lactobacillus* species make up the largest share of the functional market, rendering it the primary product responsible for the development of this sector. For instance, in Latin America alone, the market has grown 32 % between 2005 and 2007, attaining a level of about 40 % in total sales. This may be due to product improvement, which has generated more palatable yogurts (Almeida *et al.*, 2009). Emergent probiotic applications include cheeses and ice creams (Cruz *et al.*, 2009a; Cruz *et al.*, 2009b), cereal bars, breakfast cereals, infant formulas, and so on. The main determinant factors for the development of functional food market include: the level of government support and the compatibility of legislation with market growth, the presence of a mature market for processed foods, the level of consumer confidence in products, and the health benefits to the host (Granato *et al.*, 2010).

In addition to research on their probiotic properties, *Lactobacillus* strains have been extensively studied because of the important role they play in food preservation. The strains display antimicrobial activity due primarily to the production of lactic acid as well as acetic, formic, and propionic acids. These organic acids restrict growth of some spoilage and pathogenic microorganisms (Aguilar *et al.*, 2010). The potent antimicrobial activity is mainly due to the non-dissociated acids, which are able to pass through the cell membrane and, once dissociated in the cytoplasm, lower the internal pH of the cell. Other metabolites originating from various primary pathways impart different characteristics to products. The metabolites and the amounts produced depend mainly on the bacterial strain. Different strains result in products with different tastes (Zalán et. al., 2010).

Although probiotic products have a high acceptance level among consumers, there are still many questions to be answered regarding their diversity, ecology, and functionality in terms of impacts on the host GIT. Previous studies based on cultivation techniques substantially underestimated bacterial diversity (Eckburg *et al.*, 2005). Current research on intestinal microbiota has been based upon a series of modern molecular approaches, specifically those that are based on rRNA 16S, genomics, metagenomics, and metabolomics. Gut microbial ecology has been the subject of numerous studies, since the funcitonal analysis of ecosystems is essential to the improvement of functional foods (Kleerebezem and Vaughan, 2009).

Recent advances in probiogenomics have provoked deep reflection on the diversity and evolution of comensal and probiotic bacteria, in addition to deciphering the molecular basis of the bacteria's health-promoting characteristics. Thus, probiogenomics and gene expression data in host intestines are expanding the knowledge we have regarding gut microbiota (probiotic or not) beyond the relationships between microorganisms and the host (Ventura *et al.*, 2009).

Probiotic bacteria have the ability to catabolize non-digestable carbohydrates including fructooligossacharydes in humans and other microorganisms,. These sugars can selectively stimulate the growth of other probiotic LAB. Adhesion properties result from prolonged effects on GIT and inhibit pathogens, because glycosil residues present in the mucosal cell surface may be competitively linked to LAB surface proteins (Pretzer *et al.*, 2005).

By studying public genomes available in databases, LAB genome analyses and comparative genomics have been able to highlight evolutionary mechanisms involved in speciation and specific adaptation to respective ecological niches. These efforts have significantly contributed to the identification of predicted effector molecules in probiotics, which may provide the basis for determining beneficial effects associated with the ingestion of specific probiotic strains (Ventura *et al.*, 2009).

1.3. Starter Cultures

Dairy LAB play an important role in the rheological and sensorial characteristics of end products because of the proteases, peptidases, organic acids, and exopolissacharides they produce. Moreover, they contribute to microbiological safety by inhibiting growth of spoilage and/or pathogenic bacteria, as LAB produce significant amounts of substances that inhibit microbial growth. The most important genera in food processes are *Carnobacterium*, *Enterococcus*, *Lactococcus*, *Lactobacillus*, *Leuconostoc*, *Oenococcus*, *Pediococcus*, and *Streptococcus* (Klein *et al.*, 1998).

The major challenge to producing probiotic products is finding a way to successfully perform a viable cell preparation on an industrial scale (Kleerebezem and Vaughan, 2009). Probiotic LAB must survive the GIT barriers produced by singular mechanisms, such as acid and bile salts resistance, adherence, bacteriocin production and ability for growth in the intestinal mucosa (Klaenhammer, 2000). Some of these threats have already been validated at genetic levels with the genomics and funcional genomics support (Klaenhammer *et al.*, 2008).

A starter culture can be defined as a microbial cell preparation which contains a high number of viable cells and at least one type of microorganism that is added to raw materials in order to produce fermented foods. LAB play an essential role in this process, and have been historically used to produce safe food and beverages, because they are considered GRAS (*Generally Recognized as Safe*) (Leroy and de Vuyst, 2004). The microorganisms in starter cultures promote fast acidification of raw materials by producing organic acids, primarily lactic acid, as well as other substances, such as acetic acid, etanol, bacteriocins, hydrogen peroxide, aroma compounds, exopolissacharides and various enzymes. Thus, growth and metabolism of LAB usually increase the shelf life of fermented products and ensure their microbiological safety. They also contribute to the desirable sensory characteristics of the end products.

A second feature of a starter culture is that it can be added directly to the raw material, as it has a high viable cell concentration obtained through centrifugation, which must be performed before the dehydration process. Lyophilization and spray-drying are common methods for cell dehydration. The resultant cells have high viability and better uniformity between batches (Peppler and Perlman, 1979).

Mixed cultures are widely used in food fermentations and those used in yogurt production have been most extensively studied. Typically, the mixed culture used for yogurt production contains *S. thermophilus* and *L. delbrueckii* subsp. *bulgaricus* cells. These stimulate mutual growth in a process called protocooperation (Courtin and Rul, 2004). The main results from this symbiosis are (i) an increase in milk acidification by converting lactose into lactic acid; (ii) changes in texture due to exopolissacharide production and to protein coagulation; and (iii) a typical yogurt flavor production, which is the result of the combination of lactic acid with various carbonile compounds such as acetaldehyde and diacetyl (Chaves *et al.*, 2002). Indeed, it has been demonstrated that certain metabolic functions absent in *L. bulgaricus* are complemented by *S. thermophilus* and vice-versa, which corroborates the mutualistic interaction (Sakai and Ezaki, 2006). In this study, the strain interactions and analysis of both genomes suggested a quick environmental adaptation (milk), indicated by the high degree of inactivated genes.

1.4. Lactic Acid Bacteria and Flavor Production in Dairy Products

For the development of new and improved dairy products, flavor characteristics are the primary target. The process of finding suitable strains for optimal flavor production in dairy products can be accelerated if the metabolic processes which generate typical flavor compounds, the effects of different strains on flavor production, and the ideal composition of different flavor compounds leading to the desired sensory effect are all understood (Smit *et al.*, 2005; Pastink *et al.*, 2008).

The quality of fermented dairy products is largely defined by sensory perception. The balance of flavor components in a given food product determines whether that product will be liked or disliked. Many flavor molecules are found in cheeses and fermented milks (van Kranenburg *et al.*, 2002; Marilley and Casey, 2004). In order to encourage acceptance of a dairy product, it is important to understand which flavor compounds determine the overall quality of the product (Smit *et al.*, 2005).

On an industrial scale, a large variation in flavor-forming capacities exists among the different LAB used to produce fermented dairy products. The differences occur at a species level, but also among individuals of the same species, which can vary from one strain to another (Pastink *et al.*, 2008).

Flavor formation is an important feature for industrial application of LAB. The improvement of gene annotation has led to better prediction of flavor-producing pathways in specific lineages. Certain genes involved in flavor formation in different LAB strains have been identified and provide a starting point for direct selection of suitable strains for potential flavor formation. These include *pepN*, *pepX*, *pepF*, and *pepC* genes, encoding aminopeptidases, as well as the *ilvE* gene, which encodes a branched-chain amino acid aminotransferase (Liu *et al.*, 2008; do Carmo *et al.*, 2011). Several studies suggest that a comparative genomics approach may be an efficient tool to determine new biochemical routes which lead to the production of potentially useful compounds in the dairy industry (Zhu *et al.*, 2009).

Flavor compounds can be produced via the degradation of three principal constituents: lactose (glycolysis and Leloir pathway), lipids (lipolysis) and protein (proteolysis). The latter is the most complex and important process for the formation of flavor compounds in fermented foods (van Kranenburg *et al.*, 2002). Flavor production is strain-dependent, and the starter culture composition can exert an impact on specific features in the end products. Nevertheless, flavor production also depends on substrates. Differences between flavor compound precursors and their regulatory response mechanisms may result in different metabolic fluxes along flavor-forming pathways (Pastink *et al.*, 2008).

The first step in food fermentation is the catabolism of carbohydrates by LAB. As a group, LAB exhibit an enormous capacity to degrade various sugars and other related compounds. Lactose fermentation is by far the most studied disaccharide metabolism in LAB, since it is the major carbohydrate found in milk. LAB metabolize lactose by means of a lactose carrier (permease) and subsequent cleavage by β-galactosidase (β-gal), to yield glucose and galactose, which then enter two major pathways, the glycolytic or pentose phosphate pathway for glucose, or the tagatose-6-phosphate pathway or Leloir pathway for galactose (Marilley and Casey, 2004). Glycolysis is used by homofermentative LAB, with lactic acid as the major end product, while the pentose phosphate pathway is commonly used by heterofermentative LAB, producing lactate, acetate and ethanol. However, it should be noted that glycolysis may lead to heterolactic fermentation under certain conditions (significant amounts of other end-products besides lactic acid), and some LAB, regarded as homofermentative, use the pentose phosphate pathway when metabolyzing certain substrates (Axelsson, 2004; Dicks and Endo, 2009). Some thermophilic lactobacilli, such as *L. delbrueckii* subsp. *bulgaricus*, *L. delbrueckii* subsp. *lactis* and *L. acidophilus* only metabolize glucose moiety after lactose transport and cleavage by β-galactosidase, while galactose is excreted into the medium (Kandler and Weiss, 1989). *L. delbrueckii* subsp. *delbrueckii* in turn is unable to ferment lactose or degrade casein; therefore it cannot grow in milk. Besides the importance to cell viability, it has been shown that only 6 % of LAB genes encode enzymes which are involved in carbohydrate transport and metabolism (Kant *et al.*, 2010).

In Gram-positive bacteria, sugar catabolism is usually controlled by a catabolite regulatory element encoded in *cre*. However, *Lactobacillus delbrueckii* subsp. *bulgaricus* N299, a variant of type strain ATCC 11842, lost the ability to regulate the *lac* operon expression, thus exemplifying the genetic instabilities that result in large deletions of *lac* operon locus (Lapierre *et al.*, 2002). The loss was achieved through an accumulation of mutations in the *cre* region, which allowed for nonspecific carbon catabolite regulation induced by the CcpA (catabolite control protein A). As a consequence, glucose catabolic regulation is missing in *L. delbrueckii* subsp. *bulgaricus*. This event is the result of of high selective pressure for fast lactose fermentation in yogurt manufacture, which is the unique sugar found in milk (Lapierre *et al.*, 2002).

It has been shown that LAB can metabolyze different carbon sources; however, there is still missing information about how these energy sources are acquired and degraded into the cells. For instance, LAB can use casein as a source for both nitrogen and carbon, depending on the environmental conditions. Analysis of the transcriptome and proteome profiles, *Lactococcus lactis*, *S. thermophilus*, and *L. helveticus* has shown an overregulation in the proteolytic genes during growth on milk (Derzelle *et al.*, 2005; Gitton *et al.*, 2005; Smeianov *et al.*, 2007). In order to increase LAB growth yield, it is important to understand these processes, which can improve industrial applications (Zhu *et al.*, 2009). Thus, recent advances in comparative genomics of fully completed LAB genomes provide unprecedented possibilities for discovering and exploring the proteolytic systems of various strains, including those which are considered probiotic (Liu *et al.*, 2010).

Proteolysis is considered to be the most complex of the three primary events during food fermentation, the other two being carbohydrate catabolism and lipolysis. In general, most raw materials contain insufficient amounts of free amino acids, which are essential for LAB growth, and LAB have been designated as nutritionally fastidious microorganisms. In spite of this, LAB possess a complex proteolytic system capable of hydrolyzing food proteins into peptides and amino acids. Furthermore, it is generally accepted that their proteolytic system contributes to the degradation of food proteins and hence to the texture, taste and aroma of fermented products (McSweeney and Sousa, 2000; Chen *et al.*, 2003; Pastink *et al.*, 2008).

The proteolytic system of LAB converts proteins to peptides and then to amino acids, which are essential for bacterial growth and contribute significantly to the desired flavor-forming characteristics of end products (Pastink *et al.*, 2008). The *Lactococcus lactis* proteolytic system has been the best characterized among LAB. However, characteristics from lactobacilli are yet to be fully understood, though previous studies suggest that their proteolytic mechanisms are similar (Christensen *et al.*, 1999; Savijoki *et al.*, 2006).

In general, the proteolytic system of LAB comprises the following components: a cell-envelope proteinase (CEP) which initiates the casein degradation producing oligopeptides; specific peptide transport systems which absorb resulting oligopeptides into the cells; various intracellular peptidases, which act together to degrade peptides into shorter peptides and amino acids; and different enzymes which convert amino acids into flavor compounds (Christensen *et al.*, 1999; Siezen, 1999; Savijoki *et al.*, 2006). However, there are differences among the proteolytic systems of various LAB strains which reflect the environmental niches they occupy, as demonstrated by a comparative genomics study (Boekhorst *et al.*, 2004). For instance, as compared to *Lactococcus lactis*, biosynthetic capacity for most amino acids is missing in lactobacilli, but it is compensated for by important proteolytic mechanisms, where the strains are capable of producing large numbers of peptidases, amino acid permeases, and multiple oligopeptide transport (Opp) systems (Klaenhammer *et al.*, 2005).

The Opp system is capable of transporting oligopeptides containing from 4 to 18 amino acid residues, which indicates that casein must suffer a previous cleavage in order to generate oligopeptide fragments in the medium. As a consequence, the residues can be efficiently absorbed by the cells (Tynkkynen *et al.*, 1993). In *L. delbrueckii* subsp. *bulgaricus* it has been shown that a proteinase named PrtB initiates the casein hydrolysis process (Liu *et al.*, 2006), but other enzymes, such as plasmin, may also produce minor peptides from casein. In addition, several aminopeptidases and endopeptidases can be released in the medium when cell lysis occurs, which may also contribute to oligopeptide generation (El Soda *et al.*, 1995; McSweeney and Sousa, 2000).

The proteolytic enzymes catalyze transamination, desamination, decarboxylation, and cleavage of side-chain amino acids reactions. Aminotransferases usually catalyze these conversions, which are essential for synthesis and catabolism of many amino acids. Three aminotransferases commonly found in eukaryotic and prokaryotic organisms are the acidic, aromatic, and branched-chain aminotransferases (IlvE) (Atiles *et al.*, 2000). *L. lactis* is known to be auxotroph for many amino acids, such as leucine, Isoleucine and valine, but it is also known that numerous *L. lactis* strains still encode IlvE in addition to its auxotrophies. IlvE is the only enzyme able to synthesize isoleucine and valine from their biosynthetic precursor, pyruvate, as demonstrated by an IlvE-deficient strain which was able to grow in defined medium containing α-keto acid precursor (Atiles *et al.*, 2000). Thus, this enzyme is probably involved in the catabolism of branched-chain amino acids (BcAAs) (do Carmo *et al.*, 2012).

The intracellular endopeptidases are, in general, aminopeptidases, such as the metallopeptidase PepN, the cysteine-peptidase PepC and the serine peptidase X-prolyl dipeptidyl aminopeptidase, PepX. These aminopeptidases appear to be the first to catabolyze oligopeptides (Liu *et al.*, 2008). Nevertheless, a common feature shared by these enzymes is that they cannot hydrolyze intact casein, but rather perform cleavage on internal peptide bonds of casein derived peptides (Savijoki *et al.*, 2006). In addition, in *L. lactis*, another metallopeptidase named PepF is important for protein turnover under nitrogen starvation conditions, since a PepF mutant showed decreased growth on minimal medium (Monnet et al, 1994; Nardi *et al.*, 1997).

PepN and PepC are broad-specificity aminopeptidases characterized from diverse LAB strains (Klein *et al.*, 1993; Fernández *et al.*, 1994; Klein *et al.*, 1994; Christensen *et al.*, 1995; Charvagnat *et al.*, 1999). Together, these two enzymes remove the N-terminal amino acids from a peptide. The specificity depends on the peptide length and on the nature of the N-terminal amino acid residue (Christensen *et al.*, 1999).

L. delbrueckii subsp. *lactis* DSM7290 was purified and characterized a cysteine peptidase named PepG (Klein *et al.*, 1997). This enzyme probably has dipeptidyl specificity, displaying di-/tripeptides and a tetrapeptide containing leucine hydrolysis capacity, but it was unable to degrade peptides which contained proline as the sole essential amino acid, as has been shown with a leucine/proline auxotroph strain of *Salmonella* Typhimurium expressing a recombinant PepG (Klein *et al.*, 1997).

It has been shown that only 2 % of LAB genes encode enzymes involved in amino acid transport and metabolism (Kant *et al.*, 2010). Despite the small representation among these bacterial genomes, protein catabolism plays an essential role during growth in milk and other environments inhabited by these LAB.

The overall molecular stress response mechanisms in LAB have been extensively studied. Through a genomics approach, this complex process has been elucidated as a whole, which helps designate suitable strategies to improve LAB robusteness (Zhu *et al.*, 2009).

Bacterial responses to environmental stresses usually depend on the coordinated expression of genes which impart changes in cell processes, as exemplified by cell division, DNA metabolism, and transport (Serrazanetti *et al.*, 2009). The stress responses may also act in concert to improve bacterial stress tolerance, and the responses are usually integrated by regulatory networks which allow cells to react to the environmental changes (Storz and Hengge-Aronis, 2000).

Adverse conditions are frequently found during the processing of fermented foods, including extreme temperatures, acids, oxegen and osmotic stress. An ideal LAB strain with potential uses in the food industry must survive these hostile conditions. For example, LAB commonly used as starter or probiotic cultures must tolerate the spray-drying dehydration process (van de Guchte *et al.*, 2002; Carmo, 2006). However, stress response mechanisms differ according to the strain and the environmental stress in question. Various authors suggest that the involvement of metabolic process in stress response may significantly alter the fermented product, as well as the product's desirable and typical characteristics (Serrazanetti *et al.*, 2009).

The acid stress response is highly important for LAB, since lactic acid is the main metabolic end product, which promotes acidification of the medium, and, as a consequence, reduces cell growth. Where oxidative stress is concerned, LAB are facultative anaerobic microorganisms and reactive oxygen species such as superoxide and hydroxil radicals may cause significantly cell damages (Zhu *et al.*, 2009).

The introduction of probiotic LAB in starter cultures of fermented foods, in addition to meeting current consumer demands which call for innovative and differentiated products that provide health benefits at the same time, has provided significant economic return and stimulated a multimillion-dollar market worldwide (Teusink and Smid, 2006; Saxelin, 2008; Kleerebezem and Vaughan, 2009; Granato *et al.*, 2010).

Decades of experience have proven that large-scale production of cheeses with high quality and uniform characteristics is facilitated by the use of these extensively characterized starter bacteria. Thus, even if certain traditional cheese fermentations remain dependent on the microbiota present in raw milk, the vast majority of industrial processes are able to use these cultures. As the future growth and economic vitality of the cheese industry depend on starter cultures with known characteristics and stability, it is essential to understand the genetics and physiology of LAB, which is the cornerstone of the large-scale worldwide production. (van Kranenburg *et al.*, 2002; Broadbent and Steele, 2005; Smit *et al.*, 2005; Vlieg and Hugenholtz, 2007).

1.5. Lactobacillus Delbrueckii and Lactobacillus Delbrueckii UFV H2b20

Lactobacillus delbrueckii species can be represented by three main subspecies, *delbrueckii*, *bulgaricus*, and *lactis*. The latter two subspecies are of great importance to the dairy industry. These subspecies demonstrate unique evolutionary histories, as they colonize distinct habitats as well as present different fermentation patterns. The subspecies *bulgaricus* and *lactis* are almost exclusively present in milk, whereas the subspecies *delbrueckii* also colonizes vegetable sources, as it is unable to ferment lactose and to degrade casein; therefore, this subspecies cannot grow in milk (Germond *et al.*, 2003). However, *L. delbrueckii* subsp. *bulgaricus* is known to catabolyze a much lower number of carbohydrates than *L. delbrueckii* subsp. *lactis* (Dicks and Endo, 2009).

{Polymorphism among *L. delbrueckii* can be demonstrated through different approaches. For instance, Polymerase Chain Reactions (PCR's) using specific primers were conducted, enabling the distinction between *L. delbrueckii* subsp. *bulgaricus* and *L. delbrueckii* subsp. *lactis* strains (Torriani *et al.*, 1999). Moreover, the patterns obtained by the Randomly Amplified Polymorphic DNA PCR (RAPD-PCR) approach showed match results of specific PCR for all strains tested, except for *L. delbrueckii* subsp. *delbrueckii* LMG 6412, which grouped with *L. delbrueckii* subsp. *lactis* strains. Data analysis showed a similarity level of 60 % between these two subspecies, which were grouped in a well-separated cluster with two distinct subclusters, while *L. delbrueckii* subsp. *bulgaricus* grouped separately (Torriani *et al.*,

1999). However, these approaches failed to distinguish between subspecies, despite the fact that it is possible to differentiate individual *L. delbrueckii* strains.

Lactobacillus delbrueckii UFV H2b20 is a commensal bacterium isolated from faecal material of a breast fed newborn-child (Sabioni *et al.*, 1988). Since its discovery, several scientific works have been conducted in Laboratório de Microbiologia Industrial e Biotecnologia – BIOAGRO, at Universidade Federal de Viçosa, Minas Gerais, Brazil, and also at Universidade Federal de Minas Gerais, Minas Gerais, Brazil. This probiotic LAB strain demonstrates desirable physiological and functional characteristics for commercial use (Agostinho, 1988; Monteiro, 1999; Leite, 2005; Ferreira, 2006; Carmo, 2006; Leite, 2008; Leandro, 2009). Because strain H2b20 has been extensively characterized in our laboratory and because it is a probiotic, we have added detailed information on this strain to this review.

The ribosomal RNA operon of this probiotic strain has been partially characterized (Magalhães *et al.*, 2005) and DNA-DNA hybridization showed a reassociation above 70 % between *L. delbrueckii* UFV H2b20 and *L. delbrueckii* subsp. *delbrueckii* (77.4 %) and the dairy strain *L. delbrueckii* subsp. *lactis* (75.2 %). This indicates that they belong to a similar species with at least 96 % similarity in rDNA identity (Magalhães *et al.*, 2008). On the other hand, this bacterium can be compared to *L. delbrueckii* subsp. *lactis*, *L. delbrueckii* subsp. *bulgaricus* and *L. acidophilus* according to phenotypical aspects, and all are able to degrade lactose and can therefore grow on milk. About 20 % of *L. delbrueckii* strains cannot grow when lactose is the sole energy source (Kandler and Weiss, 1989; Dicks and Endo, 2009).

These results raise the question of subspecies identity: it would seem that *L. delbrueckii* UFV H2b20 cannot be considered a new species, since the DNA-DNA homology supports the idea that it in fact belongs to the *L. delbrueckii* species (Magalhães *et al.*, 2008). *Lactobacillus delbrueckii* UFV H2b20 was first identified as *L. acidophilus* based only on its sugar fermentation pattern (Sabioni et al. 1988). The identity question was first raised by Uetanabaro (1999). Based on its phenotypical aspect, the strain can be compared to *L. delbrueckii* subsp. *bulgaricus* and *L. delbrueckii* subsp. *lactis*, which are able to degrade lactose and therefore are able to grow on milk. However, its ability to metabolize galactose is unusual, which raises doubts about the subspecies in question.

The genus *Lactobacillus* is also known for its discrepancies in taxonomy due to non-correlation between phylogeny and metabolic properties (Canchaya *et al.*, 2006; Felis and Dellaglio, 2007). In a recent study, it has been observed that *L. delbrueckii* UFV H2b20 presented an evolutionary pattern closer to that of probiotic *L. acidophilus* NCFM, which corroborates the suggestion that this strain could be considered as a new and unusual subspecies among *L. delbrueckii* subspecies, and may be the first identified as a probiotic (do Carmo *et al.*, 2011a). This strain is able to survive to stress conditions of the intestinal tract in vivo, as previously demonstrated (Neumann *et al.*, 1998). Moreover, the monoassociation of germ-free mice with *L. delbrueckii* UFV H2b20 for seven days improved the host's macrophage phagocytic capacity, suggesting that this probiotic strain stimulates a nonspecific immune response, making it a strong candidate to be used as a probiotic (Neumann *et al.*, 1998). *L. delbrueckii* UFV H2b20 also has the ability to stimulate the production of type 1 cytokines in vitro and in vivo, therefore suggesting that the microoganism might have adjuvant properties for infection in which these cytokines play a major (Neumann *et al.*, 2009). As regards the antagonistic effect that probiotic microorganisms may possess, survival data showed that *L. delbrueckii* UFV H2b20 colonized the digestive tract of gnotobiotic mice, ranging from 10^9 to 10^{10} CFU. g^{-1} of viable cells in faeces. Indeed, it has been shown that this strain exhibit the ability to protect both conventional and gnotobiotic *Lactobacillus*-treated mice against *Salmonella enterica* subsp. *enterica* ser. Typhimurium (Moura *et al.*, 2001).

Possess genes which are fundamentally important for probiosis and which are absent in the pathogen *L. monocytogenes*. This is the case for most of the genes involved in the adhesion mechanism, which are of foremost importance for food health and bio-processing. A study was conducted to evaluate by means of comparative genomics the similarities and differences in the protein content of virulence-modifying genes in *Listeria monocytogenes* EGD-e and those of probiosis in probiotic *Lactobacillus* (Silva, 2007). As a consequence, phylogenetic analysis results reinforce the hypothesis that a parallel comparison may be made between the pathogenic and the probiotic responses. It sugests that probiotic bacteria and some foodborne pathogens may share genes that enable them to survive the stomach and intestinal stresses. Alternatively, oxidative stress response has been shown to be important for the development of industrial processes suitable to *L. delbrueckii* UFV H2b20. Thus, *spx*,*dps*, and *pox* genes were identified in this probiotic bacterium and the involvement of the genes in the oxidative stress response was clearly demonstrated (Leite, 2008). In previous studies, viable or inactivated cells of *L. delbrueckii* UFV H2b20 were concentrated in stabilized Minas Frescal cheese whey, from which it was possible to manufacture dairy products containing this bacterium (Leite, 2005). Subsequently, an efficient method for dehydration and maintenance of cell viability in the dehydrated state was evaluated. It followed that a probiotic starter culture with promising characteristics for use in dairy products was produced on a laboratory scale (Carmo, 2006). Afterwards, it was possible to develop a symbiotic Cottage cheese containing viable cells of *L. delbrueckii* UFV H2b20 in the end product (Araújo *et al.*, 2009). A recent study has demonstrated that *L. delbrueckii* UFV H2b20 possesses several proteins required for an efficient proteolytic system that supports growth on milk. Based on the genes identified thus far, it has been possible to propose a schematic model which explains how these proteolytic enzymes are involved in flavor formation (do Carmo *et al.*, 2011). Moreover, a separated study has identified that this strain metabolizes galactose and produces high amounts of succnic acid during growth in 10 % reconstituted skimmed milk, raising doubts about the subspecies in question (do Carmo *et al.*, 2012). The unexpected succinic acid production is probably involved in flavor formation, as it was previously suggested in the proteolytic model proposed for *L. delbrueckii* UFV H2b20 (do Carmo *et al.*, 2011). In addition, the unusual ability to metabolyze galactose os this strain might be exploited to produce low browning probiotic Mozzarella cheeses, which is a desirable property in Pizza cheeses (do Carmo *et al.*, 2012).

In summary, future scientific research using *L. delbrueckii* UFV H2b20 is needed in order to certify its probiotic potential and to demonstrate further technological potential. This research will provide important insights for the industrial application of this strain.

References

Agostinho, S. M. M. Comportamento do *L. acidophilus* H2b20 sob condições do trato digestivo *in vitro* e efeitos de métodos de preservação em sua atividade. 1988. Dissertação (Mestrado em Ciência e Tecnologia de Alimentos) – Universidade Federal de Viçosa, Viçosa, MG.

Aguilar, C., C. Venegas, B. Klotz. 2010. Antagonistic effect of *Lactobacillus* strains against *Escherichia coli* and *Listeria monocytogenes* in milk. *J. Dairy Res.* DOI. 10.1017/S0022029910000877.

Almeida, M. H. B., A. G. Cruz, J. A. F. Faria, M. R. L. Moura, L. M. J. Carvalho, M. C. J. Freitas. 2009. Effect of the açai pulp on the sensorial attributes of probiotic yogurts. *Int. J. Prob. Preb.*, 4:41-44.

Araújo, E. A., Carvalho, A. F., Leandro, E. S., Furtado, M. M., Moraes, C. A. 2009. Produção de queijo tipo Cottage simbiótico e estudo de sobrevivência das células probióticas quando expostas a diferentes estresses. *Pesq. Agropec. Trop.*, 39:111-118.

Atiles, M., E. G. Dudley, J. L. Steele. 2000. Gene cloning, sequencing and inactivation of the branched-chain aminotransferase of *Lactococcus lactis* LM0230. *Appl. Environ. Microbiol.*, 66:2325-2329.

Axelsson, L. 2004. Lactic acid bacteria: classification and physiology. Pages 1-76 in Lactic Acid Bacteria, Microbiology and Functional Aspects. S. Salminen, A. von Wright, and A. Ouwehand ed. Marcel Dekker, Inc., New York, USA.

Boekhorst, J., R. J. Siezen, M. C. Zwahlen, D. Vilanova, R. D. Pridmore, A. Mercenier, M. Kleerebezem, W. M. de Vos, H. Brüssow, F. Desiere. 2004. The complete genomes of *Lactobacillus plantarum* and *Lactobacillus johnsonii* reveal extensive differences in chromosome organization and gene content. *Microbiol.*, 150:3601-3611.

Broadbent, J., J. L. Steele. 2005. Cheese flavour and the genomics of lactic acid bacteria. *ASM News*. 71:121-128, 2005.

Cai, H., R. Thompson, M. F. Budinich, J. R. Broadbent, and J. L. Steele. 2009. Genome sequence and comparative genome analysis of *Lactobacillus casei*: insights to their niche-associated evolution. *Genome Biol. Evol.*, 1:239-257.

Callanan, M., P. Kaleta, J. O'Callaghan, O. O'Sullivan, K. Jordan, O. McAuliffe, A. Sangrador-Vegas, L. Slattery, G. F. Fitzgerald, T. Beresford, R. P. Ross. 2008. Genome sequence of *Lactobacillus helveticus*, an organism distinguished by selective gene loss and insertion sequence element expansion. *J. Bacteriol.*, 190:727-735.

Canchaya, C., M. J. Claesson, D. von S. Fitzgerald, P. O'Toole. 2006. Diversity of the genus *Lactobacillus* revealed by comparative genomics of five species. *Microbiol.*, 152:3185-3196.

Carmo, A. P. Produção de cultura DVS (*Direct Vat Set*) para *Lactobacillus delbrueckii* UFV H2b20 cultivado em soro de queijo Minas Frescal. 2006. Viçosa: UFV, 66p. Dissertação (Mestrado em Microbiologia Agrícola) – Universidade Federal de Viçosa, Viçosa, MG.

Chaves, A. C., M. Fernandez, A. L. Lerayer, I. Mireau, M. Kleerebezem, J. Hugenholtz. 2002. Metabolic engineering of acetaldehyde production by *Streptococcus thermophilus*. *Appl. Environ. Microbiol.*, 68:5656-5662.

Charvagnat, F., M. G. Casey, J. Meyer. 1999. Purification, characterization, gene cloning, sequencing, and overexpression of aminopeptidase N from *Streptococcus thermophilus* A. *Appl. Environ. Microbio*l., 65:3001-3007.

Chen, Y. -S., J. E. Christensen, J. R. Broadbent, J. L. Steele. 2003. Identification and characterization of *Lactobacillus helveticus* PepO2, an endopeptidase with post-proline specificity. *Appl. Environ. Microbiol.*, 69:1276-1282.

Christensen, J. E., D. Lin, A. Palva, J. L. Steele. 1995. Sequence analysis, distribution and expression of an aminopeptidase gene from *Lactobacillus helveticus* CN*RZ*32. *Gene.*, 155:89-93.

Christensen, J. E., E. G. Dudley, J. A. Pederson, J. L. Steele. 1999. Peptidases and amino acid catabolism in lactic acid bacteria. *Antonie Van Leeuwenhoek*. 76:217-246.

Claesson, M. J., D. Van Sideren, P.W. O'Toole. 2007. The genus *Lactobacillus* – a genomic basis for understanding its diversity. *FEMS Microbiol. Lett*., 269:22-28.

Claesson, M. J., D. van Sideren, P. W. O'Toole. 2008. *Lactobacillus* phylogenomics – towards a reclassification of the genus. *Int. J. Syst. Evol. Microbiol*., 58:2945-2954.

Courtin, P., F. Rul. 2004. Interactions between microorganisms in a simple ecosystem: Yogurt bacteria as a study model. *Lait*., 84:125-134.

Cruz, A. G., A. E. C. Antunes, A. L. O. P. S. Pilleggi, J. A. F. Faria, S. M. I. Saad. 2009a. Ice cream as probiotic food carrier. *Food Res. Int*., 42:1233-1239.

Cruz, A. G., F. C. A. Buriti, C. H. B. Souza, J. A. F. Faria, S. M. I. Saad. 2009b. Probiotic cheese: health benefits, technological and stability aspects. *Trends Food Sci. Technol*., 20:344-354.

Derzelle, S., A. Bolotin, M. Y. Mistou, F. Rul. 2005. Proteome analysis of *Streptococcus thermophilus* grown in milk reveals pyruvate formate-lyase as the major upregulated protein. *Appl. Environ. Microbiol*., 71:8597-8605.

Dicks, L. M. T., A. Endo. 2009. Taxonomic status of lactic acid bacteria in wine and key characteristics to differentiate species. *S. Afr. J. Enol. Vitric*., 30:72-90.

Do Carmo, A. P., D. F. da Silva, M. N. V. de Oliveira, A. C. Borges, A. F. de Carvalho, C. A. de Moraes. 2011. Genes involved in protein metabolism of the probiotic lactic acid bacterium *Lactobacillus delbrueckii* UFVH2b20. *Beneficial Microbes* ,2:27-38.

Eckburg, P. B., E. M. Bik, C. N. Bernstein, E. Purdom, L. Dethlefsen, M. Sargent, S. R. Gill, K. E. Nelson, D. A. Relman. 2005. Diversity of the human intestinal microbial flora. *Science*, 308:1635-1638.

El Soda, M., N. Farkye, J. C. Vuillemard, R. E. Simard, N. F. Olson, W. El Kholy, E. Dako, E. Medrano, M. Gaber, L. Lim. 1995. Autolysis of lactic acid bacteria: impact on flavor development in cheese. *Dev. Food Sci*., 37:2205-2223.

Felis, G. E., F. Dellaglio. 2007. Taxonomy of lactobacilli and bifidobacteria. *Curr. Issues Intest. Microbiol*., 8:44-61.

Fernández, L., T. Bhowmik, J. Steele. 1994. Characterization of the *Lactobacillus helveticus* CNRZ32 *pepC* gene. *Appl Environ Microbiol*., 60:333-336.

Germond, J-E., L. Lapierre, M. Delley, B. Mollet, G. E. Felis, F. Dellagio. 2003. Evolution of the bacterial species *Lactobacillus delbrueckii*: a partial genomic study with reflections on prokaryotic species concept. *Mol. Biol. Evol*., 20: 93-104.

Gitton, C., M. Meyrand, J. Wang, C. Caron, A. Trubuil, A. Guillot, M. Y. Mistou. 2005. Proteomic signature of *Lactococcus lactis* NCDO763 cultivated in milk. *Appl. Environ. Microbiol*., 71:7152-7163.

Granato, D., G. F. Branco, F. Nazzaro, A. G. Cruz, J. A. F. Faria. 2010. Functional foods and nondairy probiotic food development: trends, concepts and products. *Compr. Rev. Food Sci. Food Safety*, 9:292-302.

Hooper, L. V., J. I. Gordon. 2001. *Commensal host-bacterial relationships in the gut. Science*, 292:1115-1118.

Kandler, O., N. Weiss. 1989. Regular, nonsporeforming gram-positive rods. Pages 1209-1234 in Bergey's Manual of Systematic Bacteriology. Vol. 2. P. H. H. Sneath, N. S. Mair, M. E. Sharpe, and J. G. Holt eds. Williams and Wilkins, Baltimore, USA.

Kant, R., J. Blom, A. Palva, R. J. Siezen, W. H. de Vos. 2010. Comparative genomics of *Lactobacillus*. *Microb. Biotechnol.*, 1-10.

Klaenhammer, T. R. 2000. Probiotic bacteria: today and tomorrow. *J. Nutr.*, 130:415S-416S.

Klaenhammer, T. R., E. Altermann, F. Arigoni, A. Bolotin, F. Breidt, J. Boradbent, R. Cano, S. Chaillou, J. Deustcher, M. Gasson, M. van de Guchte, J. Guzzo, A. Hartke, T. Hawkins, P. Hols, R. Hutkins, M. Kleerebezem, J. Kokl, O. Kuipers, M. Lubbers, E. Maguin, L. Mckay, D. Mills, A. Nauta, R. Overbeek, H. Pel, D. Pridmore, M. Saier, D. van Sinderen, A. Sorokin, J. Steele, D. O'Sullivan, W. de Vos, B. Weimer, M. Zagorec, R. Siezen. 2002. Discovering lactic acid bacteria by genomics. *Antonie van Leeuwenhoek*, 82: 29-58.

Klaenhammer, T. R., R. Barrangou, B. L., Buck, M. A. Azcarate-Peril, E. Altermann. 2005. Genomic features of lactic acid bacteria effecting bioprocessing and health. *FEMS Microbiol. Lett.*, 29:393-409.

Klaenhammer, T. R., E. Altermann, E. Pfeiler, B. L. Buck, Y. J. Goh, S. O'Flaherty, R. Barrangou, T. Duong. 2008. Functional genomics of probiotic Lactobacilli. *J.Clin. Gastroenterol.*, 42:S160-S162.

Kleerebezem, M., E. E. Vaughan. 2009. Probiotic gut lactobacilli and bifidobacteria: molecular approaches to study diversity and activity. *Annu. Rev. Microbiol.*, 63:269-290.

Klein, J., U. Klein, M. Schad, R. Plapp. 1993. Cloning, DNA sequence and partial characterization of *pepN*, a lysil-aminopeptidase from *Lactobacillus delbrueckii* subsp. *lactis* DSM 7290. *Eur. J. Biochem.*, 217:105-114.

Klein, J., B. Henrich, R. Plapp. 1994. Cloning and nucleotide sequence analysis of the *Lactobacillus delbrueckii* subsp. *lactis* DSM 7290 cysteine aminopeptidase *pepC*. *FEMS Microbiol Lett.*, 124: 291-300.

Klein, J., U. Klein, J. Schick, H. Matern, B. Henrich, R. Plapp. 1997. *Lactobacillus delbrueckii* subsp. *lactis* DSM 7290 *pepG* gene encodes a novel cysteine aminopeptidase. *Microbiol.*, 143:527-537.

Klein, G., A. Pack, C. Bonaparte, G. Reuter. 1998. Taxonomy and physiology of probiotic lactic acid bacteria. Int. J. Food Microbiol. 41:103-125, 1998.

Lapierre, L., B. Mollet, J.-E. Germond. 2002. Regulation and adaptive evolution of lactose operon expression in *Lactobacillus delbrueckii*. *J. Bacteriol.*, 184:928-935.

Leandro. E. S. Tolerância ao congelamento em mutantes de *Lactobacillus delbrueckii* UFV H2b20. 2009. Viçosa: UFV, 61p. Dissertação (Mestrado em Microbiologia Agrícola) – Universidade Federal de Viçosa, Viçosa, MG.

Leite, M. C. T. Identificação, caracterização e análise da expressão de genes de resposta ao estresse oxidativo em *Lactobacillus delbrueckii* UFV H2b20. 2008. Viçosa: UFV, 127p. Tese (Doutorado em Microbiologia Agrícola) – Universidade Federal de Viçosa, Viçosa, MG.

Leite, M. O. Desenvolvimento de processos para a produção de probióticos com *Lactobacillus delbrueckii* UFV H2b20. 2005. Viçosa: UFV, 76p. Dissertação (Mestrado em Microbiologia Agrícola) – Universidade Federal de Viçosa, Viçosa, MG.

Lenoir-Wijnkoop, I., M. E. Sanders, M. D., Cabana, E. Caglar, G. Corthier, N. Rayes, P. M. Sherman, H. M. Timmerman, M. Vaneechoutte, J. van Loo, D. A. W. Wolvers. 2007. Probiotic and prebiotic influence beyond the intestinal tract. *Nutr. Rev.*, 65-469-489.

Leroy, F., L. de Vuyst. 2004. Lactic acid bacteria as funcional starter cultures for the food fermentation industry. *Trends Food Sci. Technol.*, 15: 67-78.

Liu, M., R. J. Siezen. 2006. Comparative genomics of flavour-forming pathways in lactic acid bacteria. *Aust. J. Dairy Technol.*, 61:61-68.

Liu, M., A. Nauta, C. Francke, R. J. Siezen. 2008. Comparative genomics of enzymes in flavour-forming pathways from amino acids in lactic acid bacteria. *Appl. Environ. Microbiol.*, 74:4590-4600.

Liu, M., J. R. Bayjanov, B. Reckens, A. Nauta, R. J. Siezen. 2010. The proteolytic system of lactic acid bacteria revisited: a genomic comparison. *BMC Genomics.*, 11:1-15.

McSweeney, P. L. H., M. J. Sousa. 2000. Biochemical pathways for the production of flavour compounds in cheeses during ripening: *a review*. *Lait.*, 80:293-324.

Magalhães, J. T., F. A. Floresta, C. A. Moraes. 2005. Partial characterization of ribosomal operons of *Lactobacillus* UFV H2b20. *Braz. J. Microbiol.*, 36:177-183.

Magalhães, J. T., A. P. T. Uetanabaro, C. A. Moraes. 2008. Identification of *Lactobacillus* UFV H2b20 (probiotic strain) using DNA-DNA hybridization. *Braz. J. Microbiol.*, 39:524-546.

Makarova, K., A. Slesarev, Y. Wolf, A. Sorokin, B. Mirkin, E. Koonin, A. Pavlov, N. Pavlova, V. Karamychev, N. Polouchine, V. Shakhova, I. Grigoriev, Y. Lou, D. Rohksar, S. Lucas, K. Huang, D. M. Goodstein, T. Hawkins, V. Plengvidhya, D. Welker, J. Hughes, Y. Goh, A. Benson, K. Baldwin, J.-H. Lee, I. Díaz-Muñiz, B. Dosti, V. Smeianov, W. Wechter, R. Barabote, G. Lorca, E. Alterman, R. Barrangou, B. Ganesan, Y. Xie, H. Rawsthorne, D. Tamir, C. Parker, F. Breidt, J. Broadbent, R. Hutkins, D. O'Sullivan, J. Steele, G. Unlu, M. Saier, T. Klaenhammer, P. Richardson, S. Kozyavkin, B. Weimer, D. Mills. 2006. Comparative genomics of lactic acid bacteria. *Proc. Natl. Acad. Sci. USA*, 103:15611-15616.

Makarova, K., E. V. Koonin. 2007. Evolutionary genomics of lactic acid bacteria. *J. Bacteriol.*, 189:1199-1208.

Marilley, L., M. G. Casey. 2004. Flavours of cheese products: metabolic pathways, analytical tools and identification of producing strains. *Int. J. Food Microbiol.*, 90:19-159.

Monnet, V., M. Nardi, A. Chopin, M. C. Chopin, and J. C. Gripon. 1994. Biochemical and genetic characterization of PepF, an oligopeptidase from *Lactococcus lactis*. *J. Biol. Chem.* 269:32070-32076.

Monteiro, R. C. B. Resposta ao estresse em *Lactobacillus acidophilus* UFV H2b20. 1999. Viçosa: UFV, 57p. Dissertação (Mestrado em Microbiologia Agrícola) – Universidade Federal de Viçosa, Viçosa, MG.

Moura, L. N., E. Neumann, L. Q. Vieira, J. R. Nicoli. 2001. Protection by *Lactobacillus acidophilus* UFV H2b20 against experimental oral infection with *Salmonella enteric* subsp. *enteric* ser. typhimurium in gnotobiotic and conventional mice. *Braz. J Microbiol.*, 32:66-69.

Nardi, M., P. Renault, F. Monnet. 1997. Duplication of the *pepF* gene and shuffling of DNA fragments on the lactose plasmid of *Lactococcus lactis*. *J. Bacteriol.*, 179:4164-4171.

Neumann, E., M. A. P. Oliveira, C. M. Cabral, L. N. Moura, J. R. Nicoli, E. C. Vieira, D. C. Cara, G. I. Podoprigora, L. Q. Vieira. 1998. Monoassociation with *Lactobacillus acidophilus* UFV H2b20 stimulates the immune defense mechanisms of germfree mice. *Braz. J. Med. Biol. Res.,* 31: 1565-1573.

Neumann, E., M. G. Ramos, L. M. Santos, A. C. P. Rodrigues, E. C. Vieira, L. C. C. Afonso, J. R. Nicoli, and L. Q. Vieira. 2009. *Lactobacillus delbrueckii* UFV-H2b20 induces type

identification of feasible strategies to combat this disease burden is an important public health need, we have evaluated the efficacy of *L. rhamnosus* CRL1505 in preventing respiratory infections and severe illnesses, in children aged 2–6 years. The research from our laboratory demonstrates that *L. rhamnosus* CRL1505 represent a promising resource for the development of prevention strategies against respiratory infections that could be effective tools for medical application.

Lactobacillus Rhamnosus CRL1505 and *Lactobacillus Rhamnosus* CRL1506: Differential Immunomodulatory Capacities

Fermented dairy products are widely accepted health products and included as important components of the daily diet. Therefore, every year the food industry develops new fermented milk products that are required by consumers for their health benefits. There are a variety of studies concerning the probiotic properties of microorganisms, especially referring to LAB (Rodgers 2008). However, there are several difficulties in including LAB in food products, especially because of their viability in the final product (Heller 2001). In this sense, Heller (2001) has described several factors that may influence the survival of the probiotics in a dairy product, such as the chemical composition of the food matrix and compatibility with the starter culture. Consequently, a proper evaluation of each probiotic strain is essential for inclusion as an adjunct in a fermented food (Georgieva et al., 2009) in order to be sure about the preservation of the probiotic, nutritional and technological properties of the product.

The use of milk from small ruminants alone or in combination with bacterial strains, having probiotic properties represents a technological option for the manufacture of new dairy functional foods (Gomes et al., 1998; Minervini et al., 2009). In northwestern Argentina, goats' milk production is of great importance and it could be an interesting source of LAB with probiotic properties that can be used for the development of new fermented foods. Our laboratory was interested in finding LAB strains isolated from goat's milk that possess immunomodulatory activities in order to develop a fermented goats' milk (FGM) specially designed for the recovery of immunocompromised hosts. The selection of probiotic strains is difficult since up to now no clear concepts exist with respect to adequate methodology and criteria. Although until now probiotic products for multiple applications predominated on the market (Gueimonde and Salminen, 2006), the present trend is towards the selection of strains to be made according to specific uses. There are in the literature comparative studies concerning the immunomodulatory activity between different species of LAB both *in vitro* (Shimazu et al., 2012; Fijue et al., 2011; Hosoya et al., 2011) and *in vivo* (Perdigon et al., 1995), but there are very few in which the properties of strains belonging to the same species are compared. Recently *in vitro* studies were carried out to determine the immunomodulatory properties of *Bifidobacterium longum* strains in which the importance of selection criteria according to the therapeutic use is considered (Medina et al., 2007). Some authors have proposed the study of cytokine profiles induced by probiotic bacteria in the gut (Maldonado-Galdeano et al., 2007) or in the serum (Vinderola et al., 2006) as an important criterion in the study of their probiotic properties. Bearing in mind these facts and that our laboratory had previously demonstrated that some LAB strain have the ability to stimulate distant mucosae

(Villena et al., 2005; Villena et al. 2008), we also included the study of the activation of alveolar macrophages and the cytokine profile induced by LAB in the respiratory mucosa (Salva et al., 2010).

Several Lactobacillus strains isolated from goat's milk were evaluated according to their capacity to modulate the production of serum and mucosal cytokines in mice and we found that the oral administration of two of the Lactobacillus strains evaluated, *L. rhamnosus* CRL1505 and *L. rhamnosus* CRL1506, were able to induce differential cytokine profiles, especially when their effect was assessed in distant mucosae (Salva et al., 2010). Previous studies have demonstrated that the oral administration of wild type Lactobacillus strains can induce differential cytokine profiles in the gut (Maassen et al., 2000). In our research we analyzed the cytokines induced by the two potential probiotic strains of *L. rhamnosus* and we found that both have an influence on the production of the same cytokines in the gut, with significant differences only at the levels of two of them (INF-γ and IL-4). The influence of the LAB on the cytokines levels in the gut would be related to the improvement of the resistance to infection by *S. typhimurium* observed in immunocompetent and immunocompromised LAB-treated mice (Salva et al., 2010; Salva et al., 2011). Interferon-γ, produced by activated T cells and natural killer (NK) cells, has been shown to play an important role in host defence against intracellular pathogens such as *S. typhimurium* (Benbernou and Nauciel, 1994). In our work we observed a significant increase in IFN-γ in the gut and serum of mice fed with both strains of *L. rhamnosus*. *In vitro* studies have shown that, epithelial cells and fibroblasts are resistant to *S. typhimurium* invasion in the presence of IFN-γ (Chen et al., 1989) and that IFN-γ activates mouse peritoneal macrophages, resulting in enhanced *S. typhimurium* killing (Chen et al., 1989). Therefore, the increase in IFN-γ in the intestinal fluid is probably related to the activation of peritoneal macrophages observed in the treated groups. When we analyzed the specific antibodies produced during the Salmonella infection, we observed clear differences between strains. While the administration of *L. rhamnosus* CRL1505 induced increases in intestinal IgA and serum IgG with respect to the control group, *L. rhamnosus* CRL1506 did not affect these parameters. However, both strains were able to increase resistance to the Salmonella infection, probably because even though the production of IFN-γ is essential for the resolution of enteric Salmonella infection, the antibodies have little effect on this process (Bao et al., 2000). As to the cytokines profile, we found that the group treated with *L. rhamnosus* CRL1506 showed higher levels of inflammatory cytokines in the gut, while in serum higher cytokines levels were found in the *L. rhamnosus* CRL1505 treated group. In this sense, the significant increases in specific intestinal IgA and IgG in serum observed in this group are probably related to the enhancement of IL-4 and IL-10.

Several studies have demonstrated that certain probiotic LAB strains can exert their beneficial effect on the host through their immunomudulatory activity (Perdigon et al., 2001; Alvarez et al., 2007; Alvarez et al., 2009a; Alvarez et al., 2009b). Although most research concerning LAB-mediated enhanced immune protection is focused on gastrointestinal tract pathogens, a few recent studies have centered on whether immunobiotics might sufficiently stimulate the common mucosal immune system to provide protection to other mucosal sites as well (Cross 2002; Alvarez et al., 2007; Villena et al., 2011). In this sense, we studied the profiles of cytokines in the respiratory tract of mice treated with *L. rhamnosus* CRL1505 and *L. rhamnosus* CRL1506. We observed that only the group that received *L. rhamnosus* CRL1505 showed an increase in the levels of IL-6, IL-4 and IFN-γ in broncho-alveolar lavage fluid (BAL). Moreover, both treatments with LAB induced significant increases in IL-10 levels in BAL with respect to the control. However, the group that received *L. rhamnosus* CRL1505 showed significantly higher IL-10 levels than the *L. rhamnosus* CRL1506 treated group (Salva et al., 2010).

We also studied the potential effect of the CRL1505 and CRL1506 strains on the improvement of the immune response against respiratory pathogens using an experimental model of *S. pneumoniae* infection in adult immunocompetent mice. In our experimental model, mice were infected intranasally with *S. pneumoniae* serotype 14. After the challenge, the pathogen was detected in lung and blood samples throughout the period assayed (15 days) (Racedo et al., 2006). In order to study tissue damage in lungs, biochemical and histological studies were performed in the respiratory tract of the infected mice. Challenge with pneumococci significantly increased BAL albumin concentration and lactate dehydrogenase (LDH) activity, which indicates that infection increased the permeability of the alveolar-capillarity barrier and cell damage in lungs. Moreover, lung histopathological examination revealed a gradual and intense inflammatory response with progressive parenchymal involvement, including widespread cellular infiltration, passage of blood elements from capillaries to tissues, increased fibrosis in bronchial walls and vessels, haemorrhage and reduction in the alveolar airspaces (Racedo et al., 2006).

The immunostimulating properties of LAB have been proved to be strain- and dose-dependent (Perdigon et al., 1999a; 1999b; Racedo et al., 2006; Alvarez et al., 2007). Consequently, the ability of LAB to increase resistance against pneumococcal infection was studied using several doses and periods of administration. Only one of the treatments assayed increased the resistance of the mice to challenge with the respiratory pathogen: administration of 10^8 cells of *L. rhamnosus* CRL1505 (Salva et al., 2010). Our studies demonstrated that *L. rhamnosus* CRL1505, administered by the oral route at the proper dose, was able to increase *S. pneumoniae* clearance rates in lung and blood, improved survival of infected mice and reduced lung injuries (Salva et al., 2010).

We next addressed the technological aspects of *L. rhamnosus* CRL1505 in order to include it in a FGM. The health benefit of milk fermented with probiotic LAB has been amply documented and depends on the viability of the probiotic microorganisms in the refrigerated product (Gardini et al., 1999; Patrignani et al., 2006). The number of probiotic microorganisms should be above 10^6 cfu/g according to the FAO/WHO (2003). This condition was achieved by *L. rhamnosus* CRL1505 as adjunct culture in the manufacture of FGM (Salva et al., 2011). Another technological aspect to be considered in the addition of a probiotic LAB to milk fermentation is its ability to confer good sensory properties to the final product. This aspect plays an important role in consumer acceptance (Gardini et al., 1999). In our studies, the obtained FGM was accepted by 90.48% of the panellists (Salva et al., 2011). We also evaluated the efficiency of the FGM in a mouse malnutrition experimental model due to a protein deficiency (Salva et al., 2011). Malnourished mice show deterioration of clinical, nutritional and hematological parameters. Moreover, these mice had a state of immunosuppression and, consequently, a greater susceptibility to infections (Villena et al., 2005). In our experimental model of malnutrition we demonstrated that a repletion diet supplemented with FGM was effective to accelerate the recovery of the clinical nutritional parameters as body and thymus weights and serum proteins (Salva et al., 2011). The recovery of the serum protein profile studied could be the consequence of a greater bioavailability of peptides and amino acids, which would favor its absorption at the intestinal level. In addition, the inclusion of FGM in the repletion diet improved the hematological parameters. The early normalization of leukocytes, neutrophils and lymphocytes in blood in mice treated with FGM would be important for the recovery of immunity against infections (Salva et al., 2011). In fact, we demonstrated that the supplementation of a repletion diet with an FGM containing *L. rhamnosus* CRL1505 was efficient in increasing resistance against *S. typhimurium* in malnourished mice (Salva et al., 2011).

In order to study the effect of probiotics on the respiratory tract defences of immunocompromised malnourished hosts, we developed a model of pneumococcal infection in protein-malnourished mice (Villena et al., 2005). Weaned mice were malnourished after they consumed a protein-free diet (PFD) for 21 days. Malnourished mice were repleted with a balanced conventional diet (BCD) with or without supplemental *L. rhamnosus* CRL1505 or FGM and challenged with *S. pneumoniae* at the end of the dietary treatments assayed. We studied the effect of probiotics on the recovery of the innate and adaptive immune responses against *S. pneumoniae* respiratory infection (Villena et al., 2005; Villena et al., 2006; Alvarez et al., 2007). In this experimental model, pneumococcal colonization of lung and bacteremia were significantly greater in malnourished (MNC) mice compared with the well-nourished control (WNC) group. In addition, we found that MNC mice showed an important increase in LDH and albumin in BAL with respect to the WNC group. These results and the histopathological studies evidenced severe lung injury in the MNC mice (Villena et al., 2005). Repletion of malnourished mice with the BCD for 21 days was necessary to obtain levels of infection similar to those in the WNC mice. However, treatment with BCD and supplemental *L. rhamnosus* CRL1505 or FGM reduced to seven days the time needed to normalize the immune response (Salva et al., 2011).

Improvement of Respiratory Immunity by *Lactobacillus Rhamnosus* CRL1505 in Immunocompetent Mice

In order to elucidate the immunological mechanisms involved in the increased resistance to the pneumococcal infection induced by LAB, we studied both the innate and the specific immunity against the pathogen, and we found that the effects of LAB treatments were related to an up-regulation of both types of immune response in the respiratory tract.

Innate immunity. Macrophages have a central role in the maintenance of immunological homeostasis and host defence, the key population in the lungs being composed of alveolar macrophages (AM) (Gordon et al., 2002). Although resting AM are normally maintained in a quiescent state and produce small amounts of pro-inflammatory cytokines, they maintain the capacity to be activated in response to pathogens (Mizgerd et al., 2008). The activation of AM results in increased phagocytosis and microbial killing (Wissinger et al., 2009). Moreover, in the event that the invading pathogens are too virulent or represent too large a load to be contained by AM alone, these cells are capable of generating mediators such as TNF-α, IL-1β, MIP-1, MIP-1β, IL-8 and IL-6 that recruit large numbers of neutrophils into the alveolar space. These recruited neutrophils provide auxiliary phagocytic capacities that are critical for the effective eradication of offending pathogens (Zhang et al., 2000; Kyd et al., 2001).

Mice treated orally with *L. rhamnosus* CRL1505 had significantly higher amounts of BAL TNF-α than mice in the control group after challenge with *S. pneumoniae* (Salva et al., 2010). These increased levels of BAL TNF-α allowed an improved recruitment of neutrophils from the pulmonary vasculature into the alveolar spaces and a higher activation of BAL phagocytes, as shown by the number of BAL neutrophils and the percentage of BAL nitro blue tetrazolium positive (NBT^+) cells (Salva et al., 2010). In addition, *L. rhamnosus* CRL1505 was able to induce activation of the systemic innate immune response, which was evidenced by the increase in the number and microbicidal function of blood neutrophils (Salva et al., 2010).

On the other hand, although neutrophils are a key component of the host defence response against invading pathogens, they have also been implicated as mediators of tissue injury in a variety of inflammatory disorders (Zhang et al., 2000). Consequently, regulation of the inflammatory response by anti-inflammatory cytokines prevents damage to the host. It has been suggested that LAB could participate in tissue protection against the deleterious effect of an ongoing inflammatory process (Blum et al., 1999). In our experiments, treatments with *L. rhamnosus* CRL1505 prior to pneumococcal infection induced a significant increase in IL-10 in lung and serum (Salva et al., 2010). This increase could help to reduce the production of pro-inflammatory cytokines and chemokines and to downregulate the expression of adhesion molecules (Willems et al., 1994).

Consequently, in agreement with other reports (Kerr et al., 2002), IL-10 would be valuable to attenuate inflammatory damage and pathophysiological alterations in lung infected with pneumococci. According to these results, *L. rhamnosus* CRL1505 treatment would beneficially regulate the balance between TNF-α□and IL-10, allowing a more effective inflammatory response against infection.

Humoral immunity. After exposure to pathogens there is an activation of antibody responses in the respiratory tract. The type and concentration of antibodies produced is dependent on the site of exposure. Upper airway exposure results primarily in an IgA response; however, when organisms reach the deep lung after passing through the upper airway, they induce an increased production of pathogen-specific IgG (Twigg 2005). The organized lymphoid follicles in the nasal associated lymphoid tissue (NALT) and bronchus associated lymphoid tissue (BALT) are considered important inductive sites for the respiratory mucosal immune response (Ogra 2003). Following exposure to a pathogen there is an antigen uptake by dendritic cells (DCs) of NALT or BALT, activation of $CD4^+$ T cells and generation of IgA-producing plasma cells that populate airway lamina propria (Kiyono and Fukuyama, 2004).

On the other hand, in the gut immune response induced by commensal bacteria, the antigen presentation from the luminal flora leads to the generation of large quantities of local IgA. Moreover, the increase in the number of IgA producing cells is the most remarkable property induced by probiotic microorganisms (Maldonado-Galdeano et al., 2007). It has been demonstrated that the IgA^+ cells in the lamina propria of the small intestine can be increased by orally administered LAB (Perdigon et al., 1999a, b). A common mucosal immune system exists whereby immune cells stimulated in one mucosal tissue spread and relocate to various mucosal sites. This concept implies that oral immune stimulation can induce immunity in distal intestinal mucosal sites. Our lab and others have demonstrated that the oral administration of certain LAB strains is able to induce the IgA cycle and increase the

IgA$^+$ cell population in the respiratory tract (Racedo et al., 2006; Villena et al., 2008; Racedo et al., 2009).

When studying the capacity of *L. rhamnosus* CRL1505 to stimulate the IgA cycle, we found that this oral treatment was capable of increasing the number of IgA$^+$ cells in intestine and bronchus (Salva et al., 2010). We also found that *L. rhamnosus* CRL1505 wasable to improve the production of anti-pneumococcal IgA in the airways (Salva et al., 2010). Under inflammatory conditions, cytokines in the respiratory tract change dramatically and, when a Th2 response is needed, additional sources of IL-4, IL-5, IL-6 and IL-10 contribute to stimulate B cells to proliferate and mature into polymeric IgA-producing cells and to develop specific antibodies (Kiyono and Fukuyama, 2004; Corthesy 2002). In our studies, IL-4 and IL-10 were significantly increased in the respiratory tract of animals that received LAB, which correlates with the increase in the levels of specific IgA in BAL. The production of specific IgA in the respiratory tract during an infectious process is important because it prevents colonization of mucosal tissues and subsequent spreading into the systemic circulation (Twigg 2005). Additionally, specific IgA antibodies can bind antigens and minimize their entry with a consequent reduction in inflammatory reactions, which prevents potentially harmful effects on the tissue. Thus, the stimulation of the IgA cycle and the improvement of the levels of pathogen specific IgA induced by the *L. rhamnosus* CRL1505 strain could explain at least partly the greater resistance of the treated mice to the challenge with *S. pneumoniae*.

In the deep lung, when *S. pneumoniae* reaches the alveolar space, there is a differentiation and expansion of antibody-secreting plasma cells that are committed to the production of IgG (Lambretch et al., 2001; Holt et al., 2008). These antibodies have an important role in the protection against pneumococcal infection since opsonizing IgG antibodies are important for complement fixation and for enhancing the efficiency of macrophage killing. This immune activation also induces the production at the systemic level of antibodies responsible for preventing the passage of *S. pneumoniae* to the blood and their subsequent dissemination (Moore et al., 2001).

Our experiments demonstrated that *L. rhamnosus* CRL1505 treatment improved the production of antipneumococcal IgG in BAL (Salva et al., 2010). This fact could be related to the stimulation of antigen presenting cells in the lung that induces T cell activation and B cell clonal expansion and differentiation into IgG$^+$ antibody-secreting plasma cells. When studying the activation of the recruited macrophages and the levels of TNF-α in BAL after the challenge with the pathogen, we found that mice treated orally with *L. rhamnosus* CRL1505 showed higher BAL TNF-α values and NBT$^+$ cells percentages than mice in the control group (Salva et al., 2010), which would indicate that this treatment would be capable of improving macrophage-mediated antigen presentation. In addition, during the generation of an efficient effector immune response DCs have to overcome suppression by Treg cells. In this sense, production of IL-6 by DCs can release them from the suppression of naturally occurring Tregs (Pasare et al., 2003). It has also been demonstrated that other cytokines are able to trigger DCs activation/maturation, among them proinflammatory cytokines such as TNF-α (Blanco et al., 2008). In our studies we observed that preventive treatments with *L. rhamnosus* CRL1505 was able to increase the levels of IL-6 and TNF-α in the respiratory tract after challenge with *S. pneumoniae* (Salva et al., 2010). These results would indicate that oral treatment with *L. rhamnosus* CRL1505 would be capable of improving antigen presentation mediated by pulmonary DCs. Our laboratory is working actively to demonstrate this effect.

Improvement of Respiratory Immunity by *Lactobacillus rhamnosus* CRL1505 in Immunocompromised Malnourished Mice

We also studied the immunological mechanisms involved in the improved protection against pneumococcal infection in malnourished mice by evaluating both innate and adaptive immune responses.

Innate immunity. Suppression of neutrophil recruitment or functional activity would predictably result in an increased susceptibility to pulmonary infections. Our experiments showed a decreased number of blood leukocytes and neutrophils in MNC mice suffering from pneumococcal respiratory infection (Villena et al., 2005; Salva et al., 2008). In addition, qualitative alterations of neutrophils have been described in malnourished individuals while a depressed ability to kill ingested bacteria is characteristic of neutrophils in children suffering from malnutrition (Woodward 2001). These observations agree with our investigations, which showed a decreased bactericidal function of BAL phagocytes in MNC mice. However, administration of *L. rhamnosus* CRL1505 during repletion accelerated normalization of this function (Salva et al., 2011). The neutropenia associated with infection in malnutrition is accompanied by a disproportionately large number of band cells. Therefore, the depression reported in the microbicidal activity of neutrophils from infected malnourished individuals may partly reflect the functional limitations of immature cells. In this sense, we found decreased percentages of bone marrow proliferating myeloid cells (myeloblasts, promyelocytes and myelocytes) and reduced peroxidase activity in blood and bone marrow neutrophils in MNC mice during a pneumococcal infection (Salva et al., 2008). Repletion with BCD increased the percentage of proliferating cells. However, repletion with *L. rhamnosus* CRL1505 supplementation caused an even higher proliferation (mitotic pool) than the one found in the WNC mice (Salva et al., 2012a).

Humoral immunity. The mucosal secretory IgA antibody response is impaired in malnourished hosts (Manhart et al., 2000; Gauffin Cano et al., 2002). Malnutrition causes a remarkable decrease in the number of IgA^+ cells associated with the lamina propria of the small intestine (Gauffin Cano et al., 2002) and has the same effect on the respiratory mucosa, since we found decreased numbers of IgA^+ cells in the BALT of MNC mice (Agüero et al., 2006). We also observed a significant impairment of the local humoral immune response against *S. penumoniae* infection, which was evidenced by the decreased levels of BAL IgA anti-pneumococcal antibodies (Villena et al., 2005; 2006). We found that repletion with immunobiotics not only normalized the number of IgA^+ cells in BALT but also induced a significant increase in the number of these cells compared with the WNC group (Agüero et al., 2006). This improvement in the number of IgA producing cells in the respiratory mucosa was correlated with an enhanced local production of specific antibodies after pneumococcal challenge (Villena et al., 2005; Agüero et al., 2006).

In addition, malnutrition produced a remarkable decrease in the levels of specific IgG in BAL and serum (Villena et al., 2005; 2006). It has been suggested that the impairment of the humoral response in malnourished hosts relates to the number and competence of both T and B cells (Woodward 2001). Lymphoid atrophy, evidenced by the decrease in the size and cellularity of the thymus and of the secondary lymphoid organs, significantly contributes to the alteration in the adaptive immunity in malnourished individuals. In this sense, we observed a significant decrease in blood lymphocytes and bone marrow lymphoid lineage cells in our malnutrition model (Salva et al., 2008). Cytochemical assays have been proposed to study maturation of T cells with a scheme that differentiates T cell progress from β-glucuronidase negative (β-G-), α-naphthyl butyrate esterase negative (α-NBE-) to $\beta\text{-G}^+$, α-NBE- and finally to $\beta\text{-G}^+$, $\alpha\text{-NBE}^+$ (Basso et al., 1980). Using this cytochemical scheme, we found reduced numbers of blood and bone marrow $\beta\text{-G}^+$ cells and $\alpha\text{-NBE}^+$ cells in MNC mice, which would suggest deficiencies in T cell maturation (Salva et al., 2008). Thus, part of the defect in antibody immunity in malnourished mice could be attributed to the profound effect of malnutrition on the maturation of T cells, which results in a reduction in fully functional mature T cells and an excess of poorly functional immature T cells (Keusch 2003). The administration of *L. rhamnosus* CRL1505 to malnourished mice corrected lymphopenia and improved the number of bone marrow lymphoid cells. These treatments, which induced an increase in β-G+ cells and $\alpha\text{-NBE}^+$ cells, would improve T cell maturation. Moreover, the normalization of the systemic IgG response was also achieved with the immunobiotic treatments since the levels of serum antipneumococcal IgG were similar to those found in the normal control group (Villena et al., 2005; 2006).

Recently we reported that protein malnutrition induces a significant reduction in BM cell compartments, which is reflected in a decrease in mature B cells and which was reverted by diets supplemented with probiotics (Salva et al., 2012a). It has been reported that an inadequate protein diet diminishes the immune response and increases the susceptibility to infection because immune defenses are dependent on cell replication and the production of proteins with biological activities like immunoglobulins, cytokines and acute phase proteins (Sakai et al., 2006). In that paper we showed that protein deprivation affects splenic B cell number but not their ability to produce Igs. In general, innate and cell-mediated immunity are more sensitive to malnutrition than humoral immunity (Schrimshaw et al., 1997). More recent investigations also indicate a reduced Th2 activity and a decrease of antibody production during a diet restriction (Rodriguez et al., 2005; Ishikawa et al., 2009). In addition, Ishikawa *et al.* found that the differential response in splenic cell cultures of dietary restricted animals was associated with distinct stimuli (Ishikawa et al., 2009). However, in our experimental model there were no differences in the immunoglobulin production when different stimuli were used. Then, the impairment of the humoral immune response in MNC mice observed in our experimental model (Villena et al., 2005; 2006; 2009) was probably due to quantitative rather than qualitative changes in splenic B cells.

The antibody response against infectious agents depends on the correct orchestration of B cell generation and differentiation. When we investigated the effect of nutritional deprivation on B cell progenitors by flow cytometry, we observed a significant reduction in the total B lymphocyte number with a significant decrease of immature B cells, accompanied by an increase in the percentage of mature B cells, but not by changes in the number of mature B cells. These observations agree with the results obtained by Fock *et al.* (2009) in MNC animals, describing prominent lymphopenia with depletion in the lymphoid lineage and changes in cellular development. Our results are also supported by recent evidence showing that BM B cell development is altered by nutritional deprivation in parallel with decreased serum leptin concentrations (Tanaka et al., 2011). We observed that, in addition to B cell compartment, malnutrition induces severe histological alterations in BM of MNC mice. The decrease of hematopoietic space with an increase of extracellular matrix observed in MNC mice could be responsible of the impairment of the hematopoietic microenvironment, which could affect the interactions between cells and cellular signaling (Xavier et al., 2007). Protein deprivation also generated a decrease in the number of total BM granulocytes and erythroblasts. These results agree with previous studies that showed BM atrophy with a reduced number of leukocytes, granulocytes and lymphocytes (Xavier et al., 2007; Fock et al., 2009; Ramirez Vinolo et al., 2008).

BM is the main hematopoietic organ of an adult organism, it maintains a close relationship with the immune system, and is able to respond to an emergency in immune cells such as infection. The fast and efficient recovery of the immunity is of great importance to reduce the risk of infection and mortality in malnutrition. When the repletion diets were administered, the recovery of hematopoietic spaces was observed. However, only the inclusion of *L. rhamnosus* CRL1505 or FGM in the repletion treatment was able to achieve the normalization of the cellular compartments in BM. Here, we demonstrated for the first time the effect of a probiotic in the hematopoiesis after a malnutrition state. The addition of *L. rhamnosus* or FGM to the diet led myeloid and lymphoid cells to reach normal values and restored the marrow architecture.

The T-cell populations present in murine BM have hardly been studied in detail previously, mainly because of their relative rarity. In contrast to the findings of Fock *et al.* (2009), we observed that the nutritional deprivation induced an increase of T cells in BM ($CD3^+CD4^+$ and $CD3^+CD4^-$). That increase was accompanied by a decrease of the spleen T cells. Notoriously, it was described that hematopoiesis is maintained by activated BM $CD4^+$ T cells (Monteiro et al., 2005). Moreover, recent data have shown that normal hematopoiesis is really an antigen-induced state maintained by constant activation of BM $CD4^+$ T cells. This T cell population includes a large number of recently stimulated cells in normal mice whose priming requires the presence of the cognate antigens (Monteiro and Bonomo, 2005). It is possible that the increase of BM $CD4^+$ T cells is a consequence of the stress induced by the BM hypoplasia in malnutrition in an attempt to restore the normal hematopoiesis. The decrease of spleen T cells would be a result of their mobilization to the BM. In the present paper, all the repletion diets normalized the BM T cells. These results contribute to the understanding of the effect of probiotics on the recovery of immunity after a malnutrition state.

A remarkable finding of our investigations is that the beneficial properties of *L. rhamnosus* CRL1505 were preserved in the fermented dairy product containing the immunobiotic strain. Moreover, certain properties appear to be enhanced when the probiotic strain is included in the fermented product, since FGM was the only treatment able to normalize the absolute number of immature B220 cells after 7 days of repletion. Thus, it is probable that the effect of FGM depends not only on *L. rhamnosus* but also on non-bacterial components such as bioactive peptides (Salva et al., 2011).

At the moment, patients with immune deficiencies should probably only receive probiotics in the setting of a carefully conducted clinical research protocol. However, carefully selected and fully tested probiotic strains could provide additional options to improve the effect of conventional medical therapies. The results of our investigations highlight the importance of immunobiotics in the recovery of the lymphopoyesis in immunocompromised hosts.

Improvement of Respiratory Immunity by *Lactobacillus Rhamnosus* CRL1505 in Children

Common infectious diseases continue to be a major cause of death among preschool children in developing countries (Bryce et al., 2005; Black et al., 2008; Lopez et al., 2006). The increase in antibiotic resistance and need for new and improved strategies to tackle infectious disease have led to an examination of the therapeutic potential of commensal induced modulation of the mucosal immune response.

As described above, our laboratory has demonstrated that *L. rhamnosus* CRL1505 stimulates the innate and adaptive immune response in the gut in a dose-dependent way and confers resistance to infection with *Salmonella typhimurium* in immunocompetent (Salva et al., 2010) and immunocompromised malnourished (Salva et al., 2011) mice. Moreover, we have demonstrated that *L. rahmnosus* CRL1505 is able to increase the resistance to pneumococcal infection in immunocompetent and immunocompromised mice and that this effect is associated with a differential pattern of cytokines induced in the respiratory tract (Salva et al., 2010; 2011). Although animal studies have provided clear evidence that *L. rhamnosus* CRL1505 can have profound immunoregulatory effects and regulate immune responses beyond the gastrointestinal tract, this beneficial effect should be proved in appropiate humans trials. Therefore, we conducted experiments aimed to evalute the effect of *L. rhamnosus* CRL1505 on the health of children attending pre-school dining community centers. To do this, we evaluate the impact of probiotic yogurt containing *L. rhamnosus* CRL1505 on mucosal immunity and we study the effect of this treatment on gastrointestinal and respiratory infections, in terms of number of episodes and severity (Salva et al., 2012b, *in preparation*). With the support of the Argentina and Tucuman goverments we performed a randomised controlled trial in 298 preschool infants. Healthy children, 2 to 5 years of age, attending daycare centers in Tucuman (Argentina) were recruited into the study. The purpose of the study was explained to the parents and a written informed consent was obtained. This was a double blind clinical trial in which the probiotic yogurt containing *L. rhamnosus* CRL1505 was administered to the population of children for 6 months (five times a week). The results were statistically evaluated and compared with a similar population of children from the same community center that received a placebo yogurt (without probiotic) with similar characteristics. The results showed that administration of *L. rhamnosus* CRL1505 to young children reduced the incidence of infections since we observed that 66% of children in the placebo group presented symptoms of infection while only 34% of cases were detected in the yogurt group. In addition, significant differences were detected in the incidence of intestinal infections (74% vs 26%), upper respiratory tract infections (69% vs 31%) and angina (72% vs 28%) when the placebo and the probiotic treatments were compared (Salva et al., 2012b, *in preparation*).

The study showed that children fed the probiotic yogurt experienced fewer fevers and needed fewer antibiotics than those who received placebo. Moreover, the preventive effect of yogurt was associated with increased levels of IgA in the gut and the respiratory tract (Salva et al., 2012b, *in preparation*).

Conclusion

In this review we describe several research works dealing with the possibility of using LAB for the prevention of respiratory infection. The results discussed here and the ones described by different groups using LAB for the prevention of other respiratory infections show some important coincidences. In general, LAB strains with immunomodulatory properties are able to stimulate the innate immune system. Then, when an infectious stimulus occurs, adaptive immunity is activated according to signals induced by the pathogen, which would be enhanced by the immunomodulatory effect of LAB. Another important coincidence observed in the different research works is that the ability of LAB to modulate the inflammatory response via the induction of IL-10 production is important to improve protection.

Comparative analysis of the immunomodulatory capabilities of *L. rhamnosus* CRL1505 and *L. rhamnosus* CRL1506 described in the present work allowed us to conclude that: i) certain LAB strains may share certain functional properties. However, some of them can perform a functional role better than others, so that it is important to carry out thorough studies on specific strains, according to their therapeutic use; ii) the basal cytokine profiles induced by LAB administration is a good tool for strain selection both *in vivo* and *in vitro*, and iii) it is of fundamental importance to demonstrate the effectiveness of the probiotic strain when the host's immune system must respond to an infectious challenge.We designed yogurts containing *L. rhamnosus* CRL1505 with the ability to stimulate the common mucosal immune system and to prevent respiratory and intestinal infections. We demonstrated that these functional foods were able to improve immunity in immunocompetent and immunocompromised mice. Moreover, we demonstrated that consumption of a fermented dairy product containing *L. rhamnosus* CRL1505 is associated with a significant decrease in the duration and severity of mucosal infections providing the first evidence that a dairy fermented product containing this immunobiotic strain may have a beneficial effect against respiratory infections in young children. Based on the results summarized in this work and given the high morbidity and mortality associated with infectious diseases especially of the airways in children, dietary intervention using a dairy product containing the probiotic strain *L. rhamnosus* CRL1505 could be considered as a mean to improve health status of this population.

References

Agüero G, Villena J, Racedo S, Haro C, Alvarez S. Beneficial immunomodulatory activity of *Lactobacillus casei* in malnourished mice pneumonia: effect on inflammation and coagulation. *Nutr.*, 2006;22:810–9.

Alvarez S, Villena J, Racedo S, Salva S, Agüero G. Malnutrition, probiotics and respiratory infections. In: Global Research Network, editor. *Research Advances in Nutrition*, India; 2007, 9–23.

Alvarez S, Villena J, Salva S. Humoral immunity against respiratory pathogens: can lactic acid bacteria improve it? In: Global Research Network, editor. *Research Advances in Infection and Immunity*, India; 2009a, 1–19.

Alvarez S, Villena J, Tohno M, Salva S, Kitazawa H. Modulation of innate immunity by lactic acid bacteria: impact on host response to infections. *Curr. Res. Immunol.*, 2009b;3:87–126.

Bao S, Beagley KW, France MP, Shen J, Husband AJ. Interferon-gamma plays a critical role in intestinal immunity against Salmonella typhimurium infection. *Immunol.*, 2000;99:464–472.

Basso G, Cocito M, Semenzato G, Pezzuto A, Zanesco L. Cytochemical study of thymocytes and T lymphocytes. *Br. J. Haematol.*, 1980;44:577–82.

Benbernou N, Nauciel C. Influence of mouse genotype and bacterial virulence in the generation of interferon-gamma producing cells during the early phase of *Salmonella typhimurium* infection. *Immunol.*, 1994;83:245–249.

Black RE, Allen LH, Bhutta ZA, Caulfield LE, de Onis M, Ezzati M, Mathers C, Rivera J; Maternal and Child Undernutrition Study Group. Maternal and child undernutrition: global and regional exposures and health consequences. *Lancet*, 2008;371:243–260.

Blanco P, Palucka AK, Pascual V, Banchereau J. Dendritic cells and cytokines in human inflammatory and autoimmune diseases. *Cytokine Growth Factor Rev.*, 2008;19:41–52.

Blum S, Delneste Y, Alvarez S, Haller D, Pérez PF, Bode B, et al. Interactions between commensal bacteria and mucosal immunocompetent cells. *Internat Dairy* J., 1999;9:63–8.

Bryce J, Boschi-Pinto C, Shibuya K, Black RE, WHO Child Health Epidemiology Reference Group. WHO estimates of the causes of death in children. *Lancet*, 2005;365:1147–1152.

Chen Y, Nakane A, Minagawa T. Recombinant murine gamma interferon induces enhanced resistance to Listeria monocytogenes infection in neonatal mice. *Infect. Immun.*, 1989;57:2345–2349.

Corthésy B. Recombinant immunoglobulin A: powerful tools for fundamental and applied research. *Trends Biotechnol.*, 2002;20:65–71.

Cross M. Microbes versus microbes: immune signals generated by probiotic lactobacilli and their role in protection against microbial pathogens. *FEMS Immunol. Med. Microbiol.*, 2002;34:245–53.

FAO/WHO, Standard of fermented milks. Codex standard 243. FAO/WHO, Rome, 2003;1–8.

Fock RA, Blatt SL, Beutler B, Pereira J, Tsujita M. Study of lymphocyte subpopulations in bone marrow in a model of protein-energy malnutrition. *Nutr.*, 2009;26:1021–1028.

Fujie H, Villena J, Tohno M, Morie K, Simazu T, Aso H, Suda Y, Iwabuchi N, Xiao J, Iwatsuki K, Kawai Y, Saito T, Kitazawa H. Toll-like receptor-2 activating bifidobacteria strains differentially regulate inflammatory cytokines in porcine intestinal epithelial cell culture system: finding new anti-inflammatory immunobiotics. *FEMS Immunol. Med. Microbiol.*, 2011;63:129–139.

Gardini F, Lanciotti R, Guerzoni ME, Torriani S. Evaluation of aroma production and survival of *Streptococcus thermophilus*, *Lactobacillus delbrueckii* subsp. *bulgaricus* and *Lactobacillus acidophilus* in fermented milks. *Int. Dairy J.*, 1999;9:125–134.

Gauffin Cano P, Agüero G, Perdigón G. Adjuvant effects of *Lactobacillus casei* added to a renutrition diet in a malnourished mouse model. *Biocell*, 2002;26:35–48.

Georgieva R, Iliev I, Haertlé T, Chobert JM, Ivanova I and Danova S. Technological properties of candidate probiotic *Lactobacillus plantarum* strains. *Int. Dairy J.*, 2009;19:696–702.

Gomes AMP, Malcata FX. Development of a probiotic cheese manufactured from goat milk: response surface analysis via technological manipulation. *J. Dairy Sci.*, 1998;81:1492–1507.

Gordon SB, Read RC. Macrophage defences against respiratory tract infections. *Br. Med. Bull*, 2002;61:45–61.

Gueimonde M, Salminen S. New methods for selecting and evaluating probiotics. *Dig. Liver. Dis.*, 2006;38:242S–247S.

Heller KJ. Probiotic bacteria in fermented foods: product characteristicsand starter organisms. *Am. J. Clin. Nutr.*, 2001;73:374–379.

Holt PG, Strickland DH, Wikström ME, Jahnsen FL. Regulation of immunological homeostasis in the respiratory tract. *Nat. Rev. Immunol.*, 2008;8:142–52.

Hosoya S, Villena J, Simazu T, Tohno M, Fujie H, Chiba E, Shimaosato T, Aso H, Suda Y, Kawai Y, Saito T, Alvarez S, Ikegami S, Itoh H, Kitazawa H. Immunobiotic lactic acid bacteria beneficially regulate immune response triggered by poly(I:C) in porcine intestinal epithelial cells. *Vet. Res.*, 2011;42:111.

Ishikawa LL, França TG, Chiuso-Minicucci F, Zorzella-Pezavento SF, Marra NM. Dietary restriction abrogates antibody production induced by a DNA vaccine encoding the mycobacterial 65 kDa heat shock protein. *Genet. Vaccines .Ther.*, 2009;16:7–11.

Kerr AR, Irvine JJ, Search JJ, Gingles NA, Kadioglu A, Andrew P, et al. Role of inflammatory mediators in resistance and susceptibility to pneumococcal infection. *Infect. Immun.*, 2002;70:1547–57.

Keusch GT. The history of nutrition: malnutrition, infection and immunity. *J. Nutr.*, 2003;133:336S–40S.

Kiyono H, Fukuyama S. NALT- versus Peyer's-patch-mediated mucosal immunity. *Nat. Rev. Immunol.*, 2004;4:699–710.

Kyd JM, Foxwell AR, Cripps AW. Mucosal immunity in the lung and upper airway.*Vaccine*, 2001;19:2527–33.

Lambretch BN, Prins JB, Hoogsteden HC. Lung dendritic cells and host immunity to infection. *Eur .Respir. J.*, 2001;18:692–704.

Lopez AD, Mathers CD, Ezzati M, Jamison DT, Murray CJL. Global and regional burden of disease and risk factors, 2001: systematic analysis of population health data. *Lancet*, 2006;367:1747–1757.

Maassen CBM, van Holten-Neelen, C, Balk F, Heijne den Bak-Glashouwer MJ, Leer RJ, Laman JD, Boersma WA Claassen E. Strain-dependent induction of cytokine profiles in the gut by orally administered Lactobacillus strains. *Vaccine*, 2000;18:2613–2623.

Maldonano-Galdeano C, de Moreno de LeBlanc A, Vinderola G, Bonet ME, Perdigón G. Proposed model: mechanisms of immunomodulation induced by probiotic bacteria. *Clin. Vaccine Immunol.*, 2007;14:485–92.

Manhart N, Vierlinger K, Bergmeister H, Boltz-Nitulescu G, Spittler A, Roth E. Influence of short-term protein malnutrition of mice on the phenotype and costimulatory signals of lymphocytes from spleen and Peyer's patches. *Nutr.*, 2000;16:197–201.

Medina M, Izquierdo E, Ennahar S, Sanz Y. Differential immunomodulatory properties of *Bifidobacterium logum* strains: relevance to probiotic selection and clinical applications. *Clin. Exp. Immunol.*, 2007;150:531–538.

Minervini F, Bilancia MT, Siragusa S, Gobbetti M, Caponio F. Fermented goats' milk produced with selected multiple starters as a potentially functional food. *Food Microbiol.*, 2009;26:559–564.

Mizgerd JP. Acute lower respiratory tract infection. *N. Engl. J. Med.*, 2008;358(7):716–27.

Monteiro JP, Benjamin A, Costa ES, Barcinski MA, Bonomo A. Normal hematopoiesis is maintained by activated bone marrow CD4+ T cells. *Blood*, 2005;105:1484–1491.

Monteiro JP, Bonomo A. Linking immunity and hematopoiesis by bone marrow T cell activity. *Braz. J. Med. Biol. Res.*, 2005;38:1475–1486.

Moore BB, Moore TA, Toews GB. Role of T- and B-lymphocytes in pulmonary host defences. *Eur. Respir. J.*, 2001;18:846–56.

Ogra PL. Mucosal immunity: some historical perspective on host-pathogen interactions and implications for mucosal vaccines Immunol. *Cell Biol.*, 2003;81:23–33.

Pasare C, Medzhitov R. Toll pathway-dependent blockade of CD4+CD25+ T cell-mediated suppression by dendritic cells. *Science*, 2003;299:1033–6.

Patrignani F, Lanciotti R, Mathara JM, Guerzoni ME and Holzapfel WH. Potential of functional strains, isolated from traditional Maasai milk, as starters for the production of fermented milks. *Int. J. Food Microbiol.*, 2006; 107:1–11.

Perdigon G, Agüero G, Alvarez S, Gaudiosos de Allori C, de Ruiz Pesce Holgado A. Effect of viable *Lactobacillus casei* feeding on the immunity of the mucosae and intestinal microflora in mal-nourished mice. *Milchwis*, 1995;50:251–256.

Perdigon G, Alvarez S, Medina M, Vintiñi E, Roux E. Influence of the oral administration of lactic acid bacteria on IgA producing cells associated to bronchus. *Int. J. Immunopathol. Pharmacol.*, 1999a;12:97–102.

Perdigon G, Vintiñi E, Alvarez S, Medina M, Medici M. Study of the possible mechanisms involved in the mucosal immune system activation by lactic acid bacteria. *J. Dairy. Sci.*, 1999b;82:1108–14.

Perdigon G, Fuller R, Raya R. Lactic acid bacteria and their effect on the immune system. *Curr. Issues. Intest. Microbiol.*, 2001;2(1):27–42.

Racedo S, Villena J, Salva S, Alvarez S. Influence of yogurt consumption on the respiratory immune response. *Food Agric. Immunol.*, 2009;20:231–44.

Racedo S, Villena J, Medina M, Agüero G, Rodríguez V, Alvarez S. *Lactobacillus casei* administration reduces lung injuries in a *Streptococcus pneumoniae* infection in mice. *Microbes Infect.*, 2006;8:2359–2366.

Ramirez Vinolo MA, Rabello Crisma A, Nakajima, K, Macedo Rogero M, Fock RA. (Malnourished mice display an impaired hematologic response to granulocyte colony-stimulating factor administration. *Nutr. Res.*, 2008;28:791–797.

Rodgers S. Novel applications of live bacteria in food services: probiotics and protective cultures. *Trends Food Sci. Technol.*, 2008;19:188–197.

Rodriguez L, González C, Flores L, Jiménez-Zamudio L, Graniel J. Assement by flow cytometry of cytokine production in malnourished children. *Clin. Diag. Lab. Immunol.*, 2005;12:502–507.

Sakai T, Mitsuya K, Kogiso M, Ono K, Komatsu T. Protein deficiency impairs DNA vaccine-induced antigen-specific T cell but not B cell response in C57BL/6 mice. *J. Nutr. Sci. Vitaminol.*, 2006;52:376–382.

Salva S, Villena J, Racedo S, Alvarez S, Agüero G. *Lactobacillus casei* addition to a repletion diet induced early normalization of cytokine profiles during pneumococcal infection in malnourished mice. *Food Agricult Immunol.*, 2008;19:195–211.

Salva S, Villena J, Alvarez S. Immunomodulatory activity of *Lactobacillus rhamnosus* strains isolated from goat milk: impact on intestinal and respiratory infections. *Int. J. Food Microbiol.*, 2010;141:82–9.

Salva S, Nuñez M, Villena J, Ramón A, Font G, Alvarez S. Development of a fermented goats' milk containing *Lactobacillus rhamnosus*: in vivo study of health benefits. *J. Sci. Food Agric.*, 2011;91:2355-2362.

Salva S, Merino MC, Agüero G, Gruppi A, Alvarez S. Dietary supplementation with probiotics improves hematopoiesis in malnourished mice. *PLoS ONE* 2012; *In press.*

Salva S, Villena J, Núñez M, Corzo J, Tolaba R, Faedda J, Savoy G, Font G, Alvarez S. Fermented milk containing *Lactobacillus rhamnosus* CRL1505 reduces the incidence of gastrointestinal and respiratory tract infections in preschool infants: a randomised controlled trial. 2012; *In revision.*

Schrimshaw NS, San Giovanni JP. Synergism of nutrition, infection, and immunity: an overview. *Am. J. Clin. Nutr.*, 1997;66:464S–477S.

Shimazu T, Villena J, Tohno M, Fujie H, Hosoya S, Shimosato T, Aso H, Suda Y, Kawai Y, Saito T, Makino S, Ikegami S, Itoh H, Kitazawa H. Immunobiotic *Lactobacillus jensenii* elicit anti-inflammatory activity in porcine intestinal epithelial cells by modulating negative regulators of the toll-like receptor signaling pathway. *Infect. Immun.*, 2012; *In press.*

Tanaka M, Suganami T, Kim-Saijo M, Toda C, Tsuiji M. Role of central leptin signaling in the starvation-induced alteration of B-cell development. *J. Neurosci.*, 2011;31: 8373–8380.

Twigg HL. Humoral immune defense (antibodies): recent advances. *Proc. Am. Thorac. Soc.*, 2005;2:417–21.

Villena J, Racedo S, Aguero G, Bru E, Medina M, Alvarez S. *Lactobacillus casei* improves resistance to pneumococcal respiratory infection in malnourished mice. *J. Nutr.*, 2005;135:1462–9.

Villena J, Racedo S, Aguero G, Alvarez S. Yoghurt accelerates the recovery of defence mechanisms against *Streptococcus pneumoniae* in protein-malnourished mice. *Br. J. Nutr.*, 2006;95:591–602.

Villena J, Medina M, Vintiñi E, Alvarez S. Stimulation of respiratory immunity by oral administration of *Lactococcus lactis*. *Can. J. Microbiol.*, 2008;54:630–8.

Villena J, Barbieri N, Salva S, Herrera M, Alvarez S. Enhanced immune response to pneumococcal infection in malnourished mice nasally treated with heat-killed *Lactobacillus casei*. *Microbiol. Immunol.*, 2009;53:636–646.

Villena J, Oliveira ML, Ferreira P, Salva S, Alvarez S. Lactic acid bacteria in the prevention of pneumococcal respiratory infection: future opportunities and challenges. *Int. Immunopharmacol.*, 2011;11:1633–1645.

Vinderola G, Perdigon G, Duarte J, Farnworth E, Matar CEffect of the oral administration of the exopolysaccharide produced by *Lactobacillus kefiranofaciens* on the gut mucosal immunity. *Cytokine*, 2006:36:254–260.

Willems F, Marchant A, Delville JP, Gérard C, Delvaux A, Velu T, de Boer M, Goldman M. Interleukin-10 inhibits B7 and intercellular adhesion molecule-1 expression on human monocytes *Eur. J. Immunol.*, 1994,24:1007–9.

Wissinger E, Goulding J, Hussell T. Immune homeostasis in the respiratory tract and its impact on heterologous infection. *Semin. Immunol.*, 2009;21:147–55.

Woodward B. Nutrition, immunity and infection in infants children. In: Suskind RM, Tontisirin K, editors. Nestlé Nutrition Workshop Series. *Pediatric Programme*, 2001;45:89–116.

Xavier JG, Favero ME, Vinolo MAR, Rogero MM, Dgli MLZ. Protein-energy malnutrition alters histological and ultrastructural characteristics of the bone marrow and decreases haemopoiesis in adult mice. *Histol. Histopathol.*, 2007;22:651–660.

Zhang P, Summer WR, Bagby GJ, Nelson S. Innate immunity and pulmonary host defense. *Immunol. Rev.*, 2000;173:39–51.

In: Lactobacillus: Classification, Uses and Health Implications ISBN: 978-1-62081-151-1
Editors: Alba I. Perez Campos and Arturo Leon Mena

Chapter X

Selection of LAB Strains Based on Species- and Strain-Specific Typing for Probiotic Applications

Palmiro Poltronieri,[1] *Oscar Fernando D'Urso,*[1]
Alla V. Belyakova[2] *and Alexei B. Shevelev*[2]
[1]ISPA-CNR, Lecce, Italy
[2]Chumakov Institute of Poliomyelitis and Viral Encephalitis RAMS, Moscow, Russia

Abstract

Methods for screening putative probiotic strains are based on selection of a high number of strains isolated from foods, human or animal sources. This preliminary selection can be followed by molecular analyses focusing on gene products, metabolites and proteome analysis. Candidate strains to be screened for potential probiotic properties are required to pass several tests to assess some beneficial effect, either in the gut environment, such as the binding to cell receptors on epithelial cells, competing for attachment of pathogenic strains, either for the production of specific compounds with known beneficial property. *Lactobacillus* strains have to be isolated from the environments and niches to which these strains are adapted to live, and then are needed to be molecularly assigned to species and clusters according to genomic sequences and phylogenetic analysis. Recently, methods were set up to select LAB strains resistant to upper gastro-intestinal tract, not isolated with traditional culture methods due to other major microbial populations in natural sources (environmental samples, dairy and fermented products). After isolation of strains surviving low pH, trypsin, pancreatin, and in 0.1% up to 0.3% bile salts, strains can be screened for other beneficial effects. In this chapter a review is presented on the selection of LAB strains such as the screening of *Lactobacillus* strains based on binding to CACO-2 cells or adhering to specific substrates, protein ligands as fibronectin, collagen, or immune cell surface receptors. Additional assays may include the displacement of pathogenic species from binding to cultured cells. Finally, a multiplex PCR exploring the structure of specific domains in extracellular proteins linked to strains adapted to the GI tract may provide the basis to identify the candidate strains as belonging to gastrointestinal tract niche-adapted strains. *Lactobacillus* strains thus selected may be exploited for probiotic applications in foods, or may be transformed for the expression of heterologous proteins, immunotherapy or tolerance-regulating cytokines.

1. Introduction

Lactic Acid Bacteria (LAB) grow in very different environments and contain species adapted to survive in harsh conditions, low pH, and to ferment different types of substrates. *Lactobacillus* species occur naturally in different environments, plant, meat, dairy and cereals fermentation processes, and in different animal environments (from fishes to livestock to honeybees), interacting and influencing animal health. Bio-diversity has been reported frequently, showing that subtle variations in presence or absence of proteins and functional domain composition might lead to important traits during bacterial adaptation to their living environments.

LAB possess a variety of extracellular enzymes and carbohydrate transport systems, which correlate with their ability to interact with their environment. Especially in the gastro-intestinal (GI) tract, Lactobacilli have to rely on fermentation of carbohydrates and oligosaccharides, since glucose is no more available. According to the type of fermentation ability and differences in gene content, structure and diversity of domain repeats in several genes, it is possible to differentiate within each species different strain clusters. In Bifidobacteria, it was shown that only probiotic strains adapted to survive in the GI-tract express specific ATP binding cassette (ABC)-type carbohydrate transporters: carbohydrate catabolism enables these strains to produce short chain fatty acids and acetate in the gut, that prevents enteropathogenic infections (Fukuda et al., 2011).

Lactobacilli occur in human GI-tract but form only a minor share of the human adult faecal microbiota, i.e., around 0.01% to 0.6% of total bacterial counts. *L. gasseri, L. reuteri, L. crispatus, L. salivarius*, and *L. ruminis* appear to be predominant autochthonous Lactobacillus species. *L. acidophilus, L. fermentum, L. casei, L. rhamnosus, L. johnsonii, L. plantarum, L. brevis, L. delbrueckii, L. curvatus*, and *L. sakei* can also be found in the human GI-tract at fluctuating levels (Lebeer et al., 2008). In comparison to the adult microbiota, the infant microbiota is highly unstable but also contains lactobacilli in variable amounts (Mackie et al., 1999). The number of Lactobacillus cells in neonates was found to be in the range of 10^5 CFU/g of faeces, while in infants 1 month of age and older, the counts ranged from 10^6 to 10^8 CFU/g of faeces (Mackie et al., 1999). Sequences retrieved from infant faeces grouped with *L. salivarius, L. rhamnosus*, and *L. paracasei* (Heilig et al., 2002). In contrast to the GI-tract, the presence of lactobacilli is more pronounced in the female urogenital tract, where they often dominate the healthy microbiota (Redondo-Lopez et al., 1990; Zhou et al., 2007). The most frequently occurring species found in the vagina are *L. crispatus, L. gasseri, L. iners*, and *L. jensenii* (Anukam et al., 2006; Burton et al., 2003; Burton et al., 2002; Vasquez et al., 2002).

Eighteen *Lactobacillus* species have been sequenced, some of which with several strains resequenced, showing notable differences at strain level (Zhou et al., 2010; Gross et al., 2010). Different strains were shown to contain strain-specific genomic islands, in addition to different rearrangements and numbers of domain repeats in some extracellular proteins. This high genetic diversity is highly characteristic for the LAB, especially for the species in the *L acidophilus* group (Zhou et al., 2010), for *L. plantarum* (Gross et al., 2010; Pretzer et al., 2005), and the *L casei* group (Ventura et al., 2003).

The health-promoting effects are thought to be mediated by three main mechanisms of probiotic actions, which include pathogen inhibition, restoration of the microbial balance, enhancement of epithelial barrier function, and immunomodulatory effects via interactions with immune cells such as dendritic cells and lymphocytes (Lebeer et al., 2008).

Specific metabolic and physiological characteristics of lactobacilli play a key role in the adaptation to the host and the production and availability of probiotic factors. The typical cell surface structures of lactobacilli are in direct contact with the environment and can function as both key adaptation and probiotic factors. The LAB cell wall consists of several characteristic structures: a stratified peptidoglycan (PG) sacculus decorated with proteins, teichoic acids, and polysaccharides and, in some species, surrounded by an outer shell of proteins packed in a para-crystalline layer (S-layer). Bacterial envelope of lactobacilli may comprise different cell wall-associated proteins, which are often large proteins consisting of repeating modules or particular domains. Cell surface proteins can be anchored to the cell wall by different mechanisms (Navarre et al, 1999): by single N- or C-terminal transmembrane anchors, lipoprotein anchors (lipobox), LPxTG-type anchors, or other cell wall-binding (repeated) domains such as LysM domains or glycine-tryptophan (GW) dipeptide motifs (Åvall-Jääskeläinen and Palva, 2005; Van Pijkeren et al., 2006). Other proteins are secreted in the surroundings, mediating interactions with the environment independently from direct cell contact. Some secreted proteins have also been shown to re-associate to the cell wall by electrostatic interactions (Bath et al, 2005; Åvall-Jääskeläinen and Palva, 2005). Sortase-dependent proteins (SDPs) and S-layer proteins are best characterized in lactobacilli. These proteins and respective genes may serve as key markers allowing prediction of probiotic potential of the LAB strains and their appropriate selection by a microbiological screening.

The analysis of the exoproteome in *Lactobacillus* species (Zhou et al., 2010; Klerebeezem et al., 2010) classified the extracellular space proteins in several secretome groups, according to pathways of transport and secretion (http://www.cmbi.ru.nl/lab_exoproteome/).

A classification of extracellular proteins has been made based on sortase-dependent signals for anchoring to the cell wall or other domains responsible for anchoring via protein-protein or choline interactions (Klerebeezem et al., 2010). The analysis of LAB genomes contributed to the characterization of genes encoding the Sec, FPE, peptide-efflux ABC and holin systems in LAB species. The prediction of subcellular localization (SLC) of LAB proteins supported the grouping of secreted proteins by their SLC as: (1) lipid anchored to the cytoplasmic membrane; (2) attached to the cell wall either covalently (e.g. LPxTG proteins) or non covalently (e.g. by exhibiting LysM, SLH or WXL domains/motifs); (3) anchored to the cytoplasmic membrane via the N- or the C-terminal transmembrane helix, (4) released into the extracellular medium via Sec, holin or ABC transporters; and (5) being part of the competence pseudo-pili (Klerebeezem et al., 2010). These studies showed that several groups of extracellular proteins, devoid of classic signal peptidase recognition sequence, hence not involved in the Sec secretion pathway, are translocated into the outer space through these additional secretory mechanisms. The presence of multiple peptidoglycan layers have resulted in different strategies for anchoring proteins on the cell surface to the surrounding environment (Ramiah et al., 2008).

These alternative secretory pathways may also be relevant for the export of proteins whose location is primarily intracellular but that were also found in the extracellular space, such as Elongation factor Tu, the chaperonin GroEL, ribosomal proteins (L12/L7, L29, L30), the glycolic enzymes phosphoglycerate kinase, triose-phosphate isomerase and glyceraldehyde 3-phosphate dehydrogenase, and the stress-related proteins DnaK and GrpE (Beck et al., 2009).

Furthermore, adding complexity to our understanding on the effectors produced by individual strains, is the progression on identification of LAB peptidome (Boekhorst et al., 2011). Putative small proteins encoding ORFs are widespread in bacterial genomes, and an implementation in the annotation of the *Lactobacillus plantarum* WCFS1 genome allowed to include in it some of the ORFs previously not recognised. The identification of small peptides, originated from an ORF or by proteolysis of a larger protein, is difficult, since peptides elute fast from polyacrylamide gels, and require dedicated analytical methods focusing on small compounds. In fact, several ongoing studies using Metabolomic tools applied to study the LAB supernatants could help in identifying the peptidome world.

The studies on the LAB exoproteme (Zhou et al., 2010; Klerebeezem et al., 2010) have provided new clues on the differences and functionalities associated with species and strains living in specialised niches, such as the plant-associated LAB species and plant-fermenting strains, the GI-tract associated strains, and niche-specific and *L. acidophilus* group strains, expressing specific homologous Clusters of Secreted proteins (LaCOG). The database analysed extracellular enzymes, such as transglycosylases, also referred to as aggregation-promoting factors, and several groups of proteins that may be involved in adhesion to glucans and to cellular receptors. Many extracellular proteins were shown to contain domains for binding to macromolecular substrates, often duplicated and arranged in several repeats. In addition to domains for binding to the cell wall of the producing cell, several other domains were found related to binding to host macromolecules. Some of these annotations derived from *in vitro* binding studies.

In LAB strains, mucus-binding domains (MUB, MucBP) found in many proteins with different repeat structures, are thought to play a role in binding to the host GI-tract mucus layer (Kleerebezem et al., 2010). A high variety in size of mucus-binding proteins and in the number of mucus-binding domains was found (Zhou et al., 2010). Seven different subfamilies of mucus-binding proteins were identified based on protein size, sequence homology, domain composition and phylogeny. Among three largest subfamilies of LaCOG, one contains members from different LAB but not from *L. acidophilus* group members; a second, LaCOG00885, found in many LAB, contains solely the MucBP domains as defined by Pfam (PF00746), while the proteins in the other LaCOGs possess multiple copies of the larger MUB domains (Boekhorst et al., 2006). The MUB domain is variable in size and sequence, making it difficult to determine precise domain boundaries. One family of mucus-binding proteins (LaCOG02280) is specific for GI-tract specific strains, containing 4 proteins from *L. acidophilus*, *L. gasseri* and *L. johnsonii*, that present from 5 to 11 copies of the mucus binding domain. The acidophilus "complex" that includes the species *L. acidophilus*, *L. johnsonii*, *L. gasseri*, *L. delbrueckii* subsp. *bulgaricus* and *L. helveticus* is regarded as a phylogenetic subgroup, considering that *L. acidophilus*, *L. johnsonii*, *L. gasseri*, *L. reuteri*, and *L. salivarius* strains are specifically found in the GI-tract.

MUB-associated domain (Mubad) is present in six of the proteins containing the MUB domain and the number of Mubad domains per protein varies between 1 and 18 (Boekherst et al., 2006). Roos (Roos and Jonsson, 2002), using competitive adhesion, showed that the binding of MUB to mucus was inhibited by the glycoproteins fetuin and asialofetuin as well as fucose, suggesting that MUB interacts with specific muco-oligosaccharides.

A mannose-specific adhesin protein (Msa, a MucBP domain containing protein) is responsible for the binding to mannose by *L. plantarum* (Pretzer et al., 2005). While this was initially discovered as a protein responsible for agglutinating *Saccharomyces cerevisiae* cells, the presence of *L. plantarum* in many intestinal niches suggests that the MucBP domains of Msa may also play a role in the adhesion to non-mannosylated muco-oligosaccharides as well. Within this species, the analysis of Mucus adhesion- exoproteome in the probiotic *L. plantarum* WCFS1 strain revealed at least 12 proteins, including Msa, putatively involved in the adherence to host components such as collagen and mucin (Boekhorst et al., 2006a; Boekhorst et al., 2006b). This study was integrated with a comparative analysis of seven strains from different environmental niches (Siezen and van Hylckama Vlieg, 2011).

Mucus adhesion promoting protein (MapA) is reported to mediate the binding of *L. reuteri* and *L. fermentum* to mucus. It is also degraded into an antimicrobial peptide, which supports the host anti-pathogenic properties and provides clues on how surface proteins may contribute to beneficial effects with different peptide sequences (Bøhle et al., 2010; van Tassell and Miller, 2011).

A Mub protein from *L. reuteri* can confer immunoglobulin binding (McKenzie et al., 2010). The crystallised protein showed structural similarity to the functional repeat found in a family of Immunoglobulin (Ig)-binding proteins such as Protein L (PpL) from *Peptostreptococcus magnus*. Mub repeats were shown to interact *in vitro* with a large repertoire of mammalian Igs, including secretory IgA (McKenzie et al., 2009).

Surface Layer (S-Layer) Proteins (Slps)

Many bacteria have a crystalline layer on their cell surface (S-layer), consisting of single protein or glycoprotein species with relative molecular weights of 40,000 to 200,000 Dalton, representing 10-15% of the total proteins of the bacterial cell (Sára and Sleytr, 2000). Bacteria are efficient in expressing the genes, synthesising and secreting the S-layer proteins (Boot and Pouwels, 1996). Slps form regular structures covering the cell surface and are composed of monomeric protein units known as the S-protein that crystallize into a two-dimensional structure outside the bacterial cell surface and are extracted with charged reagents such as guanidine hydrochloride and lithium chloride. The S-layer structure contains pores of identical size and morphology that comprise up to 70% of the lattice surface area (Sleytr and Beveridge, 1999). Secondary cell wall carbohydrate polymers are the anchoring structure for Spls.

S-layer proteins are hypothesised to be involved in cell protection, adhesion, trapping molecules and ions, and virulence (Åvall-Jääskeläinen and Palva, 2005). Several functions have been described for surface-bound proteins of *Lactobacillus* spp., including cell adhesion and surface recognition. However, the removal of the S-layer proteins by treatment with LiCl did not block the adhesion of some lactobacilli to Caco-2 cells.

The presence of multiple peptidoglycan layers have resulted in different strategies for anchoring proteins on the cell surface to the surrounding environment (Ramiah et al., 2008). Klerebeezem analysed the data on protein O-glycosylation mechanism in microbes and in known LAB genomes, showing that it may have relevance in the strain ability to survive and interact with the environment (Klerebeezem et al., 2010).

L. johnsonii La1 expresses two carbohydrate-binding specificities, that may be relevant for binding to the O-linked oligomannosides and a second for asialo-GM1. Similar glucan-binding properties are present in surface proteins of other LAB species, conferring a lectin–like activity, and playing a role in host–microorganism interactions, possibly mediating the beneficial effects of probiotic strains.

It is generally assumed that a good adherence capacity is a desirable trait for probiotic lactobacilli, as it can promote the gut residence time, pathogen exclusion, and interaction with host cells for the protection of epithelial cells or immune modulation (Schierack et al., 2004). The adhesive capacity of probiotic lactobacilli is usually studied *in vitro*. Mostly, adherence to epithelial cell lines (e.g., Caco-2 or HT-29 human-derived adenocarcinoma cells) or immobilized intestinal mucus or extracellular matrix molecules (e.g., collagen and fibronectin) is investigated in short-term assays (Perea Velez et al., 2007). However, in most natural niches, adherent bacteria can form microcolonies and multicellular structures, recognized as biofilm-like communities (Branda et al., 2005). This more dynamic adherence process can also be simulated *in vitro* for lactobacilli (Lebeer et al., 2007; Walter et al., 2007). In the human gut of healthy individuals, isolated bacteria and micro-colonies seem to be the predominant colonization form (MacFarlane et al., 2004).

Roos and Jonsson studied the *mub* genes encoding mucus-binding protein in *L. reuteri* ATCC 53608 (strain 1023) (Roos and Jonsson, 2002). Using antibodies against cell surface proteins of *L. reuteri* 1023, identified a number of clones that were reactive with the antiserum and adhered to mucus by screening a phage library. Subcloning resulted in the identification of a *mub* gene, encoding a very large SDP with a highly repetitive structure (≥3,000 residues). Domains with the two main types of repeats (designated Mub1 and Mub2) were shown to adhere to mucus after recombinant expression in *E. coli*. Competition experiments indicated that Mub binds to carbohydrates on the mucus components. In another *L. reuteri* strain, 100-23, a similar approach using an antiserum against the surface proteins was used to identify the *lsp* gene, which encodes a high-molecular-mass cell wall protein, Lsp (Walter et al., 2005). Mutation analysis showed a reduced ecological performance of the *lsp* mutant in the murine GI-tract. When the *lsp* mutant was administered to mice in a 1:1 ratio with the wild type, it was recovered in the faeces at rates of 1.5% of the total Lactobacillus population. Moreover, when the *lsp* mutant was administered as pure cultures to mice, scanning electron microscopic observations of fore-stomachs revealed less biofilm formation by the mutant. The complete genome sequence of *L. acidophilus* NCFM was *in silico* analysed for the presence of genes encoding proteins that were previously implicated in the adherence of other bacteria. Some of these genes were inactivated, including two streptococcal R28 homologs, a fibronectin-binding protein (FbpA), a mucin-binding protein (Mub), and a surface layer protein (SlpA) (Boot and Pouwels, 1996), of which the R28 homologs and the Mub protein (LBA1392) are SDPs (Buck et al., 2005). That study showed that the genes encoding FbpA, Mub, and SlpA all contribute to the ability of *L. acidophilus* NCFM to adhere to Caco-2 cells in vitro, confirming that adhesion is determined by multiple factors. *mub* and *fbpA* mutations resulted in 65% and 76% decreases in adherence, respectively.

LAB Strain Differentiation and Selection Methods

Methods for pre-screening of candidate probiotic strains among many non-effective strains are required to be rapid and to be based on a high-throughput format. Several groups have shown a selection scheme based on the isolation of strains from foods, or human and animal sources, the testing of survival in simulated gastric and ileum digestion assays, the attachment property to mucins and to intestinal cells surface, and displacement of pathogenic strains from binding to intestinal cells.

Candidate strains need to be screened for potential probiotic properties. *Lactobacillus* strains have to be isolated from the environments and niches to which these strains are adapted to live in livestock (possibly adapted to survive in chicken, pig, or fish intestinal tract) or for adaptation in humans, and then are molecularly assigned to a species or strain cluster according to the homologies between genomic island sequences. Molecular analysis, next-generation DNA sequencing or PCR can be applied to prove the presence of determined domains in extracellular proteins and the domain arrangements found in GI-tract adapted strains that are characteristic of the ability to survive in a gut environment. Researches on molecular mechanisms involved in strain-specific beneficial effects are necessary for a better understanding of these effects, but could also lead to the development of molecular tools and assays to help the screening of the probiotic potential of lactobacilli.

Several tests have been designed to assess some beneficial properties of environmental strains, either toward the gut environment, such as the binding to cell receptors on epithelial cells, competition for attachment with pathogenic strains, acid stress resistance, bile salt resistance, or production of specific compounds with known beneficial properties.

Several assays are required to be performed to select a suitable strain, applying "omic" technologies such as 2D electrophoresis and proteomic analysis, Metabolomic analysis, expression profiling (DNA Arrays, Real-Time PCR), high-throughput protein-protein interaction studies, and screening methods such as immunolabelling, confocal microscopy and flow cytometry. These methods are costly and time consuming, so miniaturization and rapid screening for individuation of most promising strains are required. Protocols for the screening of strain collections are available, allowing the selection of candidate strains on the basis of some characteristic or beneficial property, or metabolite products determined by HPLC-Mass Spectrometry analysis, and screening the strains for adhesion to epithelial cells or mucin or polysaccharides.

As a first line of analysis of strain collections, incubation of strains in a simulated digestive tract may be an efficient and rapid method to individuate useful candidates. Several *in vitro* milk digestion protocols have been described and shown feasible to select LAB strains resistant to upper gastro-intestinal tract, not easily isolated from milk dominant microbial populations (Baruzzi et al., 2011). The isolated strains were shown to survive low pH, trypsin, pancreatin, and in 0.1% up to 0.3% bile salts.

In a second phase, several protocols are available to explore the effects of LAB strains on cultured cells (adhesion tests, displacement of pathogenic strains), or the induction of gene expression in cells of the gut environment (epithelium and immune cells, e.g. defensins, IL-12 differential expression) (Konstantinov et al., 2008) or activation of immune cell receptors such as the Toll-Like Receptor TLR2 and of signalling pathways downstream to TLR2 (Dicks and Botes, 2010).

In addition, strains can be differentiated for their ability to secrete peptides and bioactive compounds. Using Metabolomic tools (Gas-Mass chromatography, μHPLC combined to nano-electrospray ionization-ion trap mass spectrometry (ESI-MS-MS) or MALDI-TOF mass spectrometry studies on the LAB supernatants and cell extracts are performed in several centres.

Assays to Detect Cell Surface Properties Mediating Adhesion

Adhesion of LAB to human intestinal tissue and mucus is strain dependent and depends on the number of adhesion sites on the target cell (Gross et al., 2010; Van Tassell and Miller, 2011).

Since the intestinal epithelial cells are covered by mucus, screening of adhesion *in vitro* to tissue culture cells such as Caco-2 or HT29 has a limited value in predicting *in vivo* adhesion. It was also shown that adhesion differs between the different compartments of the gastrointestinal (GI) tract so that resected animal tissues have been used in adhesion tests (Ouwehand et al., 2002). Conversely, the presence of mucus alone was shown not sufficient to explain the adhesion of certain strains (Van Tassell and Miller, 2011).

In *L. rhamnosus*, SpaC pilin, expressed only in *L. rhamnosus* GG (Kankainen et al., 2009; von Ossowski et al., 2010) locates at the pilus tip but also on the surface of the GG strain, allowing the adherence of strain GG to human intestinal mucus.

Adhesive capacities for two peculiar proteins of *L. johnsonii* NCC533 have been documented. The first corresponds to elongation factor Tu (EF-Tu), normally involved in protein synthesis in the cytoplasm. However, EF-Tu of NCC533 has been located also at the cell surface, although no secretion or cell wall-binding motifs are present to explain this observation (Granato et al, 2004). Competition experiments with the isolated protein suggested that EF-Tu has an important role in the binding of NCC533 to Caco-2 and HT-29 intestinal epithelial cells and mucins. It is noteworthy that some EF Tu possesses protein disulfide isomerase activity, reduction of disulfides and isomerisation activity, by forming intermolecular bonds and multimeric complexes (Richarme, 1998).

Another usually cytoplasmic protein has been found to be located at the cell surface of NCC533, namely, the heat shock protein GroEL (Bergonzelli et al., 2006). This protein was also shown to mediate adhesion to epithelial cells and mucins. Interestingly, the presence of normally cytoplasmic proteins at the cell surface and a role for these proteins in adhesion have also been described for other lactobacilli. In *L. plantarum* LA 318, a surface plasmon resonance BIAcore-based assay demonstrated that cell surface glyceraldehyde-3-phosphate dehydrogenase (GAPDH) adheres to human colonic mucin (Kinoshita et al., 2007). In *L. crispatus*, the glycolytic enzymes enolase and GAPDH were found to bind and activate plasminogen (Hurmalainen et al., 2007). They were localized on the cell surface at pH 5, bound to lipotheicoic acid (LTA) but released into the medium at an alkaline pH.

Several approaches have been used to study mechanisms of adherence to an epithelial surface involving both receptor-specific binding and charge and hydrophobic interaction.

LAB commonly express cell surface hydrophobicity (CSH) as measured by the Salt Aggregation Test (SAT), contact angle and adhesion to xylene.

Adhesion properties may be assessed using different substrates such as mucus, fibrinogen, and collagen, performing *in vitro* studies. Detection of adhesion may make use of different methods like radiolabelling, fluorescent antibodies or metabolic substrates detected by fluorescence. Adhesion to fibrinogen, fibronectin, laminin, and collagen can be quantified using staining with crystal violet. Overall, the assays do not provide an accurate differentiation between adherent and non-adherent strains. Adhesion using *in vitro* models with cell lines are insufficient, whereas models with fibronectin and molecules typically used by pathogens offer a different and complementary approach.

Agglutination tests quantify the variable capacity of LAB strains to agglutinate *Saccharomyces cerevisiae* in a mannose-specific manner, investigated by several researchers (Pretzer et al., 2005).

Mutational analyses on *L. plantarum* proteins showed that the persistence capacity of the lp_2940 knockout mutant was 100- to 1,000-fold decreased in the murine GI-tract, while the colonization dynamics of an lp_1403 knockout were not significantly different from those of the wild type (Bron et al., 2007). This lp_1403 protein is a putative secreted cell surface protein of *L. plantarum*, whose expression has also been shown to be induced in the murine GI-tract (Bron et al., 2004) and, more specifically, in the small intestine (Marco et al., 2007). However, the *in vivo* persistence capacity of an *srt*A mutant of *L. plantarum* was not significantly affected (Pretzer et al., 2005), which might be due to the fact that the analysis of faecal samples reflects mainly the survival of the majority of administered bacteria during GI-tract transit.

Binding affinity and protein-glucan interactions may be exploited using high-throughput assays to screen and select strains with binding specificity. Shortly, protein ligands such as vitronectin, collagen, collectins (Veldhuizen et al., 2011) or immune cell surface receptors may be bound to beads or glass slides, or protein microarrays, and let to react with *Lactobacillus* strains freely floating in a solution in a flow regulated system. The type of interaction can be differentiated either as hydrophobic or regulated by peptide domains using synthetic peptides displacement.

Binding to specific polysaccharide structures such as glycan microarrays have been performed using several different patterns of polysaccharides used to determine binding patterns of whole bacterial cells to different glycan structures (Day et al., 2009).

Cell Culture Studies

The colonic cell line HT-29 from the human small intestine shows typical characteristics of enterocytic differentiation. The HT29-MTX cell line was developed to simulate a mucus-secreting environment. Some bacterial strains show a high affinity to HT29-MTX cells or other strains to Caco-2 cells (Dicks and Botes, 2010). The *in vitro* model determines the nature of adhesion sites in the system; some cell culture models will emphasize the measurement of direct host-microbe cellular contact, whereas mucus-secreting cultures or immobilized mucus models will emphasize mucus and muco-oligosaccharide interactions more than other models (Van Tassell and Miller, 2011). Other useful cell lines include fetal I-407 and the IPEC-J2 isolated from the jejunum of piglets (Schierack et al., 2004).

Moreover, adhesion *in vitro* is influenced by changes in bacterial gene expression caused by bile salts or pancreatin, thus *in vitro* experiments with LAB strains grown in laboratory media may not be sufficiently predictive since mub protein may be up regulated 6–8-fold when incubated in the presence of mucin and up to 25-fold when exposed to physiological concentrations of pancreatin and bile (Ramiah et al., 2008). Adhesion tests using cell cultures have been carried out by many groups, but generally performed on a reduced number of strains for adhesion to the human cell lines HT-29MTX and Caco-2 (Vankerckhoven et al., 2008) due to the intrinsic complexity of the microplate handling and analysis of results.

2. Genetic Modification of Candidate LAB Strains

Exploiting recombination between homologous flanking sequences, integration vectors were constructed and tailored for *L. plantarum*, or other LAB species, leading to insertion either at the $tRNA^{ser}$ or at the L-LDH locus. Inactivation of the L-LDH does not impair the growth of *L. plantarum in vitro* or *in vivo*. As the recombinant plasmid carries the antigen encoding DNA as a transcriptional translational fusion with the L-ldh gene (complementation marker), a second *in vivo* homologous recombination event leads to excisants containing solely the heterologous gene and no antibiotic marker. Other food-grade chromosome delivery systems have now been developed for LAB based on homologous recombination (*recA*-dependent), site-specific recombination (phage-derived systems), retrotransposition (group II intron-derived systems), and transposition (insertion sequence-based systems) (Hanniffy et al., 2000).

Researches are directed towards the development of heterologous protein expression systems using the targeting regions of native lactobacilli surface proteins. One LAB species (Turner et al., 2004) *Lactobacillus fermentum* BR11, a vaginal tract isolate of a guinea pig amenable to genetic manipulation, was used to produce a secreted recombinant protein Sep with a fragment of human E-cadherin, a receptor specific for attachment of *Listeria monocytogenes*.

An approach to improve colitis in a mouse model of gut inflammation (IL-10 deficience) was attempted by engineering *L. gasseri* (Carroll et al., 2007) and *L. plantarum* (Han and Fioramonti, 2008) to express a superoxide dismutase gene. In the animal models (IL-10 deficient mice and TNBS-induced colitis, respectively), an increased expression of IL-10 in the gut cells could be correlated with the results showed in the second paper.

Other studies have concentrated on S-layer/Bet v1 allergen fusion protein expressed in *B. subtilis* (Ilk et al., 2011) used as a cell factory tool. In order to improve the tolerance to antigens exploiting the adjuvant effect of cytokines, *L. lactis* expressing the tetanus toxin was engineered to co-express IL-2 and IL-6 (Steidler et al., 1999). The same group showed an improvement in a mouse model of inflammatory bowel disease, using *L. lactis* expressing IL-10 (Steidler et al., 2000).

Once the genetically modified strain is shown viable, with a protein and metabolite profile corresponding to the strain of origin, and tested for the same responses on effector cells (i.e. immune cells), the strain may be used in animal trials (animal models of disease), or can be tested using cultured cells. However, studies are required to test the wide range of variables and possible candidate genes and their interactions, in order to establish the effective mechanism of action supposed to be at the basis of the beneficial effects of probiotic strains in the gastrointestinal tract.

Conclusion

Selection of LAB strains based on the adhesion to culture cells of epithelial origin was shown feasible. In addition, methods for screening the structure of adhesion domains within extracellular protein genes using PCR amplification will provide additional information of strains belonging to LAB species clusters adapted to specific niches. The methods for exploitation of knowledge on gene structure will allow the insertion through homologous recombination of desired sequences (i.e. genes coding for interleukins and cytokines) in strains well adapted to the gut environment for the production of desired proteins to stimulate different cell types and also regulate the host immune system.

Acknowledgments

This work was partially supported by a grants by Russian ministry of Education, Science and Technology (state contracts 14.740.11.1033 and 16.512.11.2252).

References

Anukam, K. C., Osazuwa, E. O., Ahonkhai, I., Reid, G., 2006. Assessment of Lactobacillus species colonizing the vagina of apparently healthy Nigerian women, using PCR-DGGE and 16S rRNA gene sequencing. *World Journal Microbiology Biotechnology*, 22:1055–1060.

Åvall-Jääskeläinen, S. and Palva, A., 2005. Lactobacillus surface layers and their applications. *FEMS Microbiology Reviews*, 29:511-529.

Baruzzi, F., Poltronieri, P., Quero, G. M., Morea, M., Morelli, L., 2011. An in vitro protocol for direct isolation of potential probiotic lactobacilli from raw bovine milk and traditional fermented milks. *Applied Microbiology Biotechnology,* 90: 331-342.

Bath, K., Roos, S., Wall, T., Jonsson, H., 2005. The cell surface of Lactobacillus reuteri ATCC 55730 highlighted by identification of 126 extracellular proteins from the genome sequence. *FEMS Microbiology Letters*, 253:75–82.

Beck, H. K., Madsen, S. M., Petersen, J., Israelsen, H., Norrelykke, M. R., Antonsson, M., Hansen, A. M., 2009. Proteomic analysis of cell surface-associated proteins from probiotic Lactobacillus plantarum. *FEMS Microbiology Letters,* 297:61-66.

Bergonzelli, G. E., Granato, D., Pridmore, R. D., Marvin-Guy, L. F., Donnicola, D., Corthesy-Theulaz, I. E, 2006. GroEL of Lactobacillus johnsonii La1 (NCC 533) is cell surface associated: potential role in interactions with the host and the gastric pathogen Helicobacter pylori. *Infection Immunity*, 74:425–434.

Boekhorst, J., Wels, M., Kleerebezem, M., Siezen, R. J., 2006a. The predicted secretome of Lactobacillus plantarum WCFS1 sheds light on interactions with its environment. *Microbiology*, 152:3175-3183.

Boekhorst, J., Helmer, Q., Kleerebezem, M., Siezen, R. J., 2006b. Comparative analysis of proteins with a mucus binding domain found exclusively in lactic acid bacteria. *Microbiology*, 152, 273–280.

Boekhorst, J., Wilson, G., Siezen, R. J., 2011. Searching in microbial genomes for encoded small proteins. *Microbial Biotechnology*, 4:308-313.

Bǿhle, L. A., Brede, D. A., Diep, D. B., Holo, H., Nes, I. F., 2010. Specific degradation of the mucus adhesion-promoting protein (MapA) of Lactobacillus reuteri to an antimicrobial peptide. *Applied Environmental Microbiology,* 76;7306-7309.

Bohlen, P., Esch, F., Brazeau, P., Ling, N., Guillemin, R. 1983. Isolation and characterization of the porcine hypothalamic growth hormone releasing factor. *Biochemical Biophysical Research Communications*, 116:726-734.

Boot, H. J. and Pouwels, P. H., 1996. Expression, secretion and antigenic variation of bacterial S-layer proteins. *Molecular Microbiology*, 21:1117-1123.

Branda, S. S., Vik, A., Friedman, L., Kolter, R., 2005. Biofilms: the matrix revisited. *Trends Microbiology,* 13:20–26.

Bron, P. A., Grangette, C., Mercenier, A., De Vos, W. M., Kleerebezem, M., 2004. Identification of Lactobacillus plantarum genes that are induced in the gastrointestinal tract of mice. *Journal Bacteriology*, 186:5721–5729.

Bron, P. A., Meijer, M., Bongers, R. S., De Vos, W. M., Kleerebezem, M., 2007. Dynamics of competitive population abundance of Lactobacillus plantarum ivi gene mutants in faecal samples after passage through the gastrointestinal tract of mice. *Applied Environmental Microbiology,* 103:1424–1434.

Buck, B. L., Altermann, E., Svingerud, T., Klaenhammer, T. R., 2005. Functional analysis of putative adhesion factors in Lactobacillus acidophilus NCFM. *Applied Environmental Microbiology*, 71:8344–8351.

Burton, J. P. and Reid, G., 2002. Evaluation of the bacterial vaginal flora of 20 postmenopausal women by direct (Nugent score) and molecular (polymerase chain reaction and denaturing gradient gel electrophoresis) techniques. *Journal Infection Diseases*, 186:1770–1780.

Burton, J. P., Cadieux, P. A., Reid, G., 2003. Improved understanding of the bacterial vaginal microbiota of women before and after probiotic instillation. *Applied Environmental Microbiology*, 69:97–101.

Carey, C. M. and Kostrzynska, M., 2010. Microarray analysis of probiotics effectiveness. Bioactive foods in promoting health: Probiotics and Prebiotics. *Elsevier*. Pp. 479-495.

Carroll, I. M., Andrus, J. M., Bruno-Barcena, J. M., Klaenhammer, T. R., Hassan, H. M., Threadgill, D. S., 2007. Anti-inflammatory properties of Lactobacillus gasseri expressing manganese superoxide dismutase using the interleukin 10-deficient mouse model of colitis. *American Journal Physiology Gastrointestine Liver Physiology*, 293: G729–G738.

Chernysh, S., Kim, S. I., Bekker, G., Pleskach, V. A., Filatova, N. A., Anikin, V. B., Platonov, V. G., Bulet, P. 2002. Antiviral and antitumor peptides from insects. *Proceedings National Academy Sciences USA*, 99:12628-12632.

Day, C. J., Tiralongo, J., Hartnell, R. D., Logue, C., Wilson, J. C., Von Itzstein, M., Korolik, V., 2009. Differential carbohydrate recognition by Campylobacter jejuni strain 11168: Influences of temperature and growth conditions. *PLoS One* 4:781–790.

Dicks, L. M. T. and Botes, M., 2010. Probiotic lactic acid bacteria in the gastro-intestinal tract: health benefits, safety and mode of action. *Beneficial Microbes*, 1:11-29.

Fukuda, S., Toh, H., Hase, K., Oshima, K., Nakanishi, Y., Yoshimura, K., Tobe, T., Clarke, J. M., Topping, D. L., Suzuki, T., Taylor, T. D., Itoh, K., Kikuchi, J., Morita, H., Hattori, M., Ohno, H., 2011. Bifidobacteria can protect from enteropathogenic infection through production of acetate. *Nature*, 469:543-549.

Granato, D., Bergonzelli, G. E., Pridmore, R. D., Marvin, L., Rouvet, M., Corthesy-Theulaz, I. E., 2004. Cell surface-associated elongation factor Tu mediates the attachment of Lactobacillus johnsonii NCC533 (La1) to human intestinal cells and mucins. *Infection Immunity*, 72:2160–2169.

Gross, G., Snel, J., Boekhorst, J., Smits, M. A., Kleerebezem, M., 2010. Biodiversity of mannose-specific adhesion in Lactobacillus plantarum revisited: strain-specific domain composition of the mannose-adhesin. *Beneficial Microbes*, 1:61-66.

Han, W. and Fioramonti, J., 2008. Anti-inflammatory properties of lactic acid bacteria producing superoxide dismutase. *American Journal Physiology Gastrointestinal Liver Physiology,* 294: G353.

Hanniffy, S., Wiedermann, U., Repa, A., Mercenier, A., Daniel, C., Fioramonti, J., Tlaskolova, H., Kozakova, H., Israelsen, H., Madsen, S., Vrang, A., Hols, P., Delcour, J., Bron, P., Kleerebezem, M., Wells, J., 2004. Potential and opportunities for use of recombinant lactic acid bacteria in human health. *Advances Applied Microbiology*, 56:1-64.

Heilig, H. G. H. J., Zoetendal, E. G., Vaughan, E. E., Marteau, P., Akkermans, A. D. L., De Vos, W. M., 2002. Molecular diversity of Lactobacillus spp. and other lactic acid bacteria in the human intestine as determined by specific amplification of 16S ribosomal DNA. *Environmental Microbiology*, 68:114–123.

Hurmalainen, V., Edelman, S., Antikainen, J., Baumann, M., Lahteenmaki, K., Korhonen, T. K., 2007. Extracellular proteins of Lactobacillus crispatus enhance activation of human plasminogen. *Microbiology*, 153:1112–1122.

Ilk, N., Schumi, C.-T., Bohle, B., Egelseer, E. M., Sleytr, U. B., 2011. Expression of an endotoxin-free S-layer/allergen fusion protein in gram-positive Bacillus subtilis 1012 for the potential application as vaccines for immunotherapy of atopic allergy. *Microbial Cell Factories*, 10:6.

Kankainen, M., Paulin, L., Tynkkynen, S., Von Ossowski, I., Reunanen, J., Partanen, P., Satokari, R., Vesterlund, S., Hendrickx, A. P., Lebeer, S., De Keersmaecker, S. C., Vanderleyden, J., Hämäläinen, T., Laukkanen, S., Salovuori, N., Ritari, J., Alatalo, E., Korpela, R., Mattila-Sandholm, T., Lassig, A., Hatakka, K., Kinnunen, K. T., Karjalainen, H., Saxelin, M., Laakso, K., Surakka, A., Palva, A., Salusjärvi, T., Auvinen, P., De Vos, W. M., 2009. Comparative genomic analysis of Lactobacillus rhamnosus GG reveals pili containing a human mucus binding protein. *Proceedings National Academy Sciences USA*, 106:17193-17198.

Kinoshita, H., Uchida, H., Kawasaki, T., Wakahara, N., Matuo, H., Kawai, Y., Kitazawa, H., Ohmura, S., Miura, K., Shiiba, K., Horii, A., Saito, T., 2007. Glyceraldehyde 3-phosphate dehydrogenase (GAPDH) expressed on the cell surface of Lactobacillus plantarum LA 318 mediates adhesion to human colonic mucin. *Journal Dairy Science* 90:427.

Kleerebezem, M., Hols, P., Bernard, E., Rolain, T., Zhou, M., Siezen, R. J., Bron, P. A., 2010. The extracellular biology of the lactobacilli. *FEMS Microbiology Reviews*, 34:199–230.

Konstantinov, S. R., Smidt, H., De Vos, W. M., Bruijns, S. C., Singh, S. K., Valence, F., Molle, D., Lortal, S., Altermann, E., Klaenhammer, T. R., Van Kooyk, Y., 2010. S layer protein A of Lactobacillus acidophilus NCFM regulates immature dendritic cell and T cell functions. *Proceedings National Academy Sciences USA*, 105:19474-19479.

Lebeer, S., Verhoeven, T. L. A., Perea Velez, M., Vanderleyden, J., De Keersmaecker, S. C. J., 2007. Impact of environmental and genetic factors on biofilm formation by the probiotic strain Lactobacillus rhamnosus GG. *Applied Environmental Microbiology*, 73:6768–6775.

Lebeer, S., Vanderleyden, J., De Keersmaecker, S. C., 2008. Genes and molecules of lactobacilli supporting probiotic action. *Microbiology Molecular Biology Reviews*, 72:728-764.

Lee, Y-K., Puong, K-Y., Ouwehand, A. C., Salminen S., 2003. Displacement of bacterial pathogens from mucus and Caco-2 cell surface by lactobacilli. *Journal Medical Microbiology*, 52:925-930.

Ljungh, A. and Wadström, T., 2006. Lactic Acid Bacteria as probiotics. *Current Issues Intestinal Microbiology*, 7:73–90.

MacFarlane, S., Furrie, E., Cummings, J. H., Macfarlane, G. T., 2004. Chemotaxonomic analysis of bacterial populations colonizing the rectal mucosa in patients with ulcerative colitis. *Clinical Infectious Diseases*, 38:1690–1699.

MacKenzie, D. A., Tailford, L. E., Hemmings, A. M., Juge, N., 2009. Crystal structure of a mucus-binding protein repeat reveals an unexpected functional Immunoglobulin binding activity. *Journal Biological Chemistry*, 284:32444-32453.

MacKenzie, D. A., Jeffers, F., Parker, M. L., Vibert-Vallet, A., Bongaerts, R. J., Roos, S., Walter, J., Juge, N., 2010. Strain-specific diversity of mucus-binding proteins in the adhesion and aggregation properties of *Lactobacillus reuteri. Microbiology,* **156**:3368-3378.

Mackie, R. I., Sghir, A., Gaskins, H. R., 1999. Developmental microbial ecology of the neonatal gastrointestinal tract. *American Journal Clinical Nutrition*, 69:1035S–1045S.

Marco, M. L., Bongers, R. S., De Vos, W. M., Kleerebezem, M., 2007. Spatial and temporal expression of Lactobacillus plantarum genes in the gastrointestinal tracts of mice. *Applied Environmental Microbiology*, 73:124–132.

Miyoshi, Y., Okada, S., Uchimura, T., Satoh, E., 2006. A mucus adhesion promoting protein, MapA, mediates the adhesion of Lactobacillus reuteri to Caco-2 human intestinal epithelial cells. *Bioscience Biotechnology Biochemistry*, 70:1622-1628.

Navarre, W. W. and Schneewind, O., 1999. Surface proteins of gram positive bacteria and mechanisms of their targeting to the cell wall envelope. *Microbiology Molecular Biology Reviews*, 63:174–229.

Oozeer, R., Furet, J. P., Goupil-Feuillerat, N., Anba, J., Mengaud, J., Corthier, G., 2005. Differential activities of four Lactobacillus casei promoters during bacterial transit through the gastrointestinal tracts of human-microbiota-associated mice. *Applied Environmental Microbiology,* 71:1356–1363.

Ouwehand, A. C., Salminen, S., Tölkkö, S., Roberts, P., Ovaska, J., Salminen, E., 2002. Resected human colonic tissue: new model for characterizing adhesion of lactic acid bacteria. *Clinical Diagnostic Laboratory Immunology*, 9:184-186.

Perea Velez, M., De Keersmaecker, S. C. J., Vanderleyden, J., 2007. Adherence factors of Lactobacillus in the human gastrointestinal tract. *FEMS Microbiology Letters* 276:140–148.

Pretzer, G., Snel, J., Molenaar, D., Wiersma, A., Bron, P. A., Lambert, J., De Vos, W. M., Van der Meer, R., Smits, M. A., Kleerebezem, M., 2005. Biodiversity-based identification and functional characterization of the mannose-specific adhesin of Lactobacillus plantarum. *Journal Bacteriology*, 187:6128-6136.

Ramiah, K., Van Reenen, C. A., Dicks, L. M., 2008. Surface-bound proteins of Lactobacillus plantarum 423 that contribute to adhesion of Caco-2 cells and their role in competitive exclusion and displacement of Clostridium sporogenes and Enterococcus faecalis. *Research Microbiology*, 159:470-475.

Redondo-Lopez, V., Cook, R. L., Sobel, J. D., 1990. Emerging role of lactobacilli in the control and maintenance of the vaginal bacterial microflora. *Reviews Infection Diseases*, 12:856–872.

Richarme, G., 1998. Protein disulfide isomerase activity of Elongation factor Tu. *Biochemical Biophysical Research Communications*, 252:156-161.

Roos, S. and Jonsson, H., 2002. A high-molecular-mass cell-surface protein from Lactobacillus reuteri 1063 adheres to mucus components. *Microbiology*, 148:433-442.

Sara, M. and Sleytr, U. B., 2000. S-Layer Proteins. Journal Bacteriology, 182:859–868.

Schierack, P., Nordhoff, M., Pollmann, M., Weyrauch, K. D., Amasheh, S., Lodemann, U., Jores, J., Servin, A. L., 2004. Antagonistic activities of lactobacilli and bifidobacteria against microbial pathogens. *FEMS Microbiology Reviews*, 28:405–440.

Siezen, R. J. and Van Hylckama Vlieg, J. E. T., 2011. Genomic diversity and versatility of Lactobacillus plantarum, a natural metabolic engineer. *Microbial Cell Factories*, 10(Suppl.1):S3.

Sleytr, U. B. and Beveridge, T. J., 1999. Bacterial S-layers. *Trends in Microbiology* 7: 253-260.

Steidler, L., Robinson, K., Chamberlain, L., Schofield, K. M., Remaut, E., Le Page, R. W. F., Wells, J. M., 1998. Mucosal delivery of murine interkeukin-2 (IL-2) and IL-6 by recombinant strains of Lactococcus lactis coexpressing antigen and cytokine. *Infection Immunity*, 66:3183-3189.

Steidler, L., Hans, W., Schotte, L., Neirynck, S., Obermeier, F., Falk, W., Fiers, W., Remaut, E., 2000. Treatment of murine colitis by Lactococcus lactis secreting Interleukin-10. *Science*, 289: 1352-1355.

Tachu, B., Kleta, S., Blikslager, A., Tedin, K., Wieler, L. H., 2006. Characterization of a porcine intestinal epithelial cell line for in vitro studies of microbial pathogenesis in swine. *Histochemistry Cell Biology*, 125:293–305.

Tuomola, E. M., Ouwehand, A. C., Salminen, S. J., 1999. The effect of probiotic bacteria on the adhesion of pathogens to human intestinal mucus. FEMS Immunology *Medical Microbiology*, 26:137-142.

Turner, M. S., Hafner, L. M., Walsh, T., Giffard, P. M. 2004. Identification and Characterization of the Novel LysM Domain-Containing Surface Protein Sep from

Lactobacillus fermentum BR11 and Its Use as a Peptide Fusion Partner in Lactobacillus and Lactococcus. *Applied Environmental Microbiology,* 70: 3673-3680.

Turpin, W., Humblot, C., Thomas, M., Guyot, J-P., 2010. Lactobacilli as multifaceted probiotics with poorly disclosed molecular mechanisms. International *Journal Food Microbiology*, 143: 87-102.

Vankerckhoven, V., Huys, G., Vancanneyt, M., Vael, C., Klare, I., Romonde, M-B., Entenza, J. M., Moreillon, P., Wind, R. D., Knol, J., Wiertz, E., Poti, B., Vaughan, E. E., Kahlmeter, G., Goossen, H., 2008. Biosafety assessment of probiotics used for human consumption: recommendations from the EU-PROSAFE project. *Trends Food Science Technology*, 19:102-114.

Van Pijkeren, J. P., Canchaya, C., Ryan, K. A., Li, Y., Claesson, M. J., Sheil, B., Steidler, L., O'Mahony, L., Fitzgerald, G. F., Van Sinderen, D., O'Toole, P. W., 2006. Comparative and functional analysis of sortase-dependent proteins in the predicted secretome of Lactobacillus salivarius UCC118. *Applied Environmental Microbiology*, 72:4143–4153.

Van Tassell, M. L. and Miller, M. J., 2011. Lactobacillus adhesion to mucus. *Nutrients* 3:613-636.

Veldhuizen, E. J. A., Van Eijk, M., Haagsman, H. P., 2011. The carbohydrate recognition domain of collectins. *FEBS Journal*, 278(20):3930–3941.

Vasquez, A., Jakobsson, T., Ahrne, S., Forsum, U., Molin, G., 2002. Vaginal Lactobacillus flora of healthy Swedish women. *Journal Clinical Microbiology*, 40:2746–2749.

Ventura, M., Canchaya, C., Meylan, V., Klaenhammer, T. R., Zink, R., 2003. Analysis, characterization, and loci of the tuf genes in lactobacillus and bifidobacterium species and their direct application for species identification. *Applied Environmental Microbiology*, 69:6908-6922.

Walter, J., Chagnaud, P., Tannock, G. W., Loach, D. M., Dal Bello, F., Jenkinson, H. F., Hammes, W. P., Hertel, C., 2005. A high-molecular-mass surface protein (Lsp) and methionine sulfoxide reductase B (MsrB) contribute to the ecological performance of Lactobacillus reuteri in the murine gut. *Applied Environmental Microbiology*, 71:979–986.

Walter, J., Loach, D. M., Alqumber, M., Rockel, C., Hermann, C., Pfitzenmaier, M., Tannock, G. W., 2007. D-Alanyl ester depletion of teichoic acids in Lactobacillus reuteri 100-23 results in impaired colonization of the mouse gastrointestinal tract. *Environmental Microbiology*, 9:1750 1760.

Zhou, X., Brown, C. J., Abdo, Z., Davis, C. C., Hansmann, M. A., Joyce, P., Foster, J. A., Forney, L. J., 2007. Differences in the composition of vaginal microbial communities found in healthy Caucasian and black Women. *Multidisciplinary Journal Microbial Ecology*, 1:121–133.

Zhou, M., Theunissen, D., Wels, M., Siezen, R. J., 2010. LAB-Secretome: a genome-scale comparative analysis of the predicted extracellular and surface-associated proteins of Lactic Acid Bacteria. *BMC Genomics*, 11:651.

In: Lactobacillus: Classification, Uses and Health Implications ISBN: 978-1-62081-151-1
Editors: Alba I. Perez Campos and Arturo Leon Mena

Chapter XI

Lactobacillus in Lacto-Fermented Vegetables

Zsolt Zalán*[*] *and Anna Halász
Central Food Research Institute, Budapest, Hungary

Abstract

The lactic acid fermentation of vegetables is an ancient practice. Wide range of fermented vegetable products is presented all over the world and takes a specific part in the human diet. The role of the lactic acid bacteria in this process is not only the extension of the shelf life but also make the raw material more digestible and enhance the organoleptical properties of the end product. The members of the genera of *Lactobacillus* play an important role in the vegetable fermentation and often they are the predominant microorganisms in these products. They produce beside the organic acids several other antimicrobial metabolites (e.g. hydrogen peroxide, alcohol, carbon dioxide, bacteriocins), which ensure the microbial safety of the product and in this way follow out the biopreservation. Lactobacilli produce also several aromatic compounds and by their activity can form bioactive components to enhance the nutritional value of the product. The lacto-fermented vegetables combine the vitamins, minerals and fibre of vegetables with the potential probiotic properties of *Lactobacillus*. Several products are fermented by spontaneously, which can form although unique food, but it is unrepeatable. For the constant and appropriate quality the application of selected starter *Lactobacillus* culture with good fermentation properties is necessary.

In this chapter is collected the knowledge about the lacto-fermented vegetables, report about our studies and the role and the usability of the *Lactobacillus* in this process.

Introduction

The members of the *Lactobacillus* genera occur and grow in several different natural habitats, from the soil and water to the human body. They are associated also with the surface

[*] Central Food Research Institute, Budapest, H-1022, Herman Ottó u. 15., zs.zalan@cfri.hu.

of plant and plant materials. The numbers of lactic acid bacteria on plant materials are considerably variable, but their number is usually low, however their presence is widespread (Daeschel et al. 1987). As the lactobacilli have complex nutritional requirements, the surface of the vegetables is a good habitat for them. Although on the plant could not be found all of the *Lactobacillus* species and in addition the quality and quantity of these species on the plants are altering, it is influenced by several parameters, e.g. the environment, temperature, humidity as Mundt and Hammer (1968) reported already in the late sixties. The lactic acid bacteria are present on the outer surface of the living plant with several other, simultaneously occurred microorganisms, where the number of the lactic acid bacteria is almost negligible in comparison to the total microbial population, however their number increase with the degree of plant maturity. After harvesting the number of microorganisms increases since the nutrients become available from the cells of damaged tissue. The lactobacilli, which are associated with plants, are *Lactobacillus* (*Lb.*) *arabinosus*, *Lb. brevis*, *Lb. buchneri*, *Lb. fermentum* and *Lb. plantarum* (Daeschel et al. 1987). (According to the novel taxonomy and nomenclature *Lb. arabinosus* is the synonym of *Lb. plantarum*.) During the vegetable processing the microbes could get into the end product from the surface, so it is not surprising the lactobacilli could be occurred in the plant based food product and play an important role in the vegetable fermentation. The fermentation itself as food preservation, food production technology is an ancient method, widely known and used from millennia. Following the earlier development of fermented beverages and milk products, the appearance of the vegetable fermentation may be put at the 3rd century BC in Asia and the Mediterranean region (Holzapfel et al., 1995, Mensah 1997, Hulse 2004), but the first documentation about the preservation of plant materials by lactofermentation dates back to 1000-1500 BC (Daeschel et al. 1987). The tradition of lacto fermentation was established mainly by the claim to the durability. It is true especially in the tropical climates where the lacto fermentation play a key role in the preservation of raw material and safety and acceptability of the fermented food products (Stiles and Holzapfel 1997, Nout and Sarkar 1999). However the lactic acid bacteria not only preserve the raw material, but also make some plant material more digestible thus improve the nutritional properties of the food and moreover develop the flavor/aroma and make a higher value product and moreover in some cases the fermentation reduces toxicity of the substrate (Caplice and Fitzgerald 1999). The earliest production of fermented foods was based on spontaneous fermentation, since the ancient people are not aware of the role of the microorganisms in this process of course. In this process took part and carried out the fermentation the naturally presented microflora of the raw material and the environment. The knowledge and the methodologies of the production of the ancient fermented food were handed down from generation to generation. Nevertheless this preservation process and itself the fermented vegetables have spread all over the world and also after millennia they could be found among our meal.

Lactobacillus in the Fermented Vegetables

The *Lactobacillus* strains can be found and are used in several fermented foods from the dairy products (yogurt, kefir, cheese) through the meet products (salami) to the bakery products (sourdough bread). Their role in the vegetable fermentation is not well-known than in the dairy industry, however the activity of lactobacilli is essential for the development of the fermented vegetable products. Several species of *Lactobacillus* were isolated from spontaneously fermented vegetables or are used in the vegetable fermentation process as starter culture. A range of fermented foods, their country of origin, the *Lactobacillus* species, which dominate or take part in the fermentation and the raw vegetable material, is shown in Table 1. Naturally this table could not be complete, because there are several other fermented vegetable products in the world, which are not known widely and not well studied or published. Moreover these products are prepared in various forms in different regions, often from various raw materials and in this way slightly or entirely different products are developed with different microbial population. For example there are reported about more than 150 kinds of kimchi (a traditional, well known Korean fermented vegetables) (Park and Cheigh 2004). In the Table 1 are shown also such kind of product, which based on raw material, which are not a typical vegetable, for example the cassava (also known as manioc or yucca), a woody scrub with starchy root, where this root is a major source of food carbohydrates in the tropics; the olives, a fruit of the olive tree and the bamboo shoot also not a typical vegetable. In some cases the consumption of this vegetable is not as usual, for example the cassava root was processed to flour/paste before or after the fermentation. However since the group of vegetables is not a scientific ranging and is rather based on culinary and cultural tradition the mentioned plant parts could be range into the group of vegetables. As from the collected data could be seen (Table 1), the dominant *Lactobacillus* species is the *Lactobacillus plantarum*, which could be found almost in all fermented vegetables. The other dominant species is *Lactobacillus brevis*.

Table 1. Occurrence of *Lactobacillus* species in fermented vegetables

Product	**Country/Region**	**Lactobacilli**	**Substrate**	**Reference**
Brovada	India	*Lb. plantarum, Lb. hilgardii, Lb. coryniformis*	turnip	Maifreni et al. 2004
Burong mustala	Philippines	*Lb. brevis*	mustard	Karovičová and Kohajdová 2003
Dakguadong	Thailand	*Lb. plantarum*	mustard leaf	Lee 1997
Dhamuoi	Vietnam	*Lb. plantarum*	cabbage	Lee 1997
Fermented/ pickled cucumber	International	*Lb. plantarum*	cucumber	Caplice and Fitzgerald 1999, Tamminen et al. 2004, Leroy and De Vuyst 2005, Guizani 2011
Fermented/ pickled eggplant	Spain	*Lb. plantarum, Lb. brevis, Lb. fermentum, Lb. pentosus*		Sánchez et al. 2000
Gari	West Africa	*Lb. plantarum, Lb. pentosus, Lb. fermentum*	cassava	Wankhede et al. 1998, Caplice and Fitzgerald 1999, Kostinek et al. 2007
Gundruk	India, Nepal	*Lb. cellobiosus, Lb. plantarum,*	leafy vegetables	Battcock and Azam-Ali 1998, Tamang et al., 2005

Table 1. (Continued)

Product	Country/Region	Lactobacilli	Substrate	Reference
Inziangsang	India	*Lb. plantarum, Lb. brevis*	leafy vegetables	Tamang et al., 2005
Kawal	Chad, Sudan	*Lb. plantarum*	sicklepod leaf	Singhal and Kulkarni 1998
Khalpi	India	*Lb. plantarum, Lb. brevis*	cucumber	Tamang et al., 2005
Kimchi	Korea	*Lb. plantarum, Lb. brevis, Lb. sakei, Lb. fermentum, Lb. curvatus*	cabbage, radish	Battcock and Azam-Ali 1998, Kim and Chun 2005, Guizani 2011, Rhee et al. 2011
Mesu	India	*Lb. plantarum, Lb. brevis, Lb. curvatus*	bamboo shoot	Choudhury et al. 2011, Tamang et al. 2008
Olives	Mediterranean	*Lb. plantarum Lb. brevis, Lb. casei, Lb. pentosus, Lb. paracollinoides*	green olives	Caplice and Fitzgerald 1999, Randazzo et al. 2004, Abriouel et al. 2011
Pak-Gard-Dong	Thailand	*Lb. brevis, Lb. plantarum*	mustard leaf	Battcock and Azam-Ali 1998
Pickled roots	International	*Lb. plantarum, Lb. brevis*	carrots, turnips	Guizani 2011
Sauerkraut	International	*Lb. plantarum, Lb. brevis, Lb, curvatus, Lb. sake*	cabbage	Caplice and Fitzgerald 1999, Tamminen et al. 2004, Leroy and De Vuyst 2005
Sayur asin	Indonesia	*Lb. confusus, Lb. curvatus, Lb. plantarum*	cabbage	Puspito and Fleet 1985
Shalgam	Turkey	*Lb. plantarum, Lb. paracasei subsp. paracasei, Lb. brevis, Lb. fermentum*	black carrot, turnip	Tanguler and Erten 2012
Sinki	Nepal	*Lb. fermentum, Lb. brevis, Lb. plantarum*	radish	Battcock and Azam-Ali 1998, Tamang et al., 2005, Guizani 2011
Sunki	Japan	*Lb. plantarum, Lb. brevis*	otaki-turnip	Battcock and Azam-Ali 1998, Guizani 2011

Role of Lactobacillus in the Fermentation

The original purpose of the fermentation of the vegetables as raw material was to preserve them, because the vegetables are highly perishable. The preservation effect of lactobacilli comes from the reduction of the available carbohydrates and the formation of a range organic molecules, which in turn exhibit antimicrobial activity (Ouwehand 1998). This antimicrobial activity of lactobacilli is mainly caused by production of several organic acids (lactic, acetic, formic, caproic, propionic, butyric and valeric acids (Corsetti et al. 1998)), but these bacteria also produce other inhibitory substances like hydrogen peroxide (Rodrígues et al., 1997, Ito et al., 2003), carbon dioxide, diacethyl (Ouwehand 1998, Ammor et al., 2006) and bacteriocins - thermostable low molecular weight compounds (Cleveland et al., 2001, Plocková et al., 2001). The production and the activity of their antifungal substances considerably depend on the composition of the media, on the substrate material (Tomás et al., 2002, Avonts et al., 2004, Zalán et al., 2005). The nutritional composition of vegetables depends on several factors. It is influenced by genetic and also environmental factors, such as temperature, light, moisture, nutritional status of the soil and also influenced by the agricultural practices and stage of maturity (Salunkhe and Kadam 1998). The nutritional composition of the diverse part of the vegetables is also different, since the tubers, roots and crops contain great amount carbohydrates (starch, free sugars), on the contrary the leaves of vegetables are low in fat and carbohydrates (Salunkhe and Kadam 1998). These carbohydrates are the main substrate for the lactobacilli, which could produce energy for their activity, for the propagation only through lacto-fermentation. The uptaken or intracellularly hydrolyzed carbohydrates can be fermented by the lactobacillus strains through two major pathways: the glycolysis (Embden-Meyerhof pathway) is used by the homofermentative lactobacilli, and the 6 phosphogluconate/phosphoketolase (6-PG/PK) pathway, is used by heterofermentative ones. Lactic acid bacteria during the fermentation of carbohydrate, when the sugars are converted to cellular energy, excrete lactate as metabolite byproduct. Generally the lactic acid is the predominant end-product (by the obligate homofermentative strains always), but under various conditions the amount of other byproduct (acetic and other acids, alcohol, carbon dioxide) could be increased by the facultative homofermentative strains, while the obligate heterofermentative ones always produce significant amount byproduct, other than lactic acid. In point of the metabolite products of lactobacilli the lactic acid and acetic acid are regarded as the main organic acids, which possess antimicrobial behavior. In the presence of oxygen lactobacilli are able to generate hydrogen peroxide by flavoprotein oxidase, NADH oxidase and pyruvate oxidase, which is then accumulated because of the lack of catalase in lactobacilli (Murphy and Condon 1984, Ouwehand 1998). Hydrogen peroxide is a strong oxidizing agent and thus a well-known antibacterial component (Ouwehand 1998, Ocaña et al. 1999, Ito et al. 2003). However, it has only a short-lived action because it reacts with organic materials in the environment. This oxidizing effect can cause bleaching of colored components and in this way causes undesired loss of color of the product, besides degradation of the antioxidant components of vegetables (Özkan et al. 2005). During the heterofermentative lacto-fermentation the lactobacilli form carbon-dioxide, which creates an anaerobic micro-environment and it has also an antimicrobial activity and in this way takes part in the preservation (Ouwehand 1998). Also the heterofermentative lactobacilli could form ethanol, which may play role in the preservation, since it has a well-established antimicrobial effect, but the concentration of the ethanol produced by lactobacilli is so low that the contribution to the preservation effect is minimal, only in the early stage of the fermentation may be more significant (Adams and Nicolaides 1997, Caplice and Fitzgerald 1999). Similarly, although diacetyl and acetoin are produced by many *Lactobacillus* species through the citrate metabolism and these aroma components have also antimicrobial activity, their level is low to present a major activity (Adams and Nicolaides 1997, Ouwehand 1998, Caplice and Fitzgerald 1999). *Lactobacillus* strains may produce bacteriocins, ribosomally-synthesized peptides or proteins with antimicrobial activity rather broad spectra of inhibition. Bacteriocinogenic strains can be readily isolated from fermented food (Caplice and Fitzgerald 1999). Several bacteriocin producer *Lactobacillus* strain and their bacteriocins were isolated from vegetable raw material and vegetable food matrices, for example *Lb. plantarum*, *Lb. pentosus* and plantaricin S and T from fermented green olives, *Lb. buchneri* and buchnericin from sauerkraut, *Lb. plantarum*, *Lb. paracasei* and *Lb. sakei* and their Type IIa bacteriocins from kimchi, *Lb. sakei* and sakacin from Chinese fermented cabbage (Kim et al. 2004, Settanni and Corsetti 2008, Gao et al. 2010). Naturally these antimicrobial components produced by lactobacilli act together and all of them contribute to the antimicrobial effect of the producer strain and their effect often not only additive, but synergetic. Due to the production of these antimicrobial metabolites the fermented food is safer, than the same unfermented form and the safety of the fermented product is very good even in the developing world, where the environments of the fermentation are often unhygienic and contaminated (Mensah 1997, Steinkraus 1997).

Beyond the preservation the metabolites of lactobacilli have remarkable effect also on the organoleptical properties of the product, which is influenced both by the quantity and quality of the produced organic acids and flavor components. Several organic acids produced by *Lactobacillus* strains are present in such small amount, which do not take part in the preservation, but result in a more characteristic flavor. For example among the plant-associated heterofermentative lactobacilli seem to be common the succinic acid production from citrate through the succinic acid pathway, which succinic acid is a well-known flavoring substance (Kaneuchi et al. 1988). However in the flavor formation play role several other metabolites. In lactobacilli the pyruvate is a starting molecule for the formation of short-chain flavor compounds such as acetaldehyde, acetate, acetoin, diacetyl and ethanol. Lactobacilli can also metabolize the citrate to produce acetoin, acetolactate and diacetyl (Marilley at al. 2004). As we have obtained from our investigations, the *Lactobacillus* strains can produce acetoin and diacetyl in a wide range of concentration in vegetable juice (Zalán et al. 2011). The amino acid-converting enzymes of *Lactobacillus* strains can play also an essential role in flavor development of the product. Amino acids of the vegetable substrate may contribute to the production of such flavor and aroma compounds as aldehydes, acids, alcohols, esters and sulphur compounds (Ayad et al. 2001). Depending on the enzymes present in the fermenter cultures, different flavors can develop due to the contribution of many enzymes, which lead to various flavor compounds and thus different products from the point of view of their aroma. Since the enzyme activity, the fermentation behavior of the lactobacilli is different between species, and also often between the strains of the same subspecies, this lead to several various products from the same raw material.

The activity of lactobacilli may reduce the toxicity of the raw material by the fermentation. The cassava (the raw material of fermented Gari) contains cyanogenic glucosides linamarin and lotaustralin, but during the fermentation the pH decrease inhibits the hydrolysis of this toxic compounds and stabilizes the cyanohydrins (Wankhede et al. 1998, Caplice and Fitzgerald 1999).

Plant materials, including leafy vegetables, may contain high levels of nitrate and because of it the vegetables are provide the one of the main source for the nitrate and nitrite intake in the human nutrition, since nitrite arising from the reduction of nitrate caused by plant enzymes.

The lactobacilli during the fermentation can reduce or in some cases remove completely the nitrites from the substrate (Heród-Leszczyńska and Miedzobrodzka 1992; Walkowiak-Tomczak and Zielinska 2006).

Usually in the vegetable fermentation several different microorganisms take part often other than *Lactobacillus*, especially in the spontaneous fermentation. For example in Gari *Streptococcus faecium*, *Leuconostoc mesenteroides*, *Bacillus* and *Corynebacterium* species, in pickled cucumber *Pediococcus* species, in kawal *Bacillus subtilis*, *Candida krusei*, *Propionibacterium* and *Saccharomyces* spp., *Leuconostoc mesenteroides* in sauerkraut and olives, *Leuconostoc* and *Weissella* species in kimchi (Wankhede et al. 1998, Musmade and Desai 1998, Singhal and Kulkarni 1998, Caplice and Fitzgerald 1999, Park and Cheigh 2004, Leroy and De Vuyst 2005, Kim and Chun 2005). These microorganisms develop together, simultaneously or consecutively, the end form of the fermented vegetable products. In the spontaneously fermented vegetables the strains of *Lactobacillus* become dominant in the second stage of the fermentation, since the vegetable fermentation is usually initiated by *Leuconostoc mesenteroides* and as the pH decrease to 4.0 the more acid tolerant *Lactobacillus* strains become predominant and they are followed by yeasts, mainly in substrates that are rich in fermentable sugars (Holzapfel 2002, Guizani 2011). In sauerkraut fermentation after the aerobic microbes activity the *L. mesenteroides* become dominant, which is followed by *Lb. plantarum* and then in the last period *Lb. brevis*, *Pediococcus* and *Enterococcus* strains is the dominant (Halász et al. 1999). The lactobacilli in this way play an important, essential role in the preservation and the developing of the organoleptical properties of the fermented vegetable products.

Lactobacillus as Starter Culture in Vegetable Fermentation

Most traditional fermented products are made by natural fermentation where the microorganisms, derived from the raw material and the environment, spontaneously grow and develop the end form of the product. The indigenous flora of these fermented foods varies according to the raw material and the environmental conditions. The spontaneous fermentation can form a unique product with appropriate properties, however it is unrepeatable, the quality of the product is changing, the accompanying microflora can cause spoilage and the safety of the product is questionable because of the un-, or low-controlling of the process. To the constant and appropriate quality, especially in a larger scale is absolute necessary the application of selected starter culture with good fermentation properties. Already in the ancient times were established by the artisanal and religious practices the application of starter culture. In that time the spontaneous fermentation was optimized through back slopping, when the raw material is inoculated with a small part of a previous successful fermentation. This method results the dominance of the best-adapted strains in the fermentation. In the 1st century A.D., Plinius the Elder described the preservation of white cabbage in special earthen vessels, which were used only for this purpose. Presumably the cabbage was fermented in these vessels by the microorganisms, which remained behind from a previous batch (Buckenhüskes 1993). The back slopping process, based on the continuous inoculation by foods from a previous, successful batch, is still an applied cheap and reliable preservation method in less developed countries, but in some cases it is also industrially applied to date (Buckenhüskes 1993, Holzapfel 2002, Leroy and De Vuyst 2004). However the pure culture inoculation of sauerkraut was studied already from the second decade of the 20th century and *Lb. plantarum* starter were used even from 1937 in the olive fermentation (Fleming and McFeeters 1981). The application of selected starter culture results control over the fermentation and standard product with constant quality. The selection criteria for lactic acid bacteria to be used as starter culture for vegetables were well established: The essential and most important criteria are rapid growing, fast acidification, good antimicrobial properties, good organoleptical product developing and low biogenic amine formation (Buckenhüskes 1993, Halász et al. 1999, Leroy and De Vuyst 2004). However there are several other criteria as the bacteriophage resistance, acid tolerance, which is also important in an industrial production or which is important in case of a certain product as the homofermentative metabolism or such functional criteria as probiotic property. Nevertheless during the selection of a starter culture should take into account the cultural traditions, dietary habits and the raw material, which are diverse in the different continents and regions (Holzapfel 2002).

The fast growing of the *Lactobacillus* starter culture on the raw material to reach quickly an appropriate high cell number results one hand competition with the other coexist microbes for the substrates and the niche and on the other hand removing the fermentable carbohydrates from the raw material to prevent in this way the growing of other undesirable microbes and a secondary fermentation. Most of *Lactobacillus* strains grow well on vegetables: *Lb. plantarum* on green olive reached 8.5 log cfu/ml (Leal-Sánchez 2003), both *Lb. acidophilus*, *Lb. plantarum*, *Lb. casei* and *Lb. delbrueckii* grew to 10^9 cfu/ml in red beet juice (Yoon et al. 2005), *Lb. plantarum*, *Lb. brevis* and *Lb. paracasei* subsp. *paracasei* reached in shalgam 8.4 log cfu/g, 7.5 log cfu/g and 6.78 log cfu/g, respectively (Tanguler and Erten 2012), *Lb. pentosus* in carrot juice 10^9 cfu/ml (Bergqvist et al. 2005), *Lb. sakei* cell concentration in sauerkraut fermentation reached 9.2×10^9 cfu/ml (Tolonen et al. 2004), *Lb. plantarum* in cucumber fermentation reached more than 9 log cfu/ml (McDonald et al. 1993) and even *Lb. casei* on sliced Jerusalem artichoke could grow to 10^{10} cfu/ml (Yokoi et al. 2006) and *Lb. pentosus* on beetroot mix reached 7.1×10^{10} cfu/ml (Jägerstad et al. 2004). According to our results the *Lactobacillus* viable cell count reached 10^8 cfu/ml in red beet juice and 10^9 in Jerusalem artichoke juice and there was no significant difference between the species (Baráth et al. 2004, Zalán et al. 2011). Moreover it can be established that the type of the substrate (chopped or juice) and the *Lactobacillus* strains only slightly influence the reachable cell count, rather the parameters of the incubation influence significantly the final cell number, especially the temperature and the length of the fermentation.

However the high viable cell number does not mean in itself a good fermentation and safety product, since even the acidification, the reduction of the pH of the raw material shows a more diverse picture. For example as Yoon and co-workers (2005) reported *Lb. acidophilus* and *Lb. plantarum* reduced the pH to 3.7 and 4.1, respectively, while *Lb. casei* and *Lb. delbrueckii* reduced only to 5.0 in the same red beet juice. Comparing to different *Lactobacillus* species, we have obtained, that in Jerusalem artichoke juice all strains reduced the pH to between 4.5 and 3.5 and most of them to lower than 4.0 (Zalán et al. 2011). Kostinek and co-workers (2007) observed that only the half of the strains of *Lb. plantarum* group isolated from fermented cassava reduced the pH to lower than 3.9 after 48 hours. Albeit the pH values of less than 4.2 constitutes a major safety concern in fermented foods so the ability to acidify their environment rapidly is an important property of a starter strain (Holzapfel 2002, Kostinek et al. 2007). Nevertheless several strains from *Lactobacillus* genera can reduce the pH on vegetable based product lower than this level (Karovičová et al. 1999, Baráth et al. 2004). The pH decreasing of the environment considerably depends on the organic acid production, the enzyme activity of the cells. The different quality and quantity of the produced acid can cause different decrease in the pH of the environment even at the same cell number. So the metabolic pathways, the fermentation profile of the strain plays an important role in the acidification. According to our study there may be differences in the quality and quantity of produced organic acids between species and even between the strains of the same subspecies in the same media and furthermore also the media influence significantly the organic acid production (Zalán et al. 2010). As it was mentioned earlier the quantity and quality of organic acids produced by lactobacilli influence the flavor of the fermented product. The starter culture with mixed acid fermentation results in a more characteristic flavor, which should be taken into consideration by the selection of starter culture.

It is also not negligible that pH decrease is faster in every case of the application of starter cultures than in the spontaneously fermented sample, which help to reduce the risk of spoilage or extend the lag phase of sensitive food spoilage/pathogen microorganisms (Leal-Sánchez et al. 2003, Baráth et al. 2004).

The microbial flora of living plants is dominated by Gram-negative bacteria and Gram-positive spore formers and in smaller numbers yeast and moulds, but this composition is changing during the vegetable processing. The fresh-cut vegetables typically contaminated with *Pseudomonas* spp, *Xanthomonas* spp, *Enterobacter* spp, *Janthinobacterium* spp, yeasts, lactic acid bacteria, less frequently *Aeromonas hydrophila* and occasionally *Listeria monocytogenes* (Sinigaglia et al. 1999). Moreover there are several pathogens, which are associated with vegetables (Beczner and Bata-Vidács 2009). Nevertheless many microbes could contaminate the product from the surface and may cause spoilage in the fermented vegetable products as the family *Enterobacteriaceae*, yeasts and moulds (Daeschel et al. 1987). Beside the fast growing and rapid acidification also the above-mentioned antimicrobial components produced by lactobacilli take part in the formation of microbial safety of the fermented vegetables. It is well-known and investigated that several strains of *Lactobacillus* are able to inhibit the growth of many pathogens and food spoilage microorganisms, e. g. *E. coli*, *L. monocytogenes*, *S. enteritidis* and moulds (Schillinger et al. 1991; Corsetti and others 1998; Park and others 2005). The *Lactobacillus* strains show various inhibitory activites against food borne microorganisms, however there can be found some strains, which have a relative broad one.

Beside the several positive effects of lactobacilli in food preservation their ability to decarboxylize amino acids and produce biogenic amines is also very important in food production. These biogenic amines may be formed and degraded during the normal metabolism of animals, plants and microorganisms. Depending on the spoilage association microbial deterioration of food may be accompanied by an increased production of amino acid decarboxylases. Fermented foods in which lactic acid bacteria grow contain significant amounts of putrescine, cadaverine, histamine and tyramine (Santos 1996). Experimental data indicate a correlation between bacterial numbers and the biogenic amine content of particular foods. This has been intensely studied also for vegetables (Halász et al. 1994). Concentrations of biogenic amines show extreme variations in traditional, spontaneously fermented food and beverages. Amine production of bacteria depends on the amino acid decarboxylase activity of the certain strain. As some people suffer adverse reactions after consuming amines, the production of fermented foods with predictable low levels of specific amines need to addressed by the food industry. Halász and co-workers (1994, 1999) found that the amino acid decarboxylase enzyme activity of bacteria used as starter cultures vary widely and some of them even produce histamine. In spite of this compared the biogenic amine content of spontaneously and with a potential starter culture fermented red beet, in the spontaneously fermented one the amount of total biogenic amine was 7-35 times higher and also histamine content was detected (Baráth et al. 2004). The starter cultures also successfully suppressed the biogenic amine formation in sauerkraut compared with spontaneous fermentation (Halász et al. 1999, Špička et al. 2002). The ability of lactobacilli to produce biogenic amines is highly variable, often being strain specific, therefore the selection of starter culture, which fail to produce biogenic amines is important even on strains level (Buckenhüskes 1993, Halász et al. 1994, Spano et al. 2010).

Although the spontaneous fermentation result product with an unique flavor and the indigenous flora is more adapted to the raw material and may be more stable than a starter culture, the large number of indigenous microorganisms, the variability in the nutrient content and in every case the shape of the vegetables make difficult the control of fermentation, however the use of appropriately selected starter culture results a controllable fermentation and a safety product with constant quality. Although in dairy and meat industry and also for sourdough bakery products a large number of lactic acid bacteria starters are routinely used, for vegetable fermentations only a few cultures have been used. Lactic acid fermentation of vegetables has nowadays an industrial significance only for cucumbers, cabbages and olives (Di Cagno et al. 2008). From the *Lactobacillus* genera the strains of *Lb. plantarum* is the starter, which is most frequently used in the fermentation food products of plant origin, among others in the fermentation of cucumbers, cabbages and olives also *Lb. plantarum* is the commercial starter, however sometimes *Lb. brevis* is also used (Mäyrä- Mäkinen and Bigret 1998, Musmade and Desai 1998, Di Cagno et al. 2008). Suitability and application of these *Lactobacillus* species as starter culture could take for granted, since these species are the dominant one also in the indigenous flora of spontaneously fermented products. In the production of fermented vegetables the salting of the raw material, the use of brine is generally used, which is help to reduce the growth of spoilage microorganisms and to prevent the softening. Since the *Lb. plantarum* is highly salt-tolerant (Sánchez et al. 2000, Rao et al. 2004), it may contribute to the domination of this species among the used starter cultures. Nevertheless starter culture must be selected at the strain level, since not all strains of a species are equally suitable for use as starters, they adaptation to the food substrate, their antimicrobial activity and flavor development are not equally (Giraffa 2004).

Conclusion

Vegetables play an important role in human nutrition, because they provide minerals, micronutrients, vitamins, antioxidants and dietary fibre, which contribute to the good health of the consumer. Lacto-fermentation has traditionally been applied to extend the shelf life of raw material and ensure the microbiological safety of it and in addition to make the raw material more digestible. However the fermentation is more than a preservation method, it has got several benefical health effects. The fermentation enhance the protein content or improve the balance of amino acids, increase the contents or availability of vitamins, reduce the indigestible materials (Kahajdová and Karovičová 2004). An increasing interest can be observed for lactic acid fermented vegetable products, the consumption of the lacto-fermented vegetable juices increases in a number of countries (Kahajdová and Karovičová 2004). Moreover among the lactobacilli there are several probiotic strains (Isolauri et al. 2004). After a circumspect selection the probiotic strain can be used as starter culture, however the starter culture also can show probiotic properties (Vinderola and Reinheimer 2003). Since the most of the probiotic foods are dairy product, the fermented, probiotic vegetables represent a perfect alternative to the dairy product for the consumers, who don't want to eat milk based food or are lactose intolerant or allergic to milk proteins. The food scientists agree about fermented plant products belong to the „Food of the future", in which the members of *Lactobacillus* genera play an important role (Buckenhüskes 1993).

References

Abriouel, H., Benomar, N., Lucas, R., Gálvez, A. (2011). Culture-independent study of the diversity of microbial populations in brines during fermentation of naturally-fermented Aloreña green table olives. *International Journal of Food Microbiology, 144,* 487-496.

Adams, M. R., Nicolaides, L. (1997). Review of the sensitivity of different foodborne pathogens to fermentation. *Food Control, 8,* 227-239.

Ammor, S., Tauveron, G., Dufour, E., Chevallier, I. (2006). Antibacterial activity of lactic acid bacteria against spoilage and pathogenic bacteria isolated from the same meat small-scale facility 1-Screening and characterization of the antibacterial compounds. *Food Control, 17,* 454-461.

Avonts, L., Van Uytven, E., De Vuyst, L. (2004). Cell growth and bacteriocin production of probiotic *Lactobacillus* strains in different media. *International Dairy Journal, 14,* 947-955.

Ayad, E. H. E., Verheul, A., Engels, W. J. M., Wouters, J. T. M., Smit, G. (2001). Enhanced flavour formation by combination of selected lactococci from industrial and artisanal origin with focus on completion of a metabolic pathway. *Journal of Applied Microbiology, 90,* 59-67.

Baráth, Á., Halász, A., Németh, E., Zalán, Zs. (2004). Selection of LAB strains for fermented red beet juice production. *European Food Research and Technology, 218,* 184-187.

Battcock, M., & Azam-Ali, S. (1998). *Fermented fruits and vegetables. A global perspective. FAO Agricultural Services Bulletin No. 134.* Rome, Food and Agriculture Organization of the United Nations.

Beczner, J. & Bata-Vidács, I. (2009). Microbiology of plant foods and related aspects. *Acta Alimentaria, 38,* 99-115.

Buckenhüskes, H. J. (1993). Selection criteria for lactic acid bacteria to be used as starter cultures for various food commodities. *FEMS Microbiology Reviews, 12,* 253-272.

Caplice, E. & Fitzgerald, G. F. (1999). Food fermentations: role of microorganisms in food production and preservation. *International Journal of Food Microbiology, 50,* 131-149.

Choudhury, D., Sahu, J. K., Sharma, G. D. (2011). Bamboo shoot based fermented food products: a review. *Journal of Scientific & Industrial Research, 70,* 199-203.

Cleveland, J., Montville, T. J., Nes, I. F., Chikindas, M. L. (2001). Bacteriocins: safe, natural antimicrobials for food preservation. *International Journal of Food Microbiology, 71,* 1-20.

Corsetti, A., Gobetti, M., Rossi, J., Daminiani, P. (1998). Antimould activity of sourdough lactic acid bacteria: identification of mixture of organic acids by *Lactobacillus sanfrancisco* CB1. *Applied Microbiology and Biotechnology, 50,* 253-256.

Corsetti, A., Gobetti, M., Rossi, J., Daminiani, P. (1998). Antimould activity of sourdough lactic acid bacteria: identification of mixture of organic acids by *Lactobacillus sanfrancisco* CB1. *Applied Microbiology and Biotechnology, 50,* 253-256.

Daeschel, M. A., Andersson, R. E., Fleming, H.P. (1987). Microbial ecology of fermenting plant materials. *FEMS Microbiology Reviews, 46,* 357-367.

Di Cagno, R., Surico, R. F., Siragusa, S., De Angelis, M., Paradiso, A., Minervini, F., De Gara, L., Gobbetti, M. (2008). Selection and use of autochthonous mixed starter for lactic acid fermentation of carrots, French beans or marrows. *International Journal of Food Microbiology, 127,* 220-228.

Fleming, H. P. & McFeeters, R. F. (1981). Use of microbial cultures: vegetable products. *Food Technology, 35,* 84-88.

Gao, Y., Jia, S., Gao, Q., Tan, Z. (2010). A novel bacteriocin with a broad inhibitory spectrum produced by *Lactobacillus sake* C2, isolated from traditional Chinese fermented cabbage. *Food Control, 21,* 76-81.

Giraffa G. (2004). Studying the dynamics of microbial populations during food fermentation. *FEMS Microbiology Reviews,* 28, 251-260.

Guizani, N. (2011). Vegetable Fermentation and Pickling. In N. H. Sinha (Ed.), *Handbook of Vegetables & Vegetable Processing.* (1st edition, pp. 351-368). Ames, USA: Blackwell Publishing Ltd.

Halász, A., Baráth, Á., Holzapfel, W. H. (1999). The influence of starter culture selection on sauerkraut fermentation. *Zeitschrift für Lebensmittel-Untersuchung und -Forschung A,* 208, 434-438.

Halász, A., Baráth, Á., Simon-Sarkadi, L., Holzapfel, W. H. (1994). Biogenic-amines and their production by microorganisms in food. *Trends in Food Science & Technology, 5,* 42-49.

Heród-Leszczyńska, T. & Miedzobrodzka, A. (1992). Effect of the fermentation process on levels of nitrates and nitrites in selected vegetables. *Rocz Panstw Zakl Hig, 43,* 253-258.

Holzapfel, W. H. (2002). Appropriate starter culture technologies for small-scale fermentation in developing countries. *International Journal of Food Microbiology, 75,* 197-212.

Holzapfel, W. H., Geisen, R., Schillinger, U. (1995). Biological preservation of foods with reference to protective cultures, bacteriocins and food-grade enzymes. *Food Microbiology, 24,* 343-362.

Hulse, J. H. (2004). Biotechnologies: past history, present state and future prospects. *Trends in Food Science & Technology, 15,* 3-18.

Isolauri, E., Salminen, S., Ouwehand, A. C. (2004). Probiotics. *Best Practice & Research Clinic Gastroenterology, 18,* 299-313.

Ito, A., Sato, Y., Kudo, S., Sato, S., Nakajima, H., Toba, T. (2003). The screening of hydrogen peroxide-producing lactic acid bacteria and their application to inactivating psychrotrophic food-borne pathogens. *Current Microbiology, 47,* 231-236.

Jägerstad, M., Jastrebova, J., Svensson, U. (2004). Folates in fermented vegetables-a pilot study. *Lebensmittel-Wissenschaft und -Technologie, 37,* 603-611.

Kaneuchi, C., Seki, M., Komagata, K. (1988). Production of succinic acid from citric acid and related acids by *Lactobacillus* strains. *Applied and Environmental Microbiology, 54,* 3053–3056.

Karovičová, J. & Kohajdová, Z. (2003). Lactic acid fermented vegetable juices. *Horticultural Science (Prague), 30,* 152-158.

Karovičová, J., Drdák, M., Greif, G., Hybenová, E. (1999). The choice of strains of *Lactobacillus* species for the lactic acid fermentation of vegetable juices. *European Food Research and Technology, 210,* 53-56.

Kim, H. T., Lee, K. H., Kim, J. H. (2004). Bacteriocins produced by Lactic acid bacteria isolated from Kimchi. *Journal of Agriculture & Life Sciences, 38,* 15-24.

Kim, M. & Chun J. (2005). Bacterial community structure in kimchi, a Korean fermented vegetable food, as revealed by 16S rRNA gene analysis. *International Journal of Food Microbiology, 103,* 91-96.

Kohajdová, Z. & Karovičová, J. (2004). Optimisation of method of fermentation of cabbage juice. *Czech Journal of Food Sciences, 22,* 39-50.

Kostinek, M., Specht, I., Edward, V. A., Pinto, C., Egounlety, M., Sossa, C., Mbugua S., Dortu, C., Thonart, P., Taljaard, L., Mengu, M., Franz, C. M. A. P., Holzapfel, W.H. (2007). Characterisation and biochemical properties of predominant lactic acid bacteria from fermenting cassava for selection as starter cultures. *International Journal of Food Microbiology, 114,* 342-351.

Leal-Sánchez, M. V., Ruiz-Barba, J. L., Sánchez, A. H., Rejano, L., Jiménez-Díaz, R., Garrido, A. (2003). Fermentation profile and optimization of green olive fermentation using *Lactobacillus plantarum* LPCO10 as a starter culture. *Food Microbiology, 20,* 421-430.

Lee, C-H. (1997). Lactic acid fermented foods and their benefits in Asia. *Food Control, 8,* 259-269.

Leroy, F., & De Vuyst, L. (2004). Lactic acid bacteria as functional starter cultures for the food fermentation industry. *Trends in Food Science & Technology, 15,* 67-78.

Maifreni, M., Marino, M., Conte, L. (2004). Lactic acid fermentation of Brassica rapa: chemical and microbial evaluation of a typical Italian product (brovada). *European Food Research and Technology, 218,* 469-473.

Marilley, L. & Casey, M. G. (2004). Flavours of cheese products: metabolic pathways, analytical tools and identification of producing strains. *International Journal of Food Microbiology, 90,* 39-159.

Mäyrä-Mäkinen, A., & Bigret, M. (1998). Industrial Use and production of lactic acid bacteria. In S. Salminen, & A. von Wright (Eds.), *Lactic acid bacteria: microbiology and functional aspects* (2nd edition, pp. 73-102) New York, USA: Marcel Dekker, Inc.

McDonald, L. C., Shieh, D.-H., Fleming, H. P., McFeeters, R. F., Thompson, R. L. (1993). Evaluation of malolactic-deficient strains of *Lactobacillus plantarum* for use in cucumber fermentations. *Food Microbiology, 10,* 489-499.

Mensah, P. (1997). Fermentation - the key to food safety assurance in Africa? *Food Control, 8,* 271-278.

Mundt, J. O. & Hammer, J. L. (1968). Lactobacilli on Plants. *Applied Microbiology, 16,* 1326-1330.

Murphy, M. G. & Condon, S. (1984). Correlation of oxygen utilization and hydrogen peroxide accumulation with oxygen induced enzymes in *Lactobacillus plantarum* cultures. *Archives of Microbiology, 138,* 44-48.

Musmade, A. M., & Desai, U. T. (1998). Cucumber and Melon. In D. K. Salunkhe, & S. S. Kadam (Eds.), *Handbook of Vegetable Science and Technology: Production, Composition, Storage, and Processing* (1st edition, pp. 245-272). New York, USA: Marcel Dekker, Inc.

Nout, M. J. R. & Sarkar P. K. (1999). Lactic acid food fermentation in tropical climates. *Antonie van Leeuwenhoek, 76,* 395-401.

Ocaña, V. S., de Ruiz Holgado, A. P., Nader-Macías, M E. (1999). Growth inhibition of *Staphylococcus aureus* by H2O2-producing *Lactobacillus paracasei* subsp. *paracasei* isolated from the human vagina. *FEMS Immunology & Medical Microbiology, 23,* 87-92.

Ouwehand, A. C. (1998). Antimicrobial Components from Lactic Acid Bacteria. In S. Salminen, & A. von Wright (Eds.), *Lactic acid bacteria: microbiology and functional aspects* (2nd edition, pp. 139-160) New York, USA: Marcel Dekker, Inc.

Özkan, M., Yemenicioğlu, A., Cemeroğlu, B. (2005). Degradation of various fruit juice anthocyanins by hydrogen peroxide. *Food Research International, 38,* 1015-1021.

Park, J. H., Seok, S. H., Cho, S. A., Baek, M. W., Lee, H. Y., Kim, D. J., Chung M. J., Kim, S. D., Hong, U. P., Park, J. H. (2005). Antimicrobial effect of lactic acid producing bacteria culture condensate mixture (LCCM) against *Salmonella enteritidis*. *International Journal of Food Microbiology, 101,* 111-117.

Park, K-Y., & Cheigh, H-S. (2004). Kimchi. In Y. H. Hui, S. Ghazala, D. M. Graham, K. D. Murrell, W-K. Nip (Eds.), *Handbook of Vegetable Preservation and Processing.* (1st edition, pp. 220-260) New York, USA: Marcel Dekker, Inc.

Plocková, M., Stiles, J., Chumchalová, J., Halfarová, R. (2001). Control of mould growth by *Lactobacillus rhamnosus* VT1 and *Lactobacillus reuteri* CCM 3625 on milk agar plates. *Czech Journal of Food Sciences, 19,* 46-50.

Puspito, H. & Fleet, G. H. (1985). Microbiology of sayur asin fermentation. *Applied Microbiology and Biotechnology, 22,* 442-445.

Randazzo, C. L., Restuccia, C., Romano, A. D., Caggia, C. (2004). *Lactobacillus casei,* dominant species in naturally fermented Sicilian green olives. *International Journal of Food Microbiology, 90,* 9-14.

Rao, M. S., Pintado, J., Stevens, W. F., Guyot, J. P. (2004). Kinetic growth parameters of different amylolytic and non-amylolytic *Lactobacillus* strains under various salt and pH conditions. *Bioresource Technology, 94,* 331-337.

Rhee, S. J., Lee, J.-E., Lee, C.-H. (2011). Importance of lactic acid bacteria in Asian fermented foods. *Microbial Cell Factories, 10,* S5

Rodrígues, E., Tomillo, J., Nuñez, M., Medina, M. (1997). Combined effect of bacteriocin-producing lactic acid bacteria and lactoperoxidase system activation on *Listeria monocytogenes* in refrigerated milk. *Journal of Applied and Environmental Microbiology, 83,* 389-395.

Salunkhe, D. K. & Kadam, S. S. (1998). Handbook of Vegetable Science and Technology: Production, Composition, Storage, and Processing Food (1st edition). New York, USA: Marcel Dekker, Inc.

Sánchez, I., Palop, L., Ballesteros, C. (2000). Biochemical characterization of lactic acid bacteria isolated from spontaneous fermentation of 'Almagro' eggplants. *International Journal of Food Microbiology, 59, 9-17.*

Santos, S. M. H. (1996). Biogenic amines: their importance in foods. *International Journal of Food Microbiology, 29,* 213-231.

Schillinger, U., Kaya, M., Lücke, F. K. (1991). Behaviour of *Listeria monocytogenes* in meat and its control by a bacteriocin-producing strain of *Lactobacillus sake. Journal of Applied Microbiology, 70,* 473-478.

Settanni, L. & Corsetti, A. (2008). Application of bacteriocins in vegetable food biopreservation. *International Journal of Food Microbiology, 121,* 123-138.

Singhal, R. S., & Kulkarni, P.R. (1998). Leafy Vegetables. In D. K. Salunkhe, & S. S. Kadam (Eds.), *Handbook of Vegetable Science and Technology: Production, Composition, Storage, and Processing* (1st edition, pp. 533-589). New York, USA: Marcel Dekker, Inc.

Sinigaglia, M., Albenzio, M., Rosaria Corbo, M. (1999). Influence of process operations on shelf-life and microbial population of fresh-cut vegetables. *Journal of Industrial Microbiology & Biotechnology, 23,* 484-488.

Spano, G., Russo, P., Lonvaud-Funel, A., Lucas, P., Alexandre, H., Grandvalet, C., Coton, E., Coton, M., Barnavon, L., Bach, B., Rattray, F., Bunte, A., Magni, C., Ladero, V., Alvarez, M., Fernández, M., Lopez, P., de Palencia, P. F., Corbi, A., Trip, H., Lolkema, J. S. (2010). Biogenic amines in fermented foods. *European Journal of Clinical Nutrition, 64,* S95-S100.

Špička, J., Kalač, P., Bover-Cid, S., Křížek, M. (2002). Application of lactic acid bacteria starter cultures for decreasing the biogenic amine levels in sauerkraut. *European Food Research and Technology, 215,* 509-514.

Steinkraus, K. H. (1997). Classification of fermented foods: worldwide review of household fermentation techniques. *Food Control, 8,* 311-317.

Stiles, M. E. & Holzapfel, W. H. (1997). Lactic acid bacteria of foods and their current taxonomy. *International Journal of Food Microbiology, 36,* 1-29.

Tamang, B., Tamang, J. P., Schillinger, U., Franz, C. M. A. P., Gores, M., Holzapfel, W. H. (2008). Phenotypic and genotypic identification of lactic acid bacteria isolated from ethnic fermented bamboo tender shoots of North East India. *International Journal of Food Microbiology, 121,* 35-40.

Tamang, J. P., Tamang, B., Schillinger, U., Franz, C. M. A. P., Gores, M., Holzapfel, W. H. (2005). Identification of predominant lactic acid bacteria isolated from traditionally fermented vegetable products of the Eastern Himalayas. *International Journal of Food Microbiology, 105,* 347-356.

Tamminen, M., Joutsjoki, T., Sjöblom, M., Joutsen, M., Palva, A., Ryhänen, E.-L., Joutsjoki, V. (2004). Screening of lactic acid bacteria from fermented vegetables by carbohydrate profiling and PCR–ELISA. *Letters in Applied Microbiology, 39,* 439-444.

Tanguler, H., Erten, H. (2012). Occurrence and growth of lactic acid bacteria species during the fermentation of shalgam (salgam), a traditional Turkish fermented beverage. *LWT - Food Science and Technology, 46,* 36-41.

Tomás, M. S., Bru, E., Wiese, B., de Ruiz Holgado, A. A. P., Nader-Macías, M. E. (2002). Influence of pH, temperature and culture media on the growth and bacteriocin production by vaginal *Lactobacillus salivarius* CRL 1328. *Journal of Applied Microbiology, 93,* 714-724.

Vinderola, C. G., Reinheimer, J. A. (2003). Lactic acid starter and probiotic bacteria: a comparative ''in vitro'' study of probiotic characteristics and biological barrier resistance. *Food Research International, 36,* 895-904.

Walkowiak-Tomczak, D. & Zielińska, A. (2006). Effect of fermentation conditions on red-beet leaven quality. *Polish Journal of Food and Nutrition Sciences, 56,* 437-444.

Wankhede, D. B., Satwadhar P. N., Sawate, A. R. (1998). Cassava. In D. K. Salunkhe, & S. S. Kadam (Eds.), *Handbook of Vegetable Science and Technology: Production, Composition, Storage, and Processing* (1st edition, pp. 99-118). New York, USA: Marcel Dekker, Inc.

Yokoi, K.-J., KawasakI, K.-I.., Nishitani, G., Taketo, A., Kodaira K.-I. (2006). Fermentation of Jerusalem Artichoke with or without Lactic Acid Bacteria Starter Cultures. *Food Science and Technology Research, 12,* 231-234.

Yoon, K. Y., Woodams, E. E., Hang, Y. D. (2005). Fermentation of beet juice by beneficial lactic acid bacteria. *Lebensmittel-Wissenschaft und -Technologie, 38,* 73-75.

Zalán Zs., Hudáček, J., Tóth-Markus, M., Husová, E., Solichová, K., Hegyi, F., Plocková, M., Chumchalová, J., Halász, A. (2011). Sensorically and antimicrobially active metabolite production of *Lactobacillus* strains on Jerusalem artichoke juice. *Journal of the Science of Food and Agriculture, 91,* 672-679.

Zalán, Zs., Hudáček, J., Štětina, J., Chumchalová, J., Halász, A. (2010). Production of organic acids by *Lactobacillus* strains in three different media. *European Food Research and Technology, 230,* 395-404.

Zalán, Zs., Németh, E., Baráth, Á., Halász, A. (2005). Influence of growth medium on hydrogen peroxide and bacteriocin production of *Lactobacillus* strains. *Food Technology and Biotechnology, 43,* 219-225.

In: Lactobacillus: Classification, Uses and Health Implications ISBN: 978-1-62081-151-1
Editors: Alba I. Perez Campos and Arturo Leon Mena

Chapter XII

Lactobacilli – Functional Starter Cultures for Meat Applications

Nevijo Zdolec[1]
University of Zagreb, Faculty of Veterinary Medicine, Department of Food Hygiene, Technology and Safety, Zagreb, Croatia

Abstract

Lactic acid bacteria, primarily lactobacilli are known beneficial microbes with wide agri-food applications. Their use in animal production as feed/water additives could contribute to better growth performance, feed conversion or gut health by reducing colonization of pathogens. Today, when antibiotics are banned as growth promoters in food animals, lactobacilli and other natural antimicrobial agents are promising tool for improving microbial safety in agri-food chain. Furthermore, beneficial properties of lactobacilli are also employed in processing of food for human consumption, mainly fermented meat and dairy products. Naturally fermented foods with autochthonous „wild" microbial population are rich source of lactobacilli with favorable technological and hygienic properties for food applications. For selection of potential functional starter cultures the strains must be phenotypically and genotypically characterized. Selection criteria should include technological, safety and probiotic features of the strain. Results of in vitro studies at given laboratory conditions could be promising, but performance of the culture in real food fermentation may not be relevant. This chapter summarizes results of current studies regarding selection and implementation of lactobacilli in (non)fermented meats with particular emphasis on technological and health benefits.

Introduction

Fermentation is one of the oldest processes of meat preservation, which depends on the biological activity of lactobacilli and other lactic acid bacteria (LAB) i.e. on the production of

[1] E mail: nzdolec@vef.hr; nevijo.zdolec1@gmail.com, Tel/Fax: ++38512390192.

different metabolites capable of suppressing the growth of undesirable microbiota (Ross et al., 2002; Hutkins, 2006). Knowledge about the role of microorganisms in the food fermentation dates long ago, so that individual strains, considered fit from the technological and hygienic point of view, have been introduced in the production of fermented foodstuffs in form of starter cultures. In case of fermented sausages, these are LAB, prevailingly species of the genus *Lactobacillus* and *Pediococcus* (Hutkins, 2006). Antimicrobial activity of lactobacilli is manifested by inhibition of other microorganisms through competition for nutritive ingredients and/ or production of one or more antimicrobials such as organic acids (lactic, acetic), hydrogen peroxide, antimicrobial enzymes, bacteriocins and reuterin (Holzapfel et al., 1995). The use of bacteriocinogenic strains as protective cultures and their protective substances (bio-preservatives, bacteriocins) in the production of fermented foods is a part of protective concept in the food industry (Montville and Matthews, 2005).

Lactobacilli are naturally present in various kinds of food of (non)animal origin. During the fermentation they significantly affect sensory characteristics, but also microbiological safety of products. Regarding microbiological safety, it is known that lactobacilli are dominant fermentative microbiota in many foods, and they are competitive towards many spoilage bacteria and pathogens. Thus different authors have reported on antimicrobial effect of lactobacilli toward the bacteria *Listeria monocytogenes*, *Escherichia coli*, *Salmonella* spp., *Clostridium* spp., *Staphylococcus aureus* and others (Leroy et al., 2006). The inhibitory effect toward food-related pathogens was shown under laboratory conditions (Zdolec et al., 2009), as well as in the real meat fermentation (Zdolec et al., 2007a; Zdolec et al., 2008a). *In vitro* studies are prerequisite for development of new (authothonous) functional starter cultures comprised of species with known metabolic profile and proved antimicrobial effect.

As far as in 1919 Orla-Jensen claimed that the group of LAB are gram-positive, immobile, non-sporogenous bacilli and cocci which ferment carbohydrates and higher alcohols, thus creating lactic acid. As the time passed, new species were found and taxonomic changes were intensive, which was conditioned by a constant progress in analytical methods. In food hygiene and technology, almost all the species within the LAB group have their role, as in the positive (fermentation, antimicrobial activity, sensory characteristics of a product, probiotic effect, etc.), so in the negative sense (spoilage bacteria, biogenic amines producers, carriers of resistance genes).

Lactobacilli can be found in dairy products, cereal products, meat or fish products, water, silage, beer, wine, fruit and fruit juices, marinated vegetables, sauerkraut, waste waters, sour dough; they are part of natural microflora of oral cavity and digestive tract of people and animals; they rarely show pathogen characteristics. They can be classified to three groups (Stiles and Holzapfel, 1997), and these are:

- obligatory homofermentative - they ferment glucose to lactic acid, and they don't ferment pentoses and gluconate. Some of the representatives are *Lactobacillus (Lb.) acidophilus, Lb. delbrueckii, Lb. helveticus, Lb. gasseri, Lb. ruminis;*
- facultative heterofermentative – they ferment hexoses to lactic acid and can produce gas from gluconate but not from glucose, and they ferment pentoses to lactic and acetic acid. In foodstuff we find facultative fermentative *Lb. casei, Lb. plantarum, Lb. curvatus, Lb. paracasei, Lb. pentosus, Lb. plantarum, Lb. rhamnosus* and *Lb. sakei*; and

- obligatory heterofermentative – they ferment hexoses to lactic acid, acetic acid, ethanol and CO_2. Creating gas from glucose is characteristic for this group of lactobacilli. Some of the representatives are *Lb. kefir, Lb. brevis, Lb. fermentum* and *Lb. reuteri.*

Bernardeau et al. (2008) indicate that lactobacilli are heterogeneous microbial group which consists of 135 species and 27 subspecies, and their classification keeps changing. By applying modern molecular methods it is suggested that extreme diversity of genome of this bacteria genus can lead to new sub generic classifications. The combination of genotypic and phenotypic tests e.g. DNA based techniques, and conventional tests of fermentation profiles (using carbohydrates) is necessary for determination of species. Pulsed Field Gel Electrophoresis (PFGE) has been successfully applied on species from milk, and at the same time this is the most discriminative and repeatable method. Data from professional literature support the hypothesis that taking in lactobacilli by food isn't risky since lactobacillemia caused by the food, especially by fermented dairy products, is extremely rare and appears only with predisposed patients. Some metabolic characteristics, like creating biogenic amines in fermented products, can lead to nonacceptance of some lactobacilli in food production. A small number of lactobacilli in starter cultures and probiotics can show antimicrobial resistance spread. Still, it can be underestimated since researches of spreading resistance aren't systematic. The authors consider the spread of antimicrobial resistance to be the only relevant reason to be cautious, which demands monitoring of sensitivity to antibiotics because such species could be the carriers of resistant genes and present a potential risk for spreading genes to other bacteria. Still, as a general rule, lactobacilli have high natural resistance which can't be spread, and it is resistance to most antibiotics, especially vancomycin. Safety assessment of lactobacilli which are significant from the technological aspect should be limited to antibiotic profiles and researches of transmissibility of antimicrobial resistance (Talon and Leroy, 2011).

Antimicrobial Effect of Lactobacilli

Bacteriocins are bacterial peptides or proteins of different antimicrobial spectrum and mode of activity, different molecular masses, genetic and biochemical characteristics, which in the case of lactobacilli damage the membranes of related gram- positive bacteria (Tagg et al., 1976; Jack et al., 1995). Many food-related species of lactobacilli were found as bacteriocinogenic and capable to suppress the growth of foodborne pathogens such as *L. monocytogenes, S. aureus, Clostridium* spp., *Enterococcus* spp. and *Bacillus* spp. (Holzapfel et al., 1995). Some of the bacteriocinogenic lactobacilli species are *Lb. sakei* (bacteriocin sakacin), *Lb. curvatus* (curvacin), *Lb. plantarum* (plantaricin), *Lb. acidophilus* (lactacin, acidocin) and *Lb. bavaricus* (bavaricin). Although the studies of bacteriocins and their effect on food safety have been very extensive, up to this day only one bacteriocin (nizin, E234) is used commercially in dairy industry.

It is considered that direct antimicrobial effect of organic acids as metabolitic products of lactobacilli and other LAB appears by their effect on cytoplasmatic membrane by exchanging potentials and inhibiting active transport (De Vuyst and Vandamme, 1994). Lactic acid, as the final product of homofermentative activity of lactobacilli, lowers the pH- value of the substrate (food, nutritional medium, etc.) which destroys or inhibits pathogens or spoilage bacteria. Professional literature especially emphasizes the sensitivity of gram- negative bacteria (like *Salmonella* spp. and *E. coli*) on acidification conditions (Adams and Nicolaides, 1997). Acetic acid shows stronger antimicrobial activity than lactic acid, and it appears in heterofermentative bacterial metabolism, e.g. by the activity of leuconostocs or heterofermentative lactobacilli. Larger quantities of acetic acid can be expected in products fermented by intrinsic autochthonous microbiota with predominance of heterofermentative species. Our experiences showed that, except for the desirable antimicrobial effect, acetic acid has also the unwanted effect to sensory characteristics of some foodstuff, like fermented sausages, which can be prevented by adding competitive homofermentative lactobacilli as starter cultures (Zdolec, 2007).

Hydrogen peroxide is produced by LAB in the presence of oxygen together with lactate, pyruvate and NADH over the flavine enzymes (Holzapfel et al., 1995). Its antimicrobial effect manifests in the oxidation of membrane lipids and cell proteins, but its real antibacterial activity in food is questionable because of natural decomposition by peroxidases (Caplice and Fitzgerald, 1999). Carbon dioxide is created heterofermentatively and contributes to the appearance of anaerobic conditions, by which it inhibits aerobic spoilage bacteria in interaction with the cell membrane by changing extracellular and intracellular pH (De Vuyst and Vandamme, 1994). Dyacethil is an antimicrobial product synthesized by some lactococci, pediococci and leuconostocs by metabolizing citrates. Creating dyacethil is disabled by overcoming homofermentative processes, so its antimicrobial effect is reduced in such conditions. It is believed that gram- negative bacteria, yeasts and moulds are more sensitive to dyacethil, and the mode of activity is attributed to interference in exploiting arginine (Caplice and Fitzgerald, 1999).

Certain species of lactobacilli are "specialized" for applying in some fermented dairy or meat products where they express their desirable effect regarding creating the aimed sensory characteristics of finished products. Except for that, by synthesizing substances which act competitively to other bacteria, lactobacilli improve microbiological stability of products. These mechanisms of microbial inhibition can also be used in preserving the non-fermented food, which can prolong its shelf- life.

Fermented meat products produced in the traditional way, without the addition of starter culture, are a source of lactobacilli strains with antimicrobial effect on foodborne bacteria and they may be included in new functional starter cultures if they meet the technological requirements. Recent results of identification of indigenous microbiota in Croatian traditional meat sausages are summarized in table 1. Identification of indigenous microbiota was performed by means of culture-dependent as well as culture-independent methods. In general, good correlation is found for traditional and molecular methods of identification for main LAB and coagulase- negative cocci (CNC) species in different European fermented sausages. However, molecular identification showed that phenotypic methods are unable to identify some LAB/CNC species (Iacumin et al., 2006), suggesting that culture-independent methods should be the first choice for studying bacterial diversity and ecology of fermented sausages (Rantsiou and Cocolin, 2006). Biochemical and molecular identification of intrinsic LAB or CNC in different types of Croatian sausages showed the domination of *Lb. plantarum* and *Staphylococcus xylosus*. Those species are also found as dominant in other Mediterranean countries (Albano et al., 2009; Drosinos et al., 2007; Iacumin et al., 2006). One recent study

(Babić et al., 2011) reported *Leuconostoc mesenteroides* as one of dominant LAB in Croatian dry sausage "kulen", which implicates even more the need for introducing competitive technologically suitable starter cultures, because that species is heterofermentative and can cause sensorial deviations of sausages.

Table 1. Results of determination of indigenous microbiota of different types of traditional Croatian fermented sausages

Strains selected	Sausage type	Methods of characterization	Reference
Lb. plantarum *Lb. delbrueckii* *Ln. mesenteroides* *Lb. acidophilus* *S. xylosus* *S. warneri* *S. lentus* *S. auricularis* *C. famata* *Penicillium* sp., *Aspergillus* sp.	Kulen	API 50 CHL API STAPH API 20C AUX RAPD PCR	Frece et al., 2010a Babić et al., 2011
Lb. plantarum *Lb. brevis* *Lb. curvatus* *Lb. pentosus* *Lb. fermentum* *Ln. mesenteroides* *Lc. lactis* *P. pentosaceus* *S. xylosus* *S. capitis* *S. carnosus* *S. saprophyticus*	Artisan traditional fermented sausage from North-West Croatia	API 50 CHL API STAPH PCR	Kozačinski et al., 2006 Kozačinski et al., 2008 Cvrtila et al., 2008
Lb. plantarum *Lb. fermentum* *Lb. pentosus* *Lb. delbrueckii* *W. viridescens* *W. confusa*	Artisan horse-meat dry sausage	API 50 CHL	Alagić et al., 2010
Lb. curvatus *Lb. brevis* *Ln. mesenteroides*	Homemade traditional fermented sausage from North-West Croatia	API 50 CHL	Zdolec et al., 2007b
Lb. acidophilus *S. xylosus*	Homemade traditional fermented sausage from North Croatia	API 50 CHL API STAPH	Frece et al., 2010b
Lc.lactis subsp. *lactis* *C. zeylanoides*	Horsemeat dry sausage	API 50 CHL API 20C AUX	Markov et al., 2010
S. xylosus *S. warneri* S. *lentus* *S. auricularis*	Kulen	API STAPH	Frece et al., 2010c
Lb. fermentum *Lb. curvatus* *Lc. lactis* subsp. *lactis* *S. xylosus* *S. sciuri*	Homemade Slavonian sausage	API 50 CHL API STAPH	Dobranić et al., 2011

**Lb.=Lactobacillus; Ln.=Leucoonostoc; C.=Candida; S.=Staphylococcus; P.=Pediococcus; W.=Weisella; Lc.=Lactococcus.*

The spectrum of lactobacilli antimicrobial effect depends on many intrinsic and extrinsic factors, and for practical application extended research is needed into their function under given conditions in real food fermentation. However, *in vitro* screening of antimicrobial activity of indigenous lactobacilli is the first step of providing potential protective cultures for practical implementations. Regarding to this, we preliminary tested the inhibitive effect of 11 lactobacilli strains from traditional fermented sausages on the bacteria *L. monocytogenes* NCTC 10527, 3 strains of *L. monocytogenes* from food, *Listeria welshimeri, Yersinia enterocolitica, Pseudomonas aeruginosa, S. aureus* NCBF 1499, *Salmonella* Enteritidis and *Salmonella* spp. by use of the agar spot and agar diffusion tests. The most sensitive indicator bacteria were *L. monocytogenes* 1 (inhibited from 10 of 11 strains of lactobacilli) and *L. monocytogenes* NCTC 10527 (8/11), whilst the most resistant was *L. welshimeri* (3/11). Six lactobacilli strains inhibited *Salmonella* spp., five *P. aeruginosa,* and four inhibited the growth of the bacteria *Y. enterocolitica, S. aureus* and *Salmonella* Enteritidis. The broadest spectrum of inhibition was confirmed for *Lb. paracasei* subsp. *paracasei*, and the narrowest for *Lb. curvatus.* The preliminary results indicate the significant antimicrobial activity of selected lactobacilli, which is one of the important characteristics of the potential functional starter cultures (Zdolec et al., 2009). Similar studies were recently performed in different kinds of Croatian traditionally fermented sausages (Table 2).

Resistance of Lactobacilli to Antibiotics

As already stated, transfer of antimicrobial resistance is the only relevant caution reason in food-related lactobacilli and other LAB. Prescription of antimicrobial drugs in prophylaxis, ignorance of real causative agent, improperly dosage, inappositely therapy period and other irregularities caused an appearance of antimicrobial resistance in bacteria which is nowadays one of emerging (veterinary) public health issues (Teuber, 2001; Hussein et al., 2005; Zdolec et al., 2006; *Marcinčák* et al., 2006; Roasto et al., 2009). The most important problem is increase of resistance in clinically significant bacteria, but today is clear that comensal bacteria (including lactobacilli) could also contribute to spreading resistance genes to pathogens (Mathur and Singh, 2005; Ruzauskas et al., 2009).

Recently, we tested indigenous LAB (from Croatian traditional sausages and cheese) for susceptibility/resistance towards 23 antimicrobials in order to estimate the presence of resistant or multi-resistant strains in traditional foods and a level of potential risk for consumer's health regarding the spreading of antimicrobial resistance through a food chain (Zdolec et al., 2011). All isolates of *Lactobacillus* spp. and *Leuconostoc* spp. were susceptible to tetracycline, penicillin, bacitracin, chloramphenicol, cefalexin, linkomycin, erythromycin, ampicillin, spiramycin and amoxicillin, but resistant to vancomycin, nalidixic acid and metronidazole.

Few strains were additionally resistant to nitrofurantoin, trimethoprim, oxacillin and neomycin. *Leuconostoc* spp. showed resistance to streptomycin, enrofloxacine, nitrofurantoin, sulphonamides, polymyxin B and oxacillin.

Table 2. Results of testing antimicrobial activity of selected strains from traditional Croatian fermented sausages

Strains tested for antimicrobial activity	Indicator microorganism	Strains with strongest antimicrobial activity	Test methods	Reference
Lb. paracasei *Lb. casei* *Lb. brevis* *Lb. curvatus* *Lb. fermentum*	*L. monocytogenes* *L. welshimeri* *Y. enterocolitica* *P. aeruginosa* *S. aureus* *Salmonella* spp.	*Lb. paracasei* *Lb. brevis*	Agar spot test Agar well diffusion test	Zdolec et al., 2009
Lb. plantarum *Lb. fermentum* *Lb. pentosus* *Lb. delbrueckii* *W. viridescens* *W. confusa*	*L. monocytogenes*	*Lb. plantarum*	Agar spot test Agar well diffusion test	Alagić et al., 2010
Lb. curvatus *Lb. brevis* *Ln. mesenteroides*	*L. monocytogenes*	*Ln. mesenteroides*	Agar spot test Agar well diffusion test	Zdolec et al., 2007c
Lb. plantarum *Lb. delbrueckii* *Ln. mesenteroides* *Lb. acidophilus*	*E. coli* *S. aureus* *L. monocytogenes* *Salmonella* Typhimurium	*Lb. plantarum* *Lb. delbrueckii* *Ln. mesenteroides*	Turbidimetric assay	Frece et al., 2010a Babić et al., 2011
Lb. acidophilus	*E. coli* *S. aureus* *L. monocytogenes* *Salmonella* spp.	*Lb. acidophilus KA1*	Turbidimetric assay	Frece et al., 2010b
Lc. lactis subsp. *lactis*	*E. coli* *S. aureus* *L. monocytogenes* *Salmonella* spp.	*Lc. lactis* subsp. *lactis 5K1*	Turbidimetric assay	Markov et al., 2010
S. xylosus *S. warneri* *S. lentus* *S. auricularis*	*E. coli* *S. aureus* *L. monocytogenes* *Salmonella* spp.	*S. xylosus* *S. warneri* *S. lentus*	Disk diffusion Turbidimetric assay	Frece et al., 2010c

**Lb.=Lactobacillus; Ln.=Leucoonostoc; S.=Staphylococcus; W.=Weisella; Lc.=Lactococcus.*

Lactobacilli are usually sensitive to penicillin and β-lactamase inhibitors, but more resistant to oxacillin and cephalosporins (Coppola et al., 2005; Danielsen and Wind, 2003). Our results were in agreement with those findings, but other studies reported the appearance of penicillin resistance. Thus, Yuksekdag and Beyatli (2008) isolated penicillin-resistant strains of *Lb. casei* and *Lb. helveticus* from milk products. In that study among 24 LAB isolates (*Lactobacillus* spp., *Leuconostoc* spp., *Streptococcus* spp., *Lactococcus* spp.), and 44 % were resistant to penicillin. Non-permeability of lactobacilli cell wall is the main reason of their resistance, but several non-specific mechanisms could cause differences among strains within the same species (Condon, 1983; Putman et al., 2001). Most lactobacilli are resistant to glycopeptides (vancomycin), while susceptibility to bacitracin is variable (Coppola et al., 2005; Katla et al., 2001).

Our study confirms a well-known resistance of lactobacilli to vancomycin which is considered as intrinsic and untransferable (Klein et al., 1998), but all our strains were sensitive to bacitracin. Lactobacilli are usually sensitive to inhibitors of protein synthesis such as chloramphenicol, erythromycin, clindamycin and tetracycline and more resistant to aminoglycosides (neomycin, kanamycin, streptomycin and gentamycin) (Charteris et al., 1998; Coppola et al., 2005; Zhou et al., 2005). This was partially presented in our results where all lactobacilli were inhibited by chloramphenicol, erythromycin, tetracycline and streptomycin.

On the other hand, some lactobacilli were resistant to neomycin, and especially to kanamycin and gentamycin. Antimicrobial resistance is known to be present in this group of antibiotics, in particular for chloramphenicol (Lin et al., 1996), erythromycin (Cataloluk and Gogebakan, 2004) and tetracycline (Kastner et al., 2006). Furthermore, our results showed the insensitivity of lactobacilli to antimicrobials which inhibit the synthesis of nucleic acids, as reported by others (Charteris et al., 1998; Katla et al., 2001; Coppola et al., 2005).

Antimicrobial resistance of foodborne lactobacilli is nowadays actual and emerging due to potential gene transfer to pathogenic bacteria. The goal of our preliminary study was to evaluate the presence of resistant lactobacilli in autochthonous foods such as fermented sausages and raw cheese. We have characterized just several strains, so further investigations of LAB populations are necessary to obtain relevant data to be included in risk assessment related to traditional food products.

Lactobacilli as Protective Cultures at Farm Level

Currently, human campylobacteriosis and salmonellosis are most important bacterial foodborne zoonoses. The presence of resistant strains of *Salmonella* and *Campylobacter* is also high in food-producing animals that are easily transferable through a food chain. The antimicrobial resistance in „natural“ animal gut microbiota such as lactic acid bacteria should be considered carefully, as they are able to transfer resistant genes to other bacteria, including pathogens. Strategies employed in meat production at slaughterhouse and processing level are directed to self-control programs which should be supported by preventive measures at primary production, i.e. at farm level.

As the use of antibiotics as animal growth promoters is prohibited, there are efforts to introduce new, natural agents with positive impact on animal growth, feed conversions, health and meat safety. For this purpose, we could use lactobacilli and their bacteriocins which are able to suppress undesirable gut microbiota and consequently reduce the risk of meat contamination by resistant (a)pathogens. Svetoch and Stern (2010) reported that giving bacteriocins trough feed to one day chickens artificially infected with *Campylobacter jejuni*, prevent gut colonization with the pathogen.

Further, they found that introducing bacteriocins by drinking water during last three days of fattening reduce *Campylobacter* count by 5 to 6 times. Those findings are promising for future implementation of bacteriocins or bacteriocin-producing cultures at farm level, as a part of whole food-chain approach directed to reducing microbial risks at meat producing facilities.

Meat Biopreservation

Ancient methods of food preservation, such as salting, fermentation, smoking, drying etc., have not been abandoned, but are still employed in the production of meat products. The improvement of industrial processing methods has resulted in the development of different additional protective strategies, which have the same aim as in the early days of food preservation i.e. prevention of spoilage, inhibition of growth of spoilage microorganisms and pathogens or prolongation of the product's shelf life. For this purpose, various chemical substances in the form of additives have been used, as well as protective microbial cultures, different physical procedures (irradiation, thermodynamic processes, and pulsed electric field) and packing under modified atmosphere (Devlieghere et al., 2004).

Hurdle technology, as a specific concept, includes a combination of the existing (temperature, water activity, pH, Eh, preservatives etc.) and novel techniques of preservation (packing under gas, treatment with high pressure, biopreservation), the aim of which was to establish selective protective factors in which pathogenic and spoilage microbiota could not survive (Leistner and Gorris, 1995). The idea of biopreservation comprehends the process of fermentation and preservation as a method designed for prolongation of shelf life and enhancement of food safety by the use of microorganisms and/or their metabolites (Ross et al., 2002). As regards meat and meat products, biopreservation includes the addition of bacteriocinogenic LAB cultures, non-purified bacteriocins, fermentation culture or bacteriocinogenic culture concentrate, purified or semi-purified bacteriocin, as well as the addition of mesophilic LAB as a form of protection in case of temperature abuse (Hugas, 1998). Lücke (2000) has pointed out that the production of lactic acid is the main mechanism of action of protective cultures on the competitive microbiota, while the effect of bacteriocins is reduced by their inactivation in meat and by the development of resistance. Also, the use of bacteriocinogenic LAB can not yield satisfactory results in the prevention of intestinal infections induced by gram-negative bacteria (e. g. EHEC) and in the prolongation of shelf life of meat kept under aerobic conditions. Several factors should be taken into consideration when selecting bacteriocinogenic LAB for use in a certain food product. These are GRAS status (Generally Recognised As Safe), spectrum of inhibition of microorganisms, thermal stability of bacteriocins, risk for human health, improvement of safety and quality and a highly specific activity (Holzapfel et al., 1995). Several authors have also pointed out that in the process of biopreservation attention should be paid to the cultures of LAB not producing bacteriocins. Their antagonism is based on the production of lactic acid and acidification, production of other antimicrobial substances and competition for nutritive ingredients in a substrate. However, upgrading of safety and quality by applying (non)bacteriocinogenic (non)autochtonous strains as starter cultures was reported in different kinds of fermented meat products all over the world (Zdolec et al., 2008b; Liu et al., 2010; Laukova et al., 2010; Sriphochanart and Skolpap, 2010; Bonomo et al., 2011; Jaworska et al., 2011; Vatanyoopaisarn et al., 2011). Bacteriocin-based conservation strategy in food production is intensively tested during last decades, but using bacteriocins as additives is still questionable mainly due their un-stability in reach proteolytic matrix such as fermented sausages. However, most studies showed that sausages produced with bacteriocinogenic cultures are superior to non-inoculated products or inoculated with non-bacteriocinogenic culture. A part of this product's improving should be attributed to bacteriocins produced in situ.

Effect of Lactobacilli as Protective Cultures Against L. monocytogenes in Fermented Sausages

Listeria monocytogenes is an important foodborne pathogen especially in minimaly processed food products. *L. monocytogenes* shows peculiar characteristics such as high resistance in, for many bacteria, pernicious conditions. For example, it survives freezing, multiplies at low temperatures during cooling process and survives in acid and alkaline medium, at low water activity and increased salt concentration.

Consequently, its survival and multiplication are possible in various types of foods and under different storage conditions. *L. monocytogenes* is an omnipresent organism in the environment, and has been isolated from many types of food and production facilities (Loncarevic, 1998; Kozačinski and Hadžiosmanović, 2001; Thévenot et al., 2005a). According to the report of de Valk et al. (2005), the number of food-related cases of listeriosis in humans in Europe is not negligible.

Bioconservation by applying bacteriocins or bacteriocinogenic protective cultures was shown as functional tool regarding lowering the risk related to foodborne listeriosis.Our research performed within of EU project "Safety of traditional fermented sausages: research on protective cultures and bacteriocins", have been focused on antilisterial capacity of bacteriocin-producing cultures *Lb. sakei* I151, I154 and I155 in various Central-European and Mediterranean traditionally fermented sausages – Hungarian salami, Bosnian sucuk, "Sremska" sausage (Serbia) and Croatian sausage (Hadžiosmanović et al., 2005; Čaklovica et al., 2006; Gasparik-Reichardt et al., 2006; Drosinos et al., 2006; Zdolec et al., 2007). Use of *Lb. sakei* cultures in the Hungarian salami resulted in by about 2 $\log_{10}$ lower pathogen count during 28-day-interval of ripening (from the initial 5.5 to 3.5 $\log_{10}$ CFU/g), about 1.5 $\log_{10}$ respectively in the control sausage.

A complete elimination of *L. monocytogenes* from the stuffing of the Bosnian sucuk was achieved during the ripening period (28 days) with the use of protective *Lb. sakei* strains (reduction about 5 $\log_{10}$). This, however, was not the case with the control sausage (reduction from the initial 4.85 to 3.83 $\log_{10}$ CFU/g). Similar results have been recorded in case of the "sremska" sausage. *L. monocytogenes* was not found in the finished product (after 28 days) owing to the use of protective strains (reduction 4.0 – 4.5 $\log_{10}$), but it survived in the control sausage (< 2 $\log_{10}$ CFU/g).

In Croatian sausages *L. monocytogenes* population was reduced below the detection limit in sausages with *Lb. sakei* I155 already on day 14, while in control samples and sausages with *Lb. sakei* I151 and I154 by the end of ripening (28th day). Study results suggest the contribution of bacteriocinogenic cultures of *Lb. sakei* to safety of fermented sausages due evidently enhanced reduction of inoculated *L. monocytogenes* population.

Survival and growth of *L. monocytogenes* in fermented sausages largely depend on the sausage type, conditions during fermentation (micro- and macroclimate), starter cultures used and adaptability of pathogen to meat substrate (Encinas et al., 1999; Thévenot et al. 2005ab). Within the concept of protective technologies in the production of fermented sausage, the use of protective bacteriocinogenic cultures has shown to be an additional protective factor against *L. monocytogenes* (Table 3).

Table 3. Examples of implementation of bacteriocinogenic LAB cultures and bacteriocins in fermented sausages inoculated with L. monocytogenes

Culture/bacteriocin	Purpose	Effect	Reference
Pediococcus acidilactici	Use in control of *L. monocytogenes* in dry fermented sausages	Reduction of pathogen during fermentation and drying due pediocin activity in situ	Foegeding et al. (1992)
Lactobacillus sakei and *Lactobacillus curvatus*	Implementation in two different type of sausages inoculated with *L. monocytogenes*	Reduction degree of *Listeria* depends on the production technology and strain adaptation	Hugas et al. (1997)
Pediococcus acidilactici PA-2 and *Lactobacillus bavaricus* DF126	Application in fermented sausages inoculated with different level of *L. monocytogenes*	Level of reduction is corelated to initial count of *Listeria.* LAB cultures enhanced the decrease of pathogen.	Lahti et al. (2001)
Lactobacillus plantarum 423 and *Lactobacillus curvatus* DF126	Control of *L. monocytogenes* in ostrich fermented sausages	2-3 $\log_{10}$ reduction of pathogen within 9 days, folowed with increase to initial count due loss of bacteriocin activity or resistance	Dicks et al. (2004)
Lactococcus lactis subsp. *lactis* M	Use in fermented sausages «marguez» with nitrites and *L. monocytogenes*	Enhanced reduction of pathogen with presence of culture. Nitrites reduced antilisterial activity of bacteriocin.	Benkerroum et al. (2003)
Lactococcus lactis subsp. *lactis* LMG21206 and *Lactobacillus curvatus* LBPE	Production of fermented sausages inoculated with *L. monocytogenes* (2-3 $\log_{10}$ CFU/g)	Elimination of *Listeria* in earlier phase compared with control or sausages with non-bacteriocinogenic culture	Benkerroum et al. (2005)
Enterocin CCM 4231	Use in production of Hornad salami inoculated with *L. monocytogenes*	3 $\log_{10}$ lower count of pathogen in final product than in control (4:7 $\log_{10}$ CFU/g)	Laukova et al. (1999)
Enterocin AS-48	Production of sausages inoculated with *L. monocytogenes* and different concentration of bacteriocin	Reduction degree of *Listeria* is corelated with enterocin concentration	Ananou et al. (2005)
Lactobacillus sakei	Production of traditional fermented sausage from North-West Croatia inoculated with high *L. monocytogenes* numbers (4-5 $\log_{10}$ CFU/g)	Reduction of *L. monocytogenes* by 2-3.5 $\log_{10}$	Zdolec et al. (2007a)

Conclusion

Animal and food-related lactobacilli have a significant role within the concept of integrated agri-food production. In food-producing animals they are beneficial microbes regarding probiotic properties contributing to animal's production and health. Food produced by the same animals (like milk, meat) can be processed to fermented products owing to the same bacteria – lactobacilli. Fermented food products are products with "health claims", meaning that they could provide benefits for consumer's health.

Selection of potential functional starter cultures for meat applications should include testing -both technological and safety properties of the strain. Dominant autochthonous strains in lactobacilli population which are isolated from the specific fermented meat product should be considered as potential functional starter cultures for the same product. These strains are most adapted to specific intrinsic ecology of specified product and technological procedures implemented. However, domination of some LAB species/strains could result in products of lower quality and/or safety. For this reason selection procedure of potential starters should involve all potentially "risky" criteria such as toxigenicity, acquired transmissible antimicrobial resistance or technologically unacceptable pathways (production of gases, acetic acid etc.). Regarding antimicrobial resistance, it would be of particular importance to monitor the presence of same molecular markers in animal and food lactobacilli because animals would be the first link of spreading the resistance determinants through agri-food chain. Preventing the occurrence or reducing the number of resistant commensal bacteria in animals should be the prerequisite for easier selection of potential functional starter cultures from food.

References

Adams, M. R., and Nicolaides, L. (1997). Review of the sensitivity of different foodborne pathogens to fermentation. *Food Control, 8*, 227-239.

Alagić, D., Kozačinski, L., Zdolec, N., Njari, B., Filipović, I., Ekert Kabalin, A., and Alagić-Ćorić, G. (2010). Characterisation of horse meat sausage produced in Croatia. In *5th Central European Congress on Food* (pp. 50-56). Bratislava, Slovakia, May 19– 22.

Albano, H., van Reenen, C. A., Todorov, S. D., Cruz, D., Fraga, L., Hogg, T., Dicks, L. M. T., and Teixeira, P. (2009). Phenotypic and genetic heterogeneity of lactic acid bacteria isolated from ''Alheira", a traditional fermented sausage produced in Portugal. *Meat Science, 82,* 389-398.

Ananou, S., Garriga, M., Hugas, M., Mawueda, M., Martinez-Bueno, M., Galvez, A., and Valdivia E. (2005). Control of *Listeria monocytogenes* in model sausages by enterocin AS-48. *International Journal of Food Microbiology, 103*, 179–190.

Babić, I., Markov, M., Kovačević, D., Trontel, A., Slavica, A., Đugum, J., Čvek, D., Svetec, I. K., Posavec, S., and Frece, J. (2011). Identification and characterization of potential autochthonous starter cultures from a Croatian "brand" product "slavonski kulen". *Meat Science, 88,* 517-524.

Benkerroum, N., Daoudi, A., and Kamal, M. (2003). Behaviour of *Listeria monocytogenes* in raw sausages (merguez) in presence of a bacteriocin-producing lactococcal strain as a protective culture. *Meat Science, 63*, 479–484.

Benkerroum, N., Daoudi, A., Hamraoui, T., Ghalfi, H., Thiry, C., Duroy, M., Evrart, P., Roblain, D., and Thonart, P. (2005). Lyophilized preparations of bacteriocinogenic *Lactobacillus curvatus* and *Lactococcus lactis* subsp. *lactis* as potential protective adjuncts to control *Listeria monocytogenes* in dry-fermented sausages. *Journal of Applied Microbiology, 98,* 56–63.

Bernardeau, M., Vernoux, J. P., Dubernet, S. H., and Guéguen, M. (2008). Safety assessment of dairy microorganisms: the *Lactobacillus* genus. *International Journal of Food Microbiology, 126*, 278-285.

Bonomo, M. G., Ricciardi, A., and Salzano, G. (2011). Influence of autochthonous starter cultures on microbial dynamics and chemical-physical features of traditional fermented sausages of Basilicata region. *World Journal of Microbiology and Biotechnology, 27*, 137-146.

Caplice E., and Fitzgerald, G. F. (1999). Food fermentation: role of microorganisms in food production and preservation. *International Journal of Food Microbiology, 50*, 131-149.

Cataloluk, O., Gogebakan B. (2004). Presence of drug resistance in intestinal lactobacilli of dairy and human origin in Turkey. *FEMS Microbiology Letters, 236*, 7-12.

Charteris, W. P., Kelly, P. M., Morelli, L., and Collins, J. K. (1998). Antibiotic susceptibility of potentially probiotic *Lactobacillus* species. *Journal of Food Protection, 61*, 1636-1643.

Condon, S. (1983). Aerobic metabolism of lactic acid bacteria. *Irish Journal of Food Science and Technology, 7*, 15-25.

Coppola, R., Succi, M., Tremonte, P., Reale, A., Salzano, G., and Sorrentino, E. (2005). Antibiotic susceptibility of *Lactobacillus rhamnosus* strains isolated from Parmigiano Reggiano cheese. *Lait, 85*, 193-204.

Cvrtila, Ž., Hadžiosmanović, M., Savić, V., Kozačinski, L., □dolec, N., and Filipović, I. (2008). Identification of *Lactobacillus* spp. from fermented sausages by PCR method. In *4th Central European Congress on Food* (pp. 45-49). Cavtat, Croatia, May 15 - 17.

Čaklovica, F., Alagić, D., Smajlović, M., Kozačinski, L., Cvrtila, Ž., Vesković-Moračanin, S., Gasparik Reichardt, J., and □dolec, N. (2006). Effect of selected LAB on *L. monocytogenes* during production of traditionally fermented sausages. *Meso, 8*, 140-143.

Danielsen, M., and Wind, A. (2003). Susceptibility of *Lactobacillus* spp. to antimicrobial agents. *International Journal of Food Microbiology, 82*, 1-11.

De Valk, H., Jacquet, C., Goulet, V., Vaillant, V., Perro, A., Simon, F., Desenclos, J. C., and Martin P. on behalf of the Listeria surveillance feasibility study participants (2005). Surveillance of listeria infections in Europe. *European Surveillance, 10*, 251–255.

De Vuyst, L. and Vandamme, E. J. (1994). Bacteriocins of Lactic Acid Bacteria. Blackie Academic and Professional, London.

Devlieghere, F., Vermeiren, L., and Debevere, J. (2004). New preservation technologies: Possibilities and limitations. *International Dairy Journal, 14,* 273-285.

Dicks, L. M. T., Mellett, F. D., and Hoffman L. C. (2004). Use of bacteriocin-producing starter cultures of *Lactobacillus plantarum* and *Lactobacillus curvatus* in production of ostrich meat salami. *Meat Science, 66,* 703–708.

Dobranić, V., Pleadin, J., □dolec, N., and Filipović, I. (2011). Chemical composition and microbial populations of Slavonian sausage. In *Days of veterinary medicine* (pp. 97). Ohrid, Macedonia, September 9-11.

Drosinos, E. H., Mataragas, M., Vesković-Moračanin, S., Gasparik-Reichardt, J., Hadžiosmanović, M., and Alagić, D. (2006). Quantifying nonthermal inactivation of Listeria monocytogenes in European fermented sausages using bacteriocinogenic lactic acid bacteria or their bacteriocins: a case study for risk assessment. *Journal of Food Protection, 69*, 2648–2663.

Drosinos, E. H., Paramithiotis, S., Kolovos, G., Tsikouras, I., and Metaxopoulos, I. (2007). Phenotypic and technological diversity of lactic acid bacteria and staphylococci isolated from traditionally fermented sausages in Southern Greece. *Food Microbiology, 24,* 260-270.

Encinas, J. P., Sanz, J. J., Garcia-Lopez, M. L., and Otero, A. (1999). Behavior of Listeria spp. in naturally contaminated chorizo (Spanish fermented sausage). *International Journal of Food Microbiology, 46,* 167–171.

Foegeding, P. M., Thomas, A. B., Pilkington, D. H., and Klaenhammer, T. R. (1992). Enhanced control of Listeria monocytogenes by in situproduced pediocin during dry fermented sausage production. *Applied and Environmental Microbiology, 58*, 884–890.

Frece, J., Markov, K., and Kovačević, D. (2010a). Determination of indigenous microbial populations, mycotoxins and characterization of potential starter cultures in Slavonian kulen. *Meso, 12,* 92-98.

Frece, J., Pleadin, J., Markov, K., Perši, N., Dukić, V., Čvek, D., and Delaš, F. (2010b). Microbe population, chemical composition and mycotoxins in sausages from Varaždin County. *Veterinarska stanica, 41*, 189-198.

Frece, J., Markov, K., Čvek, D. and Kovačević, D. (2010c). Staphylococci as potential indigenous starter cultures from "Slavonian kulen". *Meso, 12*, 150-155.

Gasparik-Reichardt, J., Toth, Sz., Cocolin, L., Drosinos, E.H., Hadžiosmanović, M., Čaklovica, F., Turubatović, L., and Petrohilou, I. (2006). Safety of traditional fermented sausages: research on protective cultures and bacteriocins. In *52nd ICoMST "Harnessing and exploiting global opportunities"* (pp. 335-336). Dublin, Ireland, August 13-18.

Hadžiosmanović, M., Gasparik-Reichardt, J., Smajlović, M., Vesković-Moračanin, S., and □dolec, N. (2005). Possible use of bacteriocins and starter cultures in upgrading of quality and safety of traditionally fermented sausages. *Tehnologija mesa, 46*, 194-211.

Holzapfel, W. H., Geisen, R., and Schillinger, U. (1995). Biological preservation of foods with reference to protective cultures, bacteriocins and food-grade enzymes. *International Journal of Food Microbiology, 24*, 343–362.

Hugas, M., Neumeyer, B., Pagès, F., Garriga, M., and Hammes, W.P. (1997). Comparison of bacteriocin-producing lactobacilli on Listeria growth in fermented sausages. *Fleischwirtschaft International, 76,* 649–652.

Hugas, M. (1998). Bacteriocinogenic Lactic Acid Bacteria for the Biopreservation of meat and Meat Products. *Meat Science, 49 (Suppl 1),* 139-150.

Hussein, K., Marcinčák, S., Máté, D., Kožárová, I., Sokol, J. and □dolec, N. (2005). Use of Premi®Test for detection of sulphonamide residues in chicken eggs. *Acta Veterinaria – Beograd, 55,* 493-500.

Hutkins, R. W. (2006). Meat fermentation. In R. W. Hutkins (Ed.), *Microbiology and technology of fermented foods* (pp. 207-232). Blackwell Publishing.

Iacumin, L., Comi, G., Cantoni, C., and Cocolin, L. (2006). Ecology and dynamics of coagulase-negative cocci isolated from naturally fermented Italian sausages. *Systematic and Applied Microbiology, 29,* 480-486.

Jack, R. W., Tagg, J. R. and Ray, B. (1995). Bacteriocins of Gram-positive bacteria. *Microbiology Review, 59,* 171-200.

Jaworska, D., Neffe, K., Kołozyn-Krajewska, D., and Dolatowski, □. (2011). Survival during storage and sensory effect of potential probiotic lactic acid bacteria *Lactobacillus acidophilus* Bauer and *Lactobacillus casei* Bif3'/IV in dry fermented pork loins. *International Journal of Food Science and Technology, 46,* 2491-2497.

Kastner, S., Perreten, V., Bleuler, H., Hugenschmidt, G., Lacroix, C., and Meile, L. (2006). Antibiotic susceptibility patterns and resistance genes of starter cultures and probiotic bacteria used in food. *Systematic and Applied Microbiology, 29*, 145-155.

Katla, A. K., Kruse, H., Johnsen, G., and Herikstad, H. (2001). Antimicrobial susceptibility of starter culture bacteria used in Norwegian dairy products. *International Journal of Food Microbiology, 67,* 147-152.

Klein, G., Pack, A., Bonaparte, C., and Reuter, G. (1998). Taxonomy and physiology of probiotic lactic acid bacteria. *International Journal of Food Microbiology, 41*, 103-105.

Kozačinski, L., and Hadžiosmanović, M. (2001). The occurrence of *Listeria monocytogenes* in home-made dairy products. *Tierärztlische Umschau, 56*, 590–594.

Kozačinski, L, □dolec, N., Hadžiosmanović, M., Cvrtila, Ž., Filipović, I., and Majić, T. (2006). Microbial flora of the Croatian traditionally fermented sausage. *Archiv für Lebensmittelhygiene 57*, 141-147.

Kozačinski, L., Drosinos, E. H., Čaklovica, F., Cocolin, L., Gasparik-Reichardt, J., and Vesković-Moračanin, S. (2008). Investigation of microbial association of traditionally fermented sausages. *Food Technology and Biotechnology, 46*, 93-106.

Lahti, E., Johansson, T , Honkanen Buzalski, T., Hill, P., and Nurmi, E. (2001). Survival and detection of *Escherichia coli* O157:H7 and *Listeria monocytogenes* during the manufacture of dry sausage using two different starter cultures. *Food Microbiology, 18*, 75–85.

Laukova, A., Czikkova, S., Laczkova, S., and Turek, P. (1999). Use of enterocin CCM 4231 to control Listeria monocytogenes in experimentally contaminated dry fermented Honrád salami. *International Journal of Food Microbiology, 52,* 115–119.

Laukova, A., Simonova, M., and Strompfova, V. (2010). *Staphylococcus xylosus* S03/1M/1/2, bacteriocin-producing meat starter culture or additive. *Food Control, 21*, 970-973.

Leistner, L., and Gorris, L. G. M. (1995). Food preservation by hurdle technology. *Trends in Food Science and Technology, 6*, 41-46.

Leroy, F., Verluyten, J., and De Vuyst, L. (2006). Functional meat starter cultures for improved sausage fermentation. *International Journal of Food Microbiology, 106*, 270–285.

Lin, C. F., Fung, □. F., Wu, C. L., and Chung, T. C. (1996). Molecular characterization of a plasmid-borne (pTC82) chloramphenicol resistance determinant (cat-TC) from *Lactobacillus reuteri* G4. *Plasmid, 36*, 116–124.

Liu, G., Griffiths, M. W., Shang, N., Chen, S. W., and Li, P. L. (2010). Applicability of bacteriocinogenic *Lactobacillus pentosus* 31-1 as a novel functional starter culture or coculture for fermented sausage manufacture. *Journal of Food Protection, 73*, 292-298.

Loncarevic, S. (1998). *Listeria monocytogenes* with special reference to food products and human listeriosis. Doctoral thesis. Swedish Univerity of Agricultural Sciences, Uppsala.

Lücke, F.-K. (2000). Utilization of microbes to process and preserve meat. *Meat Science, 56,* 105-115.

*Marcinčák, S., Hussein, K., Máté, D., Kožárová, I., Sokol, J., and Zdolec, N. (*2006). Premi®Test - a screening test for detecting sulphadimidine residues in eggs of laying hens. *Medycyna weterynaryjna, 62*, 159-161.

Markov, K., Frece, J., Čvek, D., Trontel, A., Slavica, A., and Kovačević, D. (2010). Dominant microflora of horse meat fermented sausages. *Meso, 12*, 217-221.

Mathur, S., and Singh, R. (2005). Antibiotic resistance in food lactic acid bacteria - a review. *International Journal of Food Microbiology, 105*, 281-295.

Montville, T. J. and Matthews, K. R. (2005). Biologically Based Preservation and Probiotic Bacteria. In *Food Microbiology – an introduction* (pp. 311-320). Washington, US: ASM Press.

Putman, M., van Veen, H. W., Degener, J. E., and Konings, W. N. (2001). The lactococcal secondary multidrug transporter LmrP confers resistance to lincosamides, macrolides, streptogramins and tetracyclines. *Microbiology, 147*, 2873-2880.

Rantsiou, K., and Cocolin, L. (2006). New developments in the study of the microbiota of naturally fermented sausages as determined by molecular methods: A review. *International Journal of Food Microbiology, 108,* 255-267.

Roasto, M., Praakle-Amin, K., Meremäe, K., Tamme, T., Hörman, A., Lillenberg, M., Hänninen, M.-L. (2009). Food control and research on *Campylobacter* spp. in Estonia. *Archiv für Lebensmittelhygiene, 60*, 109-115.

Ross, R. P., Morgan, S., and Hill, C. (2002). Preservation and fermentation: past, present and future. *International Journal of Food Microbiology, 79,* 3-16.

Ruzauskas, M., Virgailis, M., Šiugždinienė, R., Sužiedėlienė, E., Šeputienė, V., Daugelavičius, R., □ienius, D., Šengaut, J., and Pavilonis, A. (2009). Antimicrobial resistance of *Enterococcus* spp. isolated from livestock in Lithuania. *Veterinarski arhiv, 79*, 439-449.

Sriphochanart, W., and Skolpap, W. (2010). Characterization of proteolytic effect of lactic acid bacteria starter cultures on thai fermented sausages. *Food Biotechnology, 24*, 293-311.

Stiles, M.E., Holzapfel W. H. (1997). Lactic acid abcateria of foods and their current taxonomy. *International Journal of Food Microbiology, 36*, 1-19.

Svetoch, E. A., and Stern, N. J. (2010). Bacteriocins to control *Campylobacter* spp. in poultry – A review. *Poultry science, 89,* 1763–1768.

Tagg, J. R., Dajani, A. S., and Wannamaker, L. W. (1976). Bacteriocins of Gram-positive bacteria. *Bacteriology Review, 40*, 722-756.

Talon, R., and Leroy, S. (2011). Diversity and safety hazards of bacteria involved in meat fermentations. *Meat Science, 89*, 303-309.

Teuber, M. (2001). Veterinary use and antibiotic resistance. *Current Opinion in Microbiology, 4,* 493-499.

Thévenot, D., Delignette-Muller, M. L., Christieans, S., and Vernozy-Rozand, C. (2005a). Prevalence of *Listeria monocytogenes* in 13 dried sausage processing plants and their products. *International Journal of Food Microbiology, 102*, 85–94.

Thévenot, D., Delignette-Muller, M. L., Christieans, S., and Vernozy-Rozand, C. (2005b). Fate of *Listeria monocytogenes* in experimentally contaminated French sausages. *International Journal of Food Microbiology, 101*, 189–200.

Vatanyoopaisarn, S., Prapatsornwattana, K., Kuhakongkeat, T., and Phalakornkule, C. (2011). Potential use of lactic acid bacteria with bacteriocin-like activity against *Staphylococcus aureus* as dual starter cultures in thai fermented sausage "Sai Krok Prew". *International Food Research Journal, 18,* 680-687.

□dolec, N., Hadžiosmanović, M., Kozačinski, L., Cvrtila, Ž., and Filipović, I. (2006). Residual contaminants in milk. *Mljekarstvo, 56*, 191-202.

□dolec, N. (2007). Influence of protective cultures and bacteriocins on safety and quality of fermented sausages. Dissertation. Faculty of veterinary medicine, University of □agreb.

□dolec, N., Hadžiosmanović, M., Kozačinski, L., Cvrtila, Ž., Filipović, I., Marcinčak, S., Kuzmanović, Ž., and Hussein, K. (2007a). Protective effect of *Lactobacillus sakei* in fermented sausages. *Archiv für Lebensmittelhygiene, 58*, 152-155.

□dolec, N., Hadžiosmanović, M., Kozačinski, L., Cvrtila, Ž., Filipović, I., Leskovar, K., Vragović, N., and Budimir, D. (2007b). Home-made fermented sausages – microbiological quality. *Meso, 9*, 318-324.

□dolec, N., Kozačinski, L., Hadžiosmanović, M., Cvrtila, Ž., and Filipović, I. (2007c). Inhibition of *Listeria monocytogenes* growth in dry fermented sausages. *Veterinarski arhiv, 77*, 507-514.

□dolec, N., Hadžiosmanović, M., Kozačinski, L., Cvrtila, Ž., Filipović, I., and Škrivanko, M. (2008a). Influence of protective cultures on *Listeria monocytogenes* in fermented sausages: a review. *Archiv für Lebensmittelhygiene, 59*, 60-64.

□dolec, N., Hadžiosmanović, M., Kozačinski, L., Cvrtila, Ž., Filipović, I., Škrivanko, M., and Leskovar, K. (2008b). Microbial and physicochemical succession in fermented sausages produced with bacteriocinogenic culture of *Lactobacillus sakei* and semi-purified bacteriocin mesenterocin Y. *Meat Science, 80*, 480-487.

□dolec, N., Kozačinski, L., Njari, B., Filipović, I., Hadžiosmanović, M., Mioković, B., Kuzmanović, Ž., Mitak, M., and Samac, D. (2009). The antimicrobial effect of lactobacilli on some foodborne bacteria. *Archiv für Lebensmittelhygiene, 60*, 115-119.

□dolec, N., Filipović, I., Cvrtila Fleck, Ž., Marić, A., Jankuloski, D., Kozačinski, L., and Njari, B. (2011). Antimicrobial susceptibility of lactic acid bacteria isolated from fermented sausages and raw cheese. *Veterinarski arhiv, 81*, 133-141.

□hou, J. S., Pillidge, C. J., Gopal, P. K., and Gill, H. S. (2005). Antibiotic susceptibility profiles of new probiotic *Lactobacillus* and *Bifidobacterium* strains. *International Journal of Food Microbiology, 98*, 211-217.

Yuksekdag, □. N., and Beyatli, Y. (2008). Antimicrobial activity and antibiotic susceptibility of lactic acid bacteria isolated from different dairy products. *Archiv für Lebensmittelhygiene, 59*, 216-220.

In: Lactobacillus: Classification, Uses and Health Implications ISBN: 978-1-62081-151-1
Editors: Alba I. Perez Campos and Arturo Leon Mena

Chapter XIII

Environmental Applications of *Lactobacillus* for Protein Recovery and Biodegradation of Recalcitrant Chemical Compounds

Phisit Seesuriyachan,[1*] Charin Techapun[1] and Ken Sasaki[2]

[1]Division of Biotechnology, Faculty of Agro-Industry, Chiang Mai University, Chiang Ma, Thailand

[2]Materials Science and Engineering, Faculty of Engineering, Hiroshima Kokusai Gakuin University, Nakano, Akiku, Hiroshima, Japan

Abstract

In the past few decades concern about environmental issues has increased considerably and the implementation of water management has shifted towards the incorporation of new technology for the protection of ecosystems and sustainable development. Treatment of contaminated wastewater using biological processes has been widely applied. At this point, a combination of conventional knowledge and advances in technology has generated many new biological processes for wastewater treatment.

In this chapter, a biological treatment for dairy wastewater using lactic acid bacteria (LAB), *Lactobacillus* in particular, causing solid-liquid separation via biocoagulation and flocculation was introduced. A method of applying fermented lactic acid to settle protein in dairy effluent prior to the secondary treatment was demonstrated and a significant lowering in the organic load of these effluents was achieved. It was found that *Lactobacilli* are considerably more tolerant to low cytoplasmic pH than other groups of lactic acid bacteria. Sustained lactic fermentation was achieved by means of an ingenious

* Corresponding author: Dr. Phisit Seesuriyachan, Division of Biotechnology, Faculty of Agro-Industry, Chiang Mai University, Chiang Mai 50100, Thailand. Tel+66-53-948201, Fax+66-53-948201, E-mail: phisit.s@cmu.ac.th, phisit.seesuriyachan@gmail.com.

fill-react-drain-idle sequence. This is a highly efficient new semi-continuous system, with a micro-aerobic sequencing batch reactor (micro-aerobic SBR), lactic fermentation was achieved when *Lactobacillus* was used as a biocatalyst for dairy wastewater protein recovery and biodegradation of chemically recalcitrant compounds. This is alternative to other more costly chemical and biological treatments currently in use. Additionally, protein recovered from this process can be used as a probiotic feed replacing feed antibiotics and promoting health worldwide.

This chapter gives the understanding of lactic acid bacteria in environmental applications by providing data on its potential for biodegrading recalcitrant compounds and toxic substances, such as azo dyes, by *Lactobacillus*.

Keywords: Lactic acid, semi-continuous system, micro-aerobic sequencing batch reactor (micro- aerobic SBR), *Lactobacillus*, pre-treatment, decolorization

Introduction

Promoting environmental sustainability through renewable resources has increased the interest in the recovery of fermentation products from wastewater systems. In old-fashioned wastewater treatment systems, all chemical substances in wastewaters needed to be treated to convert them to carbon dioxide and water before the treated wastewater was discharged [1]. From this method, all useful organic substances are lost. However, the development of biotechnological techniques and processes of fermentation and separation-purification can now reduce the production cost while also optimizing process efficiency by recovering useful substances from wastewaters [2-5].

Raw material sources with an animal origin, such as whey or dairy process wastewater with a high lactose and protein content, are well suited to as animal feed supplements in addition to other applications. Wastewaters from the dairy industry are usually generated intermittently and have a high organic content. The flow rate of the influent fluctuates considerably and also creates severe problems for the secondary wastewater treatment plants. The dairy industry produces various products such as milk, butter, yoghurt, ice-cream and various types of desserts and cheese. The characteristics of these effluents vary greatly in terms of both organic loading rate and flow rate depending on the processes and methods of operation used [6]. There are three types of wastewater discarded from the dairy factory: processing water, cleaning wastewaters, and sanitary wastewater [7]. It should be noticed that Biological Oxygen Demand (BOD) content of dairy wastewater is more than 250 times greater than that of sewage [8]. This high organic content causes many problems in secondary wastewater treatment systems [9]. Many reports have shown that nitrogen compounds and other rich organic materials originate from dairy industry wastewater and may cause of eutrophication even if the wastewater is treated in the secondary wastewater treatment system [6, 7, 10-13]. Dairy wastewater has high carbohydrate contents, mainly in the form of lactose, as well as less biodegradable proteins, and lipids. Those compounds can cause oxygen depletion if they are discarded into natural resources. In cheese-processing wastewater 97.7 % of total Chemical Oxygen Demand (COD) is accounted for by lactose, lactate, protein and fat making dairy wastewater a complex substrate [7].

Pre-treatment of the dairy wastewater prior to the secondary wastewater treatment is necessary. Physical screening is the first step to remove large particles or debris that may cause damage to pumps and clogging downstream [6]. Next is a pH adjustment step in which the pH of dairy wastewater is adjusted to within a range of 6.5 – 8.5, which is suitable for biological treatment. In the dairy process, alkaline detergents are generally used for removing lipids and proteins, while acidic detergents are used for the removal of mineral deposits and acid-based sanitizers [14]. Then, the wastewater is usually pumped in an equalization tank for flow and composition balancing as well as to adjust volume, strength, temperature and nutrient levels. This step can be combined with pH amelioration in the same tank [15, 16]. In this step is very important to protect the wastewater treatment process from shock loading. It is often also essential to remove high amounts of fats, oil and grease from dairy wastewater of particular factories which process whole milk, cheese, butter, whey, or those that have a milk bottling line [17]. One of three systems (gravity traps, air and dissolved air floatation, or enzyme hydrolysis) is selected for pre-treatment of dairy wastewater depending on the characteristics [18-20].

A new process of biological pre-treatment for recovery of colloidal proteins from dairy wastewater using lactic acid fermentation with LAB, was introduced only in the past decade [21-23]. This new process effectively reduces organic loads in wastewater, while simultaneously producing lactic acid as a by-product. Casein is a major protein forming nearly 80% of the total protein content in milk [24]. Lactic acid bacteria is used to convert sugars to organic acids in dairy wastewater. Consequently the casein content coagulates at its isoelectric point at a pH of 4.5 – 4.7. However, this process under in a non-sterile environment has not been studied well. A synopsis of this new process for protein recovery and lactic acid production will be to clearly explain in this chapter.

Lactic acid bacteria are a group of bacteria that produce lactic acid as a product of their metabolism. They are fastidious microorganisms and require rich media for growth. Lactic acid bacteria have been used in fermented foods for preservation for over 4,000 years and are Generally Recognized as Safe (GRAS). Lactic acid bacteria can compete with other harmful microorganisms by reducing the pH to as low as to 4 with lactic acid. They are comprised of about 20 genera of which *Lactococcus, Lactobacillus, Streptococcus, Leuconostoc, Pediococcus, Aerococcus, Carnobacterium, Enterococcus, Oenococcus, Tetragenococcus, Vagococcus*, and *Weissella* are the main taxa [25, 26]. The largest genera is *Lactobacillus* comprised of over 70 species [25]. They are microaerophillic bacteria that do not need aeration during cultivation and can grow well under anaerobic conditions in a temperature range of 5 to 45 ^{o}C with an optimum pH of 5.5 – 6.5. Lactic acid bacteria can be divided into 2 groups: (1) homofermentative, which produces 85 % lactic acid from glucose via the Embden-Meyerhof-Parnas pathway, and (2) heterofermentatives, which produces only 50 % lactic acid, the rest being made up of ethanol and carbon dioxide via the phosphoketolase pathway. The proportions of these compounds depend on the redox potential in the bacterial growth environment [25, 27]. Lactic acid has properties that are toxic to microbial cells, particularly other pathogenic microorganisms. Seppo and Wright concluded that undissociated organic acids, lactic or acetic, can pass through cell membranes and dissociate inside the bacterial cell. The inhibition mechanism of lactic acid is dependent on the solubility of these undissociated acids within the cytoplasmic membrane. It is well known that bacteria attempt to maintain their cytoplasmic pH at a certain level, normally ranging from mildly acidic to neutral [27]. It has also been observed that *Lactobacilli* are considerably more tolerant to low cytoplasmic pH than other groups of lactic acid bacteria (*Enterococci, Leuconostocs, Streptococci*). Some strains of *Lactobacilli* can grow at a pH of 4.2 – 4.4, whilst *Enterococci, Lactococci,* and *Streptococci* cannot tolerate a pH below 5. The acidification of cytoplasmic pH causes the failure of proton motive forces and decreases in the energy pool inside the bacterial cell slowing the growth of lactic acid bacteria [27].

Application of *Lactobacillus* for Protein Recovery

The growth of lactic acid bacteria or even other acid producing microorganisms feature the consumption of substrate which are converted to organic acids (e.g. lactic, acetic, propionic, succinic acids etc.) The major bottlenecks in the batch fermentation are a long lag phase, slow growth strong end product inhibition. Many studies have focused on eliminating the cause of product inhibition and improving the productivity in acidic conditions by using acid tolerant strains, cells immobilization, high cell density, cell recycling, and using special bioreactors units employing simultaneous extractive fermentation. However, a simple mode of fermentation in batch-fermentation or semi-continuous fermentation with or without extractive membrane units can improve the production rate and yield of organic acids. The strategy of the semi-continuous process is to add substrates at frequent intervals with or without taking out the fermented broth from the bioreactor. In this process, substrates in the fermentation broth nearly reach to zero by the end of the first batch's fermentation before the addition of the next batch. The feeding supplies all the requirements of nutrients to microorganisms continuously. The feed can be added into the bioreactor with a constant rate or several rates when the growth rate of microorganisms is high, in order to satisfy the nutrient requirements of the fermentation system and also protect against the accumulation of inhibitory products (organic acids). Semi-continuous fermentation producing L-lactic acid has been applied to save on the cost of inoculum preparation with high cell concentrations in unsterilized conditions at mesophillic temperatures. A long term semi-continuous lactic acid fermentation process on a laboratory scale of up to 300 days. The optimum conditions of L-lactate fermentation was at 55 ^{o}C, pH 6 with hydraulic retention time (HRT) 10 days. Although, LAB generally can grow well and survive with a pH below 5, but at pH 6 it has an advantage by decreasing the usage of neutralizing agent (sodium hydroxide) to prevent unexpected bacteria germination [28].

Casein is a major protein, which comprises nearly 80 % of total protein in milk and dairy effluents [24]. Pre-treatments of dairy wastewater were introduced to remove protein by membrane [29], chemical [21, 22] and simultaneous physico-chemical [30, 31] methods thereby reducing the organic load. Nevertheless, these processes had an inherent drawback due to high operating costs from either the use of external acid sources or flocculating agents. An attempt had been made to reuse valuable components, particularly casein, from dairy wastewater, but this has not been possible from an economic point of view. However, the biological process is a suitable cost effective method for removal of organics [9]. Casein recovery from dairy wastewater was introduced and achieved using lactic acid fermentation in a continuous stirred tank reactor [23]. The acid fermentation, however, could not be achieved in an anaerobic complete mix bioreactor because the pH of wastewater could not be dropped to the isoelectric point (pH below 4.5) previous studies [32]. In combination with a biocoagulation process, chemical flocculants were also used in a separation step [33]. This promising pre-treatment process has potential applications in many products, such as animal feed from protein recovery and bioplastics from lactic acid [23, 32].

Lactic acid fermentation as a biocoagulation process in dairy wastewater treatment was studied in order to separate colloidal proteins from the wastewater stream. The use of microbial activity to convert sugars to organic acids in dairy wastewater coagulates casein in the wastewater at its isoelectric point. In table 1, lactic acid fermentation by a continuous system with and without immobilized cells is compared with a semi-continuous system with non-sterile fermentation. It is revealed that a semi-continuous system has a higher productivity and greater longevity with high efficiencies of protein, COD and sugar removal, as well as acid production (90, 75, 75 % and 389 mg/L) compared to the two types of continuous systems with and without immobilization [32]. Superior biocoagulation by LAB using a micro-aerobic sequencing batch reactor (micro-aerobic SBR) was achieved, which is based on a semi-continuous phenomenon for treating dairy wastewater effluent with low sugar content using non-sterile conditions [32, 34]. This process employed a coupled reaction of coagulation and flocculation, or a two-step mixing process by which acids were simultaneously produced. Colloidal proteins were then removed by employing a solid-liquid separation step.

The system of micro-aerobic SBR was started by inoculating 5 L of dairy wastewater in the reactor with 5 g (wet weight) of *Lactobacillus casei* TISTR 1500 in a batch process. Then, the bioreactor was mixed rapidly at the speed of 150 rpm for 5 min, followed by slow mixing at 50 rpm until the end of the experiment.

Table 1. The efficiencies of three types of bioprocess investigated performance [32]

	Continuous fermentation with non-immobilized cells	Continuous fermentation with immobilized cells	Semi – continuous system
Proteins removal (%)	74+18	64+19	90+4
COD removal (%)	59+16	44+13	75+3
Sugar removal (%)	74+15	69+19	75+4
Acid concentration (mg/L)	204+76	196+70	389+37
pH	4.6+1.0	4.9+0.9	3.3+1

p = 0.05; n = 30.

After the 3 day batch mode, the system was switched to the micro-aerobic SBR mode as shown in Figure 1, which is described below. Four liters of supernatant were withdrawn from the bioreactor and replaced with the untreated dairy wastewater. Rapid mixing at the speed of 150 rpm for 5 min, followed by slow mixing at 50 rpm for 23 hr in a lactic acid fermentation process. The treated wastewater with coagulated protein and freely suspended cells of the strain TISTR 1500 was then fed out with a control of desired solid retention time (SRT), before allowing the solution to settle for 50 min.

Finally, the treated wastewater was removed to control the HRT, and the remnant broth was kept idle for 5 min. The overall cycle operation took 24 hr and the next cycle started with restoring the reactor to 5 L of untreated dairy wastewater and the operation was repeated [34]. The SRT is expressed as:

$$SRT = \frac{V_T \cdot X}{Q_W \cdot X_W} \tag{1}$$

where SRT = solid retention time (days), Q_W = wastewater flow rate (1/days), X_W = biomass concentration of settled protein mixed with wet cells of *Lactobacillus casei* TISTR 1500 (mg/L), V_T = total reactor volume (L), X = biomass concentration of settled protein mixed with wet cells of *Lactobacillus casei* inside the fully filled reactor (mg/L).

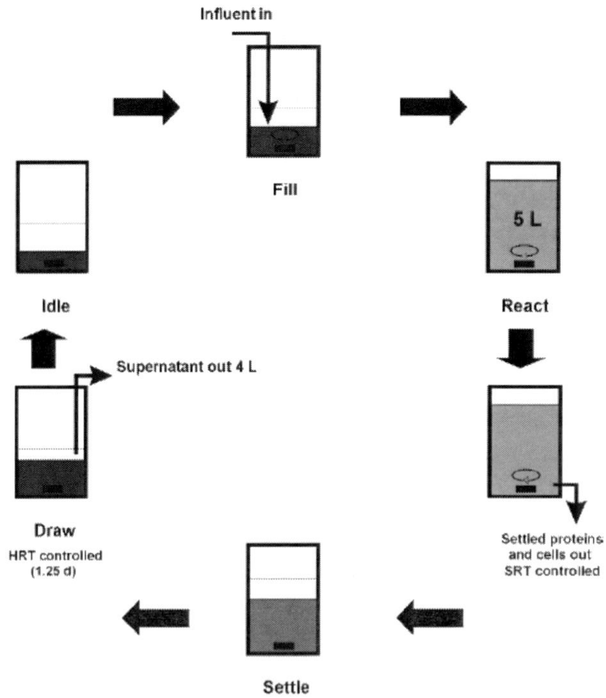

Figure 1. The schematic diagram shows fill-react-settle-draw-idle steps in one cycle of micro-aerobic SBR experiment during non-sterile acid fermentation by *Lactobacillus casei* TISTR 1500 to settle proteins in dairy effluent as a pre-treatment [32].

All sucrose was consumed within 24 hr, and over 90 % of the lactose was utilized within 3 days. Maximum total acidity of 1.16 g/L was produced after 3 days with 1.11 g/L of lactic acid (90.8 %), 0.11 g/L of acetic acid (9.2 %). This was accompanied by a decrease of the dairy process wastewater pH to 3.8, which resulted in protein coagulation and flocculation. The final effluent contained 187 mg/L of proteins, 1.83 g/L of COD, 1.11 g/L of lactic acid,

0.11 g/L of acetic acid, and 0.078 g/L of lactose [34]. High protein recovery and lactic acid production were achieved after 150 cycles for all ranges of investigated SRTs (3 – 15 days), each of which lasted for 30 days with a maintenance of HRT at 1.25 days. Depending on the SRT employed, COD removal was 44 – 74 % (Figure 2A), protein removal was 85 – 91 % (Figure 2A) and total sugar removal was 85 – 88 % (Figure 2A). Total acidity was 1.82 – 2.84 g/L and the pH level was maintained at 3.8 – 4.0 (Figure 2B and 2C, respectively).

Increasing the SRT supported lactic acid fermentation, and the highest acid production of 2.75 ± 0.06 g/L was obtained using an SRT of 15 days [34]. A long, controlled SRT and a switch in pH between acidic and neutral wastewater are critical points for successfully operating the micro-aerobic SBR process. Inoculum size can decrease the required time for achieving maximum acid formation. The restoration of culture volume to 5 L with fresh dairy wastewater with a pH from 5.5 – 6.5, provided revitalization of the slowed fermentation.

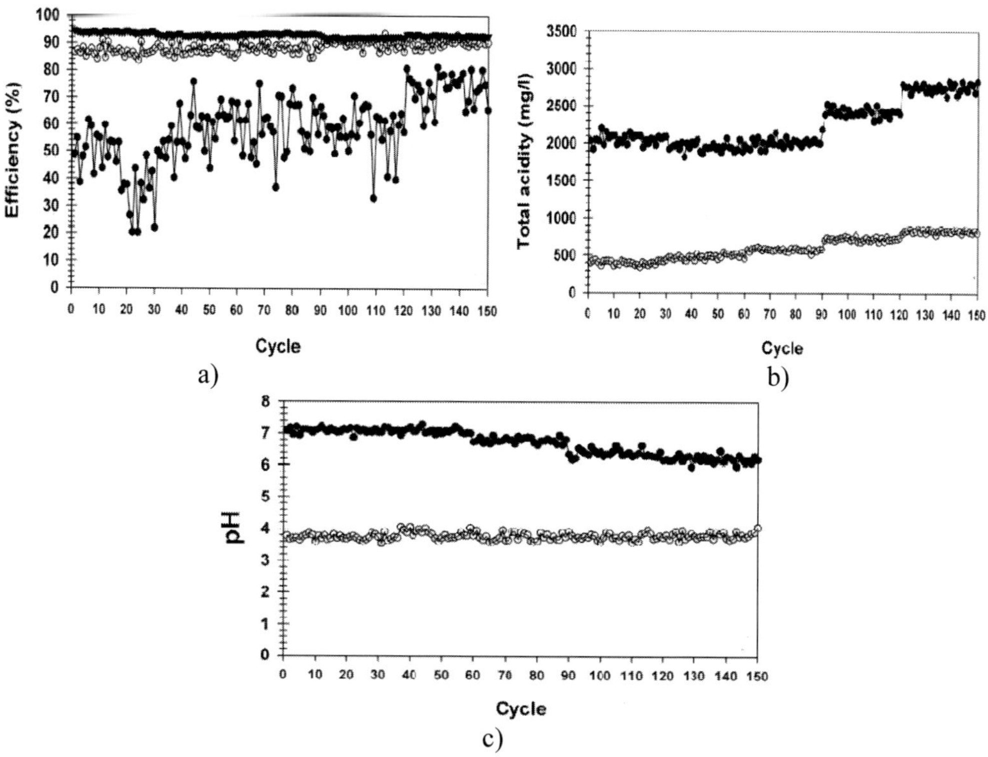

Figure 2. Performance of non-sterile acid fermentation by *Lactobacillus casei* TISTR 1500 using the micro-aerobic SBR in order to coagulate proteins in dairy process wastewater as a pre-treatment; cycles 0-30 (SRT = 3 d); 31-60 (SRT = 6 d); 61-90 (SRT = 9 d); 91-120 (SRT = 12 d), and 121-150 (SRT = 15 d), A) COD removal: (•), protein removal: (○), and sugar removal: (▼), B) total acidity: (•) Influent, (○) Effluent, and C) pH: (•) Influent, (○) Effluent [34].

A number of LAB can accumulate and survive in the system with high SRTs (i.e., 12 and 15 days), while the opposite is true for shorter SRTs. At SRTs of 12 and 15 days; the increases of LAB viability were from 6.63 and 6.83 Log CFU/mL to 7.84 and 7.89 Log CFU/mL as shown in Figure 3A, respectively. For other microorganisms at the SRT 15 days as shown in Figure 3B, the acid condition could inhibit the growth of pathogenic microorganisms in the family *Enterobacteraceae* with 1.86 Log CFU/mL [34].

Some evidence supported our application for lactic acid fermentation under non-sterile conditions if the operation was suitable to the requirements of lactic acid producing microorganisms [35-38].

Chemical coagulation of protein in dairy wastewater was tested with both inorganic acid and organic acid. Using sulfuric acid as a coagulant, 52 – 73 %, 48 – 68 % and 52 – 73 % of COD removal was obtained when Ferric chloride or Alum or Lime and Ferric chloride were used in combination after the adjustment of the wastewater to an acidity below the isoelectric point of casein [21]. In table 2, application of lactic acid with carboxymethyl cellulose (CMC) and chitosan in a continuous biological treatment process resulted in 60% of COD and protein removal at pH 4.3 using *Lactobacillus plantarum* [23]. In the case of *Lactobacillus casei* TISTR 1500, COD and protein removal were 74 %, and 90 %, respectively, which were higher than previous reports elsewhere, even though no flocculating agent was used [34].

When developing a wastewater treatment system, obtaining the lowest construction, installation and operation costs is often a concern.

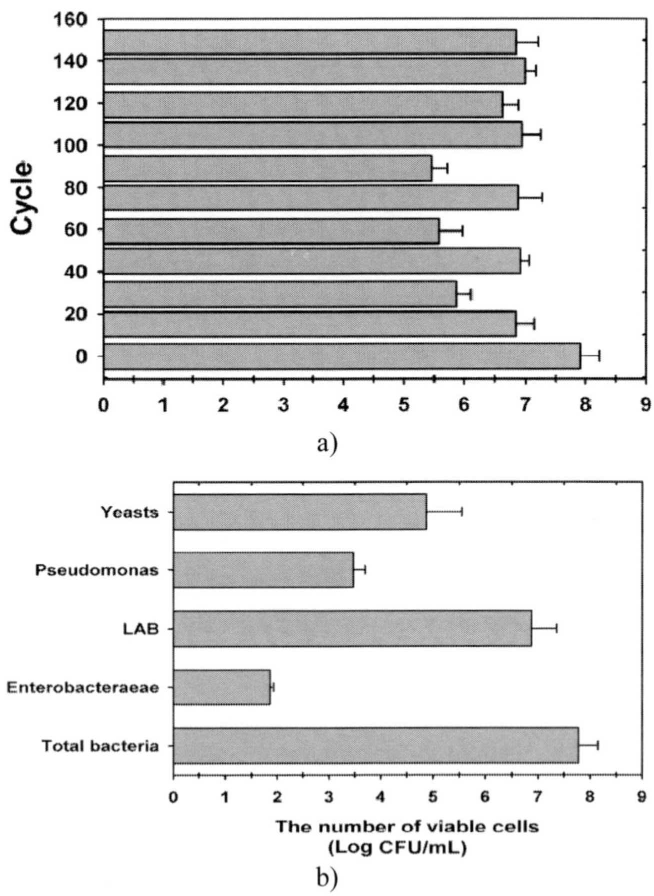

Figure 3. A) The number of viable lactic acid bacteria in the settled proteins during the micro-aerobic SBR operation using five different SRTs; cycles 0-30 (SRT = 3 d), 31-60 (SRT = 6 d), 61-90 (SRT = 9 d), 91-120 (SRT = 12 d), and 121-150 (SRT = 15 d). B) The number of viable cells of microorganisms (Total bacteria, Enterobacteraeae, Lactic acid bacteria, Pseudomonas, and yeasts) in the effluent at the end of non-sterile acid fermentation with the SRT 15 d, using the micro-aerobic SBR [34].

Table 2. Processes of solid-liquid separations of dairy wastewater in comparison to the micro-aerobic SBR process [34]

Reference	Condition of coagulation and flocculation	COD removal (%)	Protein removal (%)
Seesuriyachan et al., (2009) [34]	Non-sterile lactic acid fermentation by *L. casei* TISTR 1500, pH 3.8-4.3	74	90
Rusten et al., (1993) [22]	Lactic acid + CMC, pH 4.3	60	60
Dyrest et al., (1998) [23]	Lactic acid fermentation by *L. plantarum* INF-9F +CMC	65-78	
	Lactic acid fermentation by *L. plantarum* INF-9F +Chitosan	49-82	

It is not possible to apply high cost processes such as membrane recycling bioreactors, hollow fiber reactors or even electrodialysis.

Non-sterile lactic acid fermentation can be a low cost alternative suited to a variety of small scale applications. In batch cultivation, some disadvantages, such as a long lag phase and slow fermentation, cause higher operating costs. Whereas, semi-continuous fermentation solves problems of time labor cost. A high volume of microbial inoculum can help inhibit contamination from other microorganisms.

Applications of *Lactobacillus* for Biodegradation of Recalcitrant Chemical Compounds

As lactic acid bacteria possesses high potential of biodergradation of aromatic compounds, many research studies related to environmental issues, in particular, industrial wastewater treatment involving lactic acid bacteria, have been published through the years. Non-sterilized fermentation is necessary in most environmental applications in order to keep operating costs low.

Peddiococcus pentosacus was used for the decolorization of olive mill wastewater [39] and textile effluent decolorization was also investigated using *Lactobacillus spp*. N3A [40]. In anaerobic conditions, *Lactobacillus plantalum* can reduce quinate to dihydroxycyclohexanecarboxylate and acetic acid, and also oxidizes this compound to catechol involving two key enzymes, NAD-depent dehydrogenase and protocatechuic acid decarboxylase. Also, the strains can degrade oleuropein through hydrolysis, with the reactions of a β-glucosidase and esterase, to form aglycone, hydroxytyrosol and elenoic acid [41]. *Lactobacillus plantalum* LAC26 can hydrolyse β and α-D-glucopyranosides [42]. The degradation of tannic acid by *Lactobacillus plantalum* to methyl gallate, a simple galloyl ester of methanol, vinyl and ethyl derivatives from hydroxycinnamic acids, pyrogallol from gallic acid, has been reported [43].

Additionally, the biodegradation of oleuropein to hydroxytyrosol and protocatechuic acid to catchol with the activity of β-glucosidase and decarboxylase, has been demonstrated [44]. Moreover, the reaction of one or both enzymes (decarboxylases and reductases) activates chemical degradation of p-coumaric, caffeic, ferulic, m-coumaric, gallic and protocatechuic acids [45]. Furthermore, *Lactobacillus plantarum* No. PV-1861 is used as a biocatalyst for the decolorization of molasses wastewater [46]. *Lactobacillus paracollinoides* possesses a high potential to hydrolyse chlorogenic acid to caffeic acid and quinic acids, then caffeic acid is metabolized and forms dehydrocaffeic acid and ethyl catechol.

The strain can reduce the side chain of 3,4-hydroxy and 3-hydroxycinnamic acids by the enzyme decarboxylase to ethyl catechol and ethyl phenol [47]. It has been reported that *Lactobacillus hilgardii* can produce gallic acid, pyrogallol, catechol, protocatechuic acid, *p*-hydroxybenzoic acid, *p*-hydroxybenzldehyde, and *p*-hydroxybenzyl alcohol in a culture medium with gallic acid and also generates catechin, gallic acid, pyrogallol, catechol, *p*-hydroxybenzoic acid, acetovanillone, and homovanillic acid in a medium with catechin [48].

Lactobacillus has a biotransformation property, like a conversion of ferulic acid to vanillin and a reduction of vanillin to vanillin alcohol, p-coumaric to form 4-vinyl phenol [41, 49]. Humic acid can be degraded by *Lactococcus lactis* [50].

In addition, a new approach to bioconversion of phenolic compounds in olive wastewater with the capacities of LAB to reduce oxygen pressure, redox potential and pH, has just been introduced [51]. Wastewater in the presence of high contents of phenolic compounds is toxic and resistant to secondary wastewater treatment plants because of the auto-oxidation of phenolic compounds [51, 52]. Polyphenol hydrolysis in fresh olive mill wastewater was induced by *Lactobacillus plantarum* [53] and factors affecting the reductive decolorization and biodegradation of olive phenolic compounds were evaluated [54]. The biodegradation of high molecular weight phenolic compounds by a process of reductive depolymerization gave a better quality olive mill wastewater with 58 % of color removal, 55 % COD removal and 46 % phenolic compound removal, when the wastewater was diluted ten times [55].

The superior potential of *Lactobacillus plantarum* in biodegradation and biotransformation of recalcitrant chemical compounds is because of its enzymatic activities (benzyl alcohol dehydrogenase, phenolic acid decaboxylase; PAD, reductase, decarboxylase, β-glucosidase, esterase, and tannase) [41, 52].

As mentioned above, the potential of *Lactobacilli* biodegradation of recalcitrant chemicals is very promising as a new environmental application. A bioprocess of micro-aerobic SBR has a decreased efficiency when the system is used to treat dairy wastewater containing azo dyes and the intermediates but the efficiency of the system is compensated with high organic contents from the influent. It is almost possible to use LAB for the treatment of textile azo dyes.

A report has just been published about the treatment of azo food colorants which are normally used to enhance the appearances of food [56]. They pose a potential risk to human health, particularly if they are excessively consumed or are structurally changed through chemical or biological reactions.

The toxicity of those colorants may increase after azo bond breakage, and intermediates are generated with carcinogenic and mutagenic properties [57, 58]. Biodegradation of azo dyes occurs under both anaerobic and aerobic conditions, as well as a combination of the two when appropriate microorganisms are present. The enzymes associated with the reductive processes under aerobic conditions are more specific to the chemical structure of the azo dye than those under anaerobic conditions.

Aerobic bacteria needed to be exposed continuously to a very simple azo dye (such as methyl red) over long periods with the presence of oxygen, allowing the bacteria to synthesize a specific azoreductase for this compound. On the other hand, a universal azoreductase is produced anaerobically without regard to the azo compound involved [59].

Under a micro-aerobic condition, the strain of *Lactobacillus casei* TISTR 1500 generated two types of aromatic amine intermediates (*N'*,*N*-dimethyl-*p*-phenylenediamine and 4-aminobenzenesulfonic acid) from methyl orange under optimal conditions in pH 6 at 35 °C [60]. The rate of decolorization of methyl orange by the strain TISTR 1500 was influenced by energy sources like sugars, oligosaccharides or organic acids. The influence of sugars, oligosaccharides, organic acids, metal ions, pHs, oxygen and temperatures on the decolorization is shown in Figure 4.

From the study, sucrose is the best energy source at 35 °C with pH 6, and had a methyl orange decolorization rate of 6.1 mg/gCell/hr. These intermediates caused DNA damage and mutagenesis to microorganisms [61]. From the findings, a growing culture of *Lactobacillus casei* TISTR 1500 is more tolerant to high initial dye concentrations of methyl orange than freely suspended cells supplied with sucrose.

Furthermore, the biomass load influences the specific decolorization rate; a high biomass load increases the percentage of decolorization and the specific decolorization rate in the reaction mixture using freely suspended cells. Additionally, Na^+ ions can enhance the specific decolorization rate in the reaction mixture incubated with freely suspended cells, it cannot increase the activity of azoreductase [62]. However, the rate of decolorization can be enhanced by adding extra nitrogen supplements, like peptone, yeast extract or meat extract [63].

From this basic knowledge, the semi-continuous system for protein recovery in lactic acid fermentation was investigated [32, 34, 56]. Lower efficiencies of dye biodegradation, COD and protein removal, and lower yields of total acidities were obtained when the semi-continuous system was supplied with higher initial dye concentrations (200 and 400 mg/L) [62]. The results clearly show the inhibitory effect of ponceau 4 R and its intermediates at high concentrations even though the dye is normally recognized as safe.

The strain TISTR 1500 converted total sugar in the wastewater with efficiencies of 86 ± 0.5 % and 86 ± 0.9 %. The system deteriorated in terms of treatability when 400 mg/L of ponceau 4 R containing influent was fed into the system with the lowest number of viable lactic acid bacteria (Log 5.3).

The increase in dye concentration decreased the viability of lactic acid bacteria due to either the toxicity of the dye itself or the aromatic intermediate byproducts after decolorization, thus decreasing the efficiency of acid production. However, efficiency of the system was influenced by the sugar and protein content in dairy wastewater. Feeding low amounts of sugar into the system caused the failure of acid production and a decline in lactic acid bacteria survival. In addition, the inhibitory effect from the dye was compensated by high sugar and protein influent feeding as shown in Figure 4 A – D [56].

In addition, *Lactobacillus acidophilus* and *Lactobacillus fermentum* showed potential for biodegradation of both monoazo and diazo water-soluble dyes, similar to *Lactobacillus casei* TISTR 1500 [64]. Under anaerobic conditions, decolorization by the two strains was tested on of 6 µg/mL of various dyes (methyl red, ponceau BS, orange G, amaranth, orange II, and direct blue 15, 5 µg/mL sudan I and II, or 1.5 µg/mL sudan III and IV in MRS broth at 37°C for 36 hr.

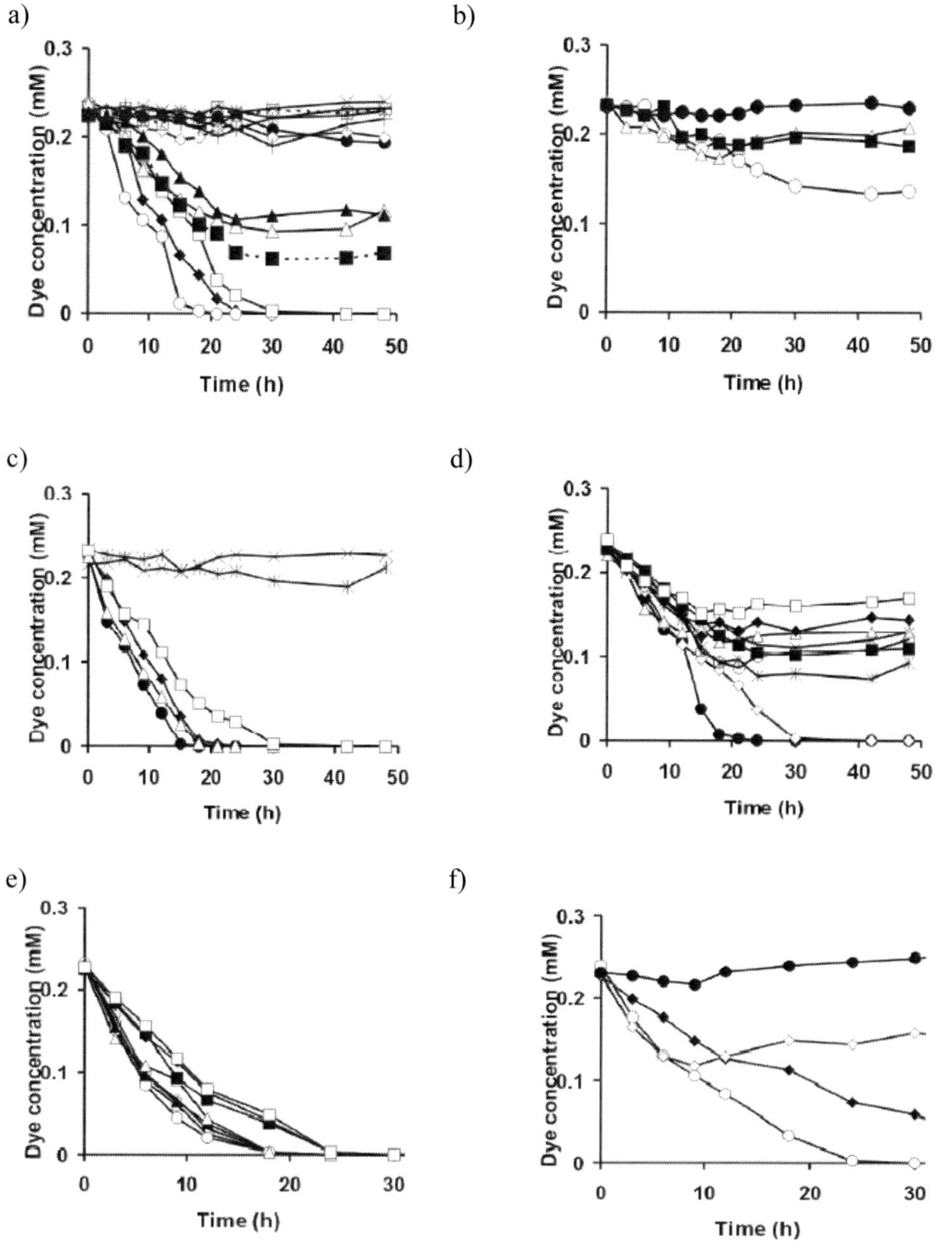

Figure 4. Influence of (A) Sugars: (●) no sugar, (◆) glucose, (○) sucrose, (✱) fructose, (◇) lactose, (✚) maltose, (□) galactose, (-) mannose, (□, dotted line) arabinose, (×) xylose, (△) mannitol, (▲) inositol, and (■) sorbitol. (B) Oligosaccharides: (●) no oligosaccharide, (○) galacto-oligosaccharide, (△) fructo-oligosaccharide, and (■) inulin. (C) Organic acids: (×) acetic acid, (◆) citric acid, (✱) lactic acid, (●) oxalic acid, (△) succinic acid, and (□) tartaric acid. (D) Metal ions: (●) no metal ions, (◇) calcium ions, (○) cobalt ions, (✱) copper ions, (△) ferric ions, (◆) ferrous ions, (■) magnesium ions, (×) manganese ions, and (□) zinc ions. (E) pH: (◆) pH 3, (◇) pH 4, (●) pH 5, (○) pH 6, (▲) pH 7, (△) pH 8, (■) pH 9, and (□) pH 10. (F) Temperature: (◆) 25 °C, (○) 35 °C, (◇) 45 °C, and (●) 55 °C on the removal of methyl orange by freely suspended cells of *Lactobacillus casei* TISTR 1500 [60].

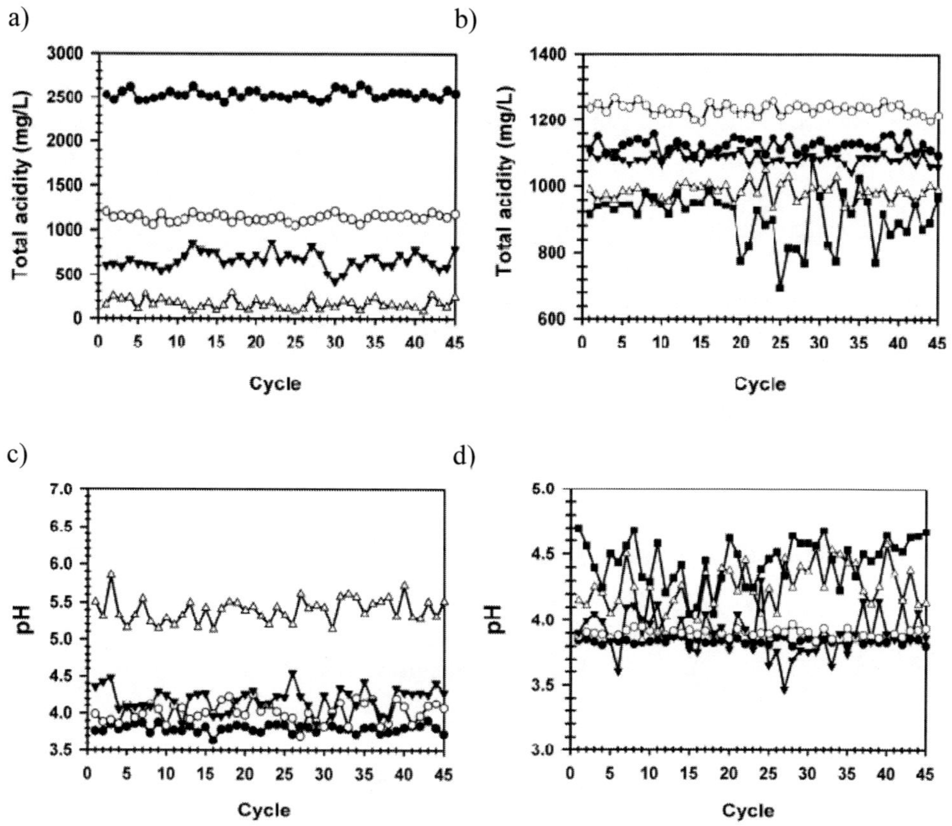

Figure 5. Total acidity and pH of micro-aerobic SBR during acid fermentation by *Lactobacillus casei* TISTR 1500 with various initial COD concentrations ((△) 3000, (▼) 6000, (○) 12000, and (●) 25000 mg/L) and azo dye concentration ((○) 25, (●) 50, (▼) 100, (△) 200, and (■) 400 mg/L) under non-sterile condition [56].

Lactobacillus acidophilus achieved 100% methyl red, orange G, sudan III and sudan IV removal, while *Lactobacillus fermentum* only completely decolorized sudan III and sudan IV at 37°C in an anaerobic chamber for 36 h without agitation. It may imply that water-soluble azo dyes can freely pass in medium through cell walls more easily than water-insoluble azo dyes, and are then degraded by azoreductases [64]. *Lactobacillus acidophilus* and *Lactobacillus fermentum* can degrade diazo dyes (sudan III and IV) but not monoazo dyes (sudan I and II). Several pieces of evidence have been described regarding the potential of lactic acid bacteria, and *Lactobacilli* in particular, for biodegradation and biotransformation. Further investigations are needed on the suitable environmental conditions for using lactic acid bacteria as a biocatalyst.

Conclusion

In this chapter, the potential applications of Lactobacilli via biocoagulation of protein in dairy wastewater under non-sterile conditions, and biodegradation and biotransformation of recalcitrant chemical compounds are demonstrated. The bioprocess of semi-continuous fermentation is employed in lactic acid fermentation to ameliorate substrate inhibition. Varying the volume of wastewater throughout the course of the treatment cycle can raise the pH within the tank to levels that allow acid producing microorganisms the chance to replenish. The recovered casein and lactic acid from the process can be used for animal feed and bioplastics, respectively. The success of the system using *Lactobacillus casei* TISTR 1500 in protein recovery is due to the influence of a long SRT control, high inoculum usage, and pH switching. Moreover, the review also shows another direction of the application for biodegradation and biotransformation of chemical compounds by lactic acid bacteria. The high potential of azo dye degradation by Lactobacilli can generate new techniques for solving the problem of dealing with these potentially dangerous textile dyes.

Acknowledgments

This work was financially supported by The Office of Higher Education Commission, part of the Ministry of Education, Thailand and the JSPS – NRCT Core University Program.

References

[1] Smith, P. G. and Scott, J. S. (2002). *Dictionary of water and waste management.* Oxford, UK: IWA Publishing, Butterworth Heinemann.

[2] Octave, S. and Thomas, D. (2009). Biorefinery: Toward an industrial metabolism. *Biochimie*, 91, 659-664.

[3] Sendich, E. D., Dale, B. E. and Kim, S. (2008). Comparison of crop and animal simulation options for integration with the biorefinery. *Biomass Bioenergy*, 32, 1162-1174.

[4] Thorsell, S., Epplin, F. M., Huhnke, R. L., and Taliaferro, C. M. (2004). Economics of a coordinated biorefinery feedstock harvest system: lignocellulosic biomass harvest cost. *Biomass Bioenergy*, 27, 327-337.

[5] Yang, S. T. (2007). Bioprocessing-from biotechnology to biorefinery. In: S. Y. Yang editor. *Bioprocessing for value-added products from renewable resources* (1st ed., p. 1-24) Amsterdam, Elsevier science.

[6] Droste, R. L. (1997). *Theory and pratice of water and wastewater treatment* New York, USA: John Wiley and Sons Inc.

[7] Britz, T. J., Van Schalkwyk, C. and Hung, Y. T. (2005). Treatment of dairy processing wastewater. In: L. K. Wang, Y. T. Hung, H. H. Lo, C. Yapijakis editors. *Waste treatment in the food processing industry.* (1st ed., p. 1-28). Boca Raton, Florida, USA: CRC press.

[8] De Man, G. and De Bekker, P. H. A. M. (1986). New technology in dairy wastewater treatment. *Dairy Ind. Int.* 51, 21-25.

[9] Gray, N. F. (2004). *Biology of Wastewater Treatment.* (2nd ed). London: Imperial College Press.

[10] Burgess, S. (1985). Anaerobic treatment of Irish creamery effluents. *Process Biochem.* 20, 6-7.

[11] Demirel, B., Yenigun, O. and Onay, T. T. (2005). Anaerobic treatment of dairy wastewaters: a review. *Process Biochem.*, 40, 2583-2595.

[12] IDF. (1990). Anaerobic treatment of dairy effluents - The present stage of development. *Bull. Int. Dairy Fed.*, 252.

[13] Li, X. and □hang, R. (2002). Aerobic treatment of dairy wastewater with sequencing batch reactor systems. *Bioprocess Biosyst. Eng.* 25, 103-109.

[14] Cramer, M. M. (2006). *Food plant sanitation: design, maintenance, and good manufacturing practices*. Boca Raton: CRC Press.

[15] Wendorff, W. L. (2001). Treatment of dairy wastes. In: E. H. Marth, J. L. Steele editors. *Applied dairy microbiology* (2nd ed., p. 681-704) New York , USA: Marcel Dekker Inc.

[16] Hemming, M. L. (1981). The treatment of dairy wastes. In: A. Herzka, R. G. Booth editors. *Food industry wastes; disposal and recovery*ed. (1st ed., p. 109-121). Essex, UK: Appiled science publishers Ltd.

[17] IDF. (1997). Removal of fats, oil, and grease in the pretreatment of dairy wastewaters. *Bull. Inter. Dairy Fed.*, Doc. No. 327.

[18] Cammarota, M. C., Teixeira, G. A. and Freire, D. M. G. (2001). Enzymeatic pre-hydrolysis and anaerobic degradation of wastewaters with high fat contents. *Biotechnol. Lett.*, 23, 1591-1595.

[19] IDF. (1997). Removal of fats, oil, and grease in the pretreatment of dairy wastewaters. *Bull. Int. Dairy Fed.*, Doc. No. 327.

[20] Leal, M. C. M. R., Cammarota, M. C., Freire, D. M. G., and Sant' Anna, Jr. G. L. (2002). Hydrolytic enzymes as coadjuvants in the anaerobic treatment of dairy wastewaters. *Brazilian J. Chem. Eng.*, 19, 175-180.

[21] Rusten, B., Eikebrokk, B. and Thorvaldsen, G. (1990). Coagulation as pretreatment of food industry wastewater. *Water Sci. Tech.*, 22, 1-8.

[22] Rusten, B., Lundar, A., Eide, O., and Ødegaard, H. (1993). Chemical pretreatment of dairy wastewater. *Water Sci. Tech.*, 28, 67-76.

[23] Dyrset, N., Selmer-Olsen, E., Havrevoll, O., Ratnaweera, H., Storro, I., and Birkeland, S. (1998). Feed supplement recovered from dairy wastewater by biological and chemical pretreatment. *J. Chem. Technol. Biotechnol.* 73, 175-182.

[24] Jones, H. R. (1974). *Pollution Control in the Dairy Industry*. Park Ridge, New Jersey, USA: Noyes Data Corp.

[25] Axelsson, L. (2004). Lactic acid bacteria: classification and physiology. In: S. Salminen, A. Von Wright, A. Ouwehand editors. *Lactic acid bacteria: microbiological and functional aspects* (3rd ed., p. 1-66) New York, USA: Marcel Dekker, Inc.

[26] Jay, J. M. (2000). *Fermentation and fermented dairy products, Modern food microbiology*. (6th ed., p. 113–130). Gaithersburg, USA: Aspen Publishers, Inc.

[27] Seppo, S. and Von Wright, A. (1993). *Lactic Acid Bacteria*. London: Marcel Dekker, Inc.

[28] Akao, S., Tsuno, H., Horie, T., and Mori, S. (2007). Effects of pH and temperature on products and bacterial community in L-lactate batch fermentation of garbage under unsterile condition. *Water Res.*, 41, 2636-2642.

[29] Chollangi, A. and Hossain, M. M. (2007). Separation of proteins and lactose from dairy wastewater. *Chem. Eng .Process*, 46, 398-404.

[30] Kasapgil, B., Anderson, G. K. and Ince, O. (1994). An investigation into the pre-treatment of dairy wastewater prior to aerobic biological treatment. *Water Sci. Tech.*, 29, 205 - 212.

[31] Suzuki, Y. and Maruyama, T. (2002). Removal of suspended solids by coagulation and foam separation using surface-active protein. *Water Res.*, 36, 2195-2204.

[32] Seesuriyachan, P., Techapun, C., Kuntiya, A., Wongroung, S., and Wungkobkiat, A. Coagulation as Pre-treatment of dairy wastewater by Lactic Acid Bacteria. *Proceedings of IWA Asia Environmental Technology* 2001, 30 Oct.- 2 Nov. Singapore; 2001. p. 391-396.

[33] Schneider, H. A. I., Neto, V. M., Soares, A., Rech, R. L., and Rubio, J. (1995). Primary treatment of a soybean proteins bearing effluent by dissolved air flotation and by sedimentation. *Water Res.*, 29, 69-75.

[34] Seesuriyachan, P., Kuntiya, A., Sasaki, K., and Techapun, C. (2009). Biocoagulation of dairy wastewater by *Lactobacillus casei* TISTR 1500 for protein recovery using micro-aerobic sequencing batch reactor (micro-aerobic SBR). *Process Biochem.*, 44, 406-411.

[35] Sakai, K. and Ezaki, Y. (2006). Open L-lactic acid fermentation of food refuse using thermophilic *Bacillus coagulans* and fluorescence in situ hybridization analysis of microflora. *J. Biosci. Bioeng.*, 101, 457-463.

[36] Sakai, K., Fujii, N. and Chukeatirote, E. (2006). Racemization of l-lactic acid in pH-swing open fermentation of kitchen refuse by selective proliferation of *Lactobacillus plantarum*. *J. Biosci. Bioeng.*, 102, 227-232.

[37] Sakai, K., Mori, M., Fujii, A., Iwami, Y., Chukeatirote, E., and Shirai, Y. (2004). Fluorescent in situ hybridization analysis of open lactic acid fermentation of kitchen refuse using rRNA-targeted oligonucleotide probes. *J. Biosci. Bioeng.*, 98, 48-56.

[38] Sakai, K., Murata, Y., Yamazumi, H., Tau, Y., Mori, M., Moriguchi, M., and Shirai, Y. (2000). Selective proliferation of lactic acid bacteria and accumulation of lactic acid during open fermentation of kitchen refuse with intermittent pH adjustment. *Food Sci. Technol. Res.*, 6, 140-145.

[39] Ben Othman, N., Ayed, L., Kachouri, F., Hammami, M., and Hamdi, M. (2008). Ecological removal of recalcitrant phenolic compounds of treated olive mill wastewater by *Pediococcus pentosaceus*. *Bioresour, Technol.*, 2996-3001.

[40] Naeem, A., Haneed, A. and Ahmed, S. (2009). Physicochemical characterrization and bioremediation perspective of textile effluent, dyes and metals by indigenous bacteria. *Bioresour. Technol.*, 164, 322-328.

[41] Rodriguez, H., Curiel, J. A., Landete, J. M., De las Rivas, B., De Felipe, F. L., Garmen-Cordovies, C., Mancheno, J. M., and Munoz, R. (2009). Food phenolics and lactic acid bacteria. *Int. J. Food Microbiol.*, 132, 79-90.

[42] Grimaldi, A., Bartowsky, E. and Jiranek, V. (2005). A survey of glycosidase activities of commercial wine strains of *Oenococcus oeni*. *Int. J. Food Microbiol.*, 105, 233-244.

[43] Osawa, R., Kuroiso, K., Goto, S., and Shimizu, A. (2000). Isolation of tannin-degrading lactobacilli from human and fermented foods. *Appl. Environ. Microbiol.*, 66, 3093-3097.

[44] Landete, J. M., Curiel, J. A., Rodríguez, H., De las Rivas, B., and Muñoz, R. (2008). Study of the inhibitory activity of phenolic compounds found in olive products and their degradation by *Lactobacillus plantarum* strains. *Food Chem.*, 107, 320-326.

[45] Landete, J. M., Rodríguez, H., De las Rivas, B., and Muñoz, R. (2008). Characterization of a benzyl alcohol dehydrogenase from *Lactobacillus plantarum* WCFS1. *J. Agric. Food Chem.*, 56, 4497-4503.

[46] Tondee, T. and Sirianuntapiboon, S. (2008). Decolorization of molasses wastewater by *Lactobacillus plantarum* No. PV71-1861. *Bioresour. Technol.*, 99, 6258-6265.

[47] Whiting, C. G. and Carr, J. G. (1959). Metabolism of cinnamic acid and hydroxycinnamic acids by *Lactobacillus pastorianus* var. *quinicus*. *Nature*, 4696, 1427-1428.

[48] Alberto, M. R., Gómez-Cordovés, M. C. and Manca De Nadra, M. C. (2004). Metabolism of gallic acid and catequin by *Lactobacillus hilgardii* from wine. *J. Agric. Food Chem.*, 52, 6465-6469.

[49] Cavin, J. F., Andioc, V., Etievant, P. X., and Divies, C. (1993). Ability of wine lactic acid bacteria to metabolize phenol carboxylic acids. *Am. J. Enol. Vitic.*, 44, 76-80.

[50] Benz, M., Schink, B. and Brune, A. (1998). Humic acid reduction by *Propionobacterium freudenreichii* and other fermenting bacteria. *Appl. Environ. Microbiol.*, 11, 4507-4512.

[51] Arvanitoyannis, I. S. and Kassaveti, A. (2007). Current and potential uses of composted olive oil waste. *Int. J. Food Sci. Technol.*, 42, 281-295.

[52] Rodriguez, H., Landete, J. M., Rivas, B. D. l., and Munoz, R. (2008). Metabolism of food phenolic acids by *Lactobacillus plantarum* CECT 748T. *Food Chem.*, 107, 1393-1398.

[53] Ayed, L. and Hamdi, M. (2003). Fermentative decolorization of olive mill wastewater by *Lactobacillus plantarum*. *Process Biochem.*, 39, 59-65.

[54] Lamia, A. and Moktar, H. (2003). Fermentative decolorization of olive mill wastewater by *Lactobacillus plantarum*. *Process Biochem.*, 39, 59-65.

[55] Kachouri, F. and Hamdi, M. (2004). Enhancement of polyphenols in olive oil by contact with fermented olive mill wastewater by *Lactobacillus plantarum*. *Process Biochem.*, 39, 841-845.

[56] Seesuriyachan, P., Chaiyaso, T., Sasaki, K., and Techapun, C. (2009). Influence of food colorant and initial COD concentration on the efficiencies of micro-aerobic sequencing batch reactor (micro-aerobic SBR) for casein recovery under non-sterile condition by *Lactobacillus casei* TISTR 1500. *Bioresour. Technol.*, 100, 4097-4103.

[57] Downham, A. and Collins, P. (2000). Colouring our foods in the last and next millennium. *Int. J. of Food Sci. Technol.*, 35, 5-22.

[58] Pandey, A., Singh, P. and Iyengar, L. (2007). Bacterial decolorization and degradation of azo dyes. *Int. Biodeter. Biodegr.*, 59, 73-84.

[59] Pearce, C. I., Lloyd, J. R. and Guthrie, J. T. (2003). The removal of colour from textile wastewater using whole bacterial cells: a review. *Dyes Pigments*, 58, 179-196.

[60] Seesuriyachan, P., Takenaka, S., Kuntiya, A., Klayraung, S., Murakami, S., and Aoki, K. (2007). Metabolism of azo dyes by *Lactobacillus casei* TISTR 1500 and effects of various factors on decolorization. *Water Res.*, 41, 985-992.

[61] Knasmüller, S., Steinkellner, H., Hirschl, A. M., Rabot, S., Nobis, E. C., and Kassie, F. (2001). Impact of bacteria in dairy products and of the intestinal microflora on the genotoxic and carcinogenic effects of heterocyclic aromatic amines. *Mutat. Res.*, 480-481, 129-138.

[62] Seesuriyachan, P., Kuntiya, A., Sasaki, K., and Techapun, C. (2009). Comparative study on methyl orange removal by growing cells and washed cell suspensions of *Lactobacillus casei* TISTR 1500 *World J. Microbiol. Biotechnol.*, 25, 973-979.

[63] Seesuriyachan, P., Kuntiya, A., Techapun, C., Chaiyaso, T., Hanmuangjai, P., and Leksawasdi, N. (2011). Nutritional requirements for methyl orange decolourisation by freely suspended cells and growing cells of *Lactobacillus casei* TISTR 1500. *Maejo Int. J. Sci. Technol.*, 5, 32-46.

[64] Chen, H., Xu, H., Heinze, T. M., and Cerniglia, C. E. (2009). Decolorization of water and oil-soluble azo dyes by *Lactobacillus acidophilus* and *Lactobacillus fermentum*. *J. Ind. Microbiol. Biotechnol.*, 36, 1459-1466.

In: Lactobacillus: Classification, Uses and Health Implications ISBN: 978-1-62081-151-1
Editors: Alba I. Perez Campos and Arturo Leon Mena

Chapter XIV

Use of Probiotics and Prebiotics on Functional Dairy Products: The Health Benefits

Svetoslav Dimitrov Todorov[1,*] and Ricardo Pinheiro de Souza Oliveira[2]
[1]Department of Food Science and Experimental Nutrition, Faculty of Pharmaceutical Sciences, Sao Paulo University, Sao Paulo, SP, Brazil
[2]Biochemical and Pharmaceutical Technology Department, Faculty of Pharmaceutical Sciences, Sao Paulo University, Sao Paulo, Brazil

Abstract

In recent years, the number of functional dairy products enriched with live probiotic microorganisms, like bifidobacteria and lactobacilli, has increased. According to FAO/WHO (2002), probiotics are live microorganisms which, when administered in adequate amounts, confer health benefits on the host. In fact, these therapeutic effects are the reason for which most probiotics are used in yoghurts, fermented milks, ice creams and pharmaceutical products. Currently, to enhance health benefits, dairy foods usually contain members of probiotic genus in association with prebiotics, i.e. non-digestible oligosaccharides that resist hydrolysis and absorption in the upper gastrointestinal tract and are metabolized selectively by at least one type of probiotic in the colon.

1. Introduction

Several health benefits of probiotic bacteria mainly belonging to *Lactobacillus* and *Bifidobacterium* genus have led to their increased incorporation in functional dairy products

[*] Corresponding author. Tel.: +55-11-3091 2191; fax: +55-11-3815 4410. E-mail address: slavi310570@abv.bg (SD Todorov).

(Lourens-Hattingh and Viljoen, 2001). Today, most probiotics are used in yoghurts, fermented milks, ice creams and pharmaceutical products, because of their therapeutic effects (Holzapfel *et al.*, 1998, Mattila-Sandholm *et al.*, 2002; Reid, 2006). For these beneficial effects, provided by functional dairy products, should be improved, are essential to adding a) lactic acid bacteria with probiotic properties, b) bioactive components and c) certain types of soluble fiber known as prebiotics (Ranadheera *et al.*, 2010, Vasiljevic and Shah, 2008).

Probiotics are microorganisms that, when used in large amounts in the preparation of foods, are able to survive the passage through the upper digestive tract and adhere to intestinal cells, helping in the intestinal balance. In particular, probiotics used in functional dairy products belong to the genera *Lactobacillus*, *Bifidobacterium*, *Streptococcus* and *Saccharomyces*. To produce the desired benefits, these bacteria should be present in the product in viable counts during their whole shelf-life (7-9 Log CFU/mL); however, their viability in commercial preparations is affected by several factors, among which the presence of other microorganisms (Kailasapathy and Rybka, 1997).

On the other hand, prebiotics are non-digestible carbohydrates that resist hydrolysis and absorption in the upper parts of the gastrointestinal tract and are metabolized selectively by at least one type of probiotic in the colon (Mattila-Sandholm *et al.*, 2002). Among these, inulin proved to exert a protective effect towards *Lactobacillus acidophilus* and *Lactobacillus casei* improving their survival and activity during storage (Donkor *et al.*, 2007). Nevertheless, other lactobacilli would deserve attention due to their ability to grow in milk and to confer functional properties and benefits to the health. In fact, some other probiotics are used for this purpose, among which are *Lactobacillus rhamnosus*, *Lactobacillus acidophilus*, *Lactobacillus johnsonii* and *Bifidobacterium lactis* (Vasiljevic and Shah, 2008).

2. Importance of Probiotic Microorganisms

Lactic **a**cid **b**acteria (LAB) comprise a heterogeneous group of bacteria having as a common metabolic property the production of lactic acid as the mainly end-product from the fermentation of sugars. Among the LAB, lactobacilli are classified as Gram-positive, non-sporulating, catalase-negative, acid-tolerant, anaerobic fermentative bacteria that have different sensitivity to oxygen. Except for a few species, LAB are nonpathogenic organisms with a reputed **G**enerally **R**ecognized **A**s **S**afe (GRAS) status (Kleerebezem and Hugenholtz, 2003). From a biochemical point of view, carbohydrates are fermented by the homofermenters mainly to lactic acid, while the heterofermenters yield, in addition to lactic acid, a large variety of fermentation products such as acetic acid, ethanol, carbon dioxide and formic acid (Gomes and Malcata, 1999; Kleerebezem and Hugenholtz, 2003). They also produce a variety of minor compounds such as diacetyl, acetoin and 2,3-butanediol from citrate metabolization, a vast array of volatile compounds, bioactive peptides from the catabolism of amino acids and exopolysaccharides from carbohydrates (Mayo *et al.*, 2010; de Vos and Hugenholtz, 2004), which provide or preserve the sensorial and nutritional properties of fermented products. In addition, bacteriocin production is common among the LAB. By definition, bacteriocins are ribosomally synthesized antimicrobial small peptides produced by bacteria that are active against others microorganisms (Todorov, 2009; Arqués *et al.*, 2011).

Nowadays, therefore, the interest in probiotics and protective cultures has constantly developed and has become important for food biotechnology (El-Dieb *et al.*, 2012). In fact, institutions and food industries were revitalized by the introduction of new products, which have quality, high nutritional value, flavor and ability to promote positive effects on consumer health (Casiraghi *et al.*, 2007).

The main health benefits associated with the ingestion of probiotic microorganisms have been described by many researchers and can be summarized as: alleviation of lactose intolerance, prevention and reduction of diarrhoea symptoms, treatment and prevention of allergy, reduction of the risk associated with mutagenicity and carcinogenicity, hypocholesterolemic effect, inhibition of *Helicobacter pylori* and intestinal pathogens, prevention of inflammatory bowel disease and modulation of the immune system (Mattila-Sandholm *et al.*, 1992; Goldin, 1998; Holzapfel *et al.*,1998; Salminen *et al.*,1998; Ouwehand *et al.*,2003; Akin *et al.*, 2007).

3. Effect of Probiotic Culture on Fermentation

A good technological performance of probiotic cultures results in a rapid growth in milk, appropriate sensory properties to the product, and survival or viability of bacteria during storage (Oliveira *et al.*, 2001; Oliveira *et al.*, 2009a). In commercial practice is very common the use of starter cultures of yoghurt, like *Streptococcus thermophilus* and *Lactobacillus delbrueckii* subsp. *bulgaricus* combined with probiotic bacteria in order to reduce the fermentation time. However, *Lactobacillus bulgaricus* subsp. *bulgaricus* favor the post-acidification in fermented milk, affecting the viability of probiotic bacteria. Thus, it is recommended the use of starter cultures without *Lactobacillus bulgaricus* subsp. *bulgaricus* (Dave and Shah, 1998).

Another technique used to increase growth and acidifying activity of probiotic microorganisms, in milk, is to increase the available source by glucose addition (Oliveira and Damin, 2003) and growth factors, i.e., yeast extract and hydrolyzed protein or appropriate antioxidants, minerals, vitamins and prebiotics (Shah and Lankaputhra, 1997; Dave and Shah, 1998; Gomes and Malcata, 1999; Naidu and Clemens, 2000; Oliveira *et al.,* 2001). In fact, the effect of culture composition and the supplementation of milk (with prebiotics) on the growth, acidification rate, survival and metabolism of probiotic bacteria in fermented milk have been studied by several researchers (Lucas *et al.,* 2004; Naidu and Clemens, 2000; Oliveira *et al.*, 2009ab, 2011ab; Shah and Lankaputhra, 1997; Sodini *et al.*, 2002).

4. Importance of Prebiotics

As industries are interested in maintaining the viability of their products, supplementation has come to enhance the probiotic effect. For example, supplementing milk with hydrolyzed protein, fructose, whey protein concentrate, pulp of tomato juice and papaya stimulates *Lactobacillus acidophilus*, while cysteine, acid hydrolyzed, tryptone, vitamins, dextrins and maltose increases the viability of *Bifidobacteria*. Prebiotic compounds also allow the growth of probiotic bacteria (Lourens-Hattingh and Viljoen, 2001).

Prebiotics are functional properties of great importance for the food industry. Both inulin and oligofructose have been used to enrich foods with fiber (Mussatto and Mancilha, 2007). Unlike other fibers, has no additional flavors, and can enrich foods without changing much of the viscosity, appearance and flavor formulations. The nutritional properties of inulin and oligofructose are similar, so the decision to make with inulin or oligofructose is a function of the attributes desired in the final product (Niness, 1999). Inulin is more appropriate when applying for products with lower fat content, while oligofructose are adequate to cover the aftertaste of high intensity sweeteners used in food preparations. Studies have shown that inulin and oligofructose have prebiotic characteristics, because they are indigestible and develop the bifidogenic effect, improving the intestinal microbiota. It is recommended that inulin and oligofructose are used in probiotic foods to obtain the synbiotic effect.

4.1. Definition and Benefits of Prebiotics

Prebiotics are nondigestible food components that beneficially affect the host by selectively stimulating the proliferation or activity of populations of bacteria that are desirable in the colon. In addition, prebiotics can inhibit the multiplication of pathogens, providing additional benefits to host health. These components act more often in the large intestine, although they may also have impact on microorganisms of the small intestine (Roberfroid, 2001; Mattila-Sandholm *et al*., 2002). Among the prebiotics, inulin and oligofructose are soluble and fermentable fiber. In particular, are known as fructans, which are not digestible by α-amylase and hydrolytic enzymes such as sucrase, maltase and iso-maltase in the upper intestinal tract (Carabin and Flamm, 1999). The quantities determined for dietary fiber, lactulose, polydextrose, inulin and oligofructossacharides (FOS) are at least 3 to 6 g fiber/100 g at solid foods and 1.5 to 3.0 g of fiber/100 g at liquid food (ANVISA, 2002).

Prebiotics, including lactulose, inulin and oligosaccharides provide various carbohydrates that bacteria of the colon are able to ferment. These functional ingredients have influence on physiological and biochemical processes in the body and resulting in the improvement of health and a reduction in the risk of developing various diseases. The main sources of inulin and oligofructose used in the food industry are chicory (*Cichorium intybus*) and Jerusalem artichoke (*Helianthus tuberosus*) (Kaur and Gupta, 2002).

4.2. Inulin and Oligofructose

Inulin is a polysaccharide found in a number of plant products, especially in the chicory root (*Cichorium intybus L.*), a plant native to Europe, Africa and America. There are aswell vegetables that contain significant concentrations of inulin and among them are quite consumed, especially Jerusalem artichoke, where concentrations can reach 20% of fresh weight (Voragen, 1998). Inulin is used in agribusiness for the production of low-calorie foods and as a functional ingredient in pharmaceutical industries. The conventional technology for the production of inulin is the aqueous extraction, which has high energy costs due to separation steps (Van Loo *et al*., 2005).

Inulin is a polymer of D-fructose (Figure 1) and a carbohydrate reserve of plants. It belongs to the fructan group and is synthesized by a variety of plants (Figueira *et al*., 2004).

This carbohydrate is a linear polymer with β(2→1) glycosidic bonds (Roberfroid and Gibson, 1998). In particular, it is a fructan consisting of a mixture of higher linear oligomers and polymers of fructose. The units of β-D fructofuranosyl are held together by bonds of the type β(2→1), and have glucose molecule in the initial portion of each linear chain of fructose (Coussement, 1997).

The inulin-type fructans are divided into two general groups: inulin and its related compounds - oligofructose and FOS. Inulin, oligofructose and FOS are chemically similar entities, with the same nutritional properties. The only difference between the inulin, oligofructose and FOS synthetic is the degree of polymerization, other words, the number of individual monosaccharide units that make up the molecule (Carabin and Flamm, 1999). The differences in size between the chains of inulin and oligofructose cause functional changes in the technological attributes of these compounds.

Oligofructose has the same structure of inulin, but the chain consists of 10 units or less of fructose. Oligofructose has approximately 30 to 50 % of sweetness. The inulin is less soluble than oligofructose and has a smooth creamy texture providing a sense of fat taste. Both inulin and oligofructose are not digested by human intestine enzymes, but are completely fermented by the intestinal microflora. The short chain fatty acids and lactate produced by the fermentation provides 1.5 Kcal per gram of inulin or oligofructose. Inulin and oligofructose are used to replace fat or sugar reducing calories from foods such as ice cream, sweets and dairy products (Coussement 1997; Akin *et al.*, 2007; Oliveira *et al.*, 2009a).

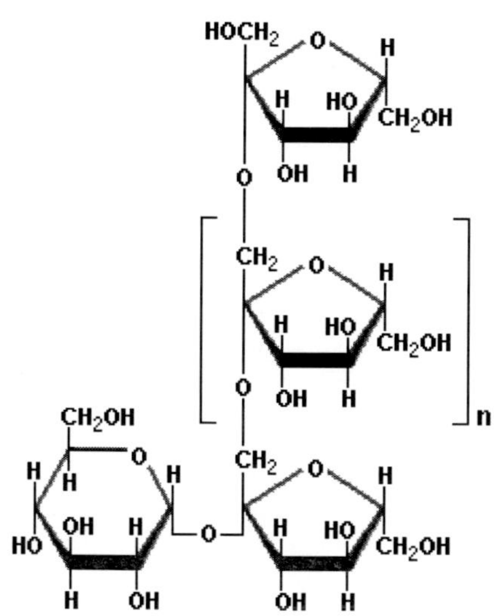

Source: Scientific Psychic, 2007.

Figure 1. Structure of inulin.

4.3. Prebiotic Effects of Inulin

Studies show that there is an interesting growth of probiotic bacteria using inulin as a substrate. Humans, who ingested inulin, had in their gastrointestinal system, high amounts of fecal bacteria such as *Bifidobacteria* spp. (Rao, 1999).

Scientific evidence indicates that the inulin and oligofructose provide an effective diet, and strengthen the gastrointestinal tract against colonization of pathogens. These prebiotics perform a selectivity favoring a competitive advantage to beneficial microorganisms such as bifidobacteria and LAB. This selective stimulation of growth and development of such groups known as "health promoters" is called prebiose, a new concept of nutrition (Marteau and Bourtron-Rualt, 2002).

The number of health-promoting microorganisms in the gastrointestinal tract may be increased by introducing probiotics in the diet or the consumption of prebiotic food supplement, which changes the composition of the microbiota by providing a competitive advantage over other bacteria in the ecosystem (Crittenden, 1999; Roberfroid, 2002).

Prebiotics such as inulin and oligofructose are resistant to digestion in the upper intestinal tract and are subsequently fermented in the colon. Prebiotics also have an effect on volume, due to increased biomass resulting from its fermentation, thus promoting an increase in stool frequency, confirming their classification as dietary fiber (Kaur and Gupta, 2002; Roberfroid, 2002).

4.4. Technological Application of Inulin

Due to the fact that it has a larger chain, inulin is less soluble than oligofructose and has the ability to form inulin microcrystals when mixed with water or milk. These microcrystals are not perceived in the mouth, but interact to form a thin creamy texture in the mouth that promotes a feeling similar to that of fat. Inulin has been used successfully as a fat substitute in various food products like cakes, chocolates, sausages and dairy products (Niness, 1999; Oliveira *et al.*, 2009b) and may also be used in ice cream, sweets and bakery products to reduce calories from these foods. However, the fact that inulin is not digested, allowed its increased use as food fiber.

Inulin is most commonly used when applying for products with lower fat content, while oligofructose are mostly used for fruit preparations for yoghurts and low fat to balance the sweetness and mask the aftertaste of sweeteners used in high intensity food preparations (Niness, 1999).

4.5. The Health Benefits of Inulin Consumption

Several experimental studies have shown the effects of inulin and oligofructose as bifidogenic factors, or that stimulate the predominance of bifidobacteria in the colon, then there is a stimulation of the host immune system, reduced levels of pathogenic bacteria in the gut, a relief in constipation, decreased risk of osteoporosis resulting from decreased absorption of minerals such as calcium. This would be a reduction in risk of atherosclerosis by decreasing the synthesis of triglycerides and fatty acids in the liver and decreased levels of these compounds in the blood (Kaur and Gupta, 2002; Fucks *et al.*, 2006).

Inulin stimulates intestinal function by increasing stool frequency, particularly in constipated patients. The increase in stool weight and bowel movements decreased fecal pH due to fermentation, have been linked to suppression of production of putrefactive substances in the colon (Menne *et al.,* 1997; Gibson and Roberfroid, 1998). Due to inulin resistance to digestion in the upper intestinal tract, its molecules reach the colon virtually intact where they are fermented by bacteria, and can then be classified as soluble food fibers (Roberfroid and Gibson, 1998).

Thus the food fibers, especially soluble fiber, effectively reduce serum cholesterol and LDL cholesterol concentrations, which plays a central role in the pathogenesis of atherosclerosis and may contribute to protection against heart disease. The effects of soluble fiber in lowering cholesterol are related to its properties of gel formation, decreased absorption of bile acids and the action of short chain fatty acids produced during fermentation on liver function. Moreover, it appears that the fibers can also act on risk factors for heart disease such as hypertension, obesity and diabetes (Anderson, 1990).

Inulin stimulates selectively the growth of bifidobacteria, and keeps at low levels the number of bacteroides, clostridia or coliforms, thus contributing to the prevention of colon cancer. The administration of inulin significantly increases fecal bifidobacteria that produce lactic acid, reducing the fecal pH and creating a microenvironment for antibacterial putrefactive bacteria, thus developing a microenvironment that may involve the modulation of bacterial enzymes such as β-glucuronidase, which can convert pro carcinogens into carcinogens (Kulkarni and Reddy, 1994; Roberfroid and Gibson, 1998).

According to Karppinen *et al.* (2000), the fermentation of inulin is faster and produces relatively more butyric acid and less propionic acid than the bran cereal. This butyrate is able to inhibit the proliferation of a large number of cells in vitro, including tumor cells. Other additional health effects such as increased absorption of minerals, especially calcium are related to the consumption of inulin. The frequent intake of inulin decreases or prevents loss of bone mass, calcium and phosphorus from the bones of gastrectomized rats and the loss of bone mineral density in ovariectomized mice (Taguchi *et al.* 1994, Ohta 1998). Heart disease, cancer, stress, high cholesterol, weight control, osteoporosis and diabetes are currently the biggest health concerns. Thus, the focus on disease prevention through food is very strong. This makes inulin appear as an important food ingredient that can be widely exploited in research and the food industry, aiming at the production of functional foods (Niness, 1999).

4.6. Other Prebiotics

4.6.1. Polydextrose and Maltodextrin

There is also a polysaccharide called polydextrose, whose physiological effects are similar to a fiber in the diet (Jie *et al.*, 2000). Polydextrose is a polysaccharide synthesized by random polymerization of glucose in the presence of minor amounts of sorbitol and a suitable acid catalyst under high temperature and partial vacuum. It is used in many countries as a body agent and as an ingredient in reduced calorie (1 kcal/g) in various foods (Hamanaka, 1987).

Polydextrose is partially fermented in the large intestine, but is neither digested nor absorbed in the small intestine, and is mostly excreted in the feces. This polysaccharide is partially fermented in the large intestine, increases the volume of the fecal mass, reduces the transit time, softens and lowers the pH of the stool. Fermentation of polydextrose leads to the growth of favorable microflora, reducing putrefactive microflora, the increased production of short chain fatty acids and suppression of production of carcinogenic metabolites, for example, indole and p-cresol (Wang and Gibson, 1993; Achour *et al.*, 1994).

From a technological point of view, the supplementation of polydextrose to a probiotic skim milk accelerated either the acidification or the fermentation, thus proving interesting prebiotics to improve the production of yoghurts (Oliveira *et al.*, 2009b). In fact, these results appear to confirm the prebiotic effect of polydextrose highlighted by Jie *et al.* (2000) and Probert *et al.* (2004).

Maltodextrins are oligosaccharides, made from starch and, are soluble in water, which popularized their use as additive in the food industry. In particular, they have dextrose equivalents of less than 20, where this term is used to describe the degree of starch hydrolysis and average molecular weight (Langrish *et al.*, 2007). In addition, maltodextrins are applied, for example, as spray-drying aids for flavors and seasonings, carriers for synthetic sweeteners, flavors enhancers, fat replacers, and bulking agents (Jaya and Das, 2004).

Oliveira *et al.* (2009b) studied the effect of maltodextrin and polydextrose on the probiotic survival in fat-free synbiotic fermented milk. Through this study, the authors concluded that all probiotic counts were stimulated by these prebiotics, and in particular, *Bifidobacterium lactis* exhibited the highest counts in all supplemented milk samples.

4.6.2. Lactulose

Lactulose is an effective drug against diseases like acute and chronic constipation (Tamura *et al.*, 1993). For this reason, this carbohydrate is widely used in pharmaceutical industry (Aider and de Halleux, 2007). Donkor *et al.* (2007) reported some promising applications in the nutraceuticals and food industries because of beneficial effects on human health provided by lactulose. From a chemical point of view, lactulose is a synthetic disaccharide obtained by isomerization of lactose, which is present in milk and whey in relatively high content, approximately 4.5 % as an average (Panesar and Kumari, 2011; □okaee *et al.*, 2002), and contains fructose instead of glucose (Strohmaier, 1998). Since it is not absorbed in the small intestine, it has the potential to function as a prebiotic (Kontula, 1999). Moreover, several studies showed the effectiveness of lactulose to stimulate the growth of bifidobacteria and lactobacilli (Olano and Corzo, 2009; Sako *et al.*, 1999; Shin *et al.*, 2000). In addition, this prebiotic can inhibit the growth of pathogenic bacteria such as *Salmonella* (Panesar and Kumari, 2011).

Taking into account all these considerations, lactulose is applied in a wide variety of functional dairy foods as a “bifidus factor” or as an important food ingredient (Oliveira *et al.*, 2011a; Panesar and Kumari, 2011). Oliveira *et al.* (2011a) explore the effects of this prebiotic carbohydrate on the acidification kinetics, growth and metabolism of binary co-cultures of *Lactobacillus acidophilus*, *Lactobacillus rhamnosus* and *Bifidobacterium lactis* with *Streptococcus thermophilus*. In particular, the use of lactulose enhanced the acidification rate and decreased the time to complete fermentation. Moreover, all probiotic cultures exhibited better growth associated with lactulose metabolization. In this study, the same authors (Oliveira *et al.*, 2011a) concluded that the viable counts of *Bifidobacterium lactis* in combination with *Streptococcus thermophilus* were significantly higher ($P < 0.05$) than those of the others *Lactobacillus* strains. According to Cardelle-Cobas *et al.* (2011), strains belonging to genera *Lactobacillus*, *Streptococcus* and *Bifidobacterium* were shown to be able to utilize lactulose and different trisaccharides derived from lactose and lactulose as carbon sources. Therefore, this above information may explain the growth of probiotic cultures in the presence of this prebiotic ingredient.

5. Influence of Prebiotics in Fermented Milk

Prebiotics and probiotics are increasingly being used to produce potentially synbiotic functional products, particularly through dairy foods as vehicles (Rodrigues *et al.*, 2012). From this, dairy industries introduced new types of dairy products in order to promote health benefits. Several authors reported that the use of prebiotics improve quality of different dairy foods fermented by probiotics cultures (Cardarelli *et al.*, 2007; Donkor *et al.*, 2007; Oliveira *et al.*, 2009b; Oliveira *et al.*, 2011)

In skim milk fermented by probiotic cultures (*Lactobacillus acidophilus*, *Lactobacillus rhamnosus*, *Bifidobacterium lactis* with *Streptococcus thermophilus*, either in binary co-cultures or in cocktail containing all microorganisms), Oliveira *et al.* (2009c) reported that the presence of inulin increased the mean acidification rate (upH/min) of binary co-cultures by approximately 6 % and that of the mixed culture or cocktail culture by about 4 %. In addition, the time to reach pH 4.5 depends of the addition of inulin. In particular, the addition of inulin reduced the fermentation time by about 10 % as an average, thus confirming its prebiotic effect already evidenced by other authors for both bifidobacteria (Akalın *et al.*, 2004; Bruno *et al.*, 2002; Donkor *et al.*, 2007; Shin *et al.*, 2000) and lactobacilli (Donkor *et al.*, 2007; Shah, 2000). In addition, Cardarelli *et al.* (2007) reported that the addition of prebiotics (inulin and oligofructose) to a probiotic *petit suisse* cheese provided a fastest fermentation and high lactic acid production, thus promoting the increase of growth rates in bifidobacteria and lactobacilli.

Bruno *et al.* (2002) and Liong and Shah (2005) suggested that the generation time (t_g) can be used as a measure of the effectiveness of a specific prebiotic in modulating the bacterial growth rate. According to Oliveira *et al.* (2011), *Streptococcus thermophilus* exhibited the shortest generation time (0.22-0.28 h) when was combined with binary co-cultures. These results were almost coincided with the value reported by Talon *et al.* (2002) for this bacterium in co-culture with *Lactobacillus* in half skim milk. The presence of inulin induced reductions of generation time by 12 % for *Streptococcus thermophilus*, 16 % for *Lactobacillus acidophilus*, 28 % for *Lactobacillus rhamnosus* and 35 % for *Bifidobacterium lactis*, as an average. The reduction of this parameter induced by inulin is a further confirmation of its prebiotic effect (Oliveira *et al.*, 2011).

Regarding to the counts, Oliveira *et al.* (2011) reported that apart from *Lactobacillus bulgaricus*, inulin has the ability to stimulate the growth of the others lactic bacteria, with particular concern to *Bifidobacterium* spp., whose counts dramatically increased, from 7.60 Log CFU/mL (after 1 day of cold storage) or 7.48 Log CFU/mL (after 7 days of cold storage) to 9.06 Log CFU/mL (after 1-7 days of cold storage). In particular, Ozer *et al.* (2005) observed a similar effect for *Bifidobacterium bifidum* BB-02 in milk supplemented with inulin or lactulose.

Conclusion

This chapter includes the knowledge of probiotics and prebiotics that spans over 100 years. It is important to note that more researches in the biotechnology and health areas should be made in order to elucidate synbiotic mechanisms involving probiotic bacteria and prebiotic ingredients that positively affect our well-being.

References

Achour, L., Flourie, B., Briet, F., Pellier, P., Marteau, P. and Rambaud, J. C. (1994). Gastrointestinal effects and energy value of polydextrose in healthy nonobese men. *American Journal of Clinical Nutrition*, *59*, 1362-1368.

Aider, M. and de Halleux, D. (2007). Isomerization of lactose and lactulose production: review. *Trends in Food Science and Technology*, *18*, 356-364.

Akalın, A. S., Fenderya, S. and Akbulut, N. (2004). Viability and activity of bifidobacteria in yoghurt containing fructooligosaccharide during refrigerated storage. *International Journal of Food Science and Technology*, *39*, 613–621.

Akın, M. B., Akın, M. S. and Kirmaci, □. (2007). Effects of inulin and sugar levels on the viability of yoghurt and probiotic bacteria and the physical and sensory characteristics in probiotic ice-cream. *Food Chemistry*, *104*, 93-99.

Anderson, J. W. (1989). Dietary fiber and human health. HortScience. 25: 1488-1495, 1990. [2nd International Symposium on Horticulture and Human Health, Alexandria, 1989, American Society for Horticultural Science].

ANVISA - Agência Nacional de Vigilância Sanitária (2002). Legislação. VisaLegis. Resolução RDC n.2, de 7 de janeiro de 2002. Aprova o regulamento técnico de substâncias bioativas e probióticos isolados com alegação de propriedades funcional e ou de saúde. Disponível em: http://e-legis.bvs.br/leisref/public/showAct.php?id=1567. Accessed on 23 November 2011.

Arqués, J. L., Rodríguez, E., Nuñez, M. and Medina, M. (2011). Combined effect of reuterin and lactic acid bacteria bacteriocins on the inactivation of food-borne pathogens in milk. *Food Control*, *22*, 457-461.

Bruno, A., Lankaputhra, W. E. W. and Shah, N. P. (2002). Growth, viability and activity of *Bifidobacterium* spp. in skim milk containing prebiotics. *Journal of Food Science*, *67*, 2740–2744.

Carabin, I. G. and Flamm, W. G. (1999). Evaluation of safety of inulin and oligofrutose as dietary fiber. *Regulatory, Toxicology and Pharmacology*, *30*, 268-282.

Cardarelli, H. R., Saad, S. M. I., Gibson, G. R. and Vulevic, J. (2007). Functional petit-suisse cheese: Measure of the prebiotic effect. *Anaerobe*, *13*, 200–207

Cardelle-Cobas, A., Corzo, N., Olano, A., Peláez, C., Requena, T. and Ávila, M. (2011). Galactooligosaccharides derived from lactose and lactulose: Influence of structure on *Lactobacillus*, *Streptococcus* and *Bifidobacterium* growth. *International Journal of Food Microbiology*, *149*, 81–87.

Casiraghi, M. C., Canzi, E., □anchi, R., Donati, E. and Villa, L. (2007). Effects of a synbiotic milk product on human intestinal ecosystem. *Journal of Applied Microbiology*, *103*, 499–506.

Coussement, P. (1997). Powerfull products. *World of Ingredients*, *17*, 12-17.

Crittenden, R. G. (1999). Prebiotics. In: Tannock, G.W. (ed). Probiotics: a critical review. Norfolk: Horizon Scientific Press, p.141-156.

Dave, R. I. and Shah, N. P. (1998). Ingredient supplementation effects on viability of probiotic bacteria in yoghurt. *Journal of Dairy Science*, *81*, 2804-2816.

De Vos, W. M. and Hugenholtz, J. (2004). Engineering metabolic highways in Lactococci and other lactic acid bacteria. *Trends in Biotechnology*, *22*, 72–79.

Donkor, O. N., Nilmini, S. L. I., Stolic, P., Vasiljevic, T. and Shah, N. P. (2007). Survival and activity of selected probiotic organisms in set-type yoghurt during cold storage. *International Dairy Journal, 17*, 657–665.

El-Dieb, S. M., Abd Rabo, F. H. R., Badran, S. M., Abd El-Fattah, A. M. and Elshaghabee, F. M. F. (2012). The growth behaviour and enhancement of probiotic viability in bioyoghurt. *International Dairy Journal*, *22*, 44-47.

FAO/WHO. (2002). Food and Agriculture Organization of the United Nations; World Health Organization. Guidelines for the evaluation of probiotics in food. London, Ontario: FAO/WHO, 2002. [Joint FAO/WHO working group report on drafting guidelines for the evaluation of probiotics in food, London, Ontario, Canada, April 30 and May 1, 2002]. Disponível em: ftp://ftp.fao.org/es/esn/food/wgreport2.pdf. Accessed on 07 November 2011.

Figueira, G. M., Park, K. J., Brod, F. P. R. and Honório, S. L. (2004). Evaluation of desorption iso-therms, drying rates and inulin concentration of chicory roots (*Cichorium intybus L.*) with and without enzymatic inactivation. *Journal of Food Engineering*, *63*, 273-280.

Fucks, R. H. B., Tanamati, A. A. C., Antonioli, C. M. and Doneda, I. (2006). Utilização de *Lactobacillus casei* e cultura iniciadora na obtenção do iogurte suplementado com inulina e oligofrutose. *Boletim do Centro de Pesquisa de Processamento de Alimentos*, *24*, 83-98.

Goldin, B. R. (1998). Health benefits of probiotics. *British Journal of Nutrition*, *80*, S203-S207.

Gomes, A. M. P. and Malcata, F. X. (1999). *Bifidobacterium* spp. and *Lactobacillus acidophilus*: biological, biochemical, technological and therapeutical properties relevant for use as probiotics. *Trends Food Science and Technology*, *10*, 139-157.

Hamanaka, M. (1987). Polydextrose as a water-soluble dietary fiber. *Syokuhin Kogyo*, *9*, 73-80.

Holzapfel, W. H., Haberer, P., Snel, J., Schillinger, U. and Huis In't Veld, J. H. J. (1998). Overview of gut flora and probiotics. *International Journal of Food Microbiology*, *41*, 85-101.

Jaya, S. and Das, H. (2004). Effect of maltodextrin, glycerol monostearate and tricalcium phosphate on vacuum dried mango powder properties. *Journal of Food Engineering, 63*, 125–134.

Jie, □., Bang-Yao, L., Ming-Jie, X., Hai-Wei, L., □u-Kang, □. and Ting-Song, W. (2000). Studies on the effects of polydextrose intake on physiologic functions in Chinese people. *American Journal of Clinical Nutrition*, *72*, 1503–1509.

Kailasapathy, K. and Rybka, S. (1997). *L. acidophilus* and *Bifidobacterium* spp.: their therapeutic potential and survival in yogurt. *Australian Journal of Dairy Technology*, *52*, 28-35.

Karppinen, S., Gibson, G. R. and Fuller, R. (2000). *In vitro* fermentation of polysaccharides of rye, wheat and oat brans and inulin by human faecal bacteria. *Journal of Science and Food Agriculture*, *80*, 1469-1476.

Kaur, N. and Gupta, A. K. (2002). Applications of inulin and oligofructose in health and nutrition. *Journal of Biosciences*, *27*, 703-714.

Kleerebezem, M. and Hugenholtz, J. (2003). Metabolic pathway engineering in lactic acid bacteria. *Current Opinions in Biotechnology*, *14*, 232–237.

Kontula, P., Suihko, M. L., von Wright, A. and Mattila-Sandholm, T. (1999). The effect of lactose derivatives on intestinal lactic acid bacteria. *Journal of Dairy Science*, *82*, 249–256.

Kulkarni, N. and Reddy, B. S. (1994). Inhibitory effect of *Bifidobacterium longum* cultures on the azoxymethane induced aberrant crypt foci formation and fecal bacterial β-glucuronidase. *Proceedings of the Society for Experimental Biology and Medicine*, *207,* 278-283.

Langrish, T. A. G., Chan, W. C. and Kota, K. (2007). Comparison of maltodextrin and skim milk wall deposition rates in a pilot-scale spray dryer. *Powder Technology*, *179*, 84-89.

Liong, M. T. and Shah, N. P. (2005). Optimization of cholesterol removal, growth and fermentation patterns of *Lactobacillus acidophilus* ATCC 4962 in the presence of mannitol, fructo-oligosaccharide and inulin: a response surface methodology approach. *Journal of Applied Microbiology*, *98*, 1115–1126.

Lourens-Hattingh, A. and Viljoen, B. C. (2001). Yogurt as probiotic carrier food. *International Dairy Journal*, *11*, 1-17.

Lucas, A., Sodini, I., Monnet, C., Jolivet, P. and Corrieu, G. (2004). Effect of milk base and starter culture on acidification, texture, and probiotic cell counts in fermented milk processing. *Journal of Dairy Science*, *85*, 2479-2488.

Marteau, P. and Bourtron-Rualt, M. C. (2002). Nutritional advantages of probiotics and prebiotics. *British Journal of Nutrition*, *87*, 153-157.

Mattila-Sandholm, T., Matto, J. and Saarela, M. (1992). Lactic acid bacteria with health claims-interations and interference with gastrointestinal flora. *International Dairy Journal*, *9*, 25-35.

Mattila-Sandholm, T., Myllärinen, P., Crittenden, R., Mogensen, G., Fondén, R. and Saarela, M. (2002). Technological challenges for future probiotic foods. *International Dairy Journal*, *12*, 173-182.

Mayo, B., Aleksandrzak-Piekarczyk, T., Fernández, M., Kowalczyk, M., Pablo Álvarez-Martín, P. and Bardowski, J. (2010). Updates in the metabolism of lactic acid bacteria, in: Mozzi, F, Raya, R. R., and Vignolo, G. M. (Ed.), Biotechnology of Lactic Acid Bacteria: Novel Applications, Wiley-Blackwell, Iowa, USA, pp. 3-33.

Menne, E., Manson, J. E. and Nilsson, U. (1997). Prebiotic effect of the fructosylfructose: a type inulin hydrolysate in humans. In: NDO SYMPOSIUM, Wageningen, 1997. Book of abstracts. Wageningen, p.164.

Mussatto, S. I. and Mancilha, I. M. (2007). Non-digestible oligosaccharides: A review. *Carbohydrate Polymers*, *68*, 587-597.

Naidu, A. S. and Clemens, R. A. (2000). Probiotics. In: NAIDU, A.S., ed. Natural food antimicrobial systems. Boca Raton: CRC Press, p.341-462.

Niness, K. R. (1999). Inulin and oligofructose: what are they? *Journal of Nutrition*, *129*, 1402-1406.

Ohta, A. (1998). Dietary fructooligosaccharides present osteopenia after gastrectomy in rats. *Journal of Nutrition*, *128*, 106-110.

Olano, A. and Corzo, N. (2009). Lactulose as a food ingredient. *Journal of the Science of Food and Agriculture*, *89*, 1987-1990.

Oliveira, M. N. and Damin, M. R. (2003). Efeito do teor de sólidos e da concentração de sacarose na acidificação, firmeza e viabilidade de bactérias do iogurte e probióticas em leite fermentado. *Ciência e Tecnologia de Alimentos*, *23*, 172-176.

Oliveira, M. N., Sodini, I., Remeuf, F. and Corrieu, G. (2001). Effect of milk supplementation and culture composition on acidification, textural properties and microbiological stability of fermented milks containing probiotic bacteria. *International Dairy Journal*, *11*, 935-942.

Oliveira, R. P. S., Perego, P., Converti, A. and Oliveira, M. N. (2009a). The effect of inulin as a prebiotic on the production of probiotic fibre-enriched fermented milk. *International Journal of Dairy Technology*, *62*, 195-203.

Oliveira, R. P. S., Florence, A. C. R., Silva, R. C., Perego, P., Converti, A., Gioielli, L. A. and Oliveira, M. N. (2009b). Effect of different prebiotics on the fermentation kinetics, probiotic survival and fatty acids profiles in nonfat symbiotic fermented milk. *International Journal of Food Microbiology*, *128*, 467-472.

Oliveira, R. P. S., Perego, P., Converti, A. and Oliveira, M. N. (2009c). Effect of inulin on growth and acidification performance of different probiotic bacteria in co-cultures and mixed culture with *Streptococcus thermophilus*. *Journal of Food Engineering*, *91*, 133–139.

Oliveira, R. P. S., Florence, A. C. R., Perego, P., Oliveira, M. N. and Converti. A. (2011a). Use of lactulose as prebiotic and its influence on the growth, acidification profile and viable counts of different probiotics in fermented skim milk. *International Journal Food Microbiology*, *145*, 22–27.

Oliveira, R. P. S., Perego, P., Oliveira, M. N. and Converti. A. (2011b). Effect of inulin as a prebiotic to improve growth and counts of a probiotic cocktail in fermented skim milk. *LWT - Food Science and Technology*, *44*, 520-523.

Ouwehand, A. C., Salvadori, B. B., Fondén, R., Mogensen, G., Salminen, S. and Sellars, R. (2003). Health effects of probiotics and cultures-containing dairy products in humans. *IDF Bulletin*, *380*, 4-19.

Ozer, D., Akın, M. S. and Ozer, B. (2005). Effect of inulin and lactulose on survival *of Lactobacillus acidophilus* LA-5 and *Bifidobacterium bifidum* BB-02 in acidophilus-bifidus yoghurt. *Food Science and Technology International*, *11*, 19-24.

Panesar, P. S. and Kumari, S. (2011). Lactulose: Production, purification and potential applications. *Biotechnology Advances*, *29*, 940-948.

Probert, H. M., Apajalahti, J. H. A., Rautonen, N., Stowell, J. and Gibson, G. R. (2004). Polydextrose, lactotol and fructo-oligosaccharide fermentation by colonic bacteria in three-stage continuous culture system. *Applied Environment Microbiology*, *70*, 4505–4511.

Ranadheera, R. D. C. S., Baines, S. K. and Adams, M. C. (2010). Importance of food in probiotic efficacy. *Food Research International*, *43*, 1-7.

Rao, A. V. (1999). Dose-response effects of inulin and oligofructose on intestinal bifidogenesis effects. *Journal of Nutrition*, *129*, 1442-1445.

Reid, G. (2006). Scientific evidence for and against the safe use of probiotics. *Trends in Microbiology*, *14*, 348–352.

Rodrigues, D., Rocha-Santos, T. A. P., Gomes, A. M., Goodfellow, B. J. and Freitas, A. C. (2012). Lipolysis in probiotic and synbiotic cheese: The influence of probiotic bacteria, prebiotic compounds and ripening time on free fatty acid profiles. *Food Chemistry*, *131*, 1414–1421.

Roberfroid, M. B. (2001). Prebiotics: preferencial substrates for specific germs? *American Journal of Clinical Nutrition*, *73*, 406-409.

Roberfroid, M. B. (2002). Funcional food concept and its application to prebiotics. *Digestive and Liver Diseases*, *34*, 105-110.

Roberfroid, M. and Gibson, G. (1998). The bifidogenic nature of chicory inulin and its hydrolysis products. *Journal Nutrition*, *128*, 11-19.

Sako, T., Matsumoto, K. and Tanaka, R. (1999). Recent progress on research and applications of non-digestible galacto-oligosaccharides. *International Dairy Journal*, *9*, 69–80.

Scientific Psychic. Disponível em: http://www.scientificpsychic.com. Accessed on 19 March 2011.

Shah, N. P. (2000). Probiotic bacteria: selective enumeration and survival in dairy foods. *Journal of Dairy Science*, *83*, 894-907.

Shah, N. P. and Lankaputhra, W. E. V. (1997). Improving viability of *Lactobacillus acidophilus* and *Bifidobacterium spp*. in yogurt. *International Dairy Journal*, *7*, 349-356.

Shin, H. S., Lee, J. H., Pestka, J. J. and Ustunol. □. (2000). Growth and viability of commercial *Bifidobacterium* spp. in skim milk containing oligosaccharides and inulin. *Journal of Food Science*, *65*, 885–887.

Sodini, I., Lucas, A., Oliveira, M. N., Remeuf, F. and Corrieu, G. (2002). Effect of milk base and starter culture on acidification, texture, and probiotic cell counts in fermented milk processing. *Journal of Dairy Science*, *85*, 2479-2488.

Strohmaier, W. (1998). Lactulose: Status of health-related applications. *International Dairy Federation, Bulletin* 9804, 262–271.

Taguchi, A., Ohta, A., Abe, M., Baba, S., Ohtsuki, M., Takizawa, T., Yuda, Y. and Adachi, T. (1994). Effects of fructooligosaccharides on the bone and mineral absorption in the rat model with ovariectomized osteoporosis. *Meiji Seika Kaisha Nenpo*, *33*, 37-44.

Talon, R., Walter, D., Viallon, C. and Berdagué, J. L. (2002). Prediction of *Streptococcus salivarius* subsp. *thermophilus* and *Lactobacillus delbrueckii* subsp. *bulgaricus* populations in yoghurt by Curie point pyrolysis-mass spectrometry. *Journal of Microbiological Methods*, *48*, 271-279.

Tamura, Y., Mizota, T., Shimamura, S. and Tomita, M. (1993). Lactulose and its application to the food and pharmaceutical industries. *International Dairy Federation Bulletin, 289*, 43-53.

Todorov, S.D. (2009). Bacteriocins from *Lactobacillus plantarum* – production, genetic organization and mode of action. A review. *Brazilian Journal of Microbiology*, 40, 209-221.

Van Loo, J., Clune, Y., Bosscher, D. and Franck, A. (2005). Prebiotic oligofructose-enriched chicory inulin combination with probiotics: in the prevention of colon cancer in experimental models and human volunteers. *Agro Food Industry Hi Tech*, *16*, 6-8.

Vasiljevic, T. and Shah, N. P. (2008). Probiotics - from Metchnikoff to bioactives. *International Dairy Journal*, *18*, 714–728.

Voragen, A. G. I. (1998). Technological aspects of functional food-related carbohydrates. *Trends in Food Science and Technology*, *9*, 328-335.

Wang, X. and Gibson, G. R. (1993). Effects of the *in vitro* fermentation of oligofrutose and inulin by bacteria growing in the human large intestine. *Journal of Applied Bacteriology*, *75*, 273-380.

□okaee, F., Kaghazchi, T., □are, A. and Soleimani, M. (2002). Isomerization of lactose to lactulose-study and comparison of three catalytic systems. *Process Biochemistry*, *37*, 629-635.

In: Lactobacillus: Classification, Uses and Health Implications ISBN: 978-1-62081-151-1
Editors: Alba I. Perez Campos and Arturo Leon Mena © 2012 Nova Science Publishers, Inc.

Chapter XV

Lactobacillus Reuteri ATCC 55730 and L22 Display Probiotic Potential In Vitro and Protect against Salmonella-Induced Pullorum Disease in a Chick Model of Infection

Zhang Dexian, LI Rui and LI Jichang*

Laboratory of Veterinary Pharmacology and Toxicology, College of Veterinary Medicine, Northeast Agricultural University, Harbin, Heilongjiang, P.R. China

Abstract

Lactobacillus reuteri ATCC 55730 (*L. reuteri* ATCC 55730) and *Lactobacillus reuteri* L22 (*L. reuteri* L22) were studied for their probiotic potential. These two strains were able to produce an antimicrobial substance, termed reuterin, the maximum production of reuterin by these two strains was detected in the late logarithmic growth phase (16h in MRS and 20h in LB broths).

These two strains could significantly reduce the growth of *S. pullorum* ATCC 9120 in MRS broth, *L. reuteri* ATCC 55730 with a reduction of 48.2±4.15% (in 5 log) and 89.7±2.59% (in 4 log) respectively, at the same time, *L. reuteri* L22 was 69.4±3.48% (in 5 log) and 80.4±3.22% respectively. *L. reuteri* ATCC 55730 was active against the majority of the pathogenic species, including *S. pullorum* ATCC 9120 and *E. coli* O_{78}, while *L. reuteri* L22 was not as effective as *L. reuteri* ATCC 55730. The two potential strains were found to survive variably at pH 2.5 and were unaffected by bile salts, while neither of the strains was haemolytic.

Moreover, *L. reuteri* ATCC 55730 exhibited variable susceptibility towards commonly used antibiotics; but *L. reuteri* L22 showed resistant to most antibiotics in this study. *L. reuteri* ATCC 55730 consequently was found to significantly increase survival rate in a *Salmonella*-induced pullorum disease model in chick. To conclude, strain *L.*

*Corresponding authors: LI Jichang. Telephone: +86 451 5519 0674; Fax: +86 451 5519 0363. E-mail: lijichang828@yahoo.cn.

reuteri ATCC 55730 possesses desirable probiotic properties, such as antimicrobial activity and immunomodulation *in vitro*, which were confirmed *in vivo* by the use of animal models.

Keywords: *Lactobacillus reuteri* ATCC 55730 and L22; Probiotics; Antimicrobial activity; Pullorum disease

Introduction

Antimicrobial agents have been widely used for extended periods to treat the diseases caused by bacteria. Unfortunately, increasing multiple drug resistance has emerged recent years (Klarin et al., 2005), many measures have been taken to reduce the risk of infection, such as applying preventive drugs, combining use of antibiotic and hygienic and/or nutritional management practices (Bernardeau et al., 2009). Recently, several studies have indicated that many species of bacteria could be used as probiotic material, including *Lcatobacillus*, *Bifidobacterium*, *Pediococcus*, *Streptococcus*, *Bacillus* and *Escherichia coli*. Among them, *Lactobacillus* and *Bacillus subtilis* have been suggested as having the greatest potential as probiotic therapies (Fuller et al., 1986; □ani et al., 1998).

There is a long history to the safe use of *Lactobacillus* by humans, particularly in the dairy industry, where it plays a crucial role in the production of fermented milk products (O'Mahony et al., 2000). Interestingly, many members belonging to the genera *Lactobacillus* and *Bifidobacterium* are being implemented as treatment and prevention strategies for immune disorders and infectious diseases in the last decade (Spinler J.K. et al., 2008). Such strains are termed as probiotic. A probiotic is defined as 'a live microbial feed supplement which beneficially affects the host animal by improving its intestinal microbial balance'(Fuller et al., 1989). They are common inhabitants in the gastrointestinal (GI) tract of humans and animals, and there are increasing evidences that they have health-promoting effects on the host (Molin, 2001; Klarin et al., 2005). Lactobacilli can confer massive beneficial effects including inhibiting pathogenic organisms; for example, lactobacilli plays an inhibiting role to *Salmonella*, *Shigella*, *Pseudomonas* and *Helicobacter* (Bernet-Camard et al., 1997; Servin et al., 2004; Aiba et al., 1998; Hammilton-Miller et al., 2003; Sgouras et al., 2004). Besides, lactobacilli possesses numerous other health benefits for both human and animal, such as reducing lactose intolerance (Adolfsson et al., 2004) and modulating immune responses (Hart et al., 2004; Mercenier et al., 2003).

Lactobacillus reuteri (*L. reuteri*) is a major component of the *Lactobacillus* populations inhabiting the GI tract of animals (Mitsuoka et al., 1992), and considered to be one of a limited number of indigenous *Lactobacillus* species in the human intestine (Casas IA et al., 2000). Cleusix et al. (2007) have indicated that the addition of *L. reuteri* ATCC 55730 and glycerol decrease *E. coli* populations in an *in vitro* colonic fermentation model. The anti-microbial effect exerted by *L. reuteri* is due to the production of a broad-spectrum antimicrobial substance, especially reuterin (3-hydroxy-propionaldehyde) which is produced during anaerobic growth in the presence of glycerol (Talarico et al., 1988). As a heterofermentative lactobacterium, *L. reuteri* decreases the intestinal pH to a level that is unfavorable to most pathogenic bacteria, thus the antimicrobial effect of *L. reuteri* is enhanced. In this case, *L. reuteri* has been shown *in vitro* a strong anti-microbial effect against a number of pathogenic bacteria, however, *L. reuteri* has no inhibitory influence on other *Lactobacillus* species and other bacteria normally residing in the GI tract (Axelsson et al, 1989; Nollet et al, 1999).

It is widely admitted that certain probiotics could produce beneficial effects, but little is known about the mechanisms of exerting these effects. The mechanisms are different from one probiotic to another, and various factors can contribute to this event. A number of modes of action have been proposed, including the production of antimicrobial substances, the competition for nutrients, the competitive exclusion of pathogen binding and the modulation of the immune system (Parvez et al., 2006). Among them, measuring the production of antimicrobial compounds has been considered the most challenge mode in recent years (Spinler J.K. et al., 2008).

The objective of this work was to investigate two, *L. reuteri* ATCC 55730 and L22, strains as potential probiotics in *in vitro* and *in vivo* studies.

2 Materials and Methods

2.1 Bacterial Strains used in this Study

The *L. reuteri* ATCC 55730, *L. reuteri* L22, *S. aureus* (SA) ATCC 6538, *E. coli* ATCC 43889 and *S. pullorum* ATCC 9120 were purchased from China Institute of Veterinary Drugs Control. De Man Rogosa, Sharp broth (MRS), glycerol, Luria-Bertani broth (LB), agar were purchased from Sigma (USA). *Lactobacillus* strains were cultured in de Man Rogosa, Sharp (MRS) broth media at 37^0C for 24h-48h in an anaerobic chamber (Model 1025 S/N, Forma Scientific, Inc., Marietta, OH). Enteric pathogen strains were cultured in LB media for 24h at 37°C. Modifications for specific assays are detailed accordingly. For long term storage, the bacteria were kept at -80°C in MRS broth containing 20% (v/v) glycerol in their cultural broths. Prior to experimentation, strains' isolates were taken out of stock and passaged once between MRS or LB broth and agar plate, and confluent cultures were ascertained by serial dilution plating.

2.2 Preparation of Cell Free Culture Supernatants (CFCSs)

CFCSs of the potential probiotic strains were collected after growth in MRS, containing 0.3% (w/v) yeast extract and 250 mM glycerol. More precisely, fresh overnight culture supernatants of the antimicrobial activity producer strains were collected by centrifugation (12000×*g*, 15min, 4°C), adjusted at pH 7.0 and filtered (0.22μm Millex Syringe Filter, Millipore, Billerica, Mass., USA).

2.3 Kinetic of Production of Reuterin by L. Reuteri

Culture of *L. reuteri* was carried out at 30 and 37℃ in 100 mL MRS and LB broths for 24h. Medium was inoculated at 10% with 10 mL 24h pre-culture ($\approx 10^4$ cfu·mL^{-1}) of *L.*

reuteri. Samples for cell counts and anti-bacterial activities assay were taken every 2h. For cell counting, 1mL of cultured broth was mixed with 9mL sterilized phosphate-buffer saline (PBS, pH 7.2) and stirred thoroughly. After serial 10 fold dilutions with PBS, 1mL of each dilution was spread on MRS agar. Cells on the agar plate were counted after anaerobically incubated at 37°C under for 48h.

To quantify the inhibitory activity, serial twofold dilutions of the supernatant were made in MRS and assayed by the well diffusion method as described by Ennahar *et al.* (1998). For serial twofold dilutions, 0.2mL CFCS was mixed with 1.8mL broth with *S. pullorum* ATCC 9120 at concentration of 10^5~10^6 cfu·mL^{-1} in a tube and serial twofold dilutions were carried out, tubes were incubated at 37°C for 24h.

The well diffusion method was performed on LB agar by supernatant of cultured broth. Briefly, 1×10^6 cfu·mL^{-1} of the target strain was incorporated into soft agar (1% w/v) plates of the appropriate for the target strain medium. CFCSs (50μL) was transferred in holes (5 mm diameter) drilled into the agar. The palates were then incubated at 30 or 37°C, depending on the target strain, and the antimicrobial activity was recorded as growth-free inhibition zones (diameter) around the well. As controls, MRS containing 0.3% (w/v) yeast extract with 250mM glycerol adjusted at pH 7.0, were used.

The titre (expressed in activity units AU·mL^{-1}) corresponding to the reciprocal of the highest dilution showing a distinct inhibition zone on the indicator lawn was determined (Ennahar *et al.* 1998). Anti-bacterial activity against *S. pullorum* ATCC 9120 was defined in AU·mL^{-1} (Atanassova et al., 2003).

2.4 Competition Experiments

Overnight cultures (≈10^7 cfu·mL^{-1}) of *L. reuteri* strains and indicator strain (*S. pullorum* ATCC 9120) in LB broth were diluted into fresh LB medium and incubated at 6 and 37°C both individually and in co-culture. In the last case, cultures of *L. reuteri* (10^5 cfu·mL^{-1}) were incubated in LB broth with two *S. pullorum* ATCC 9120 concentrations (≈10^4 and 10^5 cfu·mL^{-1}).

At intervals (2h), samples were removed and serially diluted into PBS. The appropriate dilutions were plated on triplicate LB or MRS agar plates for viable count, and the average number of colonies obtained after 24h (for *S. pullorum* ATCC 9120) or 48h (for *L. reuteri*) anaerobic incubation at 37°Cwas used to establish the growth and survival curves.

2.5 Detection of Antimicrobial Activity

CFCSs of the potential probiotic strains were screened for inhibitory activity using the well diffusion assay. The method was taken as prior described (same as 2.3).

2.6 Assay of Acid and Bile Salt Tolerance

The two *L. reuteri* strains were grown in MRS broth for 8h at 37°C anaerobically and harvested by centrifugation at 5000×*g* for 5min at 4°C.

After washing twice with phosphate-buffered saline (PBS, pH 7.4), cell pellets were either resuspended directly in fresh MRS broth at pH 2.5 and maintained at 37°C for 0, 1 and 2 h for acid tolerance assay.

Alternatively, the bacterial suspensions in PBS were inoculated into heat-sterilized fresh fowl bile and incubated at 37°C for 0, 1 and 2 h. The viable organisms from the cultures at hours 0, 1 and 2 h were enumerated on MRS agar plates. Each experiment was carried out in triplicate. The survival or growth rate was calculated by the following formula:

- Survival rate $=CFU_{tx}/CFU_{to} \times 100\%$

Where CFU_{tx} represents the number of bacteria recovered from the mixtures at different incubation time, and CFU_{t0}, the number of bacteria inoculated.

2.7 Determinations of Antibiotic Resistance

The potential probiotic strains were tested for antibiotic resistance as previously described (Maragkoudakis et al., 2006). These experiments were performed three times in duplicate.

Both *L. reuteri* ATCC 55730 and *L. reuteri* L22 were grown anaerobically for 12h in MRS broth at 37^0C and subsequently diluted to a concentration $\sim 1\times10^6$ $cfu\cdot mL^{-1}$. Their sensitivities to antibiotics (bacitracin, cloxacillin, penicillin, streptomycin, vancomycin, teicoplanin, chloramphenicol and tetracycline; Sigma) at various concentrations (2, 4, 8, 16, 32, 64 and 128 $\mu g\cdot mL^{-1}$) were measured by serial twofold dilutions, and the minimum inhibitory concentrations (MICs) were recorded following a 24 h incubation period at 37^0C.

2.8 Hemagglutination Assay

The screening system for hemagglutination against fowl red blood cells was used according to Andreu et al (1995). *L. reuteri* ATCC 55730 and *L. reuteri* L22 were grown in MRS broth, washed and suspended to 8×10^8 $cfu\cdot mL^{-1}$ in PBS.

The bacterial suspension and erythrocyte suspension (2%) were mixed in an equal volume in the wells of a 96-well microplate. PBS was used as a negative control. The experiment was carried out in triplicate. Hemagglutination was observed visually after 1h incubation at 37^0C.

2.9 Animals

Seven to fifteen days old, mixed breed broiler chicks of either sex (b.w. 50-95g) were used. They were maintained in batches of 20 chicks at a time in a room at a temperature of 30-34^0C.

2.10 In Vivo Anti-Salmonella Chick Experiment

Overnight culture of *S. pullorum* ATCC 9120 (purchased from China Institute of Veterinary Drugs Control) in LB broth (Sigma) was centrifuged (3500×*g*, 15min, 20°C), washed twice with PBS buffer (pH 7.2), and resuspended in distilled water.

Subsequently, all chicks were intragastrically challenged with 5×10^5 cfu (300 μL) of *S. pullorum* ATCC 9120. Chicks were kept in positive-pressure cabinets and poultry survival was monitored throughout the study, which lasted 3 weeks. To test the anti- *S. pullorum* ATCC 9120 activity of potential probiotic strains, *L. reuteri* ATCC 55730 or *L. reuteri* L22 was administrated once daily at a dose of 10^8 cfu by intragastrec gavage for 5 days before infection with *S. pullorum* ATCC 9120.

2.11 Statistical Analysis

In the pullorum disease model, data were analyzed by the nonparametric one-way of variance Mann-Whitney U-test (XLSTAT software: http://www.xlstat.com) and *S. pullorum* ATCC 9120 survival in fowel experiments was analyzed using the log rank test (GraphPad Prism, v. 4.0, California, USA). *P* values less than 0.05 were considered statistically significant.

3. Results

3.1. Kinetic of Production of Reuterin Produced by L. Reuteri ATCC 55730 and L22

The culture of the strain *L. reuteri* ATCC 55730 was carried out at 37°C in MRS and LB broths. Maximum of the anti- *S. pullorum* ATCC 9120 activity was observed after 18h of cultivation at 37°C in LB broth (Fig. 1A) and after 14h in MRS medium (Fig. 1B). The activity was stable until 20h and 14h in LB and MRS broth respectively, which corresponds to the stationary *L. reuteri* growth phase. The titres were equal to $340AU\cdot mL^{-1}$ in LB broth (Fig. 1A) and $1400\ AU\cdot mL^{-1}$ in MRS broth (Fig. 1B). For *L. reuteri* L22, its anti-*Salmonella* activities in LB broth after 18h of cultivation at 37°C (Fig. 1C) and after 14h in MRS medium (Fig. 1D) were also investigated.

Maximum reuterin production of *L. reuteri* ATCC 55730 ($340AU\cdot mL^{-1}$) was detected after 18h of incubation at cell count of 6.2×10^7 $cfu\cdot mL^{-1}$ in LB broth; while that of *L. reuteri* L22 ($320\ AU\cdot mL^{-1}$) was after 22h of incubation at cell count of 9.5×10^6 $cfu\cdot mL^{-1}$ in LB broth. In MRS broth, *L. reuteri* ATCC 55730 started to produce reuterin ($100\ AU\cdot mL^{-1}$) at 4h during the early logarithmic growth phase at a cell count of 3.1×10^5 $cfu\cdot mL^{-1}$; at the same time, *L. reuteri* L22 began to produce reuterin ($100\ AU\cdot mL^{-1}$), however, its cell count was merely 7×10^4 $cfu\cdot mL^{-1}$.

For *L. reuteri* ATCC 55730, reuterin reached a maximum ($1400\ AU\cdot mL^{-1}$) after 14h of incubation (early stationary phase), at a cell count of 6.4×10^8 $cfu\cdot mL^{-1}$; and *L. reuteri* L22 achieved a maximum reuterin (1400AU·mL^{-1}) after 18h of incubation with a cell count of 5.9×10^{9} cfu·mL^{-1}.

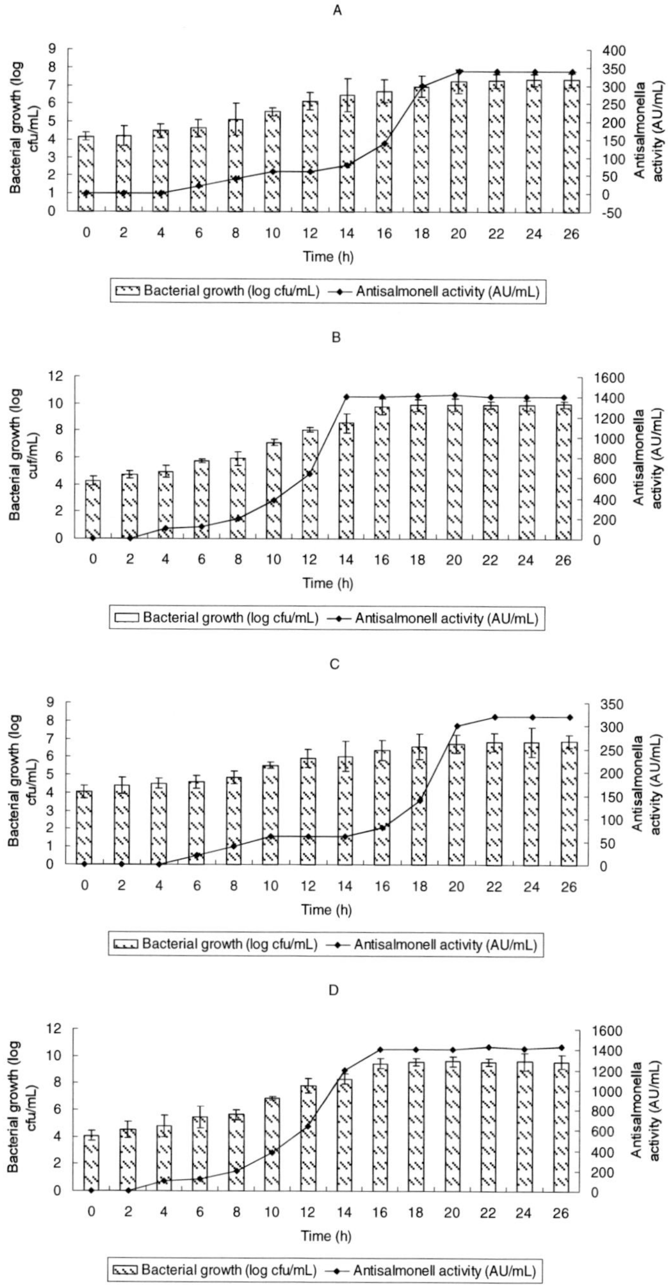

Figure 1. Growth pattern of the strain L. reuteri ATCC 55730 and L22 and kinetic of reuterin production (AU mL-1). Cultures were carried out at 37°C for 24h in LB (A and C for L. reuteri ATCC 55730 and L22 respectively) and MRS (B and D L. reuteri ATCC 55730 and L22 respectively) broths. Samples were taken every 2h for cell counts and anti-Salmonella activity assay. Results are mean ±SD of three independent determinations.

3.2 Competition Experiments

When cultures of *L. reuteri* ATCC 55730 and L22 (≈10^5 cfu·mL^{-1}) were incubated in LB broth with two *S. pullorum* ATCC 9120 concentrations (≈10^5 and 10^4 cfu·mL^{-1}) at 37℃, different antagonistic interactions were observed depending on the *S. pullorum* ATCC 9120-to-*L. reuteri* ratio as well as on the temperature of incubation. When the initial cell ratio was approximately 10 (Fig. 2B and Fig. 2D), the decrease in the number of viable *S. pullorum* ATCC 9120 coincided with the anti-bacterial activity detection in cell-free supernatants (Fig. 1B and Fig. 1D).

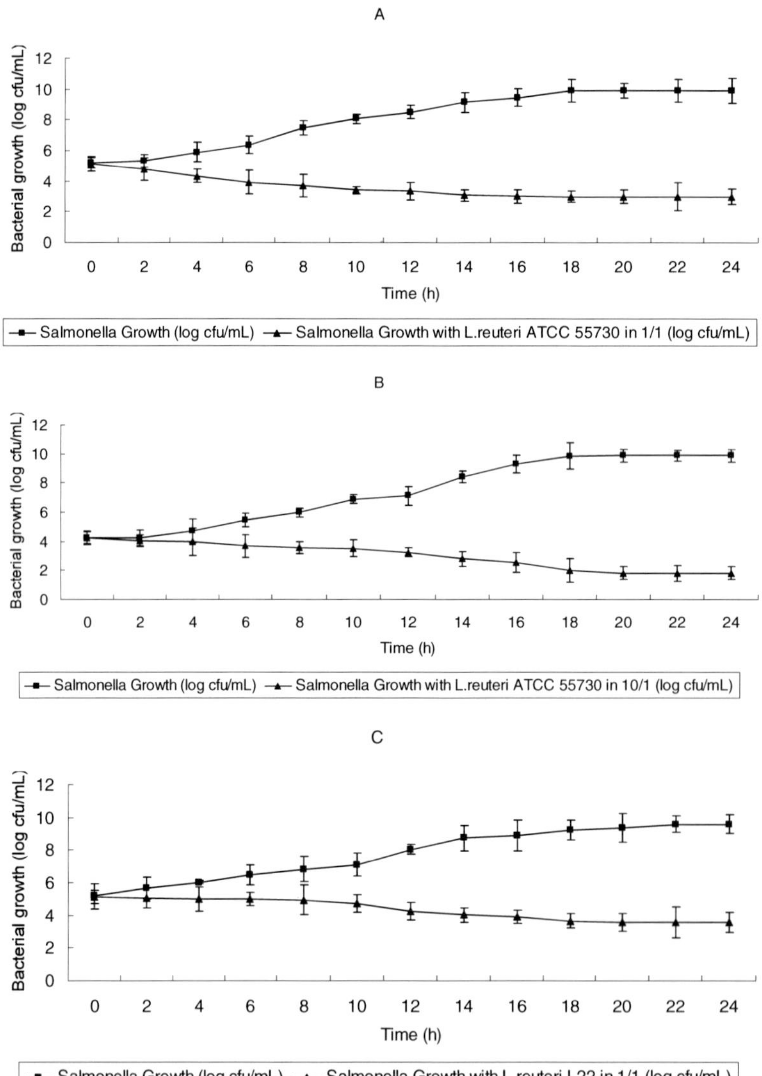

Figure 2. (Continued).

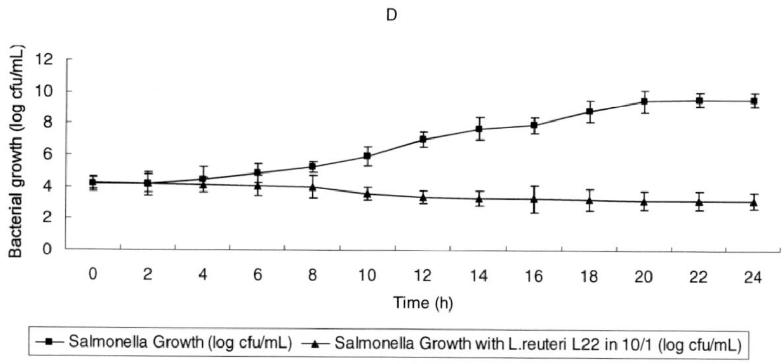

Figure 2. Effects of L. reuteri ATCC 55730 and L22 on the growth of Salmonella ATCC 9120 in co-culture at different at different ratio: for L. reuteri ATCC 55730 1/1(A), 1/10B); for L. reuteri L22 1/1(C), 1/10(D). Samples were taken each two hours for cell counts. Results are mean ±SD of three independent determinations.

However, compared with that the ratio was 1 (Fig. 2A and Fig. 2C), there was a stronger effect on *S. pullorum* ATCC 9120 when the initial cell ratio was about 10. When the initial ratio was approximately 1 (Fig. 2A and Fig. 2C), the antagonistic effect was detected much lower (after 24h at 37°C). Cocultivation at 6°C was not successful even at the highest ratios used (100). Under these conditions, no inhibition of *S. pullorum* ATCC 9120 was detected, even after prolonged incubation period for 5 days. No growth of *L. reuteri* ATCC 55730 or L22 was observed at this temperature, and no bacteriocin activity could be detected in cell-free supernatants.

3.3 Inhibitory Spectrum

L. reuteri ATCC 55730 and L22 were examined for antimicrobial activity towards 12 indicator strains. Target strains included *S. aureus, S. pullorum* ATCC 9120 and *E. coli,* which normally causing diarrhea in chick. Both MRS and mild CFCSs of the two antimicrobial producer strains were tested. Neither *L. reuteri* ATCC 55730 nor L22 was detected against strains tested in this study, while the MRS CFCSs was active against these pathogenic bacteria. The MRS CFCS of *L. reuteri* ATCC 55730 was active against *S. aureus, S. pullorum* ATCC 9120 and six *E. coli* strains, namely O_1, O_2, O_{78}, QD_1, QD_3 and S_2; at the same time, the MRS CFCS of *L. reuteri* L22 was active against *S.a pullorum* ATCC 9120 and four *E. coli* strains, which were O_2, QD_1, QD_3 and K_{12} respectively (Table 2).

3.4 Tolerance of Acid and Bile

L. reutei ATCC 55730 was resistant to both acidic condition and bile treatment during the 2h incubation at 37°C (Table 1). The result indicated that *L. reuteri* ATCC 55730 treated by the low pH and bole for 2.5h could reach 140.47±12.86% and 154.51±17.45%, respectively, in comparison with the number of bacteria inoculated.

The *L. reuteri* L22 showed substantial decrease of its viable cells in the acidic environment with about 6.65% as compared to initial counts at time zero, similarly its viability was affected by bile treatment with 97.83±12.69% in comparison with the initial number (Table 1).

Table 1. Survival rate of *L.* reuteri ATCC 55730 and L. reuteri L22 in low pH and bile. (%)

Strains	0h		1h		2h	
	Low pH	bile	Low pH	bile	Low pH	bile
L. reuteri ATCC 55730	100.64 ±14.90	100.58 ±15.81	125.58 ±15.83	120 ±15.42	140.47 ±12.86	154.51 ±17.45
L. reuteri L22	100.47 ±16.52	100.28 ±16.33	75.62 ±14.24	90.73 ±17.92	93.35 ±15.27	97.83 ±12.69

Results are mean ±SD of three independent determinations.

Table 2. Antimicrobial activity of the CFCS (skim milk + yeast extract, pH 6.5) of *L. reuteri* 55730 and L22 tested by the well diffusion assay against pathogenic bacteria

Strains	Inhibition zone (diameter, mm±SD)[a]	
	L.reuteri ATCC 55730	L.reuteri L22
Staphylococcus aureus	7.1±0.37	—[b]
Salmonella ATCC9120	12.7±0.68	6.9±0.21
E. coli		
O_1	7.6±0.77	—
O_2	11.8±0.64	9.4±0.51
O_{78}	7.9±0.44	—
QD_1	12.3±1.42	8.2±0.36
QD_2	—	—
QD_3	8.5±0.38	7.3±0.45
K_{12}	—	7.3±0.33
H_{21}	—	—
S_2	8.3±0.41	—
T_1	—	—

[a] Values are means of the results of three independent experiments.
[b] —: no inhibition.

3.5. Antibiotic Resistance and Hemagglutination Assay

L. reuteri ATCC 55730 exhibited generally no resistance to the majority of the antibiotics tested except streptomycin and vancomycin, on the contrary, *L. reuteri* L22 was highly resistant to most of the antibiotics including bacitracin, chloramphenicol, penicillin G, streptomycin, tetracycline and vancomycin (Table 3). Finally, neither strain was found to be haemolytic.

Table 3. Antibiotic resistance of strains expressed in MICs (□g mL-1) values

Strains	MICs (µg mL^{-1})[a]							
	BAC	CHL	CLO	PEN	STR	TEIC	TET	VAN
L. reutei ATCC 55730	4	2	8	4	>64	4	2	>64
L. reuteri L22	>64	16	>64	>64	>64	32	>64	>64

BAC=Bacitracin, CHL=Chloramphenicol, CLO=Cloxacillin, PEN=Penicillin G, STR=Streptomycin, TEIC=Teichoplanin, TET=Tetracycline, VAN=Vancomycin.

[a] Values are means of the results of two independent experiments.

3.6. Salmonella-Induced Pullorum Disease Model

In order to link the *in vitro* immunomodulation capacity of strains to its *in vivo* effects, we evaluated its potential to protect chick from pullorum disease induced by oral administration of *S. pullorum* ATCC 9120.

In this experiment, we comprared the development of *Salmonella*-induced pullorum disease in chick that were either not treated with any bacteria (control group) or treated with *L. reuteri* ATCC 55730 or L22 (20 chicken/group).

Characteristic features of pullorum disease were observed, including white diarrhea, liver became intumesce with punctiform and stripe bleeding, depressed, spleen is about 2-fold larger than that in normal ones with punctiform bleeding, serous exudates was observed in pericardium.

3.7. In Vivo Anti-Salmonella Chick Experiment

In our previous study, *L. reuteri* ATCC 55730 and L22 were found to exert antimicrobial activity against *S. pullorum* ATCC 9120.

More precisely, CFCSs (MRS culture supernatants, 10X concentrated by freeze-drying, pH 7.0) of the two lactobacilli decreased *S. pullorum* ATCC 9120 viability by 1.5-2.0 log after 24h of incubation with the pathogenic bacterium.

Taking into account these results, *L. reuteri* ATCC 55730 and *L. reuteri* L22 were applied in a *S. pullorum* ATCC 9120-infected chick model.

As shown in Fig. 3, survival of chicken after the oral challenge with *S. pullorum* ATCC 9120 was monitored for 20 days after infection.

Survival in the experimental group, receiving *L. reuteri* ATCC 55730, was significantly higher ($P<0.05$) when compared with the control group challenged only with *S. pullorum* ATCC 9120.

In contrast, no significant difference was observed between the group receiving *L. reuteri* L22 and the control.

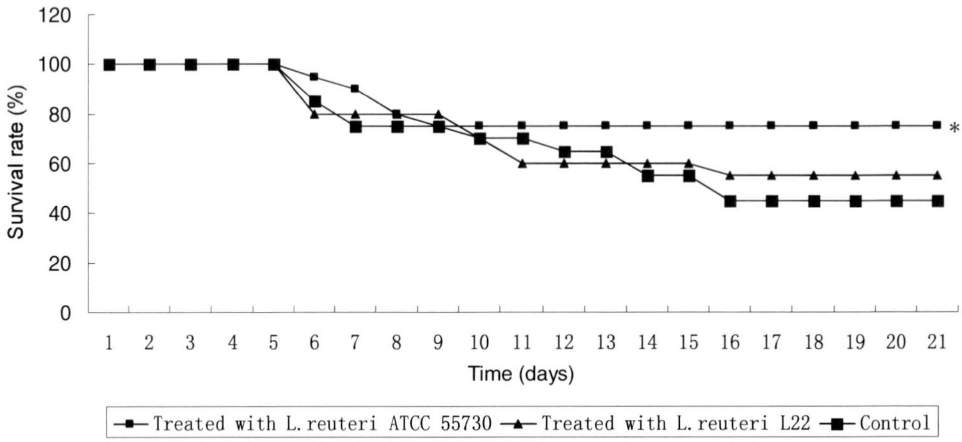

Figure 3. Survival of chicken challenged with Salmonella ATCC 9120. Salmonella ATCC 9120 was given once by oral gavage at 5×105 cfu per chick; Lactobacilli were administered once daily also by oral gavage for 5 days before infection with Salmonella ATCC 9120. Results are from one experiment (20 chicks per group). Significant difference (*$P<0.05$) are indicated in comparison with the control group challenged with Salmonella ATCC 9120 and not treated with Lactobacilli.

Discussion

The beneficial effects of live bacteria, termed probiotics, on animal health are increasingly being promoted during the last years. Most probiotics are lactic acid bacteria (LAB) isolated from the intestinal microbiota of healthy animals, or non-animal strains from fermented products. The selection criteria for LAB to be used as ‘probiotics’ in food, include the ability to produce antimicrobial substances towards pathogens, withstand transit through the GI tract, modulate immune responses and association with health benefits. In this context, the aim of the present study was to apply established *in vitro* tests and animal models to evaluate the probiotic potential of two LAB strains isolated from healthy chicken and already characterized as antimicrobial producers.

Probiotics may suppress the proliferation and virulence of bacterial pathogens by different mechanisms. Commensal-derived probioics may produce diverse antimicrobial factors including bacteriocins and non-peptide compounds, and these strains may offer therapeutic alternatives in the era of multidrug-resistant pathogens (Caglar E et al., 2006). *L. reuteri* ATCC 55730 and L22 were screened for antimicrobial activity towards several target microorganisms. These two strains demonstrated differential inhibitory activities when tested against pathogenic bacteria. Among these pathogen bacteria, *L. reuteri* ATCC 55730 showed the strongest antimicrobial activity against *S. pullorum* ATCC 9120, while *L. reuteri* L22 did not exhibit probiotic characteristics to majority of pathogenic bacteria. Interestingly, the inhibitory effect of two strains was not relative with species and/or strain. The antimicrobial effect was observed only when the strain grown in MRS, thus suggesting that this is due to the residential conditions. Besides, *L. reuteri* ATCC 55730 had showed powerful effect against the *E. coli* strains, including important pathogenic *E. coli* O_2 and O_{78}. This antimicrobial activity of *L. reuteri* ATCC 55730 can be further exploited in the field of additives in animal feeds and probiotics.

L. reuteri cultivated in anaerobic condition could convert glycerol into a broad spectrum antimicrobial substance, termed reuterin (Chung et al., 1989). Previous work has demonstrated that growing cells of *L. reuteri* had the ability of producing reuterin during glucose/glycerol cofermentation under anaerobic conditions (El-Ziney et al., 1998). The production of reuterin by *L. reuteri* ATCC 55730 when incubated in MRS containing 250mM glycerol anaerobically for 24h at 37°C was 258.53mM (Spinler J.K. et al., 2008). In our study, data imply that *L.reuteri* strains vary in their ability to convert glycerol to reuterin (Fig.1), yielding differences in reuterin synthesis between strains, and we measured the kinetic of production of reuterin produced by these two strains. The time of reuterin concentration reaching maximum was similar as observed for other bacteriocins (Toit et al., 2000). The titre of reuterin was higher in the MRS medium (both two strains were 1400 AU mL^{-1}) than in LB broth (for *L. reuteri* ATCC 55730 and L22 were 340 and 320 AU·mL^{-1} respectively), these results are not surprising since it is well known that the culture medium affects reuterin production (Parente et al., 1999). Reuterin is known to exist in three forms, with the aldehyde form being biologically relevant. The dimeric form (biologically inactive) of purified reuterin is predominant at high concentrations (4.9 M), while the hydrated molecule is prevalent from 0.030 to 1.4 M (Vollenweider S. et al., 2003), while Vollenweider et al (2003) also suggest that the biologically active monomeric form of reuterin is dominant at concentrations in the micromolar range.

Before reaching the intestinal tract where *L. reuteri* exerts its probiotic effect, cells are exposed to various environmental stresses, one of which is the detergent-like effect of bile acid, which could disrupt cell membranes (Gunn, 2000). Therefore, resistance to bile salts is considered an important property in strains envisaged as probiotics. However, there is still no consensus about the precise concentration to which the selected strain should be tolerant. The physiological concentration of bile acids in the small intestine is between 5 and 20mM (Hofmann, 1991). In this study, a concentration of 12.255mM heat-sterilized fresh fowl bile was used, which is higher than 7.353mM or 3.677μM bile salt that have previously been used by other investigators of bacterial probiotic properties (Kociubinsky et al., 1999; Chou et al., 1999; Zárate et al., 2000). As already pointed out by various workers in the field (Fernández et al., 2003; Tannock et al., 1989), in *in vitro* studies can only partially mimic the actual in situ conditions in the gut ecosystem. Nevertheless, such *in vitro* systems remain powerful tools especially for screening numerous samples. Our data imply that the survival rate of *L. reuteri* ATCC 55730 is much higher (with154.51±17.45%) than that in *L. reuteri* L22 with 97.83±12.69%. This may suggest that *L. reuteri* ATCC 55730 have a much higher number reaching the intestinal tract.

Absence of haemolytic activity and antibiotic resistance is considered safety prerequisite for the selection of a probiotic strain (Joint FAO/WHO, 2002). To ensure the safety of two *L. reuteri* strains, haemolytic activity and antibiotic resistance were carried out in our study. And the results indicated that neither strain tested exhibited haemolytic activity, confirming previous studies where strains of *Lactobacillus* were shown to be non-haemolytic (Maragkoudakis et al., 2006).

One of the required properties of probiotic strains is that they are safe for human and animal consumption. Such safety also includes that they do not harbour acquired and transferable antibiotic resistances (Pinto M.G.V. et al., 2006). It is generally known that some *Lactobacillus* species display intrinsic resistance to glycopeptide antibiotics. Also, low resistance of some lactobacilli towards ß-lactams antibiotics or towards antimicrobial agents that inhibit protein or mRNA synthesis have been reported (Danielsen et al., 2003; Mathur et al., 2005). Similar profiles were found in this study, except in the case of chloramphenicol and tetracycline, where *L. reuteri* L22 were found to be resistant. In the case of *L. reuteri* 55730, the low resistance to the most common antibiotics could be an additional criterion for selecting this strain as a potential probiotic. However, the tetracycline MIC result in our study is conflicted with the result of Rosander et al (2008), and the reason for the difference is not known. According to other researchers (Barbosa et al., 1999; Scott et al., 2000), the ribosome protection-type resistance gene *tet* (W) is responsible for the most wide spread tetracycline resistance genes in environmental samples (Aminov et al., 2001; Villedieu et al., 2003). We did not detect the ribosome protection-type resistance gene *tet*(W), which need a further study in the future.

Immune modulation by probiotics is presumed to be one of the main mechanisms of probiotic action in animal health. In previous studies, probiotic bacteria were shown to exert potent immune effects and exhibited significant differences in their capacity to induce a specific cytokine profile after stimulation of immune cells (Christensen et al., 2002).

In our study, we selected *L. reuteri* ATCC 55730 and L22 to carry out anti- *S. pullorum* ATCC 9120 experiment *in vivo*. *L. reuteri* ATCC 55730 displayed potent anti-*Salmonella* potential, both *in vitro* and *in vivo*. These results suggest that not only the antimicrobial compounds but mainly the immunomodulatory properties of *L. reuteri* 55730 are responsible for the anti-*Salmonella* activity *in vivo*. However, the underlining mechanism needs further elucidation.

In conclusion, two antimicrobial producer strains, namely *L. reuteri* ATCC 55730 and L22 were studied in order to evaluate their probiotic poetential. *L. reuteri* ATCC 55730 exhibited no haemolytic activity and antibiotic resistance; while *L. reuteri* L22 exhibited antibiotic resistance, which is not coincidence with the safety criteria.

Additionally, *L. reuteri* ATCC 55730 displayed antimicrobial activity against the majority of the target strains test, but *L. reuteri* L22 did not show the ability to these pathogenic bacteria. These two strains had no obvious difference in acid and bile tolerance, producing reuterin and competition experiments.

Moreover, *L. reuteri* ATCC 55730 was found to possess promising immunomodulatory properties that were correlated with the protection against *S. pullorum* ATCC 9120 trails and seems to be correlated also with anti-*Salmonella* activity exerted by this strain *in vivo*. Consequently, the results warrant further studies, including influence on GI tract ecology and the immunomodulatory effect of *L. reuteri* ATCC 55730 in the chick.

Acknowledgments

This study was supported by Major Program of Revitalization of Northeast Old Industrial Base of Educational Commission of Heilongjiang Province of China (1152gzd07).

References

Adolfsson, O., Meydani, S.N., Russell, R.M., 2004. *Yogurt and gut function.* American Journal of Clinical Nutrition 80, 245-256.

Aiba, Y., Suzuki, N., Kabir, A. M. A., Tagaki, A., & Koga, Y., 1998. Lactic acid mediated suppression of Helicobacter pylori by the oral administration of Lactobacillus salivarious as a probiotic in a gnotobiotic murine model. American Journal of Gastroenterology 93, 2097–2101.

Aminov, R. I., N. Garrigues-Jeanjean, and R. I. Mackie., 2001. *Molecular ecology of tetracycline resistance: development and validation of primers for detection of tetracycline resistance genes encoding ribosomal protection proteins.* Applied Environmental Microbiology 67, 22–32.

Andreu, A., Stapleton, A.E., Fennell, C.L., Hillier, S.L., Stamm, W.E., 1995. *Hemagglutination, adherence, and surface properties of vaginal Lactobacillus species.* Journal of Infectious Disease 171, 1237-1243.

Atanassova, M., Choiset, Y., Dalgalarrondo, M., Chobert, J. M., Dousset, X., &Ivanova, I., 2003. *Isolation and partial biochemical characterization of a proteinaceous anti-bacteria and anti-yeast compound produced by Lactobacillus paracasei subsp. paracasei strain 55730.* International Journal of Food Microbiology 87, 63-73.

Axelsson, L., Chung, T.L., Dobrogosz, W.J., 1989. *Production of a broad-spectrum antimicrobial substance by Lactobacillus reuteri.* Microbial Ecology Health and Disease 2, 131–136.

Barbosa, T.M., Scott K.P., Flint H.J., 1999. *Evidence for recent intergeneric transfer of a new tetracycline resistance gene, tet (W), isolated from Butyrivibrio fibrisolvens, and the occurrence of tet (O) in ruminal bacteria.* Applied Environmental Microbiology 1, 53–64.

Bernardeau, M., Gueguen, M., Smith, D.G.E., Corona-Barrera, E., Vernoux, J.P., 2009. In vitro antagonistic activities of Lactobacillus spp. against Brachyspira hyodysenteriae and Brachyspira pilosicoli. Veterinary Microbiology 138, 184-190.

Bernet-Camard, M.F., Lievin, V., Brassart, D., Neeser, J. R., Servin, A. L., Hudault, S. 1997. *The human Lactobacillus acidophilus strain LA1 secretes a nonbacteriocin antibacterial substance(s) active in vitro and in vivo.* Applied Environmental Microbiology 63, 2747–2753.

Caglar, E., Kavaloglu Cildir S., Ergeneli S., Sandalli N., Twetman S., 2006. *Salivary mutans streptococci and lactobacilli levels after ingestion of the probiotic bacterium Lactobacillus reuteri ATCC 55730 by straws or tablets.* Acta Odontol Scand 64, 314-318.

Casas, I.A., Dobrogosz, W.J., 2000. *Validation of the probiotic concept: Lactobacillus reuteri confers broad-spectrum protection against disease in humans and animals.* Microbial Ecology Health and Disease 12, 257-285.

Chou, L.,Weimer, B., 1999. *Isolation and characterisation of acid and bile-tolerant isolates from strains of Lactobacillus acidophilus*. Journal of Dairy Science 82, 23–31.

Christensen, H.R., Frokiaer, H., Pestka, J., 2002. *Lactobacilli differentially modulate expression of cytokines and maturationsurface markers in murine dendritic cells.* Journal of Immunology 168, 171-178.

Chung, T.C., Axelsson, L., Lindgren, S.E., Dobrogosz, W.J., 1989. *In vitro studies on reuterin synthesis by Lactobacillus reuteri,* Microbial Ecology in Health and Disease. 2, 137–144.

Cleusix, V., Lacroix, C., Vollenweider, S., Duboux, M., Le, B.G., 2007. *Inhibitory activity spectrum of reuterin produced by Lactobacillus reuteri against intestinal bactera.* BMC Microbiology 7, 101.

Danielsen, M., Wind, A., 2003. *Susceptibility of Lactobacillus spp. to antimicrobial agents.* International Journal of Food Microbiology 82, 1-11.

El-Ziney, M.G., Arneborg, N., Uyttendaele, M., Debevere, J., Jakobsen, M., 1998. *Characterization of growth and metabolite production of Lactobacillus reuteri during glucose/glycerol cofermentation in batch and continuous cultures.* Biotechnology Letters 20, 913–916.

Ennahar, S., Aoude-Werner, D., Assobhei, O., Hasselmann, C., 1998. *Antilisterial activity of enterocin 81, a bacteriocin produced by Enterococcus faecium WHE 81 isolated from cheese.* Journal of Applied Microbiology 85, 521-526.

Fernández, M.F., Boris, S., Barbés, C., 2003. *Probiotic properties of human lactobacilli strains to be used in the gastrointestinal tract.* Journal of Applied Microbiology 94, 449–455.

Food and Agriculture Organization of the United Nations/World Health Organization. *Guidelines for the evaluation of probiotics in food. Report of a Joint FAO/WHO Working Group on Drafting Guidelines for the Evaluation of Probiotics in Food.* London, Ontario, Canada: FAO/WHO; 2002.

Fuller, R., 1986. *Probiotics.* Appl. Bacteriol., Symposium Supplement, 1S–7S.

Fuller, R., 1989. *Probiotics in man and animals.* Journal of Applied Bacteriology 66, 365-378

Gunn, J.S., 2000. *Mechanisms of bacterial resistance and response to bile.* Microbes and Infection 2, 907-913.

Hammilton-Miller, J.M.T. 2003. *The role of probiotics in the treatment and prevention of H. pylori infection.* International Journal of Antimicrobial Agents 22, 360–366.

Hart, A.L., Lammers, K., Brigidi, P., Vitali, B., Rizello, F., Gionchetti, P., Campieri, M., Kamm, M.A., Knight, S.C., Stagg, A.J. 2004. *Modulation of human dendritic cell phenotype and function by probiotic bacteria.* Gut 53, 1602-1609.

Hofmann, A., 1991. *Enterohepatic circulation of bile acids.* In: Schultz, S.G.,Forte, J.G., Rauner, B.B. (Eds.), Handbook of Physiology. Section 6: the gastrointestinal system, vol. 3, pp. 567–580.

Klarin, B., Johansson, M.L., Molin, G., Larsson, A., Jeppsson, B., 2005 *Adhesion of the probiotic bacterium Lactobacillus plantarum 299v onto the gut mucosa in critically ill patients: a randomised open trial.* Crit Care 9, R285–93.

Kociubinsky, G., Pérez, P., Antoni, D.G., 1999. *Screening of bile resistance and bile precipitation in Lactic Acid Bacteria and Bifidobacteria.* Journal of Food Protection 62, 905–912.

Maragkoudakis, P.A., Zoumpopoulou, G., Miaris, C., Kalantzopoulos, G., Pot B., Tsakalidou, E. 2006. *Probiotic potential of Lactobacillus strains isolated from dairy products.* Inernational Dairy Journal 16, 189-199.

Mathur, S., Singh R., 2005. *Antibiotic resistance in food lactic acid bacteria-a review.* International Journal of Food Microbiology 105, 281-295.

Mercenier, A., Pavan, S., Pot, B., 2003. *Probiotics as biotherapeutic agents: present knowledge and future prospects.* Current Pharmaceutical Design 9, 175-191.

Mitsuoka, T. *The human gastrointestinal tract. Wood BJB, editor. 1992.* Lactic acid bacteria in health and disease, vol. 1. London: Elsevier, p. 76.

Molin, G. 2001 *Probiotics in foods not containing milk or milk constituents, with special reference to Lactobacillus plantarum 299v.* The American Journal of Clinical Nutrition 73, 380–5.

Nollet, L.J.A., Pereira, D.I., Verstraete, W., 1999. *Effect of a probiotic bile salt hydrolytic Lactobacillus reuteri on the human gastrointestinal microbiota as simulated in the SHIME reactor system.* Microbes Ecology Health and Disease 11, 13–21.

O'Mahony, A., O'Sullivan, T., Walsh, Y., Vaughan, A., Maher, M., Fitzgerald, G.F., Van Sinderen, D., 2000. *Characterization of antimicrobial producing lactic acid bacteria from malted barley.* Journal of Institure of Brewing 106, 403–410.

Parente, E., Ricciardi, A., 1999. *Production, recovery and purification of bacteriocins from lactic acid bacteria.* Applied Microbiology and Biotechnology 52, 628-638.

Parvez, S., Malik, K.A., Ah Kang, S., Kim, H.Y., 2006. *Probiotics and their fermented food products are beneficial for health.* Journal of Applied Microbiology 100, 1171-1185.

Pinto, M.G.V., Charles, M.A.P., Franz, Schillinger, U., Holzapfel, WH., 2006. *Lactobacillus spp. with in vitro probiotic properties from human faeces and traditional fermented products.* International Journal of Food Microbiology 109, 205-214.

Rosander, A., Connolly, E., Roos, S., 2008. *Removal of antibiotic resistance gene-carrying plasmids from Lactobacillus reuteri ATCC 55730 and characterization of the resulting daughter strain, L. reuteri DSM 17938.* Applied and Environmental Microbiology 74, 6032-6040.

Scott, K.P., Melville, C. M., Barbosa, T. M., Flint, H.J., 2000. *Occurrence of the new tetracycline resistance gene tet (W) in bacteria from the human gut.* Antimicrobial Agents and Chemother 44, 775–777.

Servin, A.L., 2004. *Antagonistic activities of lactobacilli and bifidobacteria against microbial pathogens.* FEMS Microbiology Reviews 28, 405-440.

Sgouras, D., Maragkoudakis, P., Petraki, K., Martinez-Gonzalez, B., Eriotou, E., Michopoulos, S., Kalantzopoulos, G., Tsakalidou, E., Mentis, A., 2004. *In vitro and in vivo inhibition of Helicobacter pylori by Lactobacillus casei strain Shirota.* Applied and Environmental Microbiology 70, 518-526.

Spinler, J.K., Malai, T., Cheryl, L. R., Ching, N.Ou, Somying, T., James, V., 2008. *Human-derive probiotic Lactobacillus reuteri demonstrate antimicrobial activities targeting diverse enteric bacterial pathogens.* Anaerobe 14, 166-171.

Talarico, T.L., Casas, I.A., Chung, T.C., 1988, *Production and isolation of reuterin, a growth inhibitor produced by Lactobacillus reuteri.* Antimicrobial Agents and Chemother 32, 1854–8.

Tannock, G.W., Dashkevitz, M.P., Feighner, S.D. 1989. *Lactobacilli and bile salt hydrolase in the murine intestinal tract.* Applied Environmental Microbiology 55, 1848–1851.

Toit, D.M., Franz, C.M.A.P., Dicks, L. M., Holzapfel, W.H., 2000. *Preliminary characterization of Bacteriocin produced by Enterococcus faecium and Enterococcus faecalis isolated from pig faeces.* Journal of Applied Microbiology 88, 482-494.

Villedieu, A., az-Torres, M. L., Hunt, N., McNab, R., Spratt, D.A., Wilson, M., Mullany, P. 2003. *Prevalence of tetracycline resistance genes in oral bacteria.* Antimicrobial Agents and Chemother 47, 878–882.

Vollenweider, S., Grassi, G., Konig, I., Puhan, Z., 2003. *Purification and structural characterization of 3-hydroxypropionaldehyde and its derivatives.* Journal of Agricultural Food Chemistry 51, 3287-3293.

Zani, J.L., 1998, *Effect of probiotic CenBiot on the control of diarrhea and feed efficiency in pigs.* Journal of Applied Microbiology 84, 68–71.

Zárate, G., Pérez, C.A., González, S., Oliver, G. 2000. *Viability and β-galactosidase activity of diary propionibacteria subjected to digestion by artificial gastric and intestinal fluids.* Journal of Food Protetion 63, 1214–1221.

In: Lactobacillus: Classification, Uses and Health Implications ISBN: 978-1-62081-151-1
Editors: Alba I. Perez Campos and Arturo Leon Mena © 2012 Nova Science Publishers, Inc.

Chapter XVI

Probiotics in Pediatric Diarrheal Diseases

Tapas K. Sabui[1] and Sudipta Misra[2*]
[1]North Bengal Medical College,
West Bengal, India
[2]Medical College of Georgia
Augusta, GA, US

Abstract

Probiotics have recently gained popularity as therapeutic agents in different gut and non-gut related disease conditions to prevent or to reduce the severity or to gain effective early control of the disease. However, the cost benefit analyses of probiotics treatment have not been done. We have reviewed the role of probiotic in acute and persistent diarrhea in children in this article.

Meta-analysis and evidence-based reviews suggest modest beneficial effect of probiotics, especially LGG, in infectious diarrhea in children. In a recent meta analysis that included 63 RCTs with majority (56) pediatric studies, the authors concluded that probiotics appeared to be safe and had beneficial effects in shortening the duration and reducing stool frequency in acute infectious diarrhea. The studies done in developing countries failed to show any significant benefit. The response again is dose and strain specific. LGG was found to be most effective. However, the clinical benefit of reducing duration of mild to moderate diarrhea by a day was not clear to justify the cost of probiotics. The use of probiotic for prevention of diarrhea is not recommended with the currently available data.

Meta-analysis of 4 trials of probiotics in persistent diarrhea with a total number of 464 participants showed that probiotics reduced the duration of diarrhea (mean difference 4.02 days, 95% CI 4.61 to 3.43 days, n=324, 2 trials). Stool frequency was reduced with probiotics in two trials. One trial with a small patient population reported a significantly shorter hospital stay. No adverse events were reported. Authors concluded that there is

* Tel 706 721 4724. E Mail smisra@pol.net.

limited evidence suggesting probiotics may be effective in treating persistent diarrhea in children.

The key factor in the pathogenesis of antibiotic associated diarrhea (AAD) is a disturbance in normal intestinal microflora. Therefore, the probiotics may be effective in preventing AAD. A review of six RCTs revealed that there was a moderate beneficial effect of LGG, B. Lactis, S. thermophilus and S.boulardii in preventing AAD.As AAD is uncommon in children, mostly mild in severity and responds well to discontinuation of antibiotics; its routine administration with antibiotic is not recommended.

There had been a recent surge of interest in the therapeutic use of probiotics. Probiotics are unlikely to cause harm even if ineffective in altering the course or preventing a disease condition. However, there is intense commercial pressure in promoting probiotics capitalizing on the public fascination with 'natural remedy'. As always, medical recommendation for probiotic use should be based on objective evidence including cost benefit analysis. In this chapter we have presented evidence based recommendations for probiotic use in children with diarrhea.

There is a long tradition of consuming probiotics laden fermented foods such as yoghurt as a part of normal diet in many cultures around the globe. The beneficial effects of these foods and supplements were well known to our ancestors. While the exact mechanism of this beneficial effect remains unknown, probiotics are now thought to have a modulating effect on the immune response at the gut mucosa.

Gut microflora is an organ inside an organ. There are more than one trillion bacterias in the gut outnumbering the total number of cells in a human. The single cell layer of epithelium lining the lumen of intestine acts as a physical barrier, arbiter and negotiator between luminal bacteria and sub mucosal immune system. With the help of receptors such as toll like receptors, these microorganisms can up or down regulate the inflammatory response of the gut cells. This change in inflammatory response appears to be the etiology of many gastrointestinal diseases. One of the common manifestations of disease of the gastrointestinal tract is diarrhea. For this review, we will confine ourselves to the use of probiotics in diarrheal disorders in children.

Probiotics are live microbes, bacteria or fungi that colonize the gastrointestinal tract and probably have immunomodulatory effect. These have been defined as an oral supplement or a food product that contains a sufficient number of viable microorganisms to alter the microflora of the host and has the potential for beneficial health effects [1] The criteria for a therapeutically viable probiotic are (1) human origin, (2) non pathogen, (3) resistance to processing, that is, viability in delivery vehicles, (4) stability in acid and bile, (5) adhesion to target epithelial tissue, (6) ability to persist in the gastrointestinal tract, (7) production of antimicrobial substances, (8) ability to modulate the immune system and (9) ability to influence metabolic activities [2].Though a wide variety of probiotics are available commercially, their use should be species and disease specific. The heterogeneity of probiotics should be recognized by the clinician. All probiotics are not the same. Specific probiotic or combination of probiotic may be useful in a certain disease condition.

Most of the probiotics are members of genera Lactobacillus, Bifidobacterium, and Streptococcus. In pediatric diarrhea lactobacillus rhamnosus GG, other lactobacilli and a fungus, Sachharomycis boulardii have been commonly used. Probiotics sold for medicinal purposes tend to be of higher quality than those found in dairy foods or supplements.

However, these are not regulated in most of the countries including the USA and there is no effective control on the quality of products sold for nutritional or even therapeutic purposes. Health care professionals and parents should be aware of these variations. Effective probiotics must contain an adequate number of viable colony-forming units and should be given in a timely manner for a sufficient duration. The optimal dose of probiotics is yet to be settled by pharmacokinetic studies. The minimum effective therapeutic dose is 10^6 to 10^9 CFUs/day has been cited [3].

Lactobacillus rhamnosus GG (ATCC 53103) was one of the first commercially available probiotics. It is a strain of *L. rhamnosus* first isolated in 1983 from the intestinal tract of a healthy human subject by scientists Barry Goldin and Sherwood Gorbach. [5] The "GG" suffix is derived from the last names of the discoverers who patented it in April 1985. The patent claims that the *L. rhamnosus* GG (ATCC 53103) strain is acid and bile stable, has a great avidity for human intestinal mucosal cells, and produces lactic acid. [6] In 2005, *L. rhamnosus* GG was first successfully used to treat gastrointestinal carriage of vancomycin-resistant enterococcus in renal patients [7].

Prevention of Diarrhea

Rotavirus is the most common cause of infantile diarrhea in both developed and developing countries. Vaccination against rotavirus is the best preventive strategy. However, the expensive vaccines and delivery network are beyond the reach of many children especially those in the developing countries. Probiotics have been proposed as an alternate option. LGG, *S thermophilus*, *Lactobacillus casei*, *B lactis*, or *Lactobacillus reuteri,* mixed with milk or infant formula or given as an oral supplement, have been used in randomized controlled trials (RCT). Results of published RCTs have indicated that there is a modest benefit of probiotics in preventing acute gastrointestinal tract infections in healthy infants and children. [8-11]. Rotavirus was the most common cause of acute diarrhea in these RCTs. A meta-analysis of the studies in child care centers indicated that approximately 7 children would need to be treated with LGG to prevent 1 child from developing nosocomial rotavirus gastroenteritis [12].

At the present stage of knowledge, there is not enough evidence to recommend probiotic to prevent infectious diarrhea. Rotavirus vaccination should be the first line preventive strategy. However, probiotics may be used in select vulnerable population of children.

Treatment of Acute Infectious Diarrhea

The rationale for using probiotics in acute infectious diarrhea is based on the assumption that they act against intestinal pathogens. However, the exact mechanism of their action remains unclear. The proposed hypothesis includes synthesis of antimicrobial substances,, competition for nutrients required for growth of pathogens , competitive inhibition of adhesion of pathogens , modification of toxins or toxin receptors , and stimulation of nonspecific and specific immune responses to pathogens [13]. *Lactobacillus* species (particularly LGG and *L plantarum* strain 299v) inhibit binding of *E. coli* strains to intestine derived epithelial cells grown in tissue culture by stimulation of synthesis and increased secretion of mucins. [14] However, the clinical efficacy as well as cost effectiveness of probiotics in the treatment and prevention of acute infectious diarrhea has not yet been established beyond a reasonable doubt.

Well-conducted RCTs in healthy children in developed countries have provided good data on the therapeutic benefit of probiotics in children with acute infectious diarrhea. In a randomized, double-blind, placebo-controlled trial by Szymanski et al, administration of LGG shortened the mean duration of acute rotavirus diarrhea by 40 hours, but duration of diarrhea of other etiologies was not affected. Probiotic administration also shortened mean duration intravenous rehydration by 18 hours. [15] Results of several other meta-analyses [3] and a Cochrane review points towards a modest beneficial effect of probiotics for treatment of acute infectious diarrhea in children. [4,16-18]. These reports indicate that probiotics reduce the number of diarrheal stools and the duration of the diarrhea by approximately 1 day. The benefit is strain dependent. LGG is the most effective probiotic reported to date and is dose dependent, doses greater than 10^{10} colony-forming units being most effective. It is useful mostly in rotavirus diarrhea. However, studies vary remarkably in geographical locations (developed and developing countries), presence of malnutrition, source, dose, and form of Lactobacillus used, and endpoints. Since the publication of these meta-analyses, 3 studies with LGG and related species from developing countries and another 1 published in Polish language have failed to confirm the beneficial effect. [19-21] Our own study supports these recent observations. We could not establish a clear advantage of LGG in acute infantile diarrhea irrespective of rotavirus infection status. [22] The cost benefit ratio of probiotic administration is not clear. The cost of a full course of probiotics in a developing country is equal to daily per capita income. It is not known whether probiotic treatment decreases hospital stay, treatment cost, morbidity or mortality. Most of the studies appear to include mild to moderate cases of diarrhea where shortening the duration of diarrhea by about 24 hours may not have much clinical significance. Routine administration of probiotic may put additional financial burden on the family.

Persistent Diarrhea

Viruses, parasites, and bacterial pathogens that cause acute diarrhea are capable of producing persistent diarrhea as well [23]. These microbes account for 3% to 20% of all diarrheal episodes in children aged less than five years [24]. It is also directly responsible for 36% to 54% of all diarrhea-related deaths according to two large community-based studies [25-26]. The World Health Organization (WHO) defines persistent diarrhea as an illness of proven or presumed infectious etiology that lasts 14 days or more [27]. Celiac disease, food-related enteropathies, and congenital enteropathies induced diarrheas are not included in this category. As the pathogesis of this diarrhea is not well understood, the management is mostly supportive. This includes adequate dietary management, micronutrient supplementation, adequate rehydration, and antimicrobials. As probiotics are generally safe and tried with some benefit in acute infectious diarrhea, it has also been used in children with persistent diarrhea. However, the effects of probiotics in acute diarrhea are not generalizable to persistent diarrhea, since most trials included healthy, well nourished children from developed countries. The main difficulty here is the paucity of literature. Meta-analysis of 4 trials with a total number of 464 participants showed that probiotics reduced the duration of persistent diarrhea (mean difference 4.02 days, 95% CI 4.61 to 3.43 days, n=324, 2 trials). Stool frequency was reduced with probiotics in two trials. One trial reported a shorter hospital stay, which was significant, but the number of subjects was small. No adverse events were reported. Authors concluded that there is limited evidence suggesting probiotics may be effective in treating persistent diarrhea in children [28]. Therefore, due to lack of quality data and heterogeneity of etiology of diarrhea as well as treatment, this therapy can not be recommended as a first line of therapy in children with persistent diarrhea.

Antibiotic Associated Diarrhea (AAD): Prevention

About 11% of children and adults on antibiotics develop diarrhea. The exact mechanism of this disorder is unknown. Antibiotics alter the normal balance between the large numbers of bacteria inhabitating the gut. Though it is believed that a disturbance in the balance of microflora is responsible for the pathogenesis of AAD; other causes such as increased gastrointestinal motility, osmotic diarrhea secondary to decreased digestion and absorption of carbohydrates and reduction of short-chain fatty acids (SCFAs) in feces have also been implicated. Overgrowth of toxin producing C difficile is probably the most well defined etiology of AAD and accounting for about 10 to 25% of AAD [29].Asymptomatic carriage of C. difficile has been shown to increase with antibiotic treatment. Other pathogens involved include Staphylococcus aureus, Clostridium perfringens, various Enterobacteriaceae or Candida species.

Probiotics have been used to prevent onset of AAD in children. Marginal benefit of using probiotics in AAD was noted in a meta-analysis that compared 6 RCTs. However, authors concluded that there was no evidence to support the use of probiotic to prevent the recurrence of C. difficile infection or to treat existing C. difficile diarrhea. [2] A Cochrane review published in 2007 analyzed 10 studies. Trials included treatment with either*Lactobacilli spp.*, *Bifidobacterium spp.*, *Streptococcus spp.*, or *Saccharomyces boulardii* alone or in combination. Six studies used a single strain probiotic agent and four combined two probiotic strains. The per protocol analysis for 9/10 trials reporting on the incidence of diarrhea showed statistically significant results favoring probiotics over active/non active controls (RR 0.49; 95% CI 0.32 to 0.74). However, intention to treat analysis showed non-significant results overall (RR 0.90; 95% CI 0.50 to 1.63). Five of 10 trials monitored for adverse events (n = 647); none reported a serious adverse event. *Lactobacillus GG*, *Lactobacillus sporogenes*, *Saccharomyces boulardii were found most effective* at 5 to 40 billion colony forming units/day. Authors conclude that the current data are promising, In a recent meta-analysis the risk for diarrhea was reduced (28.5% to 11.9% (relative risk [RR]: 0.44[95% CI: 0.25–77]; P _ .006).by administration of probiotics. Seven patients had to be treated with probiotics to prevent AAD in one [30]. Though effective, we think cost benefit studies are needed to justify routine administration of probiotics with antibiotic therapy. This combination is being actively promoted by the pharmaceutical industry especially in the developing countries, thereby pushing up the cost of antibiotic therapy. AAD in children is a relatively uncommon mild disease that usually responds well to discontinuation of the antibiotic. So adding expensive probiotic routinely to antibiotic therapy may not achieve significant clinical benefit but may increase the cost of treatment significantly. However, in selected vulnerable population, probiotics can be used to prevent AAD.

Antibiotic Associated Diarrhea (AAD): Treatment

Probiotics have been used in the treatment of recurrent or persistent Clostridium difficile diarrhea, mostly in adults. In general, therapeutic benefit was seen in patients with severe disease in vulnerable population [31]. A small pediatric trial with LGG showed some benefit in preventing relapsing C difficile diarrhea [32]. However, this study did not have enough power to rule out random effect. The main stay of therapy till date is withdrawal of offending antibiotic, oral metronidazole or vancomycin.

References

[1] Food and Agriculture Organization of the United Nations; World Health Organization. Guidelines for the evaluation of probiotics in food: joint FAO/WHO Working Group report on drafting guidelines for the evaluation of probiotics in food. Available at ftp://ftp.fao.org/es/esn/food/wgreport2. pdf. Accessed December 17, 2011.

[2] Szajewska H, Setty M, Mrukowicz J, Guandalini S. Probiotics in gastrointestinal diseases in children: hard and not-so-hard evidence of efficacy. J Pediatr Gastroenterol Nutr 2006;42:454-75.

[3] Lee YK, Salminen S. The coming age of probiotics. Trends Food Sci Technol 1995;6:241- 45.

[4] Van Niel C, Feudtner C, Garrison MM, Christakis DA. Lactobacillus therapy for acute infectious diarrhea in children: a meta-analysis. Pediatrics 2002; 109:678-84.

[5] Avlami A, Kordossis T, Vrizidis N, Sipsas NV. Lactobacillus rhamnosus endocarditis complicating colonoscopy. *J. Infect.2001;* 42 (4): 283–85.

[6] Silva M, Jacobus NV, Deneke C, Gorbach SL. Antimicrobial substance from a human Lactobacillus strain. *Antimicrob. Agents Chemother 1987;* 31 (8): 1231–33.

[7] Manley KJ, Fraenkel MB, Mayall BC, Power DA. Probiotic treatment of vancomycin-resistant enterococci: a randomised controlled trial. *Med. J. Aust. 2007;* 186 (9): 454–7.

[8] Oberhelman RA, Gilman RH, Sheen P, et al. A placebo-controlled trial of *Lactobacillus* GG to prevent diarrhea in undernourished Peruvian children. *J. Pediatr.* 1999;134(1): 15–20.

[9] Hatakka K, Savilahti E, Ponka A, et al. Effect of long term consumption of probiotic milk on infections in children attending day care centres: double blind, randomized trial. *BMJ.* 2001;322(7298):1327.

[10] Thibault H, Aubert-Jacquin C, Goulet O. Effects of long-term consumption of a fermented infant formula (with *Bifidobacterium breve* c50 and *Streptococcus thermophilus* 065) on acute diarrhea in healthy infants. *J. Pediatr. Gastroenterol. Nutr.* 2004;39(2):147–152.

[11] Saavedra JM, Abi-Hanna A, Moore N, Yolken RH. Long-term consumption of infant formulas containing live probiotic bacteria: tolerance and safety. *Am. J. Clin. Nutr.* 2004;79(2):261–267.

[12] Szajewska H, Kotowska M, Mrukowicz JZ, Armanska M, Mikolajczyk W. Efficacy of *Lactobacillus* GG in prevention of nosocomial diarrhea in infants. *J. Pediatr.* 2001; 138(3):361–365.

[13] Walker WA. Role of nutrients and bacterial colonisation in the development of intestinal host defense. *J. Pediatr. Gastroenterol. Nutr.* 2000;30(suppl):S2-S7.

[14] Mack DR, Michail S, Wei S, et al. Probiotics inhibit enteropathogenic E. coli adherence in vitro by inducing intestinal mucin gene expression. *Am. J. Physiol.* 1999;276:G941-G50.

[15] Szyman´ ski H, Pejcz J, Jawien´ M, Chmielarczyk A, Strus M, Heczko PB. Treatment of acute infectious diarrhoea in infants and children with a mixture of three *Lactobacıllus rhamnosus* strains: a randomized, double-blind, placebo-controlled trial. *Aliment Pharmacol. Ther.* 2006;23(2):247–253.

[16] Szajewska H, Mrukowicz JZ. Probiotics in the treatment and prevention of acute infectious diarrhea in infants and children: a systematic review of published randomized, double-blind, placebo-controlled trials. *J. Pediatr. Gastroenterol. Nutr.* 2001; 33(suppl 2):S17–S25.

[17] Huang JS, Bousvaros A, Lee JW, Diaz A, Davidson EJ. Efficacy of probiotic use in acute diarrhea in children: a meta-analysis. *Dig. Dis. Sci.* 2002;47(11):2625–2634.

[18] Allen SJ, Okoko B, Martinez E, Gregorio G, Dans LF. Probiotics for treating infectious diarrhoea. *Cochrane Database Syst. Rev.* 2004;(2):CD003048.

[19] Costa-Ribeiro H, Ribeiro TC, Mattos AP, Valois SS, Neri DA, Almeida P, et al. Limitations of probiotic therapy in acute, severe dehydrating diarrhea. *J. Pediatr. Gastroenterol. Nutr.* 2003;36:112-115.

[20] Salazar-Lindo E, Miranda-Langschwager P, Campos-Sanchez M, Chea-Woo E, Sack RB. Lactobacillus casei strain GG in the treatment of infants with acute watery diarrhea: a randomized, double-blind, placebo controlled clinical trial [ISRCTN67363048]. *BMC Pediatr* 2004;4:18.

[21] Sarker SA, Sultana S, Fuchs GJ, Alam NH, Azim T, Brussow H, et al. Lactobacillus paracasei strain ST11 has no effect on rotavirus but ameliorates the outcome of nonrotavirus diarrhea in children from Bangladesh. *Pediatrics* 2005;116:e221-8.

[22] Misra S, Sabui TK, Pal NK, A Randomized Controlled trial to evaluate the efficacy of Lactobacillus GG in infantile diarrhea. *The Journal of Pediatrics* 2009; 155: 129-32.

[23] Ochoa TJ, Salazar-Lindo E, Cleary TG. Management of children with infection-associated persistent diarrhea. *Seminars in Pediatric Infectious Diseases* 2004;15 (4):229–36.

[24] International Working Group on Persistent Diarrhoea. Evaluation of an algorithm for the treatment of persistent diarrhoea: a multicentre study. *Bulletin of the World Health Organization* 1996;74(5):479–89.

[25] Schorling JB, Wanke CA, Schorling SK, Mcauliffe JF, de Souza MA, Guerrant RL. A prospective study of persistent diarrhea among children in an urban Brazilian slum. Patterns of occurrence and etiologic agents. *American Journal of Epidemiology* 1990;132(1):144–56.

[26] Fauveau V, Henry FJ, Briend A, Yunus M, Chakraborty J. Persistent diarrhea as a cause of childhood mortality in rural Bangladesh. *Acta Paediatrica* 1992;381:12–4.
[27] Anonymous. Persistent diarrhoea in children in developing countries: memorandum from a WHO Meeting. *Bulletin of the World Health Organization* 1988;66(6):709–17.
[28] Bernaola Aponte G, Bada Mancilla CA, Carreazo Pariasca NY, Rojas Galarza RA. Probiotics for treating persistent diarrhoea in children. Cochrane Database of Systematic Reviews 2010, Issue 11. Art. No.: CD007401. DOI: 10.1002/14651858.CD007401.pub2.
[29] Turke D, Bernet JP, Marx J, Kemph H, Giard P, Welbaum O, *et al.* Incidence and risk factors of oral antibiotic-associated diarrhea in an outpatient pediatric population. *J. Pediatr. Gastroenterol. Nutr.* 2003; 37: 22–26.
[30] Johnston BC,Goldenberg JZ,Vandvik PO, SunX,GuyattGH. Probiotics for the prevention of pediatric antibiotic-associated diarrhea. *Cochrane Database of Systematic Reviews* 2011, Issue 11. Art. No.: CD004827. DOI: 10.1002/14651858.CD 004827. pub3.
[31] Dendukuri N, Costa V, McGregor M, Brophy JM. Probiotic therapy for the prevention and treatment of Clostridium difficile associated diarrhea: a systematic review. *CMAJ* 2005;173:167-70.
[32] Biller JA, Katz AJ, Flores AF, Buie TM, Gorbach SL. Treatment of recurrent Clostridium difficile colitis with Lactobacillus GG. *J. Pediatr. Gastroenterol. Nutr.* 1995;21:224-6.

Index

A

abuse, 281
acetaldehyde, 210, 217, 262
acetic acid, ix, 2, 6, 20, 37, 51, 82, 138, 177, 179, 181, 185, 186, 188, 194, 195, 196, 197, 209, 261, 274, 275, 276, 284, 296, 299, 302, 310
acidic, viii, 56, 71, 93, 94, 99, 100, 104, 106, 108, 185, 212, 293, 294, 297, 333, 334
acidity, 86, 94, 104, 120, 296, 297, 298, 303
active compound, 82
active transport, 275
adaptation, 38, 107, 168, 207, 209, 210, 242, 243, 247, 266, 283
adaptive immunity, 232, 235
additives, xii, 21, 66, 71, 84, 141, 149, 154, 155, 156, 167, 179, 200, 273, 281, 337
adenine, 202
adenocarcinoma, 246
adhesion, ix, 12, 14, 21, 25, 26, 29, 31, 39, 49, 62, 119, 120, 128, 139, 216, 221, 229, 240, 244, 245, 246, 247, 248, 249, 250, 251, 252, 253, 254, 255, 256, 344, 345
adipocyte, 53, 62
adipose tissue, 54
adsorption, 32, 69, 71, 82, 90
adults, 126, 135, 144, 347, 348
adverse conditions, 95
adverse effects, 50, 109
adverse event, xiv, 343, 347
aerobic bacteria, 82
Africa, 157, 208, 259, 270, 312
agar, 16, 49, 72, 180, 270, 278, 327, 328, 329
age, 41, 133, 224, 234, 242, 348
agglutination test, 14
aggregation, 8, 13, 14, 21, 25, 244, 254
AIDS, 143
alanine, 6, 10
albumin, 227, 228
alcohols, 262, 274
aldehydes, 262
aldolase, 160, 165
allergic reaction, 47, 178
allergic rhinitis, 136
allergy, 40, 47, 126, 136, 141, 143, 253, 311
alveolar macrophage, 226, 228
amine(s), 71, 72, 73, 85, 88, 128, 129, 171, 264, 265, 268, 271, 274, 275, 301, 307
amino acid, vii, x, 1, 31, 38, 52, 73, 74, 77, 84, 90, 155, 186, 188, 201, 204, 207, 210, 212, 213, 218, 220, 222, 227, 262, 265, 267, 310
aminoglycosides, 280
ammonium, 79
amylase, 11, 30, 312
anabolism, 199
anaerobic bacteria, 207
anchoring, 74, 243, 245, 246
angina, 234
annotation, ix, 203, 210, 244
antagonism, 75, 89, 281
antibiotic, xiv, 9, 10, 11, 22, 30, 32, 39, 42, 49, 55, 60, 61, 66, 68, 87, 126, 128, 130, 131, 133, 134, 137, 139, 140, 142, 153, 156, 174, 178, 198, 224, 234, 250, 275, 288, 289, 326, 329, 337, 338, 341, 344, 347, 348, 350
antibiotic resistance, 9, 10, 11, 22, 32, 39, 42, 55, 68, 153, 156, 224, 234, 288, 329, 337, 338, 341
antibody, 47, 229, 230, 231, 232, 237
antigen, 46, 229, 230, 233, 239, 250, 255
antimicrobial therapy, 224
antioxidant, 46, 59, 261
antitumor, 252
apoptosis, 132
aquaculture, 171
arginine, 37, 276
argyria, 146

aromatic compounds, xii, 257, 299
arteriosclerosis, 58
arthritis, 136, 141
artificial intelligence, 22
ascorbic acid, 153, 156
Asia, 157, 208, 258, 269, 306
assessment, 57, 116, 198, 256, 275, 280, 284, 285
assimilation, 120, 142
asthma, 136
atherosclerosis, 314, 315
atherosclerotic plaque, 51
atmosphere, 70, 78, 83, 91, 281
atopic dermatitis, 123, 136, 144, 145
atopic eczema, 136
ATP, 67, 242
atrophy, 232, 233
attachment, xi, 128, 139, 241, 247, 250, 253
autoimmune disease(s), 236

B

BAC, 335
Bacillus subtilis, 16, 32, 120, 174, 253, 263, 326
bacteremia, 140, 143, 145, 228
bacterial cells, 74, 99, 163, 249, 307
bacterial infection, 224
bacterial pathogens, 254, 336, 341, 346
bacterial strains, 225, 249
bacterial vaginitis, 134, 139
bacteriocinogenic LAB, viii, 66, 68, 281, 283
bacteriophage, 264
bacteriostatic, 75, 78, 79, 80
bacterium, vii, x, 1, 14, 28, 36, 60, 112, 120, 127, 128, 130, 131, 132, 133, 134, 136, 151, 168, 179, 189, 204, 215, 216, 218, 317, 335, 339, 340
barium, 132
beef, 32, 84, 87, 90, 91, 148, 174
beer, 4, 31, 94, 101, 107, 112, 114, 147, 274
beneficial effect, x, xi, xiii, xiv, 42, 44, 54, 126, 130, 132, 136, 207, 209, 223, 226, 234, 235, 241, 245, 246, 247, 251, 310, 316, 326, 327, 336, 343, 344, 346
beneficial microbes, xii, 273, 283
benefits, x, xii, xiii, 2, 19, 36, 40, 57, 61, 94, 109, 117, 126, 137, 178, 204, 207, 208, 214, 218, 222, 225, 239, 253, 269, 273, 283, 309, 310, 311, 312, 317, 319, 326, 336
benign, 104
beverages, 19, 95, 209, 258, 266
BI, 121
bile, ix, xi, xiii, 6, 8, 11, 12, 15, 17, 24, 39, 49, 52, 53, 57, 58, 59, 119, 120, 142, 164, 179, 209, 241, 247, 250, 315, 325, 329, 333, 334, 337, 338, 339, 340, 341, 344, 345
bile acids, 52, 57, 315, 337, 340
bilirubin, 44, 63
bioavailability, 227
biochemical processes, 312
biochemistry, 101
bioconversion, 300
biodegradation, xii, 292, 299, 300, 301, 303
biodiversity, 20, 72, 152, 175
bioinformatics, 62, 164
biological activity(s), 99, 101, 232, 273
biological processes, xii, 291
biomarkers, 169
biomass, ix, 70, 76, 157, 177, 179, 183, 184, 185, 186, 187, 188, 189, 190, 191, 192, 193, 194, 195, 197, 296, 301, 304, 314
biomaterials, 113
biopreservation, xi, 66, 71, 72, 73, 74, 75, 78, 79, 80, 81, 83, 91, 174, 257, 271, 281
bioremediation, 306
biosynthesis, 8, 60, 85, 169
biotechnological applications, 154
biotechnology, 169, 304, 311, 317
bladder cancer, 130, 139, 198
bleaching, 261
bleeding, 335
blood, 8, 18, 25, 41, 45, 46, 47, 50, 51, 54, 61, 128, 129, 168, 208, 227, 228, 229, 230, 231, 232, 314, 329
blood group, 25
blood pressure, 18, 51, 129, 168
body composition, 60
body fat, 53
body weight, 54, 120, 189
bonds, 98, 213, 248, 313
bone(s), 47, 161, 231, 232, 236, 238, 240, 315, 322
bone marrow, 47, 231, 232, 236, 238, 240
bone mass, 315
bowel, vii, 1, 17, 18, 19, 28, 40, 41, 43, 54, 55, 60, 61, 126, 128, 129, 130, 131, 132, 136, 142, 143, 144, 200, 208, 250, 311, 314
Brazil, 35, 75, 83, 153, 157, 203, 208, 215, 309
breakdown, 17, 98, 106, 155
breast milk, 20, 27
brevis, viii, 3, 93, 100, 108, 114, 116, 121, 122, 123, 135, 150, 151, 152, 206, 242, 258, 259, 260, 263, 264, 266, 275, 277, 279
bronchus, 229, 230, 238
Burkina Faso, 28, 38, 60
burn, 50, 54, 56, 132, 146

C

Ca^{2+}, 57
cabbage, 4, 7, 259, 260, 262, 263, 268, 269
cabinets, 330
CACO, xi, 241
calcium, 302, 314, 315
calorie, 312, 315
calorimetry, 112
cancer, 14, 16, 29, 60, 126, 128, 131, 141, 142, 146, 224, 315, 322
candidiasis, 140, 146
capillary, 96
capsule, 136
carbohydrate(s), 3, 6, 14, 15, 30, 36, 37, 51, 77, 157, 166, 209, 211, 212, 214, 242, 245, 246, 253, 256, 259, 260, 261, 264, 271, 274, 275, 292, 310, 312, 316, 322, 347
carbon, viii, xi, 2, 6, 37, 70, 93, 94, 97, 104, 115, 117, 185, 188, 189, 196, 202, 211, 257, 261, 292, 293, 310, 316
carboxylic acid(s), 51, 307
carboxymethyl cellulose, 298
carcinogenesis, 129
carcinogenicity, 40, 311
cardiovascular disease(s), viii, 28, 36, 51, 60, 62, 127, 157
caries, 48, 49, 56, 62
carnivores, 5, 22
case study, 139, 285
casein, 211, 212, 213, 214, 293, 294, 295, 298, 304, 307
catabolism, 17, 31, 199, 211, 212, 213, 218, 242, 310
catalyst, 315
catalytic activity, 62
catalytic system, 323
catheter, 133
CCA, 38
cell culture, 14, 47, 127, 232, 236, 249, 250
cell death, 97, 101, 102, 104, 105, 108, 110, 111
cell division, 213
cell invasion, 127, 139
cell line(s), 14, 45, 46, 127, 139, 246, 249, 250, 255
cell membranes, 82, 293, 337
cell size, 54
cell surface, xi, 8, 14, 25, 38, 55, 209, 241, 243, 244, 245, 246, 248, 249, 251, 254
cellular energy, 261
cellulose, 298
Central Europe, 284, 285
chain molecules, 98
challenges, 57, 111, 113, 164, 167, 239, 320
chaperones, 163, 168
cheese, x, 4, 5, 18, 21, 22, 23, 28, 30, 31, 41, 56, 61, 113, 114, 147, 153, 166, 168, 174, 179, 180, 201, 204, 214, 216, 218, 220, 221, 237, 258, 269, 278, 280, 285, 289, 292, 293, 317, 318, 322, 340
chemical(s), xii, 7, 66, 83, 87, 102, 117, 149, 150, 161, 163, 167, 225, 269, 281, 285, 286, 292, 294, 300, 303, 305, 316
chemical degradation, 300
chemokines, 130, 131, 145, 178, 229
chicken, 69, 86, 113, 247, 286, 335, 336
childhood, 48, 56, 224, 350
children, xi, xiii, xiv, 18, 19, 48, 59, 111, 126, 128, 129, 132, 133, 135, 136, 139, 140, 142, 143, 145, 198, 201, 221, 224, 225, 231, 234, 235, 236, 238, 240, 343, 344, 345, 346, 347, 348, 349, 350
China, 5, 23, 33, 63, 148, 171, 325, 327, 330, 338
chitosan, 12, 110, 114, 298, 299
chloramphenicol resistance, 287
chloroform, 14
cholesterol, viii, 16, 17, 18, 28, 31, 33, 36, 51, 52, 53, 54, 55, 59, 60, 63, 120, 127, 134, 135, 142, 315, 320
choline, 8, 173, 243
chromatography, 74, 77, 248
chromosome, 7, 55, 217, 250
chronic diseases, 56
chymotrypsin, 77
cigarette smoking, 51
circulation, 52, 230, 340
cirrhosis, 140
classification, iv, vii, ix, 7, 66, 67, 203, 205, 206, 217, 243, 275, 305, 314
climate(s), 148, 258, 270
clinical application, 238
clinical trials, 19
cloning, 23, 86, 217
cluster(s), xi, 7, 8, 30, 99, 241, 242, 251
CNS, 153
CO_2, 2, 82, 90, 104, 105, 110, 115, 275
cocoa, 20, 56, 60
coffee, 4, 22
colitis, 26, 33, 47, 126, 130, 142, 250, 252, 254, 255, 350
collagen, xi, 241, 245, 246, 249
colon, xiii, 26, 29, 51, 58, 60, 61, 63, 128, 129, 146, 309, 310, 312, 314, 315, 322
colonization, vii, xii, 1, 12, 25, 27, 44, 47, 50, 58, 63, 128, 130, 131, 133, 141, 156, 201, 228, 230, 246, 249, 256, 273, 280, 314, 349
colonoscopy, 132, 348
combined effect, 101, 104, 110

commercial, ix, 7, 40, 52, 57, 68, 74, 77, 80, 83, 94, 120, 121, 127, 137, 145, 146, 179, 215, 266, 306, 310, 311, 322, 344
communication, 28, 79, 109
community(ies), 39, 50, 111, 150, 151, 152, 170, 207, 234, 246, 256, 269, 305, 346
comparative analysis, 245, 256
compatibility, 208, 225
competition, 77, 84, 247, 264, 274, 281, 327, 338, 345
competitive advantage, 76, 314
competitiveness, 153, 154, 156, 166
complement, 230
complexity, 68, 244, 250
complications, 44
composition, 7, 13, 40, 60, 74, 79, 100, 103, 180, 199, 210, 211, 225, 242, 244, 253, 256, 261, 265, 285, 286, 293, 311, 314, 321
compression, 99, 101, 109
conditioning, 153, 169, 170
conductivity, 107
configuration, 69
conservation, 2, 72, 281
constant rate, 294
constipation, 43, 57, 130, 142, 200, 314, 316
constituents, x, 83, 100, 103, 203, 211, 341
construction, 298
consumers, viii, x, 15, 36, 66, 93, 109, 111, 136, 204, 208, 225, 267
consumption, ix, xii, 18, 48, 72, 80, 120, 126, 127, 128, 129, 130, 135, 136, 137, 148, 157, 177, 178, 182, 185, 188, 189, 192, 193, 194, 235, 238, 256, 259, 267, 273, 294, 314, 315, 337, 348, 349
contaminant, 80, 149, 153, 189
contamination, 24, 41, 71, 72, 89, 137, 280, 299
control group, 48, 51, 52, 53, 54, 133, 226, 229, 230, 232, 335, 336
controlled studies, 18
controlled trials, 126, 138, 145, 201, 345, 349
cooking, 129
cooling, 96, 106, 282
copper, 302
correlation, 14, 67, 108, 140, 215, 265, 276
cost, xiii, 83, 179, 189, 292, 294, 299, 304, 343, 344, 346, 347
cost benefit analysis, 344
cost effectiveness, 346
costimulatory signal, 237
covalent bond, 98
Croatia, 273, 277, 283, 284, 285
crop(s), 109, 261, 304
crown, 161
crystalline, 101, 243, 245
CSS, 68, 69, 70, 71, 72, 73, 74, 75, 76, 78, 79, 80, 81
CT, 345
cultivation, 38, 41, 208, 293, 299, 330
cultural tradition, 259, 264
culture conditions, 15, 73, 84
culture media, 38, 153, 179, 180, 272
culture medium, 32, 129, 179, 188, 196, 221, 300, 337
cycles, 75, 94, 99, 100, 102, 103, 105, 106, 108, 110, 192, 297, 298
cyclophosphamide, 45
cysteine, 77, 213, 219, 311
cystic fibrosis, 131
cystine, 104
cytokines, xi, 17, 18, 44, 45, 46, 47, 129, 130, 178, 215, 226, 228, 229, 230, 232, 234, 236, 242, 250, 251, 339
cytometry, 101, 113, 114, 232, 238, 247
cytoplasm, 94, 97, 104, 109, 208, 248
cytoskeleton, 14, 131, 138

D

dairy industry, ix, 203, 207, 211, 214, 259, 275, 292, 326
dairy products, vii, x, xii, xiii, 1, 15, 24, 31, 33, 35, 36, 40, 60, 99, 137, 152, 170, 203, 207, 210, 216, 225, 258, 273, 274, 275, 287, 289, 305, 307, 309, 310, 313, 314, 317, 321, 340
danger, 8
database, 205, 206, 244
death rate, 11
deaths, 51, 346
decomposition, 276
decontamination, 42, 60, 144
defects, 54, 150
defence, 54, 226, 228, 229, 239
defense mechanisms, 220
deficiency(s), 26, 131, 135, 227, 232, 233, 239
degradation, x, 17, 26, 52, 69, 74, 97, 166, 170, 203, 211, 212, 252, 261, 299, 300, 304, 305, 306, 307
dehydration, x, 204, 209, 213, 216
denaturation, 96, 100, 101, 163
dendritic cell, 47, 229, 237, 238, 243, 254, 339, 340
dental caries, 49
dental plaque, 56
depolarization, 82
depolymerization, 300
deposition, 132, 320
deposits, 293
depression, 59, 74, 231
deprivation, 232, 233
derivatives, 3, 94, 299, 320, 342

dermatitis, 123, 126, 136, 144, 145
destruction, 98, 102, 112, 113
detection, 27, 29, 39, 42, 84, 85, 101, 201, 282, 286, 287, 332, 339
detergents, 293
detoxification, 159, 164
developed countries, 264, 346, 347
developing countries, xi, xiii, 224, 234, 268, 343, 345, 346, 347, 350
deviation, 7, 184
diabetes, 45, 51, 62, 127, 140, 315
dialysis, 128
diarrhea(s), xiii, xiv, 18, 19, 43, 126, 128, 139, 142, 143, 145, 178, 198, 200, 201, 208, 333, 335, 342, 343, 344, 345, 346, 347, 348, 349, 350
diet, xi, 17, 19, 33, 41, 51, 52, 53, 54, 55, 59, 60, 62, 63, 130, 173, 189, 225, 227, 228, 232, 233, 237, 239, 257, 314, 315, 344
dietary fiber, 51, 52, 62, 312, 314, 318, 319
dietary habits, 264
dietary intake, 157
dietary supplementation, 127
differential scanning, 112
differential scanning calorimetry, 112
differentiated products, x, 204, 214
diffusion, 16, 43, 49, 72, 84, 161, 278, 279, 328, 334
diffusivity, 104
digestion, 15, 128, 129, 137, 142, 178, 201, 208, 247, 314, 315, 342, 347
digestive enzymes, 161, 162
dipoles, 97
discrimination, 62
disease model, xiii, 325, 330
diseases, vii, viii, xi, 20, 39, 51, 56, 59, 119, 120, 126, 128, 131, 133, 134, 135, 136, 138, 142, 157, 178, 224, 234, 235, 236, 312, 316, 326, 344, 348
disorder, 42, 347
displacement, xi, 23, 102, 241, 247, 249, 255
distilled water, 330
distribution, 98, 105, 115, 116, 217
diversification, 166
diversity, ix, 22, 29, 30, 51, 58, 147, 150, 162, 166, 168, 205, 208, 218, 219, 222, 242, 243, 253, 254, 255, 267, 275, 276, 285
DNA, 6, 37, 38, 61, 72, 79, 99, 100, 105, 109, 129, 164, 174, 213, 214, 215, 219, 220, 237, 239, 247, 250, 253, 275, 301
double-blind trial, 51, 139, 198
dough, 15, 274
down-regulation, 163
drinking water, 132, 280
drug resistance, 285, 326
drug therapy, 134
drugs, 120, 136, 278, 326
DSM, 16, 31, 43, 44, 48, 51, 54, 58, 88, 131, 134, 219, 341
duodenal ulcer, 128
duodenum, 15, 52
durability, 258
dyes, xiii, 292, 300, 301, 303, 304, 306, 307, 308

E

Eastern Europe, 208
E-cadherin, 250
ecology, 87, 89, 152, 168, 171, 172, 208, 254, 268, 276, 284, 338, 339
economic losses, 95
ecosystem, 48, 134, 152, 218, 314, 318, 337
eczema, 136
effluent(s), xii, 179, 291, 292, 294, 295, 296, 298, 299, 304, 305, 306
elaboration, 198
election, 268
electric field, viii, 93, 97, 105, 106, 107, 108, 110, 112, 114, 115, 116, 281
electrical resistance, 44, 58
electrodes, 105
electron(s), 103, 105, 109, 117, 128, 246
electron microscopy, 103, 105, 128
electrophoresis, 38, 74, 164, 247, 252
electroporation, 105
ELISA, 72, 271
elongation, 14, 248, 253
elucidation, 338
emergency, 233
EMS, 175, 255, 341
encapsulation, 12
encephalitis, 241
encoding, 7, 23, 30, 31, 130, 132, 178, 210, 237, 243, 244, 246, 250, 339
endocarditis, ix, 8, 39, 119, 135, 143, 348
energy, 15, 76, 97, 105, 107, 108, 109, 164, 211, 215, 236, 240, 261, 294, 301, 312, 318
energy input, 107, 108
engineering, 25, 201, 217, 250, 320
environment(s), viii, xi, 7, 12, 13, 20, 24, 36, 39, 49, 56, 62, 68, 70, 74, 94, 100, 102, 104, 106, 108, 119, 152, 153, 154, 156, 161, 163, 168, 172, 204, 213, 241, 242, 243, 244, 246, 247, 249, 251, 252, 258, 261, 262, 263, 265, 282, 293, 334
environmental change, 213
environmental conditions, ix, 68, 101, 147, 148, 150, 166, 211, 263, 303
environmental factors, 68, 70, 261
environmental impact, 111

environmental issues, xii, 291, 299
environmental stress, 213, 337
environmental sustainability, 292
enzymatic activity, 15, 17
enzyme(s), x, 7, 11, 15, 62, 72, 78, 79, 97, 99, 101, 102, 104, 114, 115, 117, 129, 141, 143, 146, 155, 161, 162, 165, 172, 173, 204, 209, 211, 212, 213, 216, 220, 242, 244, 248, 262, 265, 266, 268, 270, 274, 276, 286, 293, 299, 300, 305, 312, 313, 315
epithelial cells, vii, xi, 1, 12, 13, 14, 29, 43, 45, 132, 134, 141, 145, 202, 226, 237, 239, 241, 246, 247, 248, 254, 346
epithelium, vii, 1, 12, 133, 247, 344
EPS, 121
equilibrium, 101
ESI, 248
ester, 256, 299
Estonia, 288
ethanol, 2, 6, 15, 94, 101, 107, 164, 211, 261, 262, 275, 293, 310
ethyl acetate, 14
etiology, 344, 346, 347
eukaryotic, 212
Europe, 148, 150, 151, 152, 208, 282, 285, 312
European Union (EU), 198, 256, 282
evacuation, 43, 56
evidence, vii, xiii, xiv, 1, 19, 101, 224, 232, 234, 235, 298, 303, 314, 321, 343, 344, 345, 347, 348
evolution, 7, 72, 85, 89, 155, 205, 207, 208, 217, 219, 222
exclusion, 8, 246, 255, 327
excretion, 24, 52, 58
exopolysaccharides, 46, 59, 95, 310
experimental condition, 73
exploitation, 221, 251
exposure, 45, 48, 116, 133, 163, 229
extracellular matrix, 233, 246
extraction, 105, 182, 194, 312
extracts, 12, 27, 69, 79, 90, 91, 150, 248

F

factories, 293
families, 57, 205
family history, 51
farm environment, 172
fat, 17, 52, 53, 54, 59, 62, 70, 86, 96, 102, 148, 261, 292, 305, 312, 313, 314, 316
fatty acids, viii, 51, 65, 157, 168, 242, 313, 314, 315, 321, 347
feces, 3, 5, 6, 17, 18, 21, 22, 30, 42, 43, 53, 57, 221, 315, 347
feed additives, 141, 200
feedstock, 304
fermentable carbohydrates, 264
fermentation technology, 168
ferric ion, 302
ferritin, 164
ferrous ion, 302
fever, 133
FGM, 225, 227, 228, 233
fiber(s), 27, 51, 52, 62, 132, 144, 299, 310, 312, 314, 315, 318, 319
fibrinogen, 18, 51, 55, 249
fibroblasts, 226
fibrosis, 131, 227
fish, vii, viii, ix, 1, 5, 35, 65, 66, 67, 70, 72, 73, 74, 75, 76, 77, 78, 79, 82, 83, 85, 86, 88, 89, 90, 109, 147, 157, 161, 162, 166, 167, 169, 170, 171, 172, 173, 174, 175, 247, 274
flatulence, vii, 1, 15, 18, 42, 129, 132
flavor(our), ix, 2, 15, 94, 95, 97, 99, 147, 149, 150, 155, 172, 174, 203, 204, 210, 211, 212, 216, 217, 218, 221, 222, 258, 262, 265, 266, 311, 312
flocculation, xii, 291, 295, 296, 299
flora, 20, 23, 57, 60, 72, 76, 87, 88, 91, 135, 137, 140, 144, 150, 152, 162, 166, 171, 218, 229, 252, 256, 263, 265, 266, 287, 319, 320
flotation, 306
flour, 16, 198, 259
fluid, 86, 98, 102, 105, 145, 226, 227
fluorescence, 108, 221, 249, 306
follicles, 229
food additive, 154
food additives, 154
Food and Drug Administration (FDA), 8, 80
food chain, xii, 63, 273, 278, 280, 284
food fermentations, vii, 1, 147, 210
food industry, 7, 15, 36, 66, 78, 213, 225, 266, 274, 305, 312, 315, 316
food intake, 52
food poisoning, 31, 32
food processing industry, 304
food production, 62, 77, 80, 258, 265, 267, 275, 281, 283, 285
food products, viii, ix, 8, 19, 66, 69, 97, 120, 137, 175, 225, 258, 266, 268, 280, 282, 283, 287, 314, 341
food safety, viii, 71, 93, 95, 111, 141, 154, 157, 270, 275, 281
food security, 157, 171
food services, 238
food spoilage, 15, 16, 83, 114, 147, 265
formation, vii, ix, x, 1, 6, 38, 44, 50, 54, 55, 82, 88, 95, 106, 111, 132, 138, 163, 172, 177, 193, 195,

196, 200, 204, 210, 211, 216, 221, 222, 246, 254, 260, 262, 264, 265, 266, 267, 297, 315, 320
formula, 133, 146, 329, 345, 348
freezing, 282
frequency distribution, 115
freshwater, 75, 86, 157, 161, 167
fructose, 37, 165, 302, 311, 312, 313, 316
fruits, vii, 4, 5, 31, 35, 94, 95, 99, 109, 267
FTIR, 30
functional analysis, 256
functional changes, 106, 313
functional food, 43, 51, 136, 208, 225, 235, 238, 315, 322
fungi, 50, 82, 111, 344

G

gamma radiation, 109
gamma rays, 109
gastrectomy, 320
gastric mucosa, 130
gastritis, 128, 131, 142, 145
gastroenteritis, 345
gastrointestinal symptoms, vii, 1, 18, 43, 59, 130, 142, 200
gastrointestinal tract, vii, x, xi, xiii, 1, 3, 8, 11, 23, 27, 36, 56, 58, 63, 120, 127, 132, 135, 140, 201, 204, 223, 226, 234, 242, 251, 252, 254, 255, 256, 309, 310, 314, 340, 341, 344, 345
GC-content, 204
gel, 12, 38, 74, 165, 252, 315
gene expression, 44, 163, 208, 221, 247, 250, 349
gene transfer, 8, 11, 205, 280
genes, x, 7, 8, 10, 11, 14, 15, 17, 19, 22, 29, 32, 39, 44, 49, 54, 130, 132, 138, 170, 178, 188, 204, 205, 207, 210, 211, 213, 216, 219, 242, 243, 245, 246, 250, 251, 252, 254, 256, 274, 275, 278, 280, 286, 338, 339, 342
genetic diversity, 166, 243
genetic factors, 254
genetics, 80, 85, 151, 214
genitourinary tract, 3
genome, vii, 1, 7, 8, 19, 36, 38, 58, 152, 168, 170, 205, 206, 207, 209, 217, 222, 244, 246, 251, 256, 275
genomics, x, 25, 58, 164, 172, 204, 207, 208, 209, 210, 212, 213, 216, 217, 219, 220, 222
genotype, 236
genotyping, 144
genus, viii, ix, xiii, 2, 3, 21, 36, 61, 93, 162, 175, 178, 203, 205, 206, 215, 217, 218, 274, 275, 284, 309
geometry, 105
germination, 294
ginger, 149, 157
ginseng, 25
glucose, ix, 2, 15, 37, 51, 54, 73, 95, 149, 177, 180, 181, 182, 183, 185, 187, 188, 189, 198, 201, 204, 211, 242, 274, 275, 293, 302, 311, 313, 315, 316, 337, 340
glucose tolerance, 51
glutamate, 25, 31
glutathione, 8, 89
glycerol, 37, 163, 179, 319, 326, 327, 328, 337, 340
glycine, 8, 17, 52, 163, 243
glycolysis, x, 6, 105, 203, 211, 261
glycopeptides, 279
glycoproteins, 12, 245
glycosylation, 246
goat milk, 237, 239
grants, 251
GRAS, vii, 2, 35, 209, 281, 293, 310
grass, 98
gravity, 293
growth factor, 311
growth hormone, 252
growth models, ix, 177, 189
growth rate, 69, 75, 107, 153, 163, 192, 193, 194, 294, 317, 329
guidelines, 19, 125, 141, 171, 319, 348

H

H. pylori, 128, 130, 131, 134, 138, 340
habitat(s), ix, 3, 61, 203, 214, 257, 258
haemopoiesis, 240
harmful effects, 44, 47, 230
harvesting, 258
hazards, 71, 86, 288
healing, 45, 50, 54
health effects, 120, 135, 267, 315, 344
health status, 52, 235
heart attack, 224
heart disease, 51, 315
heat shock protein, 96, 237, 248
Helicobacter pylori, 33, 40, 128, 138, 142, 144, 145, 146, 178, 201, 251, 311, 339, 341
heme, 59
hepatocytes, 60
herpes, 15
heterogeneity, 6, 166, 284, 344, 347
high blood pressure, 51
high fat, 69, 305
highways, 319
histamine, 31, 78, 265
history, 51, 207, 222, 237, 268, 326

HIV, 132
homeostasis, 82, 94, 99, 112, 228, 237, 240
homogeneity, 74
homovanillic acid, 300
hormone, 252
host, vii, ix, x, xiii, 1, 2, 8, 12, 14, 15, 24, 28, 29, 45, 50, 52, 66, 67, 73, 119, 120, 130, 178, 207, 208, 209, 215, 218, 223, 224, 226, 228, 229, 235, 236, 237, 238, 240, 243, 244, 245, 246, 249, 251, 309, 312, 314, 326, 344, 349
house dust, 47
human body, 12, 257
human condition, 43
human genome, 207
human health, viii, x, 43, 49, 66, 71, 119, 207, 223, 253, 281, 300, 316, 318
human immunodeficiency virus, 132, 224
human subjects, 3
humidity, 258
humoral immunity, 232
Hungary, 151, 172
Hunter, 61
hybridization, 6, 7, 72, 215, 220, 221, 306
hydrocarbons, 13, 49
hydrogen, viii, ix, xi, 15, 65, 94, 119, 135, 138, 209, 257, 261, 269, 270, 272, 274
hydrogen peroxide, viii, ix, xi, 15, 65, 94, 119, 138, 209, 257, 261, 269, 270, 272, 274
hydrogen sulfide, 94
hydrolysis, xiii, 11, 37, 39, 53, 67, 74, 156, 161, 169, 212, 213, 262, 293, 299, 300, 305, 309, 310, 316, 322
hydrophobicity, 13, 14, 49, 249
hydroxide, 294
hygiene, 148, 274
hypercholesterolemia, 19, 53
hyperlipidemia, 134
hypersensitivity, 131, 138
hypertension, 315
hypoplasia, 233
hypothesis, 59, 188, 190, 216, 245, 275, 345

I

IBD, 17, 23, 41, 42, 43
identification, viii, xi, 7, 27, 30, 35, 38, 54, 58, 61, 62, 79, 112, 137, 152, 170, 174, 209, 220, 221, 224, 225, 244, 251, 255, 256, 268, 269, 271, 276
IL-13, 47
IL-8, 44, 45, 138, 228
ileum, 14, 46, 247
immobilization, 294, 295
immune activation, 230
immune defense, 129, 220, 232, 239
immune disorders, 326
immune function, 45
immune modulation, ix, 119, 246
immune reaction, 129
immune response, 24, 46, 48, 50, 63, 130, 132, 137, 140, 178, 215, 224, 227, 228, 229, 230, 231, 232, 234, 237, 238, 239, 326, 336, 344, 345
immune system, x, 18, 40, 133, 144, 223, 224, 226, 229, 233, 235, 238, 251, 311, 314, 327, 344
immunity, x, 18, 19, 20, 31, 45, 46, 47, 55, 59, 62, 67, 129, 130, 169, 201, 224, 228, 229, 231, 232, 233, 234, 235, 236, 237, 238, 239, 240
immunocompetent cells, 236
immunocompromised, ix, x, 119, 131, 132, 135, 140, 224, 225, 226, 228, 234, 235
immunodeficiency, 132, 224
immunoglobulin, 133, 232, 236, 245
immunoglobulins, 232
immunohistochemistry, 44
immunomodulation, xiii, 8, 45, 46, 57, 61, 148, 237, 326, 335
immunomodulatory, viii, x, 25, 26, 31, 35, 47, 62, 223, 224, 225, 235, 238, 243, 338, 344
immunosuppression, 227
immunosuppressive agent, 45
immunotherapy, xi, 242, 253
in situ hybridization, 221, 306
in vivo, xiii, 8, 11, 14, 16, 17, 19, 22, 24, 40, 46, 49, 50, 53, 56, 57, 120, 128, 129, 130, 131, 134, 139, 140, 145, 154, 201, 215, 221, 225, 235, 239, 248, 249, 250, 326, 327, 335, 338, 339, 341
incidence, 48, 86, 128, 130, 132, 152, 234, 239, 347
incubation period, 329, 333
incubation time, 329
India, 28, 157, 162, 170, 174, 236, 259, 260, 271, 343
indirect effect, 101, 131
Indonesia, 260
induction, 17, 18, 44, 46, 52, 137, 163, 164, 235, 237, 247
industrial processing, 281
industrialized countries, 42
industry(ies), ix, 7, 15, 36, 37, 39, 66, 78, 89, 95, 98, 117, 102, 203, 207, 208, 211, 213, 214, 219, 225, 259, 266, 269, 274, 275, 292, 304, 305, 311, 312, 315, 316, 317, 322, 326, 347
ineffectiveness, ix, 120
INF, 46, 48, 226, 299
infants, 3, 5, 24, 133, 136, 145, 146, 201, 224, 234, 239, 240, 242, 345, 348, 349
infectious agents, 232

inflammation, 17, 18, 19, 42, 46, 47, 57, 59, 130, 133, 134, 235, 250
inflammatory bowel disease, 40, 136, 208, 250, 311
inflammatory mediators, 237
influenza, 15, 18, 27, 31, 48, 59
informed consent, 234
ingestion, 11, 40, 43, 58, 130, 135, 209, 311, 339
ingredients, 84, 86, 109, 115, 167, 274, 281, 312, 318
inhibition, 23, 40, 48, 57, 69, 71, 75, 76, 77, 78, 80, 81, 83, 86, 88, 90, 94, 129, 130, 131, 144, 145, 154, 181, 185, 193, 194, 197, 201, 243, 262, 270, 274, 276, 278, 281, 293, 294, 304, 311, 328, 333, 334, 341, 345
inhibitor, 70, 80, 82, 130, 141, 341
initiation, 57
injury(ies), iv, 23, 63, 69, 82, 97, 99, 102, 103, 104, 105, 107, 110, 115, 227, 228, 229, 238
innate immunity, 20, 45, 55, 236
inoculation, 26, 72, 79, 82, 110, 150, 263
inoculum, 75, 80, 91, 107, 294, 299, 304
inositol, 37, 302
insects, 252
insertion, 67, 82, 88, 217, 250, 251
integrity, 8, 44, 50, 61, 101, 102, 107, 131, 148
intelligence, 22
intensive care unit, 42
intercellular adhesion molecule, 240
interference, 60, 62, 276, 320
interferon, 18, 27, 48, 59, 236
interleukin-8, 59, 128, 129, 139, 146
internal environment, 12
intervention, 143, 144, 235
intestinal flora, 137, 140
intestinal tract, xi, 12, 19, 21, 72, 137, 140, 162, 215, 219, 241, 247, 253, 312, 314, 315, 337, 341, 345
intestine, vii, 1, 11, 12, 15, 51, 52, 57, 170, 173, 174, 224, 229, 230, 231, 249, 253, 312, 313, 315, 316, 323, 326, 337, 344, 345
intravenously, 46
intron, 250
invertebrates, 79, 84
ionization, 248
ionizing radiation, 109
ions, 57, 97, 104, 159, 164, 245, 301, 302
irradiation, viii, 93, 97, 109, 112, 113, 117, 281
irritable bowel disease, 41, 54
irritable bowel syndrome, vii, 1, 18, 19, 28, 55, 60, 61, 126, 129, 131, 132, 142, 143, 144, 200
ischemia, 44, 63
isolation, xi, 32, 61, 62, 152, 172, 221, 241, 247, 251, 341
isoleucine, 213
isomerization, 316
isomers, 37
isotherms, 198

J

Japan, 130, 208, 260, 291
jejunum, 249
Jordan, 113, 217

K

Kenya, 5, 27
kill, 154, 155, 231
kinetic parameters, 96, 106
kinetics, ix, 95, 100, 104, 106, 112, 114, 115, 116, 155, 177, 185, 196, 198, 199, 316, 321
Korea, 260
koumiss, 33

L

lactase, 15, 120, 135, 140
lactate dehydrogenase, 185, 227
lactobacillus, vii, 2, 7, 26, 113, 169, 221, 256, 261, 344
lactoferrin, 110
lactose intolerance, 40, 126, 145, 311, 326
large intestine, 51, 174, 207, 312, 315, 323
LDL, 18, 51, 52, 53, 55, 134, 315
leakage, 79, 97, 104, 106, 110, 116
legislation, 8, 208
leptin, 51, 54, 233, 239
lesions, 45, 128, 136
leucine, 212, 213
leukocytes, 61, 231
ligand, 13
linoleic acid, 54, 59, 60
lipid metabolism, 33, 63
lipid oxidation, 113, 149, 157
lipids, x, 15, 17, 52, 61, 98, 110, 203, 211, 276, 292, 293
lipolysis, x, 174, 203, 211, 212
liquid chromatography, 77
Listeria monocytogenes, viii, 40, 46, 61, 65, 83, 84, 85, 86, 87, 88, 89, 90, 91, 116, 128, 140, 166, 173, 199, 216, 217, 236, 250, 265, 270, 271, 274, 282, 284, 285, 286, 287, 288, 289
lithium, 245
Lithuania, 288
liver, 17, 19, 20, 51, 53, 57, 131, 132, 135, 140, 144, 146, 314, 315, 335

livestock, 242, 247, 288
low temperatures, 16, 70, 282
low-density lipoprotein, 53, 127
lower respiratory tract infection, 224, 238
lumen, 344
Luo, 26, 33, 112, 144
lymphocytes, 46, 48, 131, 133, 228, 232, 233, 236, 237, 238, 243
lymphoid, 129, 133, 229, 232, 233
lysis, 79, 104, 109, 212
lysozyme, 11, 49, 110

M

macromolecules, 69, 244
macrophages, 18, 46, 144, 226, 228, 230
magnesium, 302
malabsorption, 132
malaria, 224
malate dehydrogenase, 165
malnutrition, 130, 227, 231, 232, 233, 236, 237, 240, 346
malt extract, 12
maltose, 37, 185, 200, 302, 311
management, viii, xii, 35, 41, 43, 53, 131, 141, 143, 168, 291, 304, 326, 346
manganese, 62, 86, 252, 302
mannitol, 6, 37, 302, 320
manufacturing, x, 72, 95, 150, 204, 305
marine fish, 79
marketing, 208
marketplace, 221
marrow, 47, 231, 232, 233, 236, 238, 240
mass, 73, 77, 79, 97, 98, 104, 109, 161, 182, 194, 246, 248, 255, 256, 315, 322
mass spectrometry, 77, 79, 248, 322
materials, 95, 104, 148, 166, 209, 212, 258, 259, 261, 262, 267, 268, 292
matrix(s), 12, 13, 19, 23, 31, 39, 50, 68, 69, 96, 225, 233, 246, 252, 281
maximum specific growth rate, 69, 193
measurement(s), 44, 49, 58, 249
mechanical stress, 102
media, 3, 22, 38, 70, 79, 98, 153, 179, 180, 183, 200, 202, 222, 250, 261, 265, 267, 272, 293, 327
Mediterranean, 149, 150, 157, 258, 260, 276, 282
mellitus, 45, 127
membrane permeability, 82, 98, 99, 102
membranes, 82, 96, 105, 106, 275, 293, 337
meningitis, 133, 135, 141
meta analysis, xiii, 19, 343
meta-analysis, 19, 345, 347, 348, 349
Metabolic, 25, 84, 98, 217, 221, 320
metabolic changes, 107
metabolic pathways, 37, 220, 265, 269
metabolism, 33, 36, 63, 73, 101, 111, 148, 159, 160, 164, 166, 173, 185, 199, 201, 202, 209, 211, 213, 218, 261, 264, 265, 276, 285, 293, 304, 311, 316, 320
metabolites, xi, 15, 50, 82, 101, 120, 170, 208, 241, 257, 262, 274, 281, 315
metabolized, xiii, 6, 37, 300, 309, 310
metabolizing, 276
metal ion(s), 301, 302
metals, 306
methanol, 299
methodology, 112, 161, 225, 320
Mexico, 1
mice, 15, 17, 18, 24, 26, 27, 31, 41, 45, 46, 48, 52, 53, 54, 55, 58, 59, 60, 61, 62, 128, 130, 131, 134, 145, 146, 215, 220, 226, 227, 228, 229, 230, 231, 232, 233, 234, 235, 236, 237, 238, 239, 240, 246, 250, 252, 254, 315
microbial cells, 293
microbial community(s), 151, 170, 207, 256
microbiota, 3, 9, 15, 20, 22, 48, 51, 53, 62, 150, 151, 152, 153, 161, 162, 172, 207, 208, 209, 214, 242, 252, 254, 274, 276, 277, 280, 281, 288, 312, 314, 336, 341
micronutrients, 186, 266
microorganism, vii, viii, 35, 36, 37, 38, 41, 42, 45, 46, 48, 49, 51, 93, 94, 95, 97, 99, 100, 101, 106, 109, 115, 171, 209, 246, 279
microscopy, 44, 103, 105, 128, 247
microwave heating, 98, 113
microwave radiation, 97, 113
microwaves, 97
migration, 133
mild asthma, 136
miniaturization, 247
Ministry of Education, 304
mitochondria, 87
model system, 76
models, ix, xiii, 17, 33, 62, 103, 106, 116, 138, 177, 179, 184, 185, 189, 191, 192, 195, 196, 197, 200, 201, 224, 249, 250, 322, 326, 336
modifications, ix, 177
modules, 243
moisture, 21, 157, 200, 261
molasses, 32, 112, 300, 307
molds, 15
molecular mass, 73, 77, 79, 275
molecular weight, viii, 65, 74, 82, 156, 245, 261, 300, 316
molecules, 15, 26, 67, 97, 106, 130, 159, 178, 209, 210, 229, 245, 246, 249, 254, 260, 315

monolayer, vii, 1, 14
monosaccharide, 313
morbidity, 48, 127, 235, 346
Morocco, 162
morphology, vii, 1, 2, 14, 36, 37, 99, 103, 104, 105, 108, 245
mortality, 42, 48, 60, 127, 233, 235, 346, 350
mortality rate, 42
mortality risk, 42
Moscow, 241
motif, 73, 142
mRNA(s), 45, 338
mucin, 8, 12, 245, 246, 247, 248, 250, 254, 349
mucosa, 5, 6, 12, 14, 17, 18, 20, 46, 47, 130, 144, 172, 178, 209, 226, 231, 254, 340, 344
mucus, 12, 14, 29, 120, 127, 137, 244, 245, 246, 248, 249, 252, 253, 254, 255, 256
multiple factors, 247
multiplication, 282, 312
mussels, 77
mutagenesis, 301
mutant, 60, 76, 213, 246, 249
mutation(s), 8, 211, 247
mycotoxins, 286
myeloid cells, 231
myoglobin, 149
myosin, 131

N

Na^+, 301
NaCl, 2, 6, 37, 68, 70, 73, 74, 79, 87, 149, 153, 156, 159, 160, 163, 164, 165, 181
NAD, 199, 299
NADH, 8, 199, 201, 261, 276
NAS, 127, 143
nasogastric tube, 42
natural food, 111
natural habitats, 257
natural killer (NK) cells, 226
natural resources, 292
nausea, 43
NCTC, 49, 278
Nd, 13
necrosis, 59, 126, 133, 144, 146
negative effects, viii, 18, 66, 119
neonates, 3, 242
Nepal, 157, 162, 259, 260
nervous system, 129
Netherlands, 208
neurogenic bladder, 138
neutral, 71, 100, 293, 297
neutropenia, 231
neutrophils, 45, 131, 228, 229, 231
NH_2, 173
niacin, 37
nicotinic acid, 37
Nigeria, 5
nitrates, 268
nitric oxide, 131
nitrite, 36, 59, 63, 71, 149, 154, 170, 262
nitrogen, ix, 177, 180, 181, 185, 186, 188, 189, 192, 193, 194, 196, 200, 201, 211, 213, 292, 301
nitrogen compounds, 292
NK cells, 46
nodes, 100
non-digestible oligosaccharides, xiii, 309
North America, 151, 208
Norway, 89
nucleic acid, 3, 82, 96, 159, 280
nucleotide sequence, 91, 219
nutrient(s), ix, 52, 63, 75, 76, 83, 106, 137, 150, 175, 177, 183, 184, 185, 188, 189, 194, 196, 200, 204, 258, 266, 293, 294, 327, 345, 349
nutrient concentrations, 184
nutrition, 31, 56, 57, 120, 130, 132, 136, 141, 143, 144, 198, 237, 239, 262, 266, 314, 320
nutritional status, 261

O

obesity, 51, 53, 315
ofloxacin, 9
OH, 327
oil, 70, 98, 162, 293, 305, 307, 308
oligomers, 163, 313
oligosaccharide, 249, 302, 320, 321
olive oil, 307
omega-3, 157
operating costs, 294, 299
operations, 271
operon, x, 204, 211, 215, 219
opportunities, 113, 239, 253, 286
optimization, 66, 116, 197, 269
oral antibiotic, 350
oral cavity, 3, 11, 20, 49, 54, 63, 274
oral health, viii, 36, 49, 59
orchestration, 232
organ(s), 232, 233, 344
organic compounds, 14
organism, x, 47, 68, 70, 79, 143, 171, 179, 204, 217, 233, 282
osmosis, 161
osmotic challenges, 167
osmotic pressure, 163
osmotic stress, 163, 165, 168, 213

osteoporosis, 314, 315, 322
outpatient, 350
oxidation, 15, 74, 110, 113, 149, 157, 167, 276, 300
oxidative stress, 6, 8, 214, 216
oxygen, vii, 1, 36, 75, 110, 148, 149, 150, 214, 261, 270, 276, 292, 300, 301, 310
oysters, 98
ozone, 110, 116

P

pain, vii, 1, 18, 19, 28, 42, 43, 132, 143
Pakistan, 28
pancreas, 132
pancreatitis, 132, 144
parallel, 128, 216, 232
parasites, 71, 207, 346
parents, 234, 345
participants, xiii, 42, 48, 51, 285, 343, 347
pasteurization, viii, 93, 95, 96, 105
pathogenesis, xiv, 255, 315, 344, 347
pathophysiological, 229
pathophysiology, 60
pathways, ix, 7, 37, 113, 161, 203, 207, 208, 210, 211, 220, 243, 244, 248, 261, 265, 269, 284
PBMC, 18, 47
PCR, xi, 27, 29, 38, 39, 44, 45, 53, 54, 55, 61, 79, 85, 166, 199, 214, 221, 241, 247, 251, 271, 277, 285
penicillin, 8, 40, 278, 279, 329, 334
pepsin, 11, 33
peptic ulcer, 142
peptidase, 213, 243
peptide(s), vii, x, 1, 3, 11, 38, 67, 69, 72, 73, 84, 114, 129, 141, 155, 169, 173, 204, 212, 213, 227, 233, 243, 244, 245, 248, 249, 252, 262, 275, 310, 336
per capita income, 346
pericardium, 335
peripheral blood, 45, 46
peripheral blood mononuclear cell, 46
peritonitis, 135
permeability, vii, 1, 42, 43, 44, 59, 60, 62, 82, 98, 99, 102, 104, 111, 127, 130, 131, 132, 143, 227, 279
permit, 67
peroxide, viii, ix, xi, 15, 65, 75, 94, 119, 138, 209, 257, 261, 269, 270, 272, 274, 276
personal communication, 79
Peru, 157
phage, 246, 250
phagocytosis, 131, 228
pharmaceutical, xiii, 102, 309, 310, 312, 316, 322, 347
phenol, 68, 87, 300, 307
phenolic compounds, 26, 300, 306
phenotype(s), 6, 68, 237, 340
Philippines, 5, 24, 259
phosphate(s), 25, 46, 100, 103, 104, 114, 150, 211, 244, 248, 253, 319, 328, 329
phosphoenolpyruvate, 202
phospholipids, 104
phosphorous, ix, 177, 180, 181, 185, 188, 189
phosphorus, 186, 315
phosphorylation, 14, 131
phylum, 3, 222
physical activity, 51
physical features, 285
physical inactivity, 51
physical properties, 87
physical treatments, 7
physicochemical characteristics, 72
Physiological, 36, 62, 163, 168
physiology, 2, 15, 60, 151, 162, 214, 217, 219, 287, 305
pigs, 342
pilot study, 141, 269
placebo, 18, 19, 41, 42, 43, 44, 46, 47, 48, 51, 54, 59, 125, 126, 129, 133, 134, 136, 138, 141, 142, 143, 144, 145, 200, 201, 234, 346, 348, 349
plants, 86, 110, 179, 207, 258, 265, 288, 292, 300, 312
plaque, 51, 56, 62, 132
plasma cells, 229, 230
plasmid, 10, 11, 22, 29, 33, 40, 55, 56, 220, 250, 287
plasminogen, 248, 253
plastic surgeon, 45
platelets, 8
playing, 246
pneumonia, 16, 39, 133, 135, 139, 235
polarity, 108
pollen, 47, 62, 135, 141
polyacrylamide, 244
polymerase, 38, 252
polymerase chain reaction, 38, 252
polymerization, 313, 315
polymer(s), 15, 24, 199, 245, 312, 313
polyphenols, 307
polysaccharide(s), 12, 14, 17, 50, 243, 247, 249, 312, 315, 319
polyunsaturated fat, 157, 168
polyunsaturated fatty acids, 157, 168
pools, 170
population, xii, xiv, 8, 11, 12, 42, 74, 76, 82, 94, 96, 101, 104, 105, 107, 110, 111, 120, 162, 166, 168, 169, 178, 228, 230, 233, 234, 235, 237, 246, 252, 258, 259, 271, 273, 282, 284, 286, 343, 345, 348, 350

population density, 74, 76
population group, 111
population size, 107
population structure, 168
portal vein, 129
Portugal, 5, 32, 65, 151, 152, 166, 284
potassium, 88
potential benefits, 36
poultry, 109, 148, 168, 173, 175, 288, 330
poverty, 157, 171
poverty alleviation, 157, 171
precipitation, 57, 79, 340
pregnancy, 144
preparation, iv, 40, 42, 43, 44, 127, 129, 130, 132, 134, 136, 139, 141, 161, 198, 200, 209, 234, 235, 294, 310
preschool, xi, 224, 234, 239
preschool children, xi, 224, 234
preservation, viii, 65, 66, 69, 71, 73, 78, 83, 84, 87, 93, 94, 95, 97, 98, 104, 111, 113, 114, 115, 154, 157, 174, 175, 208, 225, 258, 260, 262, 263, 265, 267, 268, 273, 281, 285, 286, 287, 293
preservative, 11, 80, 109, 154, 161
prevention, viii, ix, xi, xiii, 36, 40, 48, 51, 56, 66, 120, 126, 131, 133, 134, 136, 139, 140, 142, 145, 149, 168, 178, 201, 224, 225, 235, 239, 281, 311, 315, 322, 326, 340, 343, 346, 349, 350
priming, 233
principles, 109
probe, 61, 144
process control, 179
producers, 179, 274, 336
production costs, 109
production technology, 258, 283
professional literature, 275
professionals, 345
pro-inflammatory, 17, 18, 46, 61, 130, 178, 228, 229
project, 157, 172, 198, 205, 256, 282
proliferation, 45, 129, 231, 306, 312, 315, 336
proline, 163, 164, 213, 217
propagation, 261
prophylactic, 178
prophylactic agents, 178
prophylaxis, 132, 141, 278
proportionality, 196
protection, x, xii, 12, 18, 27, 48, 59, 68, 223, 226, 229, 230, 231, 235, 236, 245, 246, 281, 291, 315, 338, 339
protective factors, 281
protein hydrolysates, 150
protein synthesis, 39, 163, 248, 280
proteinase, 31, 77, 199, 212
proteolysis, x, 155, 203, 211, 244
proteolytic enzyme, x, 11, 72, 78, 204, 212, 216
proteome, xi, 211, 241
proteomics, 164
protons, 108
Pseudomonas aeruginosa, 45, 50, 62, 131, 139, 140, 143, 146, 278
psychosis, 19
public health, xi, 224, 225, 278
pulp, 217, 311
pumps, 102, 293
purification, 27, 74, 292, 321, 341
P-value, 184
pyrolysis, 322

Q

quantification, 29, 198

R

radiation, 97, 109, 110, 113, 117
radiation damage, 109
radicals, 214
radioisotope, 109
raw materials, 95, 148, 166, 209, 212, 259
RE, 30, 236
reaction rate, 97
reactions, x, 98, 120, 129, 132, 178, 204, 212, 221, 230, 266, 299, 300
reactive oxygen, 214
reagents, 245
real time, 38
receptors, xi, 13, 45, 69, 241, 244, 247, 248, 249, 344, 345
recognition, 73, 243, 245, 253, 256
recombination, 250, 251
recommendations, iv, 8, 137, 256, 344
recovery, xi, xii, 26, 69, 110, 224, 225, 227, 228, 233, 234, 239, 292, 293, 294, 297, 301, 304, 306, 307, 341
recurrence, 130, 136, 139, 144, 198, 347
recycling, 294, 299
red blood cells, 329
red wine, 26
redistribution, 17
refractive index, 181
regression, 96, 134
regulations, 149
regulatory systems, 129
rehydration, 129, 136, 346
relapses, 139
relevance, 20, 139, 166, 238, 246

reliability, 94
relief, 314
remission, 136, 142, 145
repair, 50, 69, 131
reparation, 107, 136, 141, 200, 294
replication, 7, 232
repression, 202
reputation, 39
requirements, 37, 109, 198, 202, 258, 276, 294, 298, 308
researchers, 14, 38, 40, 41, 43, 44, 46, 47, 48, 50, 51, 53, 54, 81, 127, 130, 131, 132, 133, 134, 135, 136, 188, 189, 249, 311, 338
resection, 144
reserves, 15
residues, 15, 73, 74, 77, 178, 202, 209, 212, 246, 286, 287
resolution, 42, 226
resources, 292, 304
restoration, 243, 297
restrictions, 190, 191, 192
retail, 86
RH, 348, 349
rheumatoid arthritis, 136, 141
rhinitis, 136
ribonucleic acid, 113
ribose, 6, 37
ribosomal RNA, 215
ribosome, 338
risk, 8, 28, 40, 42, 43, 48, 50, 51, 52, 60, 71, 72, 111, 127, 128, 129, 133, 136, 145, 233, 237, 265, 275, 278, 280, 281, 282, 285, 300, 311, 312, 314, 315, 326, 347, 350
risk assessment, 280, 285
risk factors, 28, 51, 60, 127, 129, 237, 315, 350
RNA, 215
rods, 36, 94, 218
room temperature, 83, 107, 179
root(s), 259, 260, 261, 312, 319
rotavirus, 133, 136, 145, 345, 346, 349
routes, 23, 207, 210
rowing, 323
Russia, 241

S

salinity, 162, 163, 173
saliva, 3, 6, 7
salmon, 66, 68, 69, 70, 71, 72, 73, 74, 75, 76, 77, 78, 80, 81, 82, 83, 84, 85, 86, 87, 88, 89, 90, 91, 162, 170, 171, 172, 174
Salmonella, vi, xiii, 13, 16, 22, 33, 41, 47, 56, 63, 82, 86, 127, 139, 178, 201, 213, 216, 220, 226, 234, 236, 270, 274, 276, 278, 279, 280, 316, 325, 326, 330, 331, 333, 334, 335, 336, 338
salt concentration, 2, 70, 71, 84, 103, 154, 163, 164, 166, 282
salt tolerance, 105
salts, xi, xiii, 3, 12, 15, 17, 39, 52, 53, 149, 154, 209, 241, 247, 250, 325, 337
saltwater, 157
saturated fat, 59
scanning calorimetry, 112
school, 234
science, 288, 304, 305
SDS-PAGE, 156, 170
seafood, 77, 84, 98, 171, 172
secrete, 248
secretin, 127, 230, 249
secretion, 17, 18, 44, 45, 47, 52, 59, 61, 78, 129, 139, 243, 248, 252, 346
security, 157, 171
sedimentation, 306
seed, 149
selectivity, 314
self-control, 280
senses, 145
sensitivity, 67, 75, 84, 96, 98, 111, 267, 275, 276, 284, 310
sensitization, 111, 136, 145
sepsis, 127, 132, 133, 141
sequencing, xii, 38, 79, 90, 152, 205, 217, 247, 251, 292, 295, 305, 306, 307
Serbia, 282
serine, 213
serum, 17, 18, 47, 48, 53, 54, 225, 226, 227, 229, 231, 233, 315
services, iv, 238
sewage, 292
sex, 329
shape, 98, 100, 106, 266
shelf life, xi, 12, 21, 71, 78, 149, 157, 209, 257, 266, 281
shellfish, 86
shock, 6, 38, 95, 96, 113, 237, 248, 293
shoot(s), 259, 260, 268, 271
showing, 13, 66, 72, 79, 80, 81, 96, 98, 100, 101, 103, 106, 108, 132, 208, 232, 242, 246, 328
shrimp, 79, 172
side chain, 300
side effects, 17, 42, 120, 128, 132, 140, 178
signal transduction, 44
signaling pathway, 239
signalling, ix, 119, 248
signals, 131, 235, 236, 237, 243
silver, 50, 54, 120, 132